Probability and Statistics by Example: I

Probability and statistics are as much about intuition and problem solving, as they are about theorem proving. Because of this, students can find it very difficult to make a successful transition from lectures to examinations to practice, since the problems involved can vary so much in nature. Since the subject is critical in many modern applications such as mathematical finance, quantitative management, telecommunications, signal processing, bioinformatics, as well as traditional ones such as insurance, social science and engineering, the authors have rectified deficiencies in traditional lecture-based methods by collecting together a wealth of exercises for which they've supplied complete solutions. These solutions are adapted to the needs and skills of students. To make it of broad value, the authors supply basic mathematical facts as and when they are needed, and have sprinkled some historical information throughout the text.

Probability and Statistics by Example

Volume I. Basic Probability and Statistics

Y. SUHOV
University of Cambridge

M. KELBERT
University of Wales–Swansea

CAMBRIDGE
UNIVERSITY PRESS

CAMBRIDGE UNIVERSITY PRESS
Cambridge, New York, Melbourne, Madrid, Cape Town, Singapore, São Paulo

CAMBRIDGE UNIVERSITY PRESS
The Edinburgh Building, Cambridge CB2 2RU, UK

Published in the United States of America by Cambridge University Press, New York

www.cambridge.org
Information on this title: www.cambridge.org/9780521847667

© Cambridge University Press 2005

First published 2005

Printed in the United Kingdom at the University Press, Cambridge

A catalogue record for this publication is available from the British Library

ISBN-13 978-0-521-84766-7 hardback
ISBN-10 0-521-84766-4 hardback
ISBN-13 978-0521-61233-3 paperback
ISBN-10 0-521-61233-0 paperback

Contents

Preface   *page* vii

Part I   Basic probability   1

1   Discrete outcomes   3
1.1   A uniform distribution   3
1.2   Conditional Probabilities. The Bayes Theorem. Independent trials   6
1.3   The exclusion–inclusion formula. The ballot problem   27
1.4   Random variables. Expectation and conditional expectation.
         Joint distributions   33
1.5   The binomial, Poisson and geometric distributions. Probability
         generating, moment generating and characteristic functions   54
1.6   Chebyshev's and Markov's inequalities. Jensen's inequality. The Law
         of Large Numbers and the De Moivre–Laplace Theorem   75
1.7   Branching processes   96

2   Continuous outcomes   108
2.1   Uniform distribution. Probability density functions. Random variables.
         Independence   108
2.2   Expectation, conditional expectation, variance, generating function,
         characteristic function   142
2.3   Normal distributions. Convergence of random variables
         and distributions. The Central Limit Theorem   168

Part II   Basic statistics   191

3   Parameter estimation   193
3.1   Preliminaries. Some important probability distributions   193
3.2   Estimators. Unbiasedness   204
3.3   Sufficient statistics. The factorisation criterion   209
3.4   Maximum likelihood estimators   213
3.5   Normal samples. The Fisher Theorem   215

3.6   Mean square errors. The Rao–Blackwell Theorem.
       The Cramér–Rao inequality   218
3.7   Exponential families   225
3.8   Confidence intervals   229
3.9   Bayesian estimation   233

4   Hypothesis testing   242
4.1   Type I and type II error probabilities. Most powerful tests   242
4.2   Likelihood ratio tests. The Neyman–Pearson Lemma and beyond   243
4.3   Goodness of fit. Testing normal distributions, 1: homogeneous samples   252
4.4   The Pearson χ^2 test. The Pearson Theorem   257
4.5   Generalised likelihood ratio tests. The Wilks Theorem   261
4.6   Contingency tables   270
4.7   Testing normal distributions, 2: non-homogeneous samples   276
4.8   Linear regression. The least squares estimators   289
4.9   Linear regression for normal distributions   292

**5   Cambridge University Mathematical Tripos examination questions
         in IB Statistics (1992–1999)**   298

       Appendix 1 Tables of random variables and probability distributions   346

       *Appendix 2 Index of Cambridge University Mathematical Tripos
                   examination questions in IA Probability (1992–1999)*   349

       Bibliography   352

       Index   358

Preface

The original motivation for writing this book was rather personal. The first author, in the course of his teaching career in the Department of Pure Mathematics and Mathematical Statistics (DPMMS), University of Cambridge, and St John's College, Cambridge, had many painful experiences when good (or even brilliant) students, who were interested in the subject of mathematics and its applications and who performed well during their first academic year, stumbled or nearly failed in the exams. This led to great frustration, which was very hard to overcome in subsequent undergraduate years. A conscientious tutor is always sympathetic to such misfortunes, but even pointing out a student's obvious weaknesses (if any) does not always help. For the second author, such experiences were as a parent of a Cambridge University student rather than as a teacher.

We therefore felt that a monograph focusing on Cambridge University mathematics examination questions would be beneficial for a number of students. Given our own research and teaching backgrounds, it was natural for us to select probability and statistics as the overall topic. The obvious starting point was the first-year course in probability and the second-year course in statistics. In order to cover other courses, several further volumes will be needed; for better or worse, we have decided to embark on such a project.

Thus our essential aim is to present the Cambridge University probability and statistics courses by means of examination (and examination-related) questions that have been set over a number of past years. Following the decision of the Board of the Faculty of Mathematics, University of Cambridge, we restricted our exposition to the Mathematical Tripos questions from the years 1992–1999. (The questions from 2000–2004 are available online at http://www.maths.cam.ac.uk/ppa/.) Next, we included some IA Probability regular example sheet questions from the years 1992–2003 (particularly those considered as difficult by students). Further, we included the problems from Specimen Papers issued in 1992 and used for mock examinations (mainly in the beginning of the 1990s) and selected examples from the 1992 list of so-called sample questions. A number of problems came from example sheets and examination papers from the University of Wales-Swansea.

Of course, Cambridge University examinations have never been easy. On the basis of examination results, candidates are divided into classes: first, second (divided into two categories: 2.1 and 2.2) and third; a small number of candidates fail. (In fact, a more detailed list ranking all the candidates in order is produced, but not publicly disclosed.) The examinations are officially called the 'Mathematical Tripos', after the three-legged stools on which candidates and examiners used to sit (sometimes for hours) during oral examinations in ancient times. Nowadays all examinations are written. The first-year of the three-year undergraduate course is called Part IA, the second Part IB and the third Part II.

For example, in May–June of 2003 the first-year mathematics students sat four examination papers; each lasted three hours and included 12 questions from two subjects. The following courses were examined: algebra and geometry, numbers and sets, analysis, probability, differential equations, vector calculus, and dynamics. All questions on a given course were put in a single paper, except for algebra and geometry, which appears in two papers. In each paper, four questions were classified as short (two from each of the two courses selected for the paper) and eight as long (four from each selected course). A candidate might attempt all four short questions and at most five long questions, no more than three on each course; a long question carries twice the credit of a short one. A calculation shows that if a student attempts all nine allowed questions (which is often the case), and the time is distributed evenly, a short question must be completed in 12–13 minutes and a long one in 24–25 minutes. This is not easy and usually requires special practice; one of the goals of this book is to assist with such a training programme.

The pattern of the second-year examinations has similarities but also differences. In June 2003, there were four IB Maths Tripos papers, each three hours long and containing nine or ten short and nine or ten long questions in as many subjects selected for a given paper. In particular, IB statistics was set in Papers 1, 2 and 4, giving a total of six questions. Of course, preparing for Part IB examinations is different from preparing for Part IA; we comment on some particular points in the corresponding chapters.

For a typical Cambridge University student, specific preparation for the examinations begins in earnest during the Easter (or Summer) Term (beginning in mid-April). Ideally, the work might start during the preceding five-week vacation. (Some of the examination work for Parts IB and II, the computational projects, is done mainly during the summer vacation period.) As the examinations approach, the atmosphere in Cambridge can become rather tense and nervous, although many efforts are made to diffuse the tension. Many candidates expend a great deal of effort in trying to calculate exactly how much work to put into each given subject, depending on how much examination credit it carries and how strong or weak they feel in it, in order to optimise their overall performance. One can agree or disagree with this attitude, but one thing seemed clear to us: if the students receive (and are able to digest) enough information about and insight into the level and style of the Tripos questions, they will have a much better chance of performing to the best of their abilities. At present, owing to great pressures on time and energy, most of them are not in a position to do so, and much is left to chance. We will be glad if this book helps to change this situation by alleviating pre-examination nerves and by stripping Tripos examinations of some of their mystery, at least in respect of the subjects treated here.

Thus, the first reason for this book was a desire to make life easier for the students. However, in the course of working on the text, a second motivation emerged, which we feel is of considerable professional interest to anyone teaching courses in probability and statistics. In 1991–2 there was a major change in Cambridge University to the whole approach to probabilistic and statistical courses. The most notable aspect of the new approach was that the IA Probability course and the IB Statistics course were redesigned to appeal to a wide audience (200 first-year students in the case of IA Probability and nearly the same number of the second-year students in the case of IB Statistics). For a large number of students, these are the only courses from the whole of probability and statistics which they attend during their undergraduate years. Since more and more graduates in the modern world have to deal with theoretical and (especially) applied problems of a probabilistic or statistical nature, it is important that these courses generate and maintain a strong and wide appeal. The main goal shifted, moving from an academic introduction to the subject towards a more methodological approach which equips students with the tools needed to solve reasonable practical and theoretical questions in a 'real life' situation.

Consequently, the emphasis in IA Probability moved further away from sigma-algebras, Lebesgue and Stiltjies integration and characteristic functions to a direct analysis of various models, both discrete and continuous, with the aim of preparing students both for future problems and for future courses (in particular, Part IB Statistics and Part IB/II Markov chains). In turn, in IB Statistics the focus shifted towards the most popular practical applications of estimators, hypothesis testing and regression. The principal determination of examination performance in both IA Probability and IB Statistics became students' ability to choose and analyse the right model and accurately perform a reasonable amount of calculation rather than their ability to solve theoretical problems.

Certainly such changes (and parallel developments in other courses) were not always unanimously popular among the Cambridge University Faculty of Mathematics, and provoked considerable debate at times. However, the student community was in general very much in favour of the new approach, and the 'redesigned' courses gained increased popularity both in terms of attendance and in terms of attempts at examination questions (which has become increasingly important in the life of the Faculty of Mathematics). In addition, with the ever-growing prevalence of computers, students have shown a strong preference for an 'algorithmic' style of lectures and examination questions (at least in the authors' experience).

In this respect, the following experience by the first author may be of some interest. For some time I have questioned former St John's mathematics graduates, who now have careers in a wide variety of different areas, about what parts of the Cambridge University course they now consider as most important for their present work. It turned out that the strongest impact on the majority of respondents is not related to particular facts, theorems, or proofs (although jokes by lecturers are well remembered long afterwards). Rather they appreciate the ability to construct a mathematical model which represents a real-life situation, and to solve it analytically or (more often) numerically. It must therefore be acknowledged that the new approach was rather timely. As a consequence of all this, the level and style of Maths Tripos questions underwent changes. It is strongly suggested (although perhaps it was not always achieved) that the questions should have a clear structure where candidates are led from one part to another.

The second reason described above gives us hope that the book will be interesting for an audience outside Cambridge. In this regard, there is a natural question: what is the book's place in the (long) list of textbooks on probability and statistics. Many of the references in the bibliography are books published in English after 1991, containing the terms 'probability' or 'statistics' in their titles and available at the Cambridge University Main and Departmental Libraries (we are sure that our list is not complete and apologise for any omission).

As far as basic probability is concerned, we would like to compare this book with three popular series of texts and problem books, one by S. Ross [Ros1–Ros6], another by D. Stirzaker [St1–St4], and the third by G. Grimmett and D. Stirzaker [GriS1–GriS3]. The books by Ross and Stirzaker are commonly considered as a good introduction to the basics of the subject. In fact, the style and level of exposition followed by Ross has been adopted in many American universities. On the other hand, Grimmett and Stirzaker's approach is at a much higher level and might be described as 'professional'. The level of our book is intended to be somewhere in-between. In our view, it is closer to that of Ross or Stirzaker, but quite far away from them in several important aspects. It is our feeling that the level adopted by Ross or Stirzaker is not sufficient to get through Cambridge University Mathematical Tripos examinations with Class 2.1 or above. Grimmett and Stirzaker's books are of course more than enough – but in using them to prepare for an examination the main problem would be to select the right examples from among a thousand on offer.

On the other hand, the above monographs, as well as many of the books from the bibliography, may be considered as good complementary reading for those who want to take further steps in a particular direction. We mention here just a few of them: [Chu], [Dur1], [G], [Go], [JP], [Sc] and [ChaY]. In any case, the (nostalgic) time when everyone learning probability had to read assiduously through the (excellent) two-volume Feller monograph [Fe] had long passed (though in our view, Feller has not so far been surpassed).

In statistics, the picture is more complex. Even the definition of the subject of statistics is still somewhat controversial (see Section 3.1). The style of lecturing and examining the basic statistics course (and other statistics-related courses) at Cambridge University was always rather special. This style resisted a trend of making the exposition 'fully rigorous', despite the fact that the course is taught to mathematics students. A minority of students found it difficult to follow, but for most of them this was never an issue. On the other hand, the level of rigour in the course is quite high and requires substantial mathematical knowledge. Among modern books, the closest to the Cambridge University style is perhaps [CaB]. As an example of a very different approach, we can point to [Wil] (whose style we personally admire very much but would not consider as appropriate for first reading or for preparing for Cambridge examinations).

A particular feature of this book is that it contains repetitions: certain topics and questions appear more than once, often in slightly different form, which makes it difficult to refer to previous occurrences. This is of course a pattern of the examination process which becomes apparent when one considers it over a decade or so. Our personal attitudes here followed a proverb 'Repetition is the mother of learning', popular (in various forms) in several languages. However, we apologise to those readers who may find some (and possibly many) of these repetitions excessive.

This book is organised as follows. In the first two chapters we present the material of the IA Probability course (which consists of 24 one-hour lectures). In this part the Tripos questions are placed within or immediately following the corresponding parts of the expository text. In Chapters 3 and 4 we present the material from the 16-lecture IB Statistics course. Here, the Tripos questions tend to embrace a wider range of single topics, and we decided to keep them separate from the course material. However, the various pieces of theory are always presented with a view to the rôle they play in examination questions.

Displayed equations, problems and examples are numbered by chapter: for instance, in Chapter 2 equation numbers run from (2.1) to (2.102), and there are Problems 2.1–2.55.

Symbol □ marks the end of a solution of a given problem. Symbol ■ marks the end of an example.

A special word should be said about solutions in this book. In part, we use students' solutions or our own solutions (in a few cases solutions are reduced to short answers or hints). However, a number of the so-called examiners' model solutions have also been used; these were originally set by the corresponding examiners and often altered by relevant lecturers and co-examiners. (A curious observation by many examiners is that, regardless of how perfect their model solutions are, it is rare that any of the candidates follow them.) Here, we aimed to present all solutions in a unified style; we also tried to correct mistakes occurring in these solutions. We should pay the highest credit to all past and present members of the DPMMS who contributed to the painstaking process of supplying model solutions to Tripos problems in IA Probability and IB Statistics: in our view their efforts definitely deserve the deepest appreciation, and this book should be considered as a tribute to their individual and collective work.

On the other hand, our experience shows that, curiously, students very rarely follow the ideas of model solutions proposed by lecturers, supervisors and examiners, however impeccable and elegant these solutions may be. Furthermore, students understand each other much more quickly than they understand their mentors. For that reason we tried to preserve whenever possible the style of students' solutions throughout the whole book.

Informal digressions scattered across the text have been borrowed from [Do], [Go], [Ha], the St Andrew's University website www-history.mcs.st-andrews.ac.uk/history/ and the University of Massachusetts website www.umass.edu/wsp/statistics/tales/. Conversations with H. Daniels, D.G. Kendall and C.R. Rao also provided a few subjects. However, a number of stories are just part of folklore (most of them are accessible through the Internet); any mistakes are our own responsibility. Photographs and portraits of many of the characters mentioned in this book are available on the University of York website www.york.ac.uk/depts/maths/histstat/people/ and (with biographies) on http://members.aol.com/jayKplanr/images.htm.

The advent of the World Wide Web also had another visible impact: a proliferation of humour. We confess that much of the time we enjoyed browsing (quite numerous) websites advertising jokes and amusing quotations; consequently we decided to use some of them in this book. We apologise to the authors of these jokes for not quoting them (and sometimes changing the sense of sentences).

Throughout the process of working on this book we have felt both the support and the criticism (sometimes quite sharp) of numerous members of the Faculty of Mathematics and colleagues from outside Cambridge who read some or all of the text or learned about its existence. We would like to thank all these individuals and bodies, regardless of whether they supported or rejected this project. We thank personally Charles Goldie, Oliver Johnson, James Martin, Richard Samworth and Amanda Turner, for stimulating discussions and remarks. We are particularly grateful to Alan Hawkes for the limitless patience with which he went through the preliminary version of the manuscript. As stated above, we made wide use of lecture notes, example sheets and other related texts prepared by present and former members of the Statistical Laboratory, Department of Pure Mathematics and Mathematical Statistics, University of Cambridge, and Mathematics Department and Statistics Group, EBMS, University of Wales-Swansea. In particular, a large number of problems were collected by David Kendall and put to great use in Example Sheets by Frank Kelly. We benefitted from reading excellent lecture notes produced by Richard Weber and Susan Pitts. Damon Wischik kindly provided various tables of probability distributions. Statistical tables are courtesy of R. Weber.

Finally, special thanks go to Sarah Shea-Simonds and Maureen Storey for carefully reading through parts of the book and correcting a great number of stylistic errors.

Part I
Basic probability

1 Discrete outcomes

1.1 A uniform distribution

> Lest men suspect your tale untrue,
> Keep probability in view.
> > J. Gay (1685–1732), English poet

In this section we use the simplest (and historically the earliest) probabilistic model where there are a finite number m of possibilities (often called *outcomes*) and each of them has the same probability $1/m$. A collection A of k outcomes with $k \leq m$ is called an *event* and its probability $\mathbb{P}(A)$ is calculated as k/m:

$$\mathbb{P}(A) = \frac{\text{the number of outcomes in } A}{\text{the total number of outcomes}}. \tag{1.1}$$

An empty collection has probability zero and the whole collection one. This scheme looks deceptively simple: in reality, calculating the number of outcomes in a given event (or indeed, the total number of outcomes) may be tricky.

Problem 1.1 You and I play a coin-tossing game: if the coin falls heads I score one, if tails you score one. In the beginning, the score is zero. (i) What is the probability that after $2n$ throws our scores are equal? (ii) What is the probability that after $2n+1$ throws my score is three more than yours?

Solution The outcomes in (i) are all sequences $HHH\ldots H, THH\ldots H, \ldots, TTT\ldots T$ formed by $2n$ subsequent letters H or T (or, 0 and 1). The total number of outcomes is $m = 2^{2n}$, each carries probability $1/2^{2n}$. We are looking for outcomes where the number of Hs equals that of Ts. The number k of such outcomes is $(2n)!/n!n!$ (the number of ways to choose positions for n Hs among $2n$ places available in the sequence). The probability in question is $\dfrac{(2n)!}{n!n!} \times \dfrac{1}{2^{2n}}$.

In (ii), the outcomes are the sequences of length $2n+1$, 2^{2n+1} in total. The probability equals

$$\frac{(2n+1)!}{(n+2)!(n-1)!} \times \frac{1}{2^{2n+1}}. \qquad \square$$

Problem 1.2   A tennis tournament is organised for 2^n players on a knock-out basis, with n rounds, the last round being the final. Two players are chosen at random. Calculate the probability that they meet (i) in the first or second round, (ii) in the final or semi-final, and (iii) the probability they do not meet.

Solution   The sentence 'Two players are chosen at random' is crucial. For instance, one may think that the choice has been made after the tournament when all results are known. Then there are 2^{n-1} pairs of players meeting in the first round, 2^{n-2} in the second round, two in the semi-final, one in the final and $2^{n-1} + 2^{n-2} + \cdots + 2 + 1 = 2^n - 1$ in all rounds.

The total number of player pairs is $\binom{2^n}{2} = 2^{n-1}(2^n - 1)$. Hence the answers:

(i) $\dfrac{2^{n-1} + 2^{n-2}}{2^{n-1}(2^n - 1)} = \dfrac{3}{2(2^n - 1)}$,   (ii) $\dfrac{3}{2^{n-1}(2^n - 1)}$,

and

(iii) $\dfrac{2^{n-1}(2^n - 1) - (2^n - 1)}{2^{n-1}(2^n - 1)} = 1 - \dfrac{1}{2^{n-1}}$. $\square$

Problem 1.3   There are n people gathered in a room.

(i)   What is the probability that two (at least) have the same birthday? Calculate the probability for $n = 22$ and 23.

(ii)   What is the probability that at least one has the same birthday as you? What value of n makes it close to $1/2$?

Solution   The total number of outcomes is 365^n. In (i), the number of outcomes not in the event is $365 \times 364 \times \cdots \times (365 - n + 1)$. So, the probability that all birthdays are distinct is $\big(365 \times 364 \times \cdots \times (365 - n + 1)\big)/365^n$ and that two or more people have the same birthday

$$1 - \frac{365 \times 364 \times \cdots \times (365 - n + 1)}{365^n}.$$

For $n = 22$:

$$1 - \frac{365}{365} \times \frac{364}{365} \times \cdots \times \frac{344}{365} = 0.4927,$$

and for $n = 23$:

$$1 - \frac{365}{365} \times \frac{364}{365} \times \cdots \times \frac{343}{365} = 0.5243.$$

In (ii), the number of outcomes not in the event is 364^n and the probability in question $1 - (364/365)^n$. We want it to be near $1/2$, so

$$\left(\frac{364}{365}\right)^n \approx \frac{1}{2}, \quad \text{i.e. } n \approx -\frac{1}{\log_2(364/365)} \approx 252.61. \quad \square$$

Problem 1.4   Mary tosses $n+1$ coins and John tosses n coins. What is the probability that Mary gets more heads than John?

Solution 1   We must assume that all coins are unbiased (as it was not specified otherwise). Mary has 2^{n+1} outcomes (all possible sequences of heads and tails) and John 2^n; jointly 2^{2n+1} outcomes that are equally likely. Let H_M and T_M be the number of Mary's heads and tails and H_J and T_J John's, then $H_M + T_M = n+1$ and $H_J + T_J = n$. The events $\{H_M > H_J\}$ and $\{T_M > T_J\}$ have the same number of outcomes, thus $\mathbb{P}(H_M > H_J) = \mathbb{P}(T_M > T_J)$.

On the other hand, $H_M > H_J$ if and only if $n - H_M < n - H_J$, i.e. $T_M - 1 < T_J$ or $T_M \leq T_J$. So event $H_M > H_J$ is the same as $T_M \leq T_J$, and $\mathbb{P}(T_M \leq T_J) = \mathbb{P}(H_M > H_J)$.

But for any (joint) outcome, either $T_M > T_J$ or $T_M \leq T_J$, i.e. the number of outcomes in $\{T_M > T_J\}$ equals 2^{2n+1} minus that in $\{T_M \leq T_J\}$. Therefore, $\mathbb{P}(T_M > T_J) = 1 - \mathbb{P}(T_M \leq T_J)$. To summarise:

$$\mathbb{P}(H_M > H_J) = \mathbb{P}(T_M > T_J) = 1 - \mathbb{P}(T_M \leq T_J) = 1 - \mathbb{P}(H_M > H_J),$$

whence $\mathbb{P}(H_M > H_J) = 1/2$.

Solution 2   (Fallacious, but popular with some students.) Again assume that all coins are unbiased. Consider pair (H_M, H_J), as an outcome; there are $(n+2)(n+1)$ such possible pairs, and they all are equally likely (wrong: you have to have biased coins for this!). Now count the number of pairs with $H_M > H_J$. If $H_M = n+1$, H_J can take any value $0, 1, \ldots, n$. In general, $\forall l \leq n+1$, if $H_M = l$, H_J will take values $0, \ldots, l-1$. That is, the number of outcomes where $H_M > H_J$ equals $1 + 2 + \cdots + (n+1) = \frac{1}{2}(n+1)(n+2)$. Hence, $\mathbb{P}(H_M > H_J) = 1/2$. $\square$

Problem 1.5   You throw $6n$ dice at random. Show that the probability that each number appears exactly n times is

$$\frac{(6n)!}{(n!)^6} \left(\frac{1}{6}\right)^{6n}.$$

Solution   There are 6^{6n} outcomes in total (six for each die), each has probability $1/6^{6n}$. We want n dice to show one dot, n two, and so forth. The number of such outcomes is counted by fixing first which dice show one: $(6n)!/[n!(5n)!]$. Given n dice showing one, we fix which remaining dice show two: $(5n)!/[n!(4n)!]$, etc. The total number of desired outcomes is the product that equals $(6n)!/(n!)^6$. This gives the answer. $\square$

In many problems, it is crucial to be able to spot recursive equations relating the cardinality of various events. For example, for the number f_n of ways of tossing a coin n times so that successive tails never appear: $f_n = f_{n-1} + f_{n-2}$, $n \geq 3$ (a Fibonacci equation).

Problem 1.6 (i) Determine the number g_n of ways of tossing a coin n times so that the combination HT never appears. (ii) Show that $f_n = f_{n-1} + f_{n-2} + f_{n-3}$, $n \geq 3$, is the equation for the number of ways of tossing a coin n times so that three successive heads never appear.

Solution (i) $g_n = 1 + n$; 1 for the sequence $HH \ldots H$, n for the sequences $T \ldots TH \ldots H$ (which includes $T \ldots T$).

(ii) The outcomes are 2^n sequences $(y_1, \ldots, y_n)$ of H and T. Let A_n be the event {no three successive heads appeared after n tosses}, then f_n is the cardinality $\#A_n$. Split: $A_n = B_n^{(1)} \cup B_n^{(2)} \cup B_n^{(3)}$, where $B_n^{(1)}$ is the event {no three successive heads appeared after n tosses, and the last toss was a tail}, $B_n^{(2)} = $ {no three successive heads appeared after n tosses, and the last two tosses were TH} and $B_n^{(3)} = $ {no three successive heads appeared after n tosses, and the last three tosses were THH}.

Clearly, $B_n^{(i)} \cap B_n^{(j)} = \emptyset$, $1 \leq i \neq j \leq 3$, and so $f_n = \#B_n^{(1)} + \#B_n^{(2)} + \#B_n^{(3)}$.

Now drop the last digit y_n: $(y_1, \ldots y_n) \in B_n^{(1)}$ iff $y_n = T$, $(y_1, \ldots y_{n-1}) \in A_{n-1}$, i.e. $\#B_{n-1}^{(1)} = f_{n-1}$. Also, $(y_1, \ldots y_n) \in B_n^{(2)}$ iff $y_{n-1} = T$, $y_n = H$, and $(y_1, \ldots y_{n-2}) \in A_{n-2}$. This allows us to drop the two last digits, yielding $\#B_n^{(2)} = f_{n-2}$. Similarly, $\#B_n^{(3)} = f_{n-3}$. The equation then follows. □

1.2   Conditional Probabilities. The Bayes Theorem. Independent trials

> Probability theory is nothing but common sense
> reduced to calculation.
>> P.-S. Laplace (1749–1827), French mathematician

> Clockwork Omega
>> (From the series '*Movies that never made it to the Big Screen*'.)

From now on we adopt a more general setting: our outcomes do not necessarily have equal probabilities $p_1, \ldots, p_m$, with $p_i > 0$ and $p_1 + \cdots + p_m = 1$.

As before, an *event* A is a collection of outcomes (possibly empty); the *probability* $\mathbb{P}(A)$ of event A is now given by

$$\mathbb{P}(A) = \sum_{\text{outcome } i \in A} p_i = \sum_{\text{outcome } i} p_i I(i \in A). \tag{1.2}$$

($\mathbb{P}(A) = 0$ for $A = \emptyset$.) Here and below, I stands for the *indicator function*, viz.:

$$I(i \in A) = \begin{cases} 1, & \text{if } i \in A, \\ 0, & \text{otherwise.} \end{cases}$$

The probability of the total set of outcomes is 1. The total set of outcomes is also called the whole, or full, event and is often denoted by Ω, so $\mathbb{P}(\Omega) = 1$. An outcome is often denoted by ω, and if $p(\omega)$ is its probability, then

$$\mathbb{P}(A) = \sum_{\omega \in A} p(\omega) = \sum_{\omega \in \Omega} p(\omega) I(\omega \in A). \tag{1.3}$$

As follows from this definition, the probability of the union

$$\mathbb{P}(A_1 \cup A_2) = \mathbb{P}(A_1) + \mathbb{P}(A_2) \tag{1.4}$$

for any pair of disjoint events A_1, A_2 (with $A_1 \cap A_2 = \emptyset$). More generally,

$$\mathbb{P}(A_1 \cup \cdots \cup A_n) = \mathbb{P}(A_1) + \cdots + \mathbb{P}(A_n) \tag{1.5}$$

for any collection of pair-wise disjoint events (with $A_j \cap A_{j'} = \emptyset \ \forall j \neq j'$). Consequently, (i) the probability $\mathbb{P}(A^c)$ of the complement $A^c = \Omega \setminus A$ is $1 - \mathbb{P}(A)$, (ii) if $B \subseteq A$, then $\mathbb{P}(B) \leq \mathbb{P}(A)$ and $\mathbb{P}(A) - \mathbb{P}(B) = \mathbb{P}(A \setminus B)$, and (iii) for a general pair of events A, B: $\mathbb{P}(A \setminus B) = \mathbb{P}(A \setminus (A \cap B)) = \mathbb{P}(A) - \mathbb{P}(A \cap B)$.

Furthermore, for a general (not necessarily disjoint) union:

$$\mathbb{P}(A_1 \cup \cdots \cup A_n) \leq \sum_{i=1}^{n} \mathbb{P}(A_i);$$

a more detailed analysis of the probability $\mathbb{P}(\cup A_i)$ is provided by the exclusion–inclusion formula (1.12); see below.

Given two events A and B with $\mathbb{P}(B) > 0$, the *conditional probability* $\mathbb{P}(A|B)$ of A given B is defined as the ratio

$$\mathbb{P}(A|B) = \frac{\mathbb{P}(A \cap B)}{\mathbb{P}(B)}. \tag{1.6}$$

At this stage, the conditional probabilities are important for us because of two formulas. One is the formula of complete probability: if $B_1, \ldots, B_n$ are pair-wise disjoint events partitioning the whole event Ω, i.e. have $B_i \cap B_j = \emptyset$ for $1 \leq i < j \leq n$ and $B_1 \cup B_2 \cup \cdots \cup B_n = \Omega$, and in addition $\mathbb{P}(B_i) > 0$ for $1 \leq i \leq n$, then

$$\mathbb{P}(A) = \mathbb{P}(A|B_1)\mathbb{P}(B_1) + \mathbb{P}(A|B_2)\mathbb{P}(B_2) + \cdots + \mathbb{P}(A|B_n)\mathbb{P}(B_n). \tag{1.7}$$

The proof is straightforward:

$$\mathbb{P}(A) = \sum_{1 \leq i \leq n} \mathbb{P}(A \cap B_i) = \sum_{1 \leq i \leq n} \frac{\mathbb{P}(A \cap B_i)}{\mathbb{P}(B_i)} \mathbb{P}(B_i) = \sum_{1 \leq i \leq n} \mathbb{P}(A|B_i)\mathbb{P}(B_i).$$

The point is that often it is conditional probabilities that are given, and we are required to find unconditional ones; also, the formula of complete probability is useful to clarify the nature of (unconditional) probability $\mathbb{P}(A)$. Despite its simple character, this formula is an extremely powerful tool in literally all areas dealing with probabilities. In particular, a large portion of the theory of Markov chains is based on its skilful application.

Representing $\mathbb{P}(A)$ in the form of the right-hand side (RHS) of (1.7) is called conditioning (on the collection of events $B_1, \ldots, B_n$).

Another formula is the *Bayes formula* (or the *Bayes Theorem*) named after T. Bayes (1702–1761), an English mathematician and cleric. It states that *under the same assumptions as above, if in addition* $\mathbb{P}(A) > 0$, *then the conditional probability* $\mathbb{P}(B_i|A)$ *can be expressed in terms of probabilities* $\mathbb{P}(B_1), \ldots, \mathbb{P}(B_n)$ *and conditional probabilities* $\mathbb{P}(A|B_1), \ldots, \mathbb{P}(A|B_n)$ *as*

$$\mathbb{P}(B_i|A) = \frac{\mathbb{P}(A|B_i)\mathbb{P}(B_i)}{\sum\limits_{1 \le j \le n} \mathbb{P}(A|B_j)\mathbb{P}(B_j)}. \tag{1.8}$$

The proof is the direct application of the definition and the formula of complete probability:

$$\mathbb{P}(B_i|A) = \frac{\mathbb{P}(A \cap B_i)}{\mathbb{P}(A)}, \quad \mathbb{P}(A \cap B_i) = \mathbb{P}(A|B_i)\mathbb{P}(B_i)$$

and

$$\mathbb{P}(A) = \sum_j \mathbb{P}(A|B_j)\mathbb{P}(B_j).$$

A standard interpretation of equation (1.8) is that it relates the *posterior probability* $\mathbb{P}(B_i|A)$ (conditional on A) with *prior probabilities* $\{\mathbb{P}(B_j)\}$ (valid before one knew that event A occurred).

In his lifetime, Bayes finished only two papers: one in theology and one called 'Essay towards solving a problem in the doctrine of chances'; the latter contained the Bayes Theorem and was published two years after his death. Nevertheless he was elected a Fellow of The Royal Society. Bayes' theory (of which the above theorem is an important part) was for a long time subject to controversy. His views were fully accepted (after considerable theoretical clarifications) only at the end of the nineteenth century.

Problem 1.7   Four mice are chosen (without replacement) from a litter containing two white mice. The probability that both white mice are chosen is twice the probability that neither is chosen. How many mice are there in the litter?

Solution   Let the number of mice in the litter be n. We use the notation $\mathbb{P}(2) = \mathbb{P}(\text{two white chosen})$ and $\mathbb{P}(0) = \mathbb{P}(\text{no white chosen})$. Then

$$\mathbb{P}(2) = \binom{n-2}{2} \bigg/ \binom{n}{4}.$$

Otherwise, $\mathbb{P}(2)$ could be computed as:

$$\frac{2}{n}\frac{1}{n-1} + \frac{2}{n}\frac{n-2}{n-1}\frac{1}{n-2} + \frac{2}{n}\frac{n-2}{n-1}\frac{n-3}{n-2}\frac{1}{n-3} + \frac{n-2}{n}\frac{2}{n-1}\frac{1}{n-2}$$

$$+ \frac{n-2}{n}\frac{n-3}{n-1}\frac{2}{n-2}\frac{1}{n-3} + \frac{n-2}{n}\frac{2}{n-1}\frac{n-3}{n-2}\frac{1}{n-3} = \frac{12}{n(n-1)}.$$

On the other hand,

$$\mathbb{P}(0) = \binom{n-2}{4} \bigg/ \binom{n}{4}.$$

Otherwise, $\mathbb{P}(0)$ could be computed as follows:

$$\mathbb{P}(0) = \frac{n-2}{n}\frac{n-3}{n-1}\frac{n-4}{n-2}\frac{n-5}{n-3} = \frac{(n-4)(n-5)}{n(n-1)}.$$

Solving the equation

$$\frac{12}{n(n-1)} = 2\frac{(n-4)(n-5)}{n(n-1)},$$

we get $n = (9 \pm 5)/2$; $n = 2$ is discarded as $n \geq 6$ (otherwise the second probability is 0). Hence, $n = 7$.   □

Problem 1.8   Lord Vile drinks his whisky randomly, and the probability that, on a given day, he has n glasses equals $e^{-1}/n!$, $n = 0, 1, \ldots$ Yesterday his wife Lady Vile, his son Liddell and his butler decided to murder him. If he had no whisky that day, Lady Vile was to kill him; if he had exactly one glass, the task would fall to Liddell, otherwise the butler would do it. Lady Vile is twice as likely to poison as to strangle, the butler twice as likely to strangle as to poison, and Liddell just as likely to use either method. Despite their efforts, Lord Vile is not guaranteed to die from any of their attempts, though he is three times as likely to succumb to strangulation as to poisoning.
    Today Lord Vile is dead. What is the probability that the butler did it?

Solution   Write $\mathbb{P}(\text{dead}|\text{strangle}) = 3r$, $\mathbb{P}(\text{dead}|\text{poison}) = r$, and

$$\mathbb{P}(\text{drinks no whisky}) = \mathbb{P}(\text{drinks one glass}) = \frac{1}{e},$$

$$\mathbb{P}(\text{drinks two glasses or more}) = 1 - \frac{2}{e}.$$

Next:

$$\mathbb{P}(\text{strangle}|\text{Lady V}) = \frac{1}{3}, \quad \mathbb{P}(\text{poison}|\text{Lady V}) = \frac{2}{3},$$

$$\mathbb{P}(\text{strangle}|\text{butler}) = \frac{2}{3}, \quad \mathbb{P}(\text{poison}|\text{butler}) = \frac{1}{3},$$

and

$$\mathbb{P}(\text{strangle}|\text{Liddell}) = \mathbb{P}(\text{poison}|\text{Liddell}) = \frac{1}{2}.$$

Then the conditional probability $\mathbb{P}(\text{butler}|\text{dead})$ is

$$\frac{\mathbb{P}(d|b)\mathbb{P}(b)}{\mathbb{P}(d|b)\mathbb{P}(b) + \mathbb{P}(d|LV)\mathbb{P}(LV) + \mathbb{P}(d|Lddl)\mathbb{P}(Lddl)}$$

$$= \frac{\left(1 - \dfrac{2}{e}\right)\left(\dfrac{3r \times 2}{3} + \dfrac{r}{3}\right)}{\left(1 - \dfrac{2}{e}\right)\left(\dfrac{3r \times 2}{3} + \dfrac{r}{3}\right) + \dfrac{1}{e}\left(\dfrac{3r}{3} + \dfrac{r \times 2}{3}\right) + \dfrac{1}{e}\left(\dfrac{3r}{2} + \dfrac{r}{2}\right)}$$

$$= \frac{e - 2}{e - 3/7} \approx 0.3137. \quad \square$$

Problem 1.9   At the station there are three payphones which accept 20p pieces. One never works, another always works, while the third works with probability $1/2$. On my way to the metropolis for the day, I wish to identify the reliable phone, so that I can use it on my return. The station is empty and I have just three 20p pieces. I try one phone and it does not work. I try another twice in succession and it works both times. What is the probability that this second phone is the reliable one?

Solution   Let A be the event in the question: the first phone tried did not work and second worked twice. Clearly:

$$\mathbb{P}(A|\text{1st reliable}) = 0,$$

$$\mathbb{P}(A|\text{2nd reliable}) = \mathbb{P}(\text{1st never works}|\text{2nd reliable})$$

$$+ \frac{1}{2} \times \mathbb{P}(\text{1st works half-time}|\text{2nd reliable})$$

$$= \frac{1}{2} + \frac{1}{2} \times \frac{1}{2} = \frac{3}{4},$$

and the probability $\mathbb{P}(A|\text{3rd reliable})$ equals

$$\frac{1}{2} \times \frac{1}{2} \times \mathbb{P}(\text{2nd works half-time}|\text{3rd reliable}) = \frac{1}{8}.$$

The required probability $\mathbb{P}(\text{2nd reliable})$ is then

$$\frac{1/3 \times 3/4}{1/3 \times (0 + 3/4 + 1/8)} = \frac{6}{7}. \quad \square$$

Problem 1.10   Parliament contains a proportion p of Labour Party members, incapable of changing their opinions about anything, and $1 - p$ of Tory Party members changing their minds at random, with probability r, between subsequent votes on the same issue. A randomly chosen parliamentarian is noticed to have voted twice in succession in the same way. Find the probability that he or she will vote in the same way next time.

Solution Set

$$A_1 = \{\text{Labour chosen}\}, \quad A_2 = \{\text{Tory chosen}\},$$

$$B = \{\text{the member chosen voted twice in the same way}\}.$$

We have $\mathbb{P}(A_1) = p$, $\mathbb{P}(A_2) = 1 - p$, $\mathbb{P}(B|A_1) = 1$, $\mathbb{P}(B|A_2) = 1 - r$. We want to calculate

$$\mathbb{P}(A_1|B) = \frac{\mathbb{P}(A_1 \cap B)}{\mathbb{P}(B)} = \frac{\mathbb{P}(A_1)\mathbb{P}(B|A_1)}{\mathbb{P}(B)}$$

and $\mathbb{P}(A_2|B) = 1 - \mathbb{P}(A_1|B)$. Write

$$\mathbb{P}(B) = \mathbb{P}(A_1)\mathbb{P}(B|A_1) + \mathbb{P}(A_2)\mathbb{P}(B|A_2) = p \cdot 1 + (1 - p)(1 - r).$$

Then

$$\mathbb{P}(A_1|B) = \frac{p}{p + (1 - r)(1 - p)}, \quad \mathbb{P}(A_2|B) = \frac{(1 - r)(1 - p)}{p + (1 - r)(1 - p)},$$

and the answer is given by

$$\mathbb{P}\big(\text{the member will vote in the same way}\big|B\big) = \frac{p + (1 - r)^2(1 - p)}{p + (1 - r)(1 - p)}. \quad \square$$

Problem 1.11 The Polya urn model is as follows. We start with an urn which contains one white ball and one black ball. At each second we choose a ball at random from the urn and replace it together with one more ball of the same colour. Calculate the probability that when n balls are in the urn, i of them are white.

Solution Denote by $\mathbb{P}_n$ the conditional probability given that there are n balls in the urn. For $n = 2$ and 3

$$\mathbb{P}_n(\text{one white ball}) = \begin{cases} 1, & n = 2 \\ \frac{1}{2}, & n = 3, \end{cases}$$

and

$$\mathbb{P}_n(\text{two white balls}) = \tfrac{1}{2}, \quad n = 3.$$

Make the induction hypothesis

$$\mathbb{P}_k(i \text{ white balls}) = \frac{1}{k - 1},$$

$\forall \ k = 2, \ldots, n - 1$ and $i = 1, \ldots, k - 1$. Then, after $n - 1$ trials (when the number of balls is n),

$$\mathbb{P}_n(i \text{ white balls})$$
$$= \mathbb{P}_{n-1}(i - 1 \text{ white balls}) \times \frac{i - 1}{n - 1} + \mathbb{P}_{n-1}(i \text{ white balls}) \times \frac{n - 1 - i}{n - 1}$$
$$= \frac{1}{n - 1}, \quad i = 1, \ldots, n - 1.$$

Hence,

$$\mathbb{P}_n(i \text{ white balls}) = \frac{1}{n-1}, \quad i = 1, \ldots, n-1. \quad \square$$

Problem 1.12   You have n urns, the rth of which contains $r-1$ red balls and $n-r$ blue balls, $r = 1, \ldots, n$. You pick an urn at random and remove two balls from it without replacement. Find the probability that the two balls are of different colours. Find the same probability when you put back a removed ball.

Solution   The totals of blue and red balls in all urns are equal. Hence, the first ball is equally likely to be any ball. So

$$\mathbb{P}(\text{1st blue}) = \frac{1}{2} = \mathbb{P}(\text{1st red}).$$

Now,

$$\mathbb{P}(\text{1st red, 2nd blue}) = \sum_{k=1}^{n} \mathbb{P}(\text{1st red, 2nd blue} \mid \text{urn } k \text{ chosen}) \times \frac{1}{n}$$

$$= \frac{1}{n} \sum_{k} \frac{(k-1)(n-k)}{(n-1)(n-2)}$$

$$= \frac{1}{n(n-1)(n-2)} \left[n \sum_{k=1}^{n} (k-1) - \sum_{k=1}^{n} k(k-1) \right]$$

$$= \frac{1}{n(n-1)(n-2)} \left[\frac{n(n-1)n}{2} - \frac{(n+1)n(n-1)}{3} \right]$$

$$= \frac{n(n-1)}{n(n-1)(n-2)} \left(\frac{n}{2} - \frac{n+1}{3} \right) = \frac{1}{6}.$$

We used here the following well-known identity:

$$\sum_{i=1}^{n} i(i-1) = \frac{1}{3}(n+1)n(n-1).$$

By symmetry:

$$\mathbb{P}(\text{different colours}) = 2 \times \frac{1}{6} = \frac{1}{3}.$$

If you return a removed ball, the probability that the two ball are of different colours becomes $1/2$. $\quad \square$

Problem 1.13   You are on a game show and given a choice of three doors. Behind one is a car; behind the two others are a goat and a pig. You pick door 1, and the host opens door 3, with a pig. The host asks if you want to pick door 2 instead. Should you switch? What if instead of a goat and a pig there were two goats?

Solution A popular solution of this problem always assumes that the host knows behind which door the car is, and takes care not to open this door rather than doing so by chance. (It is assumed that the host never opens the door picked by you.) In fact, it is instructive to consider two cases, depending on whether the host does or does not know the door with the car. If he doesn't, your chances are unaffected, otherwise you should switch. Indeed, consider the events

$$Y_i = \{\text{you pick door } i\}, \ C_i = \{\text{the car is behind door } i\},$$

$$H_i = \{\text{the host opens door } i\}, \ G_i/P_i = \{\text{a goat/pig is behind door } i\},$$

with $\mathbb{P}(Y_i) = \mathbb{P}(C_i) = \mathbb{P}(G_i) = \mathbb{P}(P_i) = 1/3, i = 1, 2, 3$. Obviously, event Y_i is independent of any of the events C_j, G_j and P_j, $i, j = 1, 2, 3$.
   You want to calculate

$$\mathbb{P}(C_1 | Y_1 \cap H_3 \cap P_3) = \frac{\mathbb{P}(C_1 \cap Y_1 \cap H_3 \cap P_3)}{\mathbb{P}(Y_1 \cap H_3 \cap P_3)}.$$

In the numerator:

$$\mathbb{P}(C_1 \cap Y_1 \cap H_3 \cap P_3)$$
$$= \mathbb{P}(C_1)\mathbb{P}(Y_1 | C_1)\mathbb{P}(P_3 | C_1 \cap Y_1)\mathbb{P}(H_3 | C_1 \cap Y_1 \cap P_3)$$
$$= \frac{1}{3} \times \frac{1}{3} \times \frac{1}{2} \times \frac{1}{2} = \frac{1}{36}.$$

If the host doesn't know where the car is, then

$$\mathbb{P}(Y_1 \cap H_3 \cap P_3)$$
$$= \mathbb{P}(Y_1)\mathbb{P}(P_3 | Y_1)\mathbb{P}(H_3 | Y_1 \cap P_3)$$
$$= \frac{1}{3} \times \frac{1}{3} \times \frac{1}{2} = \frac{1}{18},$$

and $\mathbb{P}(C_1 | Y_1 \cap H_3 \cap P_3) = 1/2$. But if he does then

$$\mathbb{P}(Y_1 \cap H_3 \cap P_3)$$
$$= \mathbb{P}(Y_1 \cap C_1 \cap H_3 \cap P_3) + \mathbb{P}(Y_1 \cap C_2 \cap H_3 \cap P_3)$$
$$= \frac{1}{3} \times \frac{1}{3} \times \frac{1}{2} \times \frac{1}{2} \times \frac{1}{3} \times \frac{1}{3} \times \frac{1}{2} \times 1 = \frac{1}{12},$$

and $\mathbb{P}(C_1 | Y_1 \cap H_3 \cap P_3) = 1/3$.
   The answer remains the same if there were two goats instead of a goat and a pig. Another useful exercise is to consider the case where the host has some 'preference' choosing a door with the goat with probability p_g and that with the pig with probability $p_p = 1 - p_g$. □

   We continue our study by introducing the definition of independent events. The concept of independence was an important invention in probability theory. It shaped the theory at an early stage and is considered one of the main features specifying the place of probability theory within more general measure theory.

We say that events A and B are *independent* if

$$\mathbb{P}(A \cap B) = \mathbb{P}(A)\mathbb{P}(B). \tag{1.9}$$

A convenient criterion of independence is: events A and B, where say $\mathbb{P}(B) > 0$ are independent iff $\mathbb{P}(A|B) = \mathbb{P}(A)$, i.e. knowledge that B occurred does not change the probability of A.

Trivial examples are the empty event $\emptyset$ and the whole set Ω: they are independent of any event. The next example we consider is when each of the four outcomes $00, 01, 10,$ and 11 have probability $1/4$. Here the events

$$A = \{1\text{st digit is } 1\} \text{ and } B = \{2\text{nd digit is } 0\}$$

are independent:

$$\mathbb{P}(A) = p_{10} + p_{11} = \frac{1}{2} = p_{10} + p_{00} = \mathbb{P}(B), \ \ \mathbb{P}(A \cap B) = p_{10} = \frac{1}{4} = \frac{1}{2} \times \frac{1}{2}.$$

Also, the events

$$\{1\text{st digit is } 0\} \text{ and } \{\text{both digits are the same}\}$$

are independent, while the events

$$\{1\text{st digit is } 0\} \text{ and } \{\text{the sum of digits is } > 0\}$$

are dependent.

These examples can be easily re-formulated in terms of two unbiased coin-tossings. An important fact is that if A, B are independent then A^c and B are independent:

$$\mathbb{P}(A^c \cap B) = \mathbb{P}(B \backslash (A \cap B)) = \mathbb{P}(B) - \mathbb{P}(A \cap B)$$

$$= \mathbb{P}(B) - \mathbb{P}(A)\mathbb{P}(B) \ \text{(by independence)}$$

$$= [1 - \mathbb{P}(A)]\mathbb{P}(B) = \mathbb{P}(A^c)\mathbb{P}(B).$$

Next, if (i) A_1 and B are independent, (ii) A_2 and B are independent, and (iii) A_1 and A_2 are disjoint, then $A_1 \cup A_2$ and B are independent. If (i) and (ii) hold and $A_1 \subset A_2$ then B and $A_2 \backslash A_1$ are also independent.

Intuitively, independence is often associated with an 'absence of any connection' between events. There is a famous joke about A.N. Kolmogorov (1903–1987), the renowned Russian mathematician considered the father of the modern probability theory. His monograph [Ko], which originally appeared in German in 1933, was revolutionary in understanding the basics of probability theory and its rôle in mathematics and its applications. When in the 1930s this monograph was translated into Russian, the Soviet government enquired about the nature of the concept of independent events. A senior minister asked if this concept was consistent with materialistic determinism, the core of Marxist–Leninist philosophy, and about examples of such events. Kolmogorov had to answer on the spot, and he had to be cautious as subsequent events showed, such as the infamous condemnation by the authorities of genetics as a 'reactionary bourgeois pseudo-science'. The legend is that he did not hesitate for a second, and said: 'Look, imagine a remote village where there has been a long drought. One day, local peasants in desperation go to the church, and the priest says a prayer for rain. And the next day the rain arrives! These are independent events.'

In reality, the situation is more complex. A helpful view is that independence is a geometric property. In the above example, the four probabilities

$$p_{00}, \ p_{01}, \ p_{10} \text{ and } p_{11}$$

can be assigned to the vertices

$$(0; 0), \ (0; 1), \ (1; 1), \ \text{and } (1; 0)$$

of a unit square. See Figure 1.1. Each of these four points has a projection onto the horizontal and the vertical line. The projections are points 0 and 1 on each of these lines, and a vertex is uniquely determined by its projections. If the projection points have probability mass $1/2$ on each line then each vertex has

$$p_{ij} = \frac{1}{2} \times \frac{1}{2} = \frac{1}{4}, \ i, j = 0, 1.$$

In this situation one says that the four-point probability distribution

$$\left\{ \frac{1}{4}, \frac{1}{4}, \frac{1}{4}, \frac{1}{4} \right\}$$

is a product of two two-point distributions

$$\left\{ \frac{1}{2}, \frac{1}{2} \right\}.$$

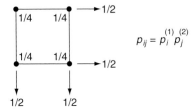

Figure 1.1

It is easy to imagine a similar picture where there are m points along the horizontal and n along the vertical line: we would then have mn pairs (i, j) (lattice sites) where $i = 0, \ldots, m - 1$, $j = 0, \ldots, n - 1$ and each pair will receive probability mass $1/mn$. Moreover, the equidistribution can be replaced by a more general law:

$$p_{ij} = p_i^{(1)} p_j^{(2)},$$

where

$$p_i^{(1)}, i = 0, \ldots, m - 1, \text{ and } p_j^{(2)}, j = 0, \ldots, n - 1,$$

are probability distributions for the two projections. Then any event that is expressed in terms of the horizontal projection (e.g., {digit i is divisible by 3}) is independent of any event expressed in terms of the vertical projection (e.g., {digit j is $\leq n/2$}). This is a basic (and in a sense the only) example of independence.

More generally, we say that events $A_1, \ldots, A_n$ are mutually independent (shortly, independent) if $\forall$ subcollections $A_{i_1}, \ldots, A_{i_l}$,

$$\mathbb{P}(A_{i_1} \cap \cdots \cap A_{i_l}) = \mathbb{P}(A_{i_1}) \ldots \mathbb{P}(A_{i_l}). \tag{1.10}$$

This includes the whole collection $A_1, \ldots, A_n$. It is important to distinguish this situation from the one where (1.10) holds only for some subcollections; say pairs A_i, A_j, $1 \leq i < j \leq n$, or only for the whole collection $A_1, \ldots, A_n$. See Problems 1.20 and 1.21.

This gives rise to an important model: a sequence of independent trials, each with two or more outcomes. Such a model is behind many problems, and it is essential to familiarize yourself with it.

So far, we assumed that the total number of outcomes ω is finite, but the material in this section can be extended to the case where Ω is a countable set, consisting of points $x_1, x_2, \ldots$, say, with assigned probabilities $p_i = \mathbb{P}(\{x_i\})$, $i = 1, 2, \ldots$ Of course, the labelling of the outcomes can be different, for instance, by $i \in \mathbb{Z}$, the set of integers. The requirements are as before: each $p_i \geq 0$ and $\sum_i p_i = 1$.

We can also work with infinite sequences of events. For example, equations (1.7) and (1.8) do not change form:

$$\mathbb{P}(A) = \sum_{1 \leq j < \infty} \mathbb{P}(A|B_j)\mathbb{P}(B_j), \quad \mathbb{P}(B_i|A) = \frac{\mathbb{P}(A|B_i)\mathbb{P}(B_i)}{\sum_{1 \leq j < \infty} \mathbb{P}(A|B_j)\mathbb{P}(B_j)}. \tag{1.11}$$

Problem 1.14   A coin shows heads with probability p on each toss. Let π_n be the probability that the number of heads after n tosses is even. By showing that $\pi_{n+1} = (1 - p)\pi_n + p(1 - \pi_n)$, $n \geq 1$, or otherwise, find π_n. (The number 0 is considered even.)

Solution 1   As always in the coin-tossing models, we assume that outcomes of different throws are independent. Set $A_n = \{n\text{th toss is a head}\}$, with $\mathbb{P}(A_n) = p$ and

$B_n = \{$even number of heads after n tosses$\}$, with $\pi_n = \mathbb{P}(B_n)$. Then, by conditioning on A_{n+1} and A^c_{n+1}:

$$\mathbb{P}(B_{n+1}) = \mathbb{P}(B_{n+1} \cap A_{n+1}) + \mathbb{P}(B_{n+1} \cap A^c_{n+1})$$
$$= \mathbb{P}(B_{n+1}|A_{n+1})\mathbb{P}(A_{n+1}) + \mathbb{P}(B_{n+1}|A^c_{n+1})\mathbb{P}(A^c_{n+1}).$$

Next, $B_{n+1} \cap A_{n+1} = B^c_n \cap A_{n+1}$ and $B_{n+1} \cap A^c_{n+1} = B_n \cap A^c_{n+1}$. In view of independence,

$$\mathbb{P}(B^c_n \cap A_{n+1}) = \mathbb{P}(B^c_n)\mathbb{P}(A_{n+1}), \text{ and } \mathbb{P}(B_n \cap A^c_{n+1}) = \mathbb{P}(B_n)\mathbb{P}(A^c_{n+1}),$$

which implies

$$\mathbb{P}(B_{n+1}) = \mathbb{P}(B^c_n)\mathbb{P}(A_{n+1}) + \mathbb{P}(B_n)\mathbb{P}(A^c_{n+1}) = (1 - \mathbb{P}(B_n))p + \mathbb{P}(B_n)(1-p),$$

with $\mathbb{P}(B_0) = 1$. That is,

$$\pi_{n+1} = (1-p)\pi_n + p(1 - \pi_n) = (1-2p)\pi_n + p.$$

Substituting $\pi_n = a(1-2p)^n + b$ gives that $b = 1/2$, and the condition $\pi_0 = 1$ that $a = 1/2$. Then $\pi_n = [(1-2p)^n + 1]/2$.

A shorter way to derive the recursion is by conditioning on B_n and B^c_n:

$$\pi_{n+1} = \mathbb{P}(B_{n+1}) = \mathbb{P}(B_{n+1} \cap B_n) + \mathbb{P}(B_{n+1} \cap B^c_n)$$
$$= \mathbb{P}(A^c_{n+1}|B_n)\mathbb{P}(B_n) + \mathbb{P}(A_{n+1}|B^c_n)\mathbb{P}(B^c_n)$$
$$= (1-p)\pi_n + p(1 - \pi_n).$$

Writing recursive equations like the one in the statement of the current problem is a convenient instrument used in a great many situations.

Solution 2 (Look at this solution after reading Section 1.5.) Let $X_i = 0$ or 1 be the outcome of the ith trial, and $Y_n = X_1 + \cdots + X_n$ the total number of successes in n trials. Then the probability generating function of Y_n

$$\psi(s) = [ps + (1-p)]^n.$$

Then the probability that n trials result in an even number of successes is

$$\frac{1}{2}[\psi(1) + \psi(-1)] = \frac{1}{2}[1 + (1-2p)^n]. \quad \square$$

Problem 1.15 My Aunt Agatha has given me a clockwork orange for my birthday. I place it in the middle of my dining table which happens to be exactly 2 metres long. One minute after I place it on the table it makes a loud whirring sound, emits a puff of green smoke and moves 10 cm towards the left-hand end of the table with probability 3/5, or 10 cm towards the right with probability 2/5. It continues in this manner (the direction of each move being independent of what has gone before) at one minute intervals until it reaches the edge of the table where it promptly falls off. If it falls off the left-hand end it will break my Ming vase (also a present from Aunt Agatha). If it falls off the right-hand end it will land safely in a bucket of water. What is the probability that the Ming vase will survive?

Solution   Set p_ℓ to be

$$\mathbb{P}(\text{falls at RH end}|\text{was at distance } \ell \times 10\,\text{cm from the LH end}),$$

then $1 - p_\ell$ equals

$$\mathbb{P}(\text{falls at LH end}|\text{was at distance } \ell \times 10\,\text{cm from LH end}).$$

We have $p_0 = 0$, $p_{20} = 1$ and $p_\ell = \frac{3}{5}p_{\ell-1} + \frac{2}{5}p_{\ell+1}$ or

$$p_{\ell+1} = \frac{5}{2}p_\ell - \frac{3}{2}p_{\ell-1}.$$

In other words, vector $(p_\ell, p_{\ell+1}) = (p_{\ell-1}, p_\ell)A$ with

$$A = \begin{pmatrix} 0 & -\frac{3}{2} \\ 1 & \frac{5}{2} \end{pmatrix}.$$

This yields

$$(p_\ell, p_{\ell+1}) = (p_0, p_1)A^\ell = (0, p_1)A^\ell,$$

i.e. p_ℓ should be a linear combination of the ℓth powers of the eigenvalues of A. The eigenvalues are $\lambda_1 = \frac{3}{2}$ and $\lambda_2 = 1$ and so:

$$p_\ell = b_1 \left(\frac{3}{2}\right)^\ell + b_2.$$

We have the equations

$$b_1 + b_2 = 0, \quad 1 = b_1 \left(\frac{3}{2}\right)^{20} + b_2,$$

whence

$$b_1 = -b_2 = \left[\left(\frac{3}{2}\right)^{20} - 1\right]^{-1}$$

and

$$p_{10} = \left(\frac{3}{2}\right)^{10} \frac{1}{(3/2)^{20} - 1} - \frac{1}{(3/2)^{20} - 1} = \frac{1}{(3/2)^{10} + 1}. \quad \square$$

Problem 1.16   Dubrovsky sits down to a night of gambling with his fellow officers. Each time he stakes u roubles there is a probability r that he will win and receive back $2u$ roubles (including his stake). At the beginning of the night he has 8000 roubles. If ever he has 256 000 roubles he will marry the beautiful Natasha and retire to his estate in the country. Otherwise, he will commit suicide. He decides to follow one of two courses of action:

(i)   to stake 1000 roubles each time until the issue is decided;
(ii)   to stake everything each time until the issue is decided.

Advise him (a) if $r = 1/4$ and (b) if $r = 3/4$. What are the chances of a happy ending in each case if he follows your advice?

Solution   Let p_ℓ be the probability that Dubrovsky wins 256 000 with the starting capital ℓ thousands while following strategy (i). Reasoning as in Problem 1.15 yields that

$$p_\ell = b_1 \lambda_1^\ell + b_2 \lambda_2^\ell$$

where $\lambda_1 = (1 - r)/r$ and $\lambda_2 = 1$ are the eigenvalues of the matrix

$$A = \begin{pmatrix} 0 & (1-r)/r \\ 1 & 1/r \end{pmatrix}.$$

The boundary conditions $p_0 = 0$ and $p_{256} = 1$ yield

$$b_1 = -b_2 = \left[\left(\frac{1-r}{r} \right)^{256} - 1 \right]^{-1}.$$

For $r = 1/4$, $(1 - r)/r = 3$. Then he should choose (ii) as

$$p_8 = \frac{3^8 - 1}{3^{256} - 1},$$

which is tiny compared with $(1/4)^5$, the chance to win 256 000 in five successful rounds by gambling on everything he obtains.
    For $r = 3/4$, $(1 - r)/r = 1/3$. Then

$$p_8 = \frac{1 - (1/3)^8}{1 - (1/3)^{256}}$$

which is much larger than $(3/4)^5$. Therefore, he should choose (i).   □

Remark   In both Problems 1.15 and 1.16 one of the eigenvalues of the recursion matrix A equals one. This is not accidental and is due to the fact that in equation (1.7) (which gives rise to the recursive equations under consideration) the sum $\sum_j \mathbb{P}(B_j) = 1$.

Problem 1.17   I play the dice game 'craps' against 'Lucky' Pete Jay as follows. On each throw I throw two dice. If my first throw is 7 or 11, then I win and if it is 2, 3 or 12, then I lose. If my first throw is none of these, I throw repeatedly until I score the same as my first throw, in which case I win, or I throw a 7, in which case I lose. What is the probability that I win?

Solution   Write

$$\mathbb{P}(\text{I win}) = \mathbb{P}(\text{I win at the 1st throw}) + \mathbb{P}(\text{I win, but not at the 1st throw}).$$

The probability $\mathbb{P}(\text{I win at the 1st throw})$ is straightforward and equals

$$\sum_{i,j=1}^{6} \frac{1}{36} I(i+j=7) + \sum_{i,j=1}^{6} \frac{1}{36} I(i+j=11) = \frac{6}{36} + \frac{2}{36} = \frac{2}{9}.$$

Here

$$I(i+j=7) = \begin{cases} 1, & \text{if } i+j=7, \\ 0, & \text{otherwise.} \end{cases}$$

For the second probability we have

$$\mathbb{P}(\text{I win, but not at the 1st throw}) = \sum_{i=4,5,6,8,9,10} p_i q_i.$$

Here

$$p_i = \mathbb{P}(\text{the 1st score is } i)$$

and

$$q_i = \mathbb{P}(\text{get } i \text{ before 7 in the course of repeated throws} | \text{the 1st score is } i)$$

$$= \mathbb{P}(\text{get } i \text{ before 7 in the course of repeated throws}).$$

Then for q_i, by conditioning on the result of the first throw:

$$q_i = p_i + (1 - p_i - p_7)q_i, \text{ i.e. } q_i = \frac{p_i}{p_i + p_7}.$$

Equivalently,

$$q_i = p_i + (1 - p_i - p_7)p_i + (1 - p_i - p_7)^2 p_i + \cdots,$$

with the same result.

Now

$$q_4 = \frac{3/36}{3/36 + 6/36} = \frac{3}{3+6} = \frac{1}{3}, \ q_5 = \frac{4/36}{4/36 + 6/36} = \frac{4}{4+6} = \frac{2}{5},$$

and likewise,

$$q_6 = \frac{5/36}{5/36 + 6/36} = \frac{5}{5+6} = \frac{5}{11}, \ q_8 = \frac{5}{11}, \ q_9 = \frac{2}{5}, \ q_{10} = \frac{1}{3},$$

giving for $\mathbb{P}(\text{I win, but not at the 1st throw})$ the value

$$\frac{1}{12} \times \frac{1}{3} + \frac{1}{9} \times \frac{2}{5} + \frac{5}{36} \times \frac{5}{11} + \frac{5}{36} \times \frac{5}{11} + \frac{1}{9} \times \frac{2}{5} + \frac{1}{12} \times \frac{1}{3} = \frac{134}{495}.$$

Then

$$\mathbb{P}(\text{I win}) = \frac{2}{9} + \frac{134}{495} = \frac{244}{495}. \quad \square$$

Problem 1.18   Two darts players throw alternately at a board and the first to score a bull wins. On each of their throws player A has probability p_A and player B p_B of success; the results of different throws are independent. If A starts, calculate the probability that he/she wins.

Solution   Consider the diagram below.

$$
\begin{array}{cccc}
1-p_A & 1-p_B & 1-p_A & 1-p_B \\
\bullet \longrightarrow & \bullet \longrightarrow & \bullet \longrightarrow & \bullet \longrightarrow \quad \cdots \\
& & & \qquad \ddots \\
p_A \searrow & p_B \searrow & p_A \searrow & p_B \searrow \\
\text{A wins} & \text{B wins} & \text{A wins} & \text{B wins}
\end{array}
$$

If $q = \mathbb{P}(A\text{ wins})$, then

$$q = p_A + (1-p_A)(1-p_B)p_A + (1-p_A)^2(1-p_B)^2 p_A + \cdots$$

$$= \frac{p_A}{1-(1-p_A)(1-p_B)} = \frac{p_A}{p_A + p_B - p_A p_B}.$$

Equivalently, by conditioning on the first and the second throw, one gets the equation

$$\mathbb{P}(A\text{ wins}) = p_A + (1-p_A)(1-p_B)\mathbb{P}(A\text{ wins}),$$

which is immediately solved to give the required result. $\square$

Remark   In Problems 1.17 and 1.18 we used an equation for the probabilities q and q_i that was equivalent to their representation as series. This is another useful idea; for example, it allowed us to avoid the use of infinite outcome spaces. However, we will not be able to avoid it much longer.

Problem 1.19   A fair coin is tossed until either the sequence *HHH* occurs in which case I win or the sequence *THH* occurs, when you win. What is the probability that you win?

Solution   In principle, the results of the game could be I win, you win and the game lasts forever. Observe that I win only if *HHH* occurs at the beginning: the probability is $(1/2)^3 = 1/8$. Indeed, if *HHH* occurs but not at the beginning then *THH* should have occurred before then you will have already won. But *HHH* will appear sooner or later, with probability 1. In fact, $\forall\, N$, the event

$$A = \{HHH \text{ never occurs}\}$$

is contained in the event

$$A_N = \{HHH \text{ does not occur among the first } N \text{ subsequent triples}\}$$

(we partition first $3N$ trials into N subsequent triples). So $\mathbb{P}(A) \leq \mathbb{P}(A_N)$. But the probability $\mathbb{P}(A_N) = (1 - 1/8)^N \to 0$ as $N \to \infty$. Hence, $\mathbb{P}(A) = 0$. Therefore, the game cannot continue forever, and the probability that you win is $1 - 1/8 = 7/8$. $\square$

Problem 1.20   (i) Give examples of the following phenomena:
(a) three events A, B, C that are pair-wise independent but not independent;
(b) three events that are not independent, but such that the probability of the intersection of all three is equal to the product of the probabilities.
(ii) Three coins each show heads with probability $3/5$ and tails otherwise. The first counts 10 points for a head and 2 for a tail, the second counts 4 points for a head and tail, and the third counts 3 points for a head and 20 for a tail.
You and your opponent each choose a coin; you cannot choose the same coin. Each of you tosses your coin once and the person with the larger score wins 10^{10} points. Would you prefer to be the first or the second to choose a coin?

Solution   (i) (a) Toss two unbiased coins, with

$$A = \{1\text{st toss shows } H\}, \quad B = \{2\text{nd toss shows } H\},$$

and

$$C = \{\text{both tosses show the same side}\}.$$

Then

$$\mathbb{P}(A \cap B) = p_{HH} = \frac{1}{4} = \mathbb{P}(A)\mathbb{P}(B), \quad \mathbb{P}(A \cap C) = p_{HH} = \frac{1}{4} = \mathbb{P}(A)\mathbb{P}(C),$$

$$\mathbb{P}(B \cap C) = p_{HH} = \frac{1}{4} = \mathbb{P}(B)\mathbb{P}(C),$$

and

$$\mathbb{P}(A \cap B \cap C) = p_{HH} = \frac{1}{4} \neq \mathbb{P}(A)\mathbb{P}(B)\mathbb{P}(C).$$

Or throw three dice, with

$$A = \{\text{dice one shows an odd score}\}, \quad B = \{\text{dice two shows an odd score}\},$$

$$C = \{\text{overall score odd}\}$$

and $\mathbb{P}(A) = \mathbb{P}(B) = \mathbb{P}(C) = 1/2$. Then

$$\mathbb{P}(A \cap B) = \mathbb{P}(A \cap C) = \mathbb{P}(B \cap C) = \frac{1}{4}, \quad \text{but } \mathbb{P}(A \cap B \cap C) = 0.$$

(b) Toss three coins, with

$$A = \{\text{1st toss shows } H\}, \quad B = \{\text{3rd toss shows } H\},$$

$$C = \{HHH, HHT, HTT, TTT\} = \{\text{no subsequent } TH\}.$$

Then $\mathbb{P}(A) = \mathbb{P}(B) = \mathbb{P}(C) = 1/2$,

$$A \cap B \cap C = \{HHH\}, \quad \text{and} \quad \mathbb{P}(A \cap B \cap C) = \frac{1}{8} = \left(\frac{1}{2}\right)^3.$$

But

$$A \cap C = \{HHH, HHT, HTT\}, \quad \text{and} \quad \mathbb{P}(A \cap C) = \frac{3}{8},$$

while

$$B \cap C = \{HHH\}, \quad \text{and} \quad \mathbb{P}(B \cap C) = \frac{1}{8}.$$

Or (as the majority of students' attempts did so far), take $A = \emptyset$, and any dependent pair B, C (say, $B = C$ with $0 < \mathbb{P}(B) < 1$). Then $A \cap B \cap C = \emptyset$ and

$$\mathbb{P}(A \cap B \cap C) = 0 = \mathbb{P}(A)\mathbb{P}(B)\mathbb{P}(C), \quad \text{but} \quad \mathbb{P}(B \cap C) \neq \mathbb{P}(B)\mathbb{P}(C).$$

(ii) Suppose I choose coin 1 and you coin 2, then $\mathbb{P}(\text{you win}) = 2/5$. But if you choose 3 then

$$\mathbb{P}(\text{you win}) = \frac{2}{5} + \frac{3}{5} \times \frac{2}{5} = \frac{16}{25} > \frac{1}{2}.$$

Similarly, if I choose 2 and you choose 1, $\mathbb{P}(\text{you win}) = 3/5 > 1/2$. Finally, if I choose 3 and you choose 2 then $\mathbb{P}(\text{you win}) = 3/5 > 1/2$. Thus, it's always better to be the second. $\square$

Problem 1.21 Let $A_1, \ldots, A_n$ be independent events, with $\mathbb{P}(A_i) < 1$. Prove that there exists an event B with $\mathbb{P}(B) > 0$ such that $B \cap A_i = \emptyset$, for $1 \leq i \leq n$.

Give an example of three events A_1, A_2, A_3 which are not independent, but are such that for each $i \neq j$ the pair A_i, A_j is independent.

Solution If A^c denotes the complement of event A, then

$$\mathbb{P}(A_1 \cup \cdots \cup A_n) = 1 - \mathbb{P}\left(\bigcap_{i=1}^{n} A_i^c\right) = 1 - \prod_{i=1}^{n} \mathbb{P}(A_i^c) < 1,$$

as $\mathbb{P}(A_i^c) > 0 \; \forall \; i$. So, if $B = (\bigcup A_i)^c$, then $\mathbb{P}(B) > 0$ and $B \cap A_i = \emptyset \; \forall \; i$.

Next, take events

$$A_1 = \{1, 4\}, \; A_1 = \{2, 4\}, \; A_3 = \{3, 4\},$$

where the probability assigned to each outcome $k = 1, 2, 3, 4$ equals $1/4$. Then $\mathbb{P}(A_i) = 1/2$, $i = 1, 2, 3$, and

$$\mathbb{P}(A_i \cap A_j) = \mathbb{P}\{4\} = \left(\frac{1}{2}\right)^2 = \mathbb{P}(A_i)\mathbb{P}(A_j), \quad 1 \le i < j \le 3.$$

However, the intersection $A_1 \cap A_2 \cap A_3$ consists of a single outcome 4. Hence

$$\mathbb{P}(A_1 \cap A_2 \cap A_3) = \frac{1}{4} \ne \left(\frac{1}{2}\right)^3 = \mathbb{P}(A_1)\mathbb{P}(A_2)\mathbb{P}(A_3). \quad \square$$

Problem 1.22 n balls are placed at random into n cells. Find the probability p_n that exactly two cells remain empty.

Solution   'At random' means here that each ball is put in a cell with probability $1/n$, independently of other balls. First, consider the cases $n = 3$ and $n = 4$. If $n = 3$ we have one cell with three balls and two empty cells. Hence,

$$p_3 = \binom{n}{2} \times \frac{1}{n^n} = \frac{1}{9}.$$

If $n = 4$ we either have two cells with two balls each (probability p_4') or one cell with one ball and one cell with three balls (probability p_4''). Hence,

$$p_4 = p_4' + p_4'' = \binom{n}{2} \times \frac{(n-2)}{n^n} \times \left[4 + \frac{n(n-1)}{4}\right] = \frac{21}{64}.$$

Here 4 stands for a number of ways to select three balls that will go to one cell, and $n(n-1)/4$ stands for the number of ways to select two pairs of balls that will go to two prescribed cells.

For $n \ge 5$, to have exactly two empty cells means that either there are exactly two cells with two balls in them and $n - 4$ with a single ball, or there is one cell with three balls and $n - 3$ with a single ball. Denote probabilities in question by p_n' and p_n'', respectively. Then $p_n = p_n' + p_n''$. Further,

$$p_n' = \binom{n}{2} \times \frac{1}{2}\binom{n}{2}\binom{n-2}{2} \times \frac{n-2}{n}\frac{n-3}{n}\cdots\frac{1}{n} \times \left(\frac{1}{n}\right)^2$$

$$= \frac{1}{4}\frac{n!}{n^n}\binom{n}{2}\binom{n-2}{2}.$$

Here the first factor, $\binom{n}{2}$, is responsible for the number of ways of choosing two empty cells among n. The second factor,

$$\frac{1}{2}\binom{n}{2}\binom{n-2}{2},$$

accounts for choosing which balls 'decide' to fall in cells that will contain two balls and also which cells will contain two balls. Finally, the third factor,

$$\frac{n-2}{n}\frac{n-3}{n}\cdots\frac{1}{n},$$

gives the probability that $n-2$ balls fall in $n-2$ cells, one in each, and the last $(1/n)^2$ that two pairs of balls go into the cells marked for two-ball occupancy. Next,

$$p_n'' = \binom{n}{2} \times (n-2) \times \binom{n}{3} \frac{(n-3)!}{n^n}.$$

Here the first factor $\binom{n}{2}$ is responsible for the number of ways of choosing two empty cells among n, the second $(n-2)$ is responsible for the number of ways of choosing the cell with three balls, the third $\binom{n}{3}$ is responsible for the number of ways of choosing three balls to go into this cell, and the last factor $\frac{(n-3)!}{n^n}$ describes the distribution of all balls into the respective cells.   □

Problem 1.23   You play a match against an opponent in which at each point either you or he/she serves. If you serve you win the point with probability p_1, but if your opponent serves you win the point with probability p_2. There are two possible conventions for serving:

(i) alternate serves;
(ii) the player serving continues until he/she loses a point.

You serve first and the first player to reach n points wins the match. Show that your probability of winning the match does not depend on the serving convention adopted.

Solution   Both systems give you equal probabilities of winning. In fact, suppose we extend the match beyond the result achieved, until you have served n and your opponent $n-1$ times. (Under rule (i) you just continue the alternating services and under (ii) the loser is given the remaining number of serves.) Then, under either rule, if you win the actual game, you also win the extended one, and vice versa (as your opponent won't have enough points to catch up with you). So it suffices to check the extended matches.

An outcome of an extended match is $\omega = (\omega_1, \ldots, \omega_{2n-1})$, a sequence of $2n-1$ subsequent values, say zero (you lose a point) and one (you gain one). You may think that $\omega_1, \ldots, \omega_n$ represent the results of your serves and $\omega_{n+1}, \ldots, \omega_{2n-1}$ those of your opponent. Define events

$$A_i = \{\text{you win your } i\text{th service}\}, \text{ and } B_j = \{\text{you win his } j\text{th service}\}.$$

Their respective indicator functions are

$$I(\omega \in A_i) = \begin{cases} 1, & \omega \in A_i \\ 0, & \omega \notin A_i, \ 1 \le i \le n, \end{cases}$$

and

$$I(\omega \in B_j) = \begin{cases} 1, & \omega \in B_j \\ 0, & \omega \notin B_j, \ 1 \le j \le n-1. \end{cases}$$

Under both rules, the event that you win the extended match is

$$\left\{ \omega = (\omega_1, \ldots, \omega_{2n-1}) : \sum_{1 \le i \le 2n-1} \omega_i \ge n \right\},$$

and the probability of outcome ω is

$$p_1^{\sum_i I_{A_i}(\omega)} (1 - p_1)^{n - \sum_i I_{A_i}(\omega)} p_2^{\sum_j I_{B_j}(\omega)} (1 - p_2)^{n - \sum_j I_{B_j}(\omega)}.$$

Because they do not depend on the choice of the rule, the probabilities are the same. □

Remark   The ω-notation was quite handy in this solution. We will use it repeatedly in future problems.

Problem 1.24   The departmental photocopier has three parts A, B and C which can go wrong. The probability that A will fail during a copying session is 10^{-5}. The probability that B will fail is 10^{-1} if A fails and 10^{-5} otherwise. The probability that C will fail is 10^{-1} if B fails and 10^{-5} otherwise. The 'Call Engineer' sign lights up if two or three parts fail. If only two parts have failed I can repair the machine myself but if all three parts have failed my attempts will only make matters worse. If the 'Call Engineer' sign lights up and I am willing to run a risk of no greater than 1 per cent of making matters worse, should I try to repair the machine, and why?

Solution   The final outcomes are

$$\begin{aligned} A\,f\,B\,f\,C\,f, & \quad \text{probability } 10^{-5} \times 10^{-1} \times 10^{-1} = 10^{-7}, \\ A\,f\,B\,w\,C\,f, & \quad \text{probability } 10^{-5} \times 9 \cdot 10^{-1} \times 10^{-5} = 9 \cdot 10^{-11}, \\ A\,f\,B\,f\,C\,w, & \quad \text{probability } 10^{-5} \times 10^{-1} \times 9 \cdot 10^{-1} = 9 \cdot 10^{-7}, \\ A\,w\,B\,f\,C\,f, & \quad \text{probability } (1 - 10^{-5}) \times 10^{-5} \times 10^{-1} = (1 - 10^{-5}) \cdot 10^{-6}. \end{aligned}$$

So,

$$\mathbb{P}\left(3 \text{ parts fail} \,|\, 2 \text{ or } 3 \text{ fail}\right) = \frac{\mathbb{P}\,(3 \text{ fail})}{\mathbb{P}\,(2 \text{ or } 3 \text{ fail})}$$

$$= \frac{\mathbb{P}(A, B, C\,f)}{\mathbb{P}(A, B\,f\,C\,w) + \mathbb{P}(A, C\,f\,B\,w) + \mathbb{P}(B, C\,f\,A\,w) + \mathbb{P}(A, B, C\,f)}$$

$$= \frac{10^{-7}}{9 \cdot 10^{-7} + 9 \times 10^{-11} + (1 - 10^{-5}) \cdot 10^{-6} + 10^{-7}}$$

$$\approx \frac{10^{-7}}{9 \cdot 10^{-7} + 10^{-6} + 10^{-7}} = \frac{1}{20} > \frac{1}{100}.$$

Thus, you should not attempt to mend the photocopier: the chances of making things worse are $1/20$. □

1.3   The exclusion–inclusion formula. The ballot problem

> Natural selection is a mechanism
> for generating an exceedingly high degree of improbability.
>> R.A. Fisher (1890–1962), British statistician

The *exclusion–inclusion formula* helps to calculate the probability $\mathbb{P}(A)$, where $A = A_1 \cup A_2 \cup \cdots \cup A_n$ is the union of a given collection of events $A_1, \ldots, A_n$. We know (see Section 1.2) that if events are pair-wise disjoint,

$$\mathbb{P}(A) = \sum_{1 \le i \le n} \mathbb{P}(A_i).$$

In general, the formula is more complicated:

$$\begin{aligned}
\mathbb{P}(A) &= \mathbb{P}(A_1) + \cdots + \mathbb{P}(A_n) \\
&\quad - \mathbb{P}(A_1 \cap A_2) - \mathbb{P}(A_1 \cap A_3) - \cdots - \mathbb{P}(A_{n-1} \cap A_n) \\
&\quad + \mathbb{P}(A_1 \cap A_2 \cap A_3) + \cdots + \mathbb{P}(A_{n-2} \cap A_{n-1} \cap A_n) \\
&\quad + \cdots + (-1)^{n+1} \mathbb{P}(A_1 \cap \cdots \cap A_n) \\
&= \sum_{k=1}^{n} (-1)^{k-1} \sum_{1 \le i_1 < \cdots < i_k \le n} \mathbb{P}\left(\cap_1^k A_{i_j} \right).
\end{aligned} \tag{1.12}$$

The proof is reduced to a counting argument: on the left-hand side (LHS) we count each outcome from $\bigcup_{1 \le i \le n} A_i$ once. But if we take the sum $\sum_{1 \le i \le n} \mathbb{P}(A_i)$ then those outcomes that enter more than one event among $A_1, \ldots, A_n$ will be counted more than once. Now if we subtract the sum $\sum_{1 \le i < j \le n} \mathbb{P}(A_i \cap A_j)$ we will count precisely once the outcomes that enter exactly two events $A_1, \ldots, A_n$, but will be in trouble with the outcomes that enter three or more events. Therefore we have to add $\sum_{1 \le i < j < k \le n} \mathbb{P}(A_i \cap A_j \cap A_k)$, and so on.

A formal proof can be carried by induction in n. It is convenient to begin the induction with $n = 2$ (for $n = 1$ the formula is trivial). For two events A and B, $A \cup B$ coincides with $(A \setminus (A \cap B)) \cup (B \setminus (A \cap B)) \cup (A \cap B)$, the union of non-intersecting events. Hence, $\mathbb{P}(A \cup B)$ can be written as

$$\begin{aligned}
&\mathbb{P}(A \setminus (A \cap B)) + \mathbb{P}(B \setminus (A \cap B)) + \mathbb{P}(A \cap B) \\
&= \mathbb{P}(A) - \mathbb{P}(A \cap B) + \mathbb{P}(B) - \mathbb{P}(A \cap B) + \mathbb{P}(A \cap B) \\
&= \mathbb{P}(A) + \mathbb{P}(B) - \mathbb{P}(A \cap B).
\end{aligned}$$

The induction hypothesis is that the formula holds for any collection of n or less events. Then for any collection $A_1, \ldots, A_{n+1}$ of $n+1$ events, the probability $\mathbb{P}\left(\cup_1^{n+1} A_i \right)$ equals:

$$\mathbb{P}\left(\left(\bigcup_1^n A_i \right) \cup A_{n+1} \right) = \mathbb{P}\left(\bigcup_1^n A_i \right) + \mathbb{P}\left(A_{n+1} \right) - \mathbb{P}\left(\left(\bigcup_1^n A_i \right) \cap A_{n+1} \right)$$

$$= \sum_{k=1}^{n} (-1)^{k-1} \sum_{1 \le i_1 < \ldots < i_k \le n} \mathbb{P}\left(\cap_1^k A_{i_j} \right) + \mathbb{P}\left(A_{n+1} \right) - \mathbb{P}\left(\bigcup_1^n (A_i \cap A_{n+1}) \right).$$

For the last term we have, again by the induction hypothesis:

$$-\mathbb{P}\left(\bigcup_1^n (A_i \cap A_{n+1})\right) = \sum_{k=1}^{n}(-1)^k \sum_{1 \le i_1 < \ldots < i_k \le n} \mathbb{P}\left(\bigcap_1^k \left(A_{i_j} \cap A_{n+1}\right)\right)$$

$$= \sum_{k=1}^{n}(-1)^k \sum_{1 \le i_1 < \ldots < i_k \le n} \mathbb{P}\left(\left(\bigcap_1^k A_{i_j}\right) \cap A_{n+1}\right).$$

We see that the whole sum in the expansion for $\mathbb{P}\left(\bigcup_1^{n+1} A_i\right)$ includes all possible terms identified on the RHS of formula (1.12) for $n+1$, with correct signs. This completes the induction.

An alternative proof (which is instructive as it shows relations between various concepts of probability theory) will be given in the next section, after we introduce random variables and expectations.

The exclusion–inclusion formula is particularly efficient under assumptions of independence and symmetry. It also provides an interesting asymptotical insight into various probabilities.

Example 1.1   An example of using the exclusion–inclusion formula is the following *matching problem*. An absent-minded person (some authors prefer talking about a secretary) has to put n personal letters in n addressed envelopes, and he does it at random. What is the probability $p_{m,n}$ that exactly m letters will be put correctly in their envelopes? Verify the limit

$$\lim_{n \to \infty} p_{m,n} = \frac{1}{em!}.$$

The solution is as follows. The set of outcomes consists of $n!$ possible matchings of the letters to envelopes. Let $A_k = \{\text{letter } k \text{ in correct envelope}\}$.
Then

$$\mathbb{P}(A_{i_1} \cap A_{i_2} \cap \cdots \cap A_{i_r}) = \frac{(n-r)!}{n!},$$

and so

$$\sum_{i_1 < i_2 < \cdots < i_r} \mathbb{P}(A_{i_1} \cap A_{i_2} \cap \cdots \cap A_{i_r}) = \binom{n}{r} \frac{(n-r)!}{n!} = \frac{1}{r!}.$$

Thus,

$$\mathbb{P}(\text{at least one letter in the correct envelope})$$

$$= \mathbb{P}\left(\bigcup_{i=1}^n A_i\right) = 1 - \frac{1}{2!} + \frac{1}{3!} - \cdots + (-1)^{n-1}\frac{1}{n!},$$

and

$$\mathbb{P}(\text{no letter in the correct envelope})$$

$$= 1 - \mathbb{P}\left(\bigcup_{i=1}^n A_i\right) = 1 - 1 + \frac{1}{2!} + \frac{1}{3!} + \cdots + (-1)^n\frac{1}{n!}.$$

which tends to e^{-1} as $n \to \infty$. The number of outcomes in the event {no letter put in the correct envelope} equals $= n! p_{0,n}$, and so

$$P_{m,n} = \binom{n}{m} \frac{(n-m)! p(0, n-m)}{n!}.$$

Therefore,

$$P_{m,n} = \frac{1}{m!} p(0, n-m) = \frac{1}{m!} \left[1 - \frac{1}{1!} + \cdots + (-1)^{n-m} \frac{1}{(n-m)!} \right]$$

which approaches $e^{-1}/m!$ as $n \to \infty$. ∎

Problem 1.25   A total of n male psychologists remembered to attend a meeting about absentmindness. After the meeting, none could recognise his own hat so they took hats at random. Furthermore, each was liable, with probability p and independently of the others, to lose the hat on the way home. Assuming, optimistically, that all arrived home, find the probability that none had his own hat with him, and deduce that it is approximately $e^{-(1-p)}$.

Solution   Set $A_j = \{$psychologist j had his own hat$\}$, then the event $A = \{$none had his own hat$\}$ is the complement of $\bigcup_{1 \le j \le n} A_j$. By the exclusion–inclusion formula and the symmetry:

$$\mathbb{P}(A) = 1 - n\mathbb{P}(A_1) + \frac{n(n-1)}{2} \mathbb{P}(A_1 \cap A_2) - \cdots + (-1)^n \mathbb{P}(A_1 \cap \cdots \cap A_n).$$

Next,

$$\mathbb{P}(A_1) = (1-p) \frac{(n-1)!}{n!},$$

$$\mathbb{P}(A_1 \cap A_2) = (1-p)^2 \frac{(n-2)!}{n!},$$

$$\cdots$$

$$\mathbb{P}(A_1 \cap \cdots \cap A_n) = (1-p)^n \frac{1}{n!}.$$

Then

$$\mathbb{P}(A) = 1 - (1-p) + \frac{(1-p)^2}{2!} - \cdots + (-1)^n \frac{(1-p)^n}{n!},$$

which is the partial sum of $e^{-(1-p)}$. □

Problem 1.26   It is certain that at least one, but no more than two of the events $A_1, \ldots, A_n$ occur. Given that $\mathbb{P}(A_1) = p_1$ for all i, and $\mathbb{P}(A_i \cap A_j) = p_2$ for all i, j $(i \ne j)$, show that

$$1 = np_1 - \frac{n(n-1)}{2} p_2.$$

Deduce that $p_1 \ge 1/n$ and $p_2 \le 2/n$.

Solution   As $\mathbb{P}(A_{i_1} \cap \cdots \cap A_{i_k}) = 0$ for $k > 2$, the exclusion–inclusion formula gives $1 = np_1 - n(n-1)p_2/2$. Rearranging:

$$p_1 = \frac{1}{n} + \frac{n-1}{2}p_2 \geq \frac{1}{n},$$

as $p_2 \geq 0$. Finally,

$$p_2 = \frac{2}{n-1}\left(p_1 - \frac{1}{n}\right) \leq \frac{2}{n-1}\left(1 - \frac{1}{n}\right) = \frac{2}{n}. \quad \square$$

Problem 1.27   State the exclusion–inclusion formula for $\mathbb{P}\left(\bigcup_{i=1}^{n} A_i\right)$.

A large and cheerful crowd of n junior wizards leave their staff in the Porter's Lodge on the way to a long night in the Mended Drum. On returning, each collects a staff at random from a pile, return to his room and attempts to cast a spell against hangovers. If a junior wizard attempts this spell with his own staff, there is a probability p that he will turn into a bullfrog. If he attempts it with someone else's staff, he is certain to turn into a bullfrog. Show that the probability that in the morning we will find n very surprised bullfrogs is approximately e^{p-1}.

Solution   Set $A_i (= A_i(n)) = \{$wizard i gets his own stuff$\}$. Then $\forall r = 1, \ldots, n$ and $1 \leq i_1 < \ldots < i_r \leq n$:

$$\mathbb{P}(A_{i_1} \cap A_{i_2} \cap \cdots \cap A_{i_r}) = \frac{(n-r)!}{n!},$$

as there are $(n-r)!$ ways of distributing the remaining stuff. So, by the exclusion–inclusion formula, the probability $\mathbb{P}\left(\bigcup_{i=1}^{n} A_i\right)$ is equal to

$$\sum_{r=1}^{n}\binom{n}{r}(-1)^{r-1}\frac{(n-r)!}{n!} = \sum_{r=1}^{n}(-1)^{r-1}\frac{1}{r!} = 1 - \sum_{r=0}^{n}(-1)^{r}\frac{1}{r!}$$

which tends to $1 - e^{-1}$ as $n \to \infty$. We can also consider similar events $A_i(k)$ defined for a given subset of k wizards, $1 \leq k \leq n$.

Further, if $P_r(k) = \mathbb{P}($precisely r out of given k get their own staff$)$ then

$$P_r(k) = \binom{k}{r}\frac{(k-r)!}{k!}P_0(k-r) = \frac{1}{r!}P_0(k-r), \quad 0 \leq r \leq k.$$

Also:

$$P_0(k-r) = 1 - \mathbb{P}(\bigcup_{i=1}^{k-r}A_i(k-r)) = \sum_{i=1}^{k-r}(-1)^{i}\frac{1}{i!} \to e^{-1}$$

as $k \to \infty$. So, $P_r(k) \approx e^{-1}/r!$ for k sufficiently large. Finally

$$\mathbb{P}(\text{all turn into bullfrogs}) = \sum_{r=0}^{n}P_r(n)p^r \approx \sum_{r=0}^{n}e^{-1}\frac{p^r}{r!} = e^{-1}e^p = e^{p-1}$$

for n large enough. $\square$

Remark   To formally prove the convergence $\sum_{r=0}^{n} P_r(n)p^r \to e^{p-1}$ one needs an assertion guaranteeing that the term-wise convergence (in our case $P_r(n)p^r \to e^{-1}p^r/r! \ \forall \ r$)
implies the convergence of the sum of the series. For example, the following theorem
will do:

> If $a_n(m) \to a_n \forall \ n$ as $m \to \infty$ and $|a_n(m)| \leq b_n$, where $\sum_n b_n < \infty$, then the sum
> $S(m) = \sum_n a_n(m) \to S = \sum_n a_n$.

In fact, $P_r(n)p^r = P_0(n-r)p^r/r! \leq p^r/r! \ (= b_r)$, and the series $\sum_{r\geq 0} p^r/r!$ converges to
e^p for all p (i.e. $\sum_r b_r < \infty$).

The remaining part of this section addresses the so-called *ballot problem*. Its original
formulation is: a community of voters contains m conservatives and n anarchists voting
for their respective candidates, where $m \geq n$. What is the probability that in the process
of counting the secret ballot papers the conservative candidate will never be behind the
anarchist? This question has emerged in many situations (but, strangely, not in Cambridge
University IA Mathematical Tripos papers so far). However, at the University of Wales-
Swansea it has been actively discussed, in the slightly modified form presented below.

We start with a particular case $m = n$. A series of $2n$ drinks is on offer, n of them are
gin and n tonic. In a popular local game, a blindfolded participant drinks all $2n$ glasses,
one at a time, selected in a random order. The participant is declared a winner if the
volume of gin drunk is always not more than that of tonic. We will check that this occurs
with probability $1/(n+1)$.

Consider a random walk on the set $\{-n, -n+1, \ldots, n\}$ where a particle moves one
step up if a glass of tonic was selected and one down if it was gin. The walk begins at
the origin (no drink consumed) and after $2n$ steps always comes back to it (the number
of gins = the number of tonics). On Figure 1.2 that includes time, every path $X(t)$ of the
walk begins at $(0,0)$ and ends at $(2n,0)$ and each time jumps up and to the right or down
and to the right. We look for the probability that the path remains above the line $X = -1$.

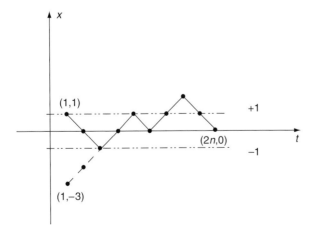

Figure 1.2

The total number of paths from $(0,0)$ to $(2n,0)$ is $(2n)!/n!n!$. The number of paths staying above the line is the same as the total number of paths from $(1,1)$ to $(2n,0)$ less the total number of paths from $(1,-3)$ to $(2n,0)$. In fact, the first step from $(0,0)$ must be up. Next, if a path $(0,0)$ to $(2n,0)$ touches or crosses line $X=-1$, then we can reflect its initial bit and obtain a path from $(1,-3)$ to $(2n,0)$. This is sometimes called the *reflection principle*.

Hence, the probability of winning is

$$\left[\frac{(2n-1)!}{n!(n-1)!}-\frac{(2n-1)!}{(n+1)!(n-2)!}\right]\bigg/\frac{(2n)!}{n!n!}=\frac{1}{2n}\left[n-\frac{n(n-1)}{n+1}\right]=\frac{1}{n+1}.$$

Now assume that the number m of tonics is $> n$, the number of gins. As before, winning the game means that at each time the number of consumed tonics is not less than that of gins. Then the total number of paths from $(0,0)$ to $(m+n, m-n)$ equals $(m+n)!/m!n!$. Again, the first step of a winning path is always up. The total number of paths from $(1,1)$ to $(m+n, m-n)$ equals $(m+n-1)!/(m-1)!n!$. Using the reflection principle, we see that the number of losing paths equals the total number of paths from $(1,-3)$ to $(m+n, m-n)$, which is $(m+n-1)!/(m+1)!(n-2)!$. Finally, the winning probability is

$$\left[\frac{(m+n-1)!}{(m-1)!n!}-\frac{(m+n-1)!}{(m+1)!(n-2)!}\right]\bigg/\frac{(m+n)!}{m!n!}=\frac{m-n+1}{m+1}.$$

We apply these results to the following.

Problem 1.28 n married couples have to cross from the left to the right bank of a river via a narrow bridge, one by one. They decided that at any time on the left bank the number of men should be no less than that of women; apart from this the order can be arbitrary. Find the probability that every man will cross the river after his own wife.

Solution   Set

$$A = \{\text{every man crosses after his own wife}\},$$

$$B = \{\text{at all times, the \# of men on the left bank} \geq \text{that of women}\}.$$

Then

$$P(A|B) = P(A\cap B)/P(B) = P(A)/P(B) = \frac{n+1}{2^n}. \quad \square$$

We conclude this section with another story about Kolmogorov. He began his academic studies as a historian, and at the age of 19 finished a work focused on the principles of distribution and taxation of arable land in fifteenth and sixteenth century Novgorod, an ancient Russian state (a principality and a republic at different periods of its existence, fully or partially independent, until it was annexed by Moscow in 1478). In this work, he used mathematical arguments to answer the following question: was it (i) a village that was taxed in the first place, and then the tax was divided between households, or (ii) the other way around, where it was a household that was originally taxed, and then the sum represented the total to be paid by the village? The sources were ancient cadastres and other official manuscripts of the period. Because the totals received from the villages were always an integer number of (changing) monetary units, Kolmogorov proved that it was rule (ii) that was adopted.

Kolmogorov reported his findings at a history seminar at Moscow University in 1922. However, the head of the seminar, a well-known professor (a street in Moscow was later named after him) commented that the conclusions of his young colleague could not be considered final, because 'in history, every statement must be supported by several proofs'. Kolmogorov then decided to move to a discipline where 'a single proof would be sufficient for a statement to be considered correct', i.e. mathematics.

1.4   Random variables. Expectation and conditional expectation. Joint distributions

> He who has heard the same thing told by 12 000 eye-witnesses has only 12 000 probabilities, which are equal to one strong probability, which is far from certain.
> F.M.A. Voltaire (1694–1778), French philosopher

This section is considerably larger than the previous ones: the wealth of problems generated here is such that getting through it allows a student to secure in principle at least the first third of the Cambridge University IA Probability course.

The definition of a *random variable* (RV) is that it is a function X on the total set of outcomes Ω, usually with real, sometimes with complex values, $X(\omega)$, $\omega \in \Omega$ (in the complex case we may consider a pair formed by the real and imaginary parts of X). A simple (and important) example of an RV is the indicator function of an event:

$$\omega \in \Omega \mapsto I(\omega \in A) = \begin{cases} 1, & \text{if } \omega \in A, \\ 0, & \text{if } \omega \notin A. \end{cases} \tag{1.13}$$

It is obvious that the product $I(\omega \in A_1)I(\omega \in A_2)$ equals 1 iff $\omega \in A_1 \cap A_2$, i.e.

$$I(\omega \in A_1)I(\omega \in A_2) = I(\omega \in A_1 \cap A_2).$$

On the other hand,

$$I(\omega \in A_1 \cup A_2) = I(\omega \in A_1) + I(\omega \in A_2) - I(\omega \in A_1)I(\omega \in A_2).$$

In the case of finitely many outcomes, every RV is a finite linear combination of indicator functions; if there are countably many outcomes, then infinite combinations will be needed.

The *expected value* (or the *expectation*, or the *mean value*, or simply the *mean*) of a RV X taking values $x_1, \ldots, x_m$ with probabilities $p_1, \ldots, p_m$ is defined as the sum

$$\mathbb{E}X = \sum_{1 \le i \le m} p_i x_i = \sum_{1 \le i \le m} x_i \mathbb{P}(X = x_i); \tag{1.14}$$

in the ω-notation:

$$\mathbb{E}X = \sum_{\omega \in \Omega} p(\omega)X(\omega). \tag{1.15}$$

If $X \equiv b$ is a constant RV, then $\mathbb{E}X = b$.

This definition is meaningful also for RVs taking countably many values $x_1, x_2, \ldots$ with probabilities $p_1, p_2, \ldots$:

$$\mathbb{E}X = \sum_i p_i x_i,$$

provided that the series converges absolutely: $\sum_i p_i |x_i| < \infty$. If $\sum_i p_i |x_i| = \infty$, one says that X does not have a finite expected value.

In many applications, it is helpful to treat the expected value $\mathbb{E}X$ as the position of the centre of mass for the system of massive points $x_1, x_2, \ldots$ with masses $p_1, p_2, \ldots$.

The first (and a very useful) observation is that the expected value of the indicator $I_A(\omega) = I(\omega \in A)$ of an event equals the probability:

$$\mathbb{E}I_A = \sum_{\omega \in \Omega} p(\omega)I(\omega \in A) = \sum_{\omega \in A} p(\omega) = \mathbb{P}(A). \tag{1.16}$$

Furthermore, if RVs X, Y have $X \leq Y$, then $\mathbb{E}X \leq \mathbb{E}Y$.

Next, the expectation of a linear combination of RVs equals the linear combination of expectations. The shortest proof is in the ω-notation:

$$\mathbb{E}(c_1 X_1 + c_2 X_2) = \sum_\omega p(\omega)\big[c_1 X_1(\omega) + c_2 X_2(\omega)\big]$$

$$= c_1 \sum_\omega p(\omega)X_1(\omega) + c_2 \sum_\omega p(\omega)X_2(\omega) = c_1 \mathbb{E}X_1 + c_2 \mathbb{E}X_2.$$

This fact (called the linearity of the expectation) can be easily extended to n summands:

$$\mathbb{E}\left(\sum_{1 \leq k \leq n} c_k X_k\right) = \sum_{1 \leq k \leq n} c_k \mathbb{E}X_k;$$

in particular if $\mathbb{E}X_k = \mu$ for every k, then $\mathbb{E}\sum_{1 \leq k \leq n} X_k = \mu n$. A similar property also holds for an infinite sequence of RVs $X_1, X_2, \ldots$:

$$\mathbb{E}\left(\sum_k c_k X_k\right) = \sum_k c_k \mathbb{E}X_k, \tag{1.17}$$

provided that the series on the RHS converges absolutely: $\sum_k |c_k \mathbb{E}X_k| < \infty$. (The precise statement is that if $\sum_k |c_k \mathbb{E}X_k| < \infty$, then the series $\sum_{k \geq 1} c_k X_k$ defines an RV with a finite mean equal to the sum $\sum_{k \geq 1} c_k \mathbb{E}X_k$.)

Also, for a given function $g : \mathbb{R} \to \mathbb{R}$

$$\mathbb{E}g(X) = \sum_{\omega \in \Omega} p(\omega)g(X(\omega)) = \sum_i p_i g(x_i), \tag{1.18}$$

provided that the sum $\sum_i p_i |g(x_i)| < \infty$. In fact, writing $Y = g(X)$ and denoting the values of Y by $y_1, y_2, \ldots$, we have

$$\mathbb{E}Y = \sum_j y_j P(g(X) = y_j) = \sum_j y_j \sum_i I(g(x_i) = y_j) \mathbb{P}(X = x_i)$$

which is simply $\sum_i p_i g(x_i)$ under the condition that $\sum_i p_i |g(x_i)| < \infty$, as we then can do summation by grouping terms.

Remark Formula (1.18) is a discrete version of what is known as the *Law of the Unconscious Statistician*. See equation (2.69).

Given two RVs, X and Y, with values $X(\omega)$ and $Y(\omega)$, we can consider events $\{\omega : X(\omega) = x_i, \ Y(\omega) = y_j\}$ (shortly, $\{X = x_i, \ Y = y_j\}$) for any pair of their values $x_i, \ y_j$. The probabilities $\mathbb{P}(X = x_i, \ Y = y_j)$ will specify the *joint distribution* of the pair $X, \ Y$. The 'marginal' probabilities $\mathbb{P}(X = x_i)$ and $\mathbb{P}(Y = y_j)$ are obtained by summation over all possible values of the complementary RV:

$$\begin{aligned} \mathbb{P}(X = x_i) &= \sum_{y_j} \mathbb{P}(X = x_i, \ Y = y_j), \\ \mathbb{P}(Y = y_j) &= \sum_{x_i} \mathbb{P}(X = x_i, \ Y = y_j). \end{aligned} \tag{1.19}$$

We can also consider the conditional probabilities

$$\mathbb{P}(X = x_i | Y = y_j) = \frac{\mathbb{P}(X = x_i, \ Y = y_j)}{\mathbb{P}(Y = y_j)}. \tag{1.20}$$

Similar concepts are applicable in the case of several RVs $X_1, \ldots, X_n$.

For example, for the sum $X + Y$ of RVs X and Y with joint probabilities $\mathbb{P}(X = x_i, \ Y = y_j)$:

$$\begin{aligned} \mathbb{P}(X + Y = u) &= \sum_{x_i, y_j : x_i + y_j = u} \mathbb{P}(X = x_i, \ Y = y_j) \\ &= \sum_x \mathbb{P}(X = x, \ Y = u - x) \\ &= \sum_y \mathbb{P}(X = u - y, \ Y = y). \end{aligned} \tag{1.21}$$

Similarly, for the product XY:

$$\begin{aligned} \mathbb{P}(XY = u) &= \sum_{x_i, y_j : x_i y_j = u} \mathbb{P}(X = x_i, \ Y = y_j) \\ &= \sum_x \mathbb{P}(X = x, \ Y = u/x) \\ &= \sum_y \mathbb{P}(X = u/y, \ Y = y). \end{aligned} \tag{1.22}$$

A powerful tool is the formula of conditional expectation: if X and N are two RVs then

$$\mathbb{E}X = \mathbb{E}[\mathbb{E}(X|N)]. \tag{1.23}$$

Here, on the RHS, the external expectation is relative to the probabilities $\mathbb{P}(N=n_j)$ with which RV N takes its values n_j:

$$\mathbb{E}[\mathbb{E}(X|N)] = \sum_j \mathbb{P}(N=n_j)\mathbb{E}(X|N=n_j).$$

The internal expectation is relative to the conditional probabilities $\mathbb{P}(X=x_i|N=n_j)$ of values x_i of RV X, given that $N=n_j$:

$$\mathbb{E}(X|N=n_j) = \sum_i x_i\mathbb{P}(X=x_i|N=n_j).$$

To prove formula (1.23), we simply substitute and use the definition of the conditional probability:

$$\mathbb{E}[\mathbb{E}(X|N)] = \sum_j \mathbb{P}(N=n_j)\sum_i x_i\mathbb{P}(X=x_i|N=n_j)$$

$$= \sum_{x_i,n_j} x_i\mathbb{P}\left(X=x_i, N=n_j\right) = \sum_{x_i} x_i\mathbb{P}(X=x_i) = \mathbb{E}X.$$

We see that formula (1.23) is merely a paraphrase of (2.6). We will say again that it is the result of conditioning by RV N.

A handy formula is that for N with non-negative integer values

$$\mathbb{E}N = \sum_{n\geq 1}\mathbb{P}(N\geq n). \tag{1.24}$$

In fact,

$$\sum_{n\geq 1}\mathbb{P}(N\geq n) = \sum_{n\geq 1}\sum_{k\geq n}\mathbb{P}(N=k) = \sum_{k\geq 1}\mathbb{P}(N=k)\sum_{1\leq n\leq k}1$$

$$= \sum_{k\geq 1}\mathbb{P}(N=k)k = \mathbb{E}N.$$

Of course, in formula (1.23) the rôles of X and N can be swapped. For example, in Problem 1.17, the expected duration of the game is

$$\mathbb{E}N = \mathbb{E}[\mathbb{E}(N|X)] = \sum_{i=2}^{12} p_i\mathbb{E}(N|X=i),$$

where X is the combined outcome of the first throw, and $p_i = \mathbb{P}(X=i)$. We have:

$$\mathbb{E}(N|X=i) = 1, \ \ i+2, 3, 7, 11, 12,$$

$$\mathbb{E}(N|X=i) = 1 + \frac{1}{p_i+p_7}, \ \ i=4,5,6,8,9,10.$$

Substituting the values of p_i we get $\mathbb{E}N = 557/165 \approx 3.38$.

Further, we say that RVs X and Y with values x_i and y_j are *independent* if for any pair of values

$$\mathbb{P}(X = x_i, Y = y_j) = \mathbb{P}(X = x_i)\mathbb{P}(Y = y_j). \tag{1.25}$$

For a triple of variables X, Y, Z, we require that for any triple of values $\mathbb{P}(X = x_i, Y = y_j, Z = z_k) = \mathbb{P}(X = x_i)\mathbb{P}(Y = y_j)\mathbb{P}(Z = z_k)$. This definition is extended to the case of any number of RVs $X_1, \ldots, X_n$: $\forall x_1, \ldots, x_n$:

$$\mathbb{P}(X_1 = x_1, X_2 = x_2, \ldots, X_n = x_n) = \prod_{i=1}^{n} \mathbb{P}(X_i = x_i). \tag{1.26}$$

Furthermore, an infinite sequence of RVs X_1, X_2, $\ldots$ is called independent if $\forall n$ the variables $X_1, \ldots, X_n$ are independent.

One may ask here why it suffices to consider in equation (1.26) the 'full' product $\prod_{i=1}^{n}$ while in the definition of independent events it was carefully stated that $\mathbb{P}\left(\cap A_{i_k}\right) = \prod \mathbb{P}(A_{i_k})$ for any subcollection (see equation (1.10)). The answer is: because we require it for any collection of values $x_1, \ldots, x_n$ of our RVs $X_1, \ldots, X_n$. In fact, when some of the RVs are omitted, we should take the summation over their values, viz.

$$\mathbb{P}(X_2 = x_2, \ldots, X_n = x_n) = \sum_{x_1} \mathbb{P}(X_1 = x_1, X_2 = x_2, \ldots, X_n = x_n),$$

as the events $\{X_1 = x_1, \ldots, X_{n-1} = x_{n-1}, X_n = x_n\}$ are pair-wise disjoint for different x_1s, and partition the event $\{X_2 = x_2, \ldots, X_n = x_n\}$. So, if

$$\mathbb{P}(X_1 = x_1, \ldots, X_n = x_n) = \prod_{i=1}^{n} \mathbb{P}(X_i = x_i),$$

then

$$\mathbb{P}(X_2 = x_2, \ldots, X_n = x_n) = \sum_{x_1} \mathbb{P}(X_1 = x_1)\mathbb{P}(X_2 = x_2) \ldots \mathbb{P}(X_n = x_n)$$

$$= \mathbb{P}(X_2 = x_2) \ldots \mathbb{P}(X_n = x_n),$$

i.e. the subcollection $X_2, \ldots, X_n$ is automatically independent.

However, the inverse is not true. For instance, if each of (X, Y), (X, Z) and (Y, Z) is a pair of independent RVs, it does not necessarily mean that the triple X, Y, Z is independent. An example is produced from Problem 1.21: you simply take I_{A_1}, I_{A_2} and I_{A_3} from the solution.

An important concept that now emerges and will be employed throughout the rest of this volume is a sequence of *independent, identically distributed* (IID) RVs X_1, X_2, $\ldots$. Here, in addition to the independence, it is assumed that the probability $\mathbb{P}(X_i = x)$ is the same for each $i = 1, 2, \ldots$. A good model here is coin-tossing: X_n may be a function of the result of the nth tossing, viz.

$$X_n = I(n\text{th toss a head}) = \begin{cases} 1, & \text{if the } n\text{th toss produces a head,} \\ 0, & \text{if the } n\text{th toss produces a tail.} \end{cases}$$

More generally, you could divide trials into disjoint 'blocks' of some given length l and think of X_n as a function of the outcomes in the nth block.

An immediate consequence of the definition is that for independent RVs X, Y, the expected value of the product equals the product of the expected values:

$$\mathbb{E}(XY) = \sum_{x_i, y_j} x_i y_j \mathbb{P}(X = x_i, Y = y_j)$$

$$= \sum_{x_i, y_j} x_i y_j \mathbb{P}(X = x_i) \mathbb{P}(Y = y_j)$$

$$= \left[\sum_{x_i} x_i \mathbb{P}(X = x_i) \right] \left[\sum_{y_j} y_j \mathbb{P}(Y = y_j) \right] = \mathbb{E}X \mathbb{E}Y. \tag{1.27}$$

However, the inverse assertion is not true: there are RVs with $\mathbb{E}(XY) = \mathbb{E}X\mathbb{E}Y$ which are not independent. A simple example is the following.

Example 1.2

$$X = \begin{cases} -1, & \text{probability } \dfrac{1}{3}, \\ 0, & \text{probability } \dfrac{1}{3}, \\ 1, & \text{probability } \dfrac{1}{3}, \end{cases} \qquad Y = -\frac{2}{3} + X^2.$$

Here,

$$\mathbb{E}X = 0, \ \mathbb{E}Y = 0, \ \mathbb{E}(XY) = -\frac{2}{3}\mathbb{E}X + \mathbb{E}X^3 = 0.$$

But, obviously, X and Y are dependent.

On the other hand, it is impossible to construct an example of dependent RVs X and Y taking two values each such that $\mathbb{E}(XY) = \mathbb{E}X\mathbb{E}Y$. In fact, let X, $Y = 0, 1$, with

$$\mathbb{P}(X = Y = 0) = w, \ \ P(X = 1, Y = 0) = x,$$

$$\mathbb{P}(X = 0, Y = 1) = y, \ \ P(X = Y = 1) = z.$$

Here

$$\mathbb{P}(X = 0) = w + y, \ \ \mathbb{P}(Y = 0) = w + x,$$

$$\mathbb{P}(X = 1) = x + z, \ \ P(Y = 1) = y + z,$$

and independence occurs iff $z = (x + z)(y + z)$. Next,

$$\mathbb{E}(XY) = z, \ \ \mathbb{E}X\mathbb{E}Y = (x + z)(y + z),$$

and $\mathbb{E}(XY) = \mathbb{E}X\mathbb{E}Y$ iff $z = (x + z)(y + z)$. ∎

The *variance* of RV X with real values and mean $\mathbb{E}X = \mu$ is defined as

$$\text{Var } X = \mathbb{E}(X - \mu)^2; \tag{1.28}$$

by expanding the square and using linearity of expectation and the fact that the expected value of a constant equals this constant we have

$$\text{Var } X = \mathbb{E}(X^2 - 2X\mu + \mu^2) = \mathbb{E}X^2 - 2\mu\mathbb{E}X + \mu^2$$
$$= \mathbb{E}X^2 - 2\mu^2 + \mu^2 = \mathbb{E}X^2 - (\mathbb{E}X)^2. \tag{1.29}$$

In particular, if X is a constant RV: $X \equiv b$ then $\text{Var } X = b^2 - b^2 = 0$.

The variance is considered as a measure of 'deviation' of RV X from its mean. (The square root $\sqrt{\text{Var } X}$ is called the standard deviation.)

From the first definition we see that $\text{Var } X \geq 0$, i.e. $(\mathbb{E}X)^2 \leq \mathbb{E}X^2$. This is a particular case of the so-called *Cauchy–Schwarz* (CS) *inequality*

$$|\mathbb{E}(X\overline{Y})|^2 \leq \mathbb{E}|X|^2 \mathbb{E}|Y|^2, \tag{1.30}$$

valid for any pair of real or complex RVs X and Y, with $\overline{Y}$ standing for the complex conjugation. (This inequality implies that if $|X|^2$ and $|Y|^2$ have finite expected values then XY also has a finite expected value.) The CS inequality is named after two famous mathematicians: the Frenchman A.-L. Cauchy (1789–1857) who is credited with its discovery in the discrete case and the German H.A. Schwarz (1843–1921) who proposed it in the continuous case.

The proof of the CS inequality provides an elegant algebraic exercise (and digression); for simplicity, we will conduct it for real RVs. Observe that $\forall \lambda \in \mathbb{R}$, the RV $(X + \lambda Y)^2$ is ≥ 0 and hence has $\mathbb{E}(X + \lambda Y)^2 \geq 0$. As a function of λ it gives an expression

$$\mathbb{E}(X + \lambda Y)^2 = \mathbb{E}X^2 + 2\lambda\mathbb{E}XY + \lambda^2\mathbb{E}Y^2,$$

which is a quadratic polynomial. To be ≥ 0 for $\forall \lambda \in \mathbb{R}$, it must have a non-positive discriminant, i.e.

$$4\left(\mathbb{E}XY\right)^2 \leq 4\mathbb{E}X^2\mathbb{E}Y^2.$$

A concept closely related to the variance is the *covariance* of two RVs X, Y defined as

$$\text{Cov}(X, Y) = \mathbb{E}(X - \mathbb{E}X)(Y - \mathbb{E}Y) = \mathbb{E}(XY) - \mathbb{E}X\mathbb{E}Y. \tag{1.31}$$

By the CS inequality, $|\text{Cov}(X, Y)|^2 \leq (\text{Var } X)(\text{Var } Y)$.

For independent variables, $\text{Cov}(X, Y) = 0$. Again, the inverse assertion does not hold: there are non-independent variables X, Y for which $\text{Cov}(X, Y) = 0$. See Example 1.2. However, as we checked before, if X and Y take two values (or one) and $\text{Cov}(X, Y) = 0$, they will be independent.

RVs with $\text{Cov}(X, Y) = 0$ are called *uncorrelated*.

For the variance Var $(X+Y)$ of the sum $X+Y$ we have the following representation:

$$\text{Var}\,(X+Y) = \text{Var}\,X + \text{Var}\,Y + 2\text{Cov}\,(X, Y). \tag{1.32}$$

In fact, Var $(X+Y)$ equals

$$\mathbb{E}\big((X-\mathbb{E}X)+(Y-\mathbb{E}Y)\big)^2$$
$$= \mathbb{E}(X-\mathbb{E}X)^2 + 2\mathbb{E}(X-\mathbb{E}X)(Y-\mathbb{E}Y) + \mathbb{E}(Y-\mathbb{E}Y)^2,$$

which is the RHS of equation (1.32).

An important corollary is that the variance of the sum of independent variables equals the sum of the variances:

$$\text{Var}\,(X+Y) = \text{Var}\,X + \text{Var}\,Y.$$

This fact is easily extended to any number of independent summands:

$$\text{Var}\,(X_1 + \cdots + X_n) = \text{Var}\,X_1 + \cdots + \text{Var}\,X_n. \tag{1.33}$$

In the case of IID RVs, Var X_i does not depend on i, and if Var $X_i = \sigma^2$ (a standard probabilistic notation) then Var $\left(\sum_{1 \le j \le n} X_j\right) = n\sigma^2$.

On the other hand, if c is a real constant then Var $(cX) = \mathbb{E}(cX)^2 - (\mathbb{E}(cX))^2 = c^2(\mathbb{E}X^2 - (\mathbb{E}X)^2) = c^2\text{Var}\,X$. Hence, Var $nX = n^2\text{Var}\,X$. That is summing identical RVs produces a quadratic growth in the variance, whereas summing IID RVs produces only a linear one. At the level of mean values both options give the same (linear) growth.

A constant RV taking a single value $(X \equiv b)$ is independent of any other RV (and group of RVs). Therefore, Var $(X+b) = \text{Var}\,X + \text{Var}\,b = \text{Var}\,X$.

Summarising, for independent RVs X_1, $X_2, \ldots$ and real coefficients c_1, $c_2, \ldots$,

$$\text{Var}\,\sum_i c_i X_i = \sum_i c_i^2 \text{Var}\,X_i,$$

provided that the series in the RHS converges absolutely. More precisely, if $\sum_i c_i^2 \text{Var}\,X_i < \infty$, then the series $\sum_i c_i X_i$ defines an RV with finite variance $\sum_i c_i^2 \text{Var}\,X_i$.

Finally, formulas (1.21), (1.22) in the case of independent RVs become

$$\mathbb{P}(X+Y=u) = \sum_y \mathbb{P}(X=u-y)\mathbb{P}(Y=y)$$

$$= \sum_y \mathbb{P}(X=x)\mathbb{P}(Y=u-x) \tag{1.34}$$

and

$$\mathbb{P}(XY=u) = \sum_y \mathbb{P}(X=u/y)\mathbb{P}(Y=y)$$

$$= \sum_x \mathbb{P}(X=x)\mathbb{P}(Y=u/x). \tag{1.35}$$

Equation (1.34) is often referred to as the *convolution formula*.

Remark There is variation in the notation: many authors write $\mathbb{E}[X]$, Var $[X]$ and/or Var (X). As we usually follow the style of the Tripos questions, such notation will also occasionally appear in our text.

Concluding this section, we give an alternative proof of the exclusion–inclusion formula (1.12) for probability $\mathbb{P}(A)$ of a union $A = \bigcup_{1 \le i \le n} A_i$. By using the indicators I_{A_i} of events A_i, the formula becomes

$$\mathbb{E}I_A = \sum_{1 \le k \le n} (-1)^{k+1} \sum_{1 \le i_1 < \ldots < i_k \le n} \mathbb{E}\left(I_{A_{i_1}} \cdots I_{A_{i_k}} \right).$$

Given an outcome $\omega \in \bigcup_{1 \le i \le n} A_i$, let $A_{j_1}, \ldots, A_{j_s}$ be the list of events containing ω: $\omega \in A_{j_1} \cap \cdots \cap A_{j_s}$ and $\omega \notin A_j \; \forall \; A_j$ not in the list. Without loss of generality, assume that $\{j_1, \ldots, j_s\} = \{1, \ldots, s\}$; otherwise relabel events accordingly. Then for this ω the RHS must contain terms $\mathbb{E}(I_{A_{i_1}} \cdots I_{A_{i_k}})$ with $1 \le k \le s$ and $1 \le i_1 < \ldots < i_k \le s$ only. More precisely we want to check that

$$1 = I_{\bigcup_{1 \le i \le n} A_i}(\omega) = \sum_{1 \le k \le s} (-1)^{k+1} \sum_{1 \le i_1 < \ldots < i_k \le s} I_{A_{i_1}}(\omega) \ldots I_{A_{i_k}}(\omega).$$

But the RHS is

$$\sum_{1 \le k \le s} (-1)^{k+1} \binom{s}{k} = 1 - \left(1 + \sum_{1 \le k \le s} (-1)^k \binom{s}{k} \right) = 1 - (1-1)^s = 1.$$

Taking the expectation yields the result.

Problem 1.29 I arrive home from a feast and attempt to open my front door with one of the three keys in my pocket. (You may assume that exactly one key will open the door and that if I use it I will be successful.) Find the expected number a of tries that I will need if I take the keys at random from my pocket but drop any key that fails onto the ground. Find the expected number b of tries that I will need if I take the keys at random from my pocket and immediately put back into my pocket any key that fails. Find a and b and check that $b - a = 1$.

Solution Label the keys 1, 2 and 3 and assume that keys 2 and 3 are wrong. Consider the following cases (i) the first chosen key is right, (ii) the second key is right, and (iii) the third key is right. Then for a we have

$$a = \left(\frac{1}{3} \times 1 \right) + \left(2 \times \frac{1}{3} \times \frac{1}{2} \times 2 \right) + \left(2 \times \frac{1}{3} \times \frac{1}{2} \times 3 \right) = 2.$$

For b, use conditioning by the result of the first attempt:

$$b = \frac{1}{3} \times 1 + \frac{2}{3} \times (1 + b), \quad \text{whence } b = 3.$$

Here factor $(1 + b)$ reflects the fact that when the first attempt fails, we spend one try, and the problem starts again, by independence. $\square$

Remark   The equation for b is similar to earlier equations for probabilities q and q_i; see Problems 1.17 and 1.18.

Problem 1.30   Let N be a non-negative integer-valued RV with mean μ_1 and variance σ_1^2, and $X_1, X_2, \ldots$ be identically distributed RVs with mean μ_2 and variance σ_2^2; furthermore, assume that $N, X_1, X_2, \ldots$ are independent. Calculate the mean and variance of the RV $S_N = X_1 + X_2 + \cdots + X_N$.

Solution 1   By conditioning,

$$\mathbb{E}S_N = \mathbb{E}[\mathbb{E}(S_N|N)] = \sum_{n=0}^{N} \mathbb{P}(N=n)\mathbb{E}\left(\sum_{i=1}^{n} X_i\right)$$

$$= \sum_{n=0}^{N} \mathbb{P}(N=n)\, n\mu_2 = \mu_1\mu_2.$$

Next,

$$\mathbb{E}S_N^2 = \mathbb{E}\left[\mathbb{E}\left(S_N^2|N\right)\right] = \sum_{n=0}^{N} \mathbb{P}(N=n)\mathbb{E}\left(\sum_{i=1}^{n} X_i\right)^2$$

$$= \sum_{n=0}^{N} \mathbb{P}(N=n)\left[\mathrm{Var}\left(\sum_{i=1}^{n} X_i\right) + \left(\mathbb{E}\sum_{i=1}^{n} X_i\right)^2\right]$$

$$= \sum_{n=0}^{N} \mathbb{P}(N=n)[n\sigma_2^2 + (n\mu_2)^2]$$

$$= \mu_1\sigma_2^2 + \mu_2^2\mathbb{E}N^2 = \mu_1\sigma_2^2 + \mu_2^2(\mu_1^2 + \sigma_1^2).$$

Therefore,

$$\mathrm{Var}\,(S_N) = \mathbb{E}S_N^2 - (\mathbb{E}S_N)^2 = \mu_1\sigma_2^2 + \mu_2^2\sigma_1^2.$$

Solution 2   (Look at this solution after you have learnt about probability generating functions in Section 1.5.) Write

$$\phi(s) = \mathbb{E}s^{S_n} = g(h(s)),$$

where $h(s) = \mathbb{E}s^{X_1}$, $g(s) = \mathbb{E}s^N$. Differentiating yields $\phi'(s) = g'(h(s))h'(s)$, and so

$$\mathbb{E}(S_N) = \phi'(1) = g'(h(1))h'(1) = g'(1)h'(1) = \mu_1\mu_2.$$

Further, $\phi''(s) = g''(h(s))h'(s)^2 + g'(h(s))h''(1)$, and so

$$\mathbb{E}(S_N(S_N - 1)) = \phi''(1) = g''(1)h'(1)^2 + g'(1)h''(1)$$

and

$$\mathrm{Var}\,(S_N) = \phi''(1) + \phi'(1) - (\phi'(1))^2 = \mu_1\sigma_2^2 + \mu_2^2\sigma_1^2. \quad \square$$

Problem 1.31   Define the variance Var X of a random variable X and the covariance Cov (X, Y) of the pair X, Y.

Let $X_1, X_2, \ldots, X_n$ be random variables. Show that

$$\mathrm{Var}\,(X_1 + X_2 + \cdots + X_n) = \sum_{i,j=1}^{n} \mathrm{Cov}\,(X_i, X_j).$$

Ten people sit in a circle, and each tosses a fair coin. Let N be the number of people whose coin shows the same side as both of the coins tossed by the two neighbouring people. Find $\mathbb{P}(N = 9)$ and $\mathbb{P}(N = 10)$. By representing N as a sum of random variables, or otherwise, find the mean and variance of N.

Solution   Let $Y_i = X_i - \mathbb{E}X_i$. Then

$$\mathrm{Var}\,\left(\sum_{i=1}^{n} X_i\right) = \mathbb{E}\left(\sum_{i=1}^{n} Y_i\right)^2 = \mathbb{E}\left(\sum_{i,j=1}^{n} Y_i Y_j\right) = \sum_{i,j=1}^{n} \mathbb{E}(Y_i Y_j)\,.$$

But $\mathbb{E}(Y_i Y_j) = \mathrm{Cov}\,(X_i, X_j)$, hence the result. Next observe that $\mathbb{P}(N = 9) = 0$. Indeed, you can't have nine people out of ten in agreement with their neighbours and just one person in disagreement, as his neighbours must quote him. Further, $\mathbb{P}(N = 10) = 2 \times (1/2)^{10}$: there are two ways to fix the side, and $(1/2)^{10}$ is the probability that all ten coins show this particular side.

Finally, write $N = I_{A_1} + \cdots + I_{A_n}$, where I_{A_i} is the indicator of the event

$$A_i = \{i\text{th person has two neighbours with the same side showing}\}.$$

By symmetry,

$$\mathbb{E}N = 10\mathbb{P}(A_1) = 10 \times \frac{1}{4} = 2.5.$$

Further, RVs I_{A_i}, I_{A_j} are pair-wise independent if i is not next to j. In fact, when i and j do not have a common neighbour, this is obvious, as each RV depends on disjoint collection of independent trials. Next, assume that i and j have a (single) common neighbour. Then

$$\mathbb{P}(I_{A_i} = 1, I_{A_j} = 1) = \frac{1}{32} + \frac{1}{32} = \frac{1}{16}.$$

At the same time,

$$\mathbb{P}(I_{A_i} = 1) = \mathbb{P}(I_{A_j} = 1) = \frac{1}{4}.$$

Thus,

$$\mathrm{Var}\,(N) = 10\,\mathrm{Var}\,(I_{A_1}) + 20\,\mathrm{Cov}\,(I_{A_1}, I_{A_2})$$

$$= 10\left[\frac{1}{4} - \left(\frac{1}{4}\right)^2\right] + 20\left[\mathbb{E}(I_{A_1} I_{A_2}) - \frac{1}{16}\right].$$

Further,

$$\mathbb{E}(I_{A_1} I_{A_2}) = \mathbb{P}(I_{A_1} = 1 | I_{A_2} = 1)\mathbb{P}(I_{A_2} = 1) = \frac{1}{2} \times \frac{1}{4} = \frac{1}{8}.$$

Thus,

$$\text{Var } N = 10 \times \frac{3}{16} + 20\left(\frac{1}{8} - \frac{1}{16}\right) = 3.125. \quad \square$$

Problem 1.32 $X_1, \ldots, X_n$ are independent, identically distributed RVs with mean μ and variance σ^2. Find the mean of

$$S = \sum_{i=1}^{n}(X_i - \overline{X})^2, \text{ where } \overline{X} = \frac{1}{n}\sum_{1}^{n}X_i.$$

Solution   First, consider the mean value and the variance of $\overline{X}$:

$$\mathbb{E}\overline{X} = \frac{1}{n}\mathbb{E}\sum_{i=1}^{n}X_i = \frac{1}{n}\sum_{i=1}^{n}\mathbb{E}X_i = \frac{1}{n} \cdot n\mu = \mu,$$

and

$$\text{Var }\overline{X} = \frac{1}{n^2}\sum_{i=1}^{n}\text{Var }X_i = \frac{\sigma^2}{n}.$$

Next,

$$\mathbb{E}S = \mathbb{E}\left(\sum_{i=1}^{n}X_i^2 - 2\overline{X}\sum_{i=1}^{n}X_i + n\overline{X}^2\right) = \mathbb{E}\left(\sum_{i=1}^{n}X_i^2 - 2n\overline{X}^2 + n\overline{X}^2\right)$$

$$= \mathbb{E}\left(\sum_{i=1}^{n}x_i^2 - n\overline{X}^2\right) = \sum_{i=1}^{n}\mathbb{E}X_i^2 - n\mathbb{E}\overline{X}^2$$

$$= n(\mu^2 + \sigma^2) - n[(\mathbb{E}\overline{X})^2 + \text{Var }\overline{X}]$$

$$= n(\mu^2 + \sigma^2) - n(\mu^2 + \frac{\sigma^2}{n}) = \sigma^2(n - 1). \quad \square$$

Problem 1.33 (X_k) is a sequence of independent identically distributed positive RVs, where $\mathbb{E}(X_k) = a$ and $\mathbb{E}X_k^{-1} = b$ exist. Let $S_n = \sum_{i=1}^{n}X_k$. Show that $\mathbb{E}(S_m/S_n) = m/n$ if $m \leq n$, and $\mathbb{E}(S_m/S_n) = 1 + (m - n)a\mathbb{E}\ (S_n^{-1})$ if $m > n$.

Solution   For $m \leq n$ write

$$\mathbb{E}\left(\frac{S_m}{S_n}\right) = \mathbb{E}\frac{X_1 + \cdots + X_m}{S_n} = \sum_{i=1}^{m}\mathbb{E}\left(\frac{X_i}{S_n}\right) = m\mathbb{E}\left(\frac{X_1}{S_n}\right).$$

But for $m = n$:

$$1 = n\mathbb{E}\left(\frac{X_1}{S_n}\right).$$

Hence $\mathbb{E}(S_m/S_n) = m/n.$

For $m > n$,

$$\mathbb{E}\left(\frac{S_m}{S_n}\right) = \mathbb{E}\left(\frac{S_n}{S_n}\right) + \mathbb{E}\sum_{j=m+1}^{n}\frac{X_j}{S_n} = 1 + \sum_{j=n+1}^{m}\mathbb{E}X_j\mathbb{E}\left(\frac{1}{S_n}\right)$$

$$= 1 + (m-n)a\mathbb{E}\left(\frac{1}{S_n}\right).$$

The mean value $\mathbb{E}(1/S_n)$ is finite (and is $\le b$) since $1/S_n \le 1/X_1$.
It is important to stress that S_n and S_m are not independent.   ☐

Problem 1.34   Suppose that X and Y are independent real random variables with $|X|$, $|Y| \le K$ for some constant K. If $Z = XY$ show that the variance of Z is given by

$$\text{Var } Z = \text{Var } X\text{Var } Y + \text{Var } Y(\mathbb{E}X)^2 + \text{Var } X(\mathbb{E}Y)^2,$$

stating the properties of expectation that you use.

Solution   Write $\text{Var}(XY) = \mathbb{E}(XY)^2 - (\mathbb{E}(XY))^2 = \mathbb{E}X^2\mathbb{E}Y^2 - (\mathbb{E}X)^2(\mathbb{E}Y)^2$ and continue

$$= [\text{Var } X + (\mathbb{E}X)^2][\text{Var } Y + (\mathbb{E}Y)^2] - (\mathbb{E}X)^2(\mathbb{E}Y)^2$$
$$= \text{Var } X\text{Var } Y + (\mathbb{E}X)^2\text{Var } Y + (\mathbb{E}Y)^2\text{Var } X.$$

The facts used are: linearity of expectation, the equations $\mathbb{E}Z = \mathbb{E}X\mathbb{E}Y$ and $\mathbb{E}Z^2 = \mathbb{E}X^2\mathbb{E}Y^2$ that hold due to independence, and finiteness of all mean values because of the boundedness of random variables X and Y.   ☐

Problem 1.35   Let $a_1, a_2, \ldots, a_n$ be yearly rainfalls in Cambridge over the past n years: assume $a_1, a_2, \ldots, a_n$ are IID RVs. Say that k is a record year if $a_k > a_i$ for all $i < k$ (thus the first-year is always a record year). Let $Y_i = 1$ if i is a record year and 0 otherwise. Find the distribution of Y_i and show that $Y_1, Y_2, \ldots, Y_n$ are independent. Calculate the mean and variance of the number of record years in the next n years.

Set $N = j$ if j is the first record year after year 1, $1 \le j < n$, and $N = n$ if a_1 or a_n are maximal among $a_1, \ldots, a_n$ (i.e. the first or the last year produced the absolute record). Show that $\mathbb{E}N \to \infty$ as $n \to \infty$.

Solution   Ranking the a_is in a non-increasing order generates a random permutation of $1, 2, \ldots, n$; by symmetry, all such permutations will be equiprobable. Ranking the first i rainfalls yields a random permutation of $1, 2, \ldots, i$. Hence,

$$\mathbb{P}(Y_i = 1) = \frac{1}{i}, \ \mathbb{P}(Y_i = 0) = \frac{i-1}{i}, \ \mathbb{E}Y_i = \frac{1}{i}, \ \text{Var } Y_i = \frac{1}{i} - \frac{1}{i^2}.$$

Observe that RVs $Y_1, \ldots, Y_n$ are independent. In fact, set

$$X_i = \text{the ranking of year } i \text{ among } 1, \ldots, i,$$

with

$$P(X_i = l) = \frac{1}{i}, \quad 1 \le l \le i.$$

Then Y_i is a function of X_i only. So, it is enough to check that the X_i are independent. In fact, for any collection of pair-wise distinct values l_i, $i = 1, \ldots, n$, with $l_i \le i$, the event $(X_i = l_i, \ 1 \le i \le n)$ is realised for a unique permutation out of $n!$. Thus,

$$\frac{1}{n!} = P(X_i = x_i, 1 \le i \le n) = \prod_{1 \le i \le n} P(X_i = x_i).$$

Next,

$$\mathbb{E} \sum_1^n Y_i = \sum_1^n \frac{1}{i},$$

and

$$\mathrm{Var}\left(\sum_1^n Y_i\right) = \sum_{i=1}^n \mathrm{Var}\, Y_i = \sum_1^n \left(\frac{1}{i} - \frac{1}{i^2}\right).$$

Finally, for $m = 2, \ldots, n$, the probability $P(N \ge m)$ equals

$$P(2, \ldots, m-1 \text{ are not records}) = 1 \times \frac{1}{2} \times \frac{2}{3} \cdots \times \frac{m-2}{m-1} = \frac{1}{m-1}.$$

Hence, $\mathbb{E}N \ge \sum_{m=2}^n 1/(m-1)$ which tends to ∞ as $n \to \infty$. Observe, however, that with probability 1 there will be infinitely many record years if observations are continued indefinitely. □

Problem 1.36   An expedition is sent to the Himalayas with the objective of catching a pair of wild yaks for breeding. Assume yaks are loners and roam about the Himalayas at random. The probability $p \in (0, 1)$ that a given trapped yak is male is independent of prior outcomes. Let N be the number of yaks that must be caught until a pair is obtained.
(i) Show that the expected value of N is $1 + p/q + q/p$, where $q = 1 - p$.
(ii) Find the variance of N.

Solution   (i) Clearly, $P(N = n) = p^{n-1}q + q^{n-1}p$ for $n \ge 2$. Then $\mathbb{E}N$ equals

$$q \sum_{n=2}^\infty np^{n-1} + p \sum_{n=2}^\infty nq^{n-1} = pq \left[\frac{1}{1-p} + \frac{1}{1-q} + \frac{1}{(1-p)^2} + \frac{1}{(1-q)^2}\right],$$

which gives $1 + (p/q) + (q/p)$.
(ii) Further,

$$\mathrm{Var}\, N = \mathbb{E}N(N-1) + \mathbb{E}N - (\mathbb{E}N)^2$$

and

$$\mathbb{E}N(N-1) = pq\sum_{n=2}^{\infty} n(n-1)p^{n-2} + pq\sum_{n=2}^{\infty} n(n-1)q^{n-2}$$
$$= \frac{2pq}{(1-p)^3} + \frac{2pq}{(1-q)^3} = \frac{2p}{q^2} + \frac{2q}{p^2}.$$

So, the variance of N equals

$$\frac{2p}{q^2} + \frac{2q}{p^2} + \frac{p}{q} + \frac{q}{p} + 1 - \left(\frac{p}{q} + \frac{q}{p} + 1\right)^2 = \frac{1}{q^2} + \frac{1}{p^2} - \frac{p}{q} - \frac{q}{p} - 4. \quad \square$$

Problem 1.37   Liam's bowl of spaghetti contains n strands. He selects two ends at random and joins them together. He does this until there are no ends left. What is the expected number of spaghetti hoops in the bowl?

Solution   Setting

$$X_i = \begin{cases} 1, & \text{if the } i\text{th join makes a loop,} \\ 0, & \text{otherwise,} \end{cases}$$

find

$$\mathbb{P}(X_i = 1) = \frac{1}{2n - 2i + 1}.$$

(By the ith join you have $2n - 2(i-1)$ ends untied; for an end chosen there are $2n - 2i + 1$ possibilities to choose the second end, and only one of them leads to a hoop.) Thus

$$\mathbb{E}\sum_{i=1}^{n} X_i = \sum_{i=1}^{n} \mathbb{E}X_i = \sum_{i=1}^{n} \mathbb{P}(X_i = 1) = \sum_{i=1}^{n} \frac{1}{2n - 2i + 1} = \sum_{i=0}^{n-1} \frac{1}{2i + 1}. \quad \square$$

Problem 1.38   Sarah collects figures from cornflakes packets. Each packet contains one figure, and n distinct figures make a complete set. Show that the expected number of packets Sarah needs to collect a complete set is

$$n\sum_{i=1}^{n} \frac{1}{i}.$$

Solution   The number of packets needed is

$$N = 1 + Y_1 + \cdots + Y_{n-1}.$$

Here Y_1 represents the number of packets needed for collecting the second figure, Y_3 the third figure, and so on. Each Y_j has a geometric distribution:

$$\mathbb{P}(Y_j = s) = \left(\frac{j}{n}\right)^{s-1} \frac{n-j}{n}.$$

Hence, $\mathbb{E}Y_j = n/(n-j)$. Then

$$\mathbb{E}N = n\sum_{j=0}^{n-1}\frac{1}{n-j} = n\sum_{1}^{n}\frac{1}{j} \approx n\ln n. \quad \square$$

Problem 1.39   The RV N takes values in the non-negative integers. Show that N has mean satisfying

$$\mathbb{E}(N) = \sum_{k=0}^{\infty}\mathbb{P}(N>k),$$

whenever this series converges.

Each packet of the breakfast cereal Soggies contains exactly one token, and tokens are available in each of the three colours blue, white and red. You may assume that each token obtained is equally likely to be of the three available colours, and that the (random) colours of different tokens are independent. Find the probability that, having searched the contents of k packets of Soggies, you have not yet obtained tokens of every colour.

Let N be the number of packets required until you have obtained tokens of every colour. Show that $\mathbb{E}(N) = 11/2$.

Solution   Write

$$\sum_{n=1}^{\infty}\mathbb{P}(N=n)\sum_{r=1}^{n}1 = \sum_{n=1}^{\infty}\sum_{r=n}^{\infty}\mathbb{P}(N=r) = \sum_{n=1}^{\infty}\mathbb{P}(N\ge n) = \sum_{k=0}^{\infty}\mathbb{P}(N>k).$$

Further, the probability $\mathbb{P}(N>k)$ equals

$$\mathbb{P}(\text{not yet obtained after } k \text{ trials}) = \mathbb{P}(\text{all red or blue after } k \text{ trials})$$
$$+ \mathbb{P}(\text{all red or white after } k \text{ trials}) + \mathbb{P}(\text{all blue or white after } k \text{ trials})$$
$$- \mathbb{P}(\text{all red}) - \mathbb{P}(\text{all blue}) - \mathbb{P}(\text{all white}) = 3\left[(2/3)^k - (1/3)^k\right].$$

Then the expected value of N equals

$$3+3\left[\sum_{k=3}^{\infty}\left(\frac{2}{3}\right)^k - \sum_{k=3}^{\infty}\left(\frac{1}{3}\right)^k\right] = 3+3\left[3\left(\frac{2}{3}\right)^3 - \frac{3}{2}\left(\frac{1}{3}\right)^3\right] = \frac{11}{2}. \quad \square$$

Example 1.3   A useful exercise is to prove formulas for the mean values of $R = \min[X, Y]$ and $S = \max[X, Y]$, where X and Y are independent RVs with non-negative integer values and finite means. Here

$$\mathbb{E}R = \sum_{n\ge1}\mathbb{P}(\min[X, Y]\ge n) = \sum_{n\ge1}\mathbb{P}(X\ge n)\mathbb{P}(X\ge n).$$

Next, as $R+S = X+Y$, the mean value $\mathbb{E}S = \mathbb{E}X + \mathbb{E}Y - \mathbb{E}R$, which is equal to

$$\sum_{n\ge1}[\mathbb{P}(X\ge n)+\mathbb{P}(Y\ge n) - \mathbb{P}(X\ge n)\mathbb{P}(Y\ge n)]. \quad \blacksquare$$

Remark   Independence plays an important rôle in the analysis of gambling and betting (which was a strong motivation for developing probabilistic concepts in the sixteenth–nineteenth centuries). For example, the following strategy is related to a concept of a *martingale*. (This term will appear many times in the future volumes.) A gambler bets on a sequence of independent events each of probability 1/2. First, he bets $1 on the first event. If he wins he quits, if he loses, he bets $2 on the next event. Again, if he wins he quits, otherwise he bets $4 and then quits anyway. In principle, one could stick to the same strategy for any given number of rounds, or even wait for the success however long it took. The point here is that success will eventually occur with probability 1 (as an infinite series of subsequent failures has probability 0). In this case the gain of $1 is guaranteed: if success occurs at trial k, the profit is 2^k and the total loss $1 + 2 + \cdots + 2^{k-1} = 2^k - 1$.

In general, the gambler could also bet different amounts $S_1, S_2, \ldots$ on different rounds. It is easy to see that the expected gain $\mathbb{E}X_n$ will be zero after any given number of rounds n. For instance, after three betting rounds the expected gain is

$$\frac{S_1}{2} + \frac{S_2 - S_1}{4} + \frac{S_3 - S_2 - S_1}{8} - \frac{S_1 + S_2 + S_3}{8} = 0.$$

In general, this fact is checked by induction in n. It seems that it contradicts with the previous argument that amount $1 could be obtained with probability 1. However, that will occur at a random trial! And although the expected number of trials until the first success is 2, the expected capital the gambler will need with doubled bets $S_i = 2^i$ is

$$\sum_{k=1}^{\infty} \left(\frac{1}{2}\right)^k 2^{k-1} = \sum_{k=1}^{\infty} \frac{1}{2} = \infty.$$

(This is known as St Petersburg's *gambling paradox*: it caused consternation in the Russian high society in the early nineteenth century.)

It would be natural for the gambler to try to minimize the variance Var X_n of his loss after n rounds. Again, a straightforward calculation shows that for $n = 3$:

$$\text{Var } X_3 = \frac{S_1^2}{2} + \frac{(S_2 - S_1)^2}{4} + \frac{(S_3 - S_2 - S_1)^2}{8} + \frac{(S_1 + S_2 + S_3)^2}{8}$$
$$= S_1^2 + \frac{S_2^2}{2} + \frac{S_3^2}{4}.$$

This is minimized when the bets are placed in increasing order, i.e. $S_1 \leq S_2 \leq S_3$. The same is true for any n.

Problem 1.40   Hamlet, Rosencrantz and Guildenstern are flipping coins. The odd man wins the coins of the others; if all coins appear alike, no coins change hands. Find the expected number of throws required to force one man out of the game provided Hamlet has 14 coins to start with and Rosencrantz and Guildenstern have 6 coins each.

Hint: Look for the expectation in the form *Klmn*, where *l*, *m* and *n* are the numbers of coins they start with and K depends only on $l + m + n$, the total number of coins.

Solution   The conditions of the game are that the total number of coins is constant. Hence, if H denotes the number of Hamlet's coins, R of Rosencrantz's and G Guildenstern's, then

$$H + R + G = 26. \tag{1.36}$$

The game corresponds to a random walk on integer points of a three-dimensional lattice satisfying equation (1.36), with $H \geq 0$, $R \geq 0$ and $G \geq 0$. The game ends when at least one of the inequalities becomes equality. The jump probabilities are

$$(H, R, G) \to (H+2, R-1, G-1), \quad \text{probability } \frac{1}{4},$$

$$(H, R, G) \to (H-1, R+2, G-1), \quad \text{probability } \frac{1}{4},$$

$$(H, R, G) \to (H-1, R-1, G+2), \quad \text{probability } \frac{1}{4},$$

$$(H, R, G) \to (H, R, G) \quad \text{probability } \frac{1}{4}.$$

The walk starts at $H = 14$, $R = G = 6$.

Let $E_{H,R,G}$ be the expected number of throws to reach the end if the starting amounts are H, R, G. Then, by the formula of conditional expectation,

$$E_{H,R,G} = \frac{1}{4}(1 + E_{H,R,G}) + \frac{1}{4}(1 + \mathbb{E}_{H+2,R-1,G-1})$$

$$+ \frac{1}{4}(1 + E_{H-1,R-1,G+2}) + \frac{1}{4}(1 + E_{H-1,R+2,G-1}),$$

where we condition on the first throw. That is

$$\frac{3}{4}E_{H,R,G} - 1 = \frac{1}{4}(E_{H+2,R-1,G-1} + E_{H-1,R+2,G-1} + E_{H-1,R-1,G+2}),$$

with the boundary conditions $E_{0,R,G} = E_{H,0,G} = E_{H,R,0} = 0$.

The conjectured form is $E_{H,R,G} = K(H + G + R)HGR$ whence

$$K = \frac{4}{3(H + G + R - 2)}$$

and

$$E_{14,6,6} = \frac{4 \times 14 \times 6 \times 6}{3 \times (26 - 2)} = \frac{336}{12} = 28. \quad \square$$

Remark   A suggestive guess is as follows: $E_{H,R,G}$ is a symmetric function of H, R, G. Hence, it must be a function of symmetric polynomials in H, R, G, viz.

$$H + R + G, \quad HG + HR + GR, \quad HRG,$$

etc. The boundary condition yields that HRG should appear as a factor.

Problem 1.41   We play the following coin-tossing game. Each of us tosses one (unbiased) coin; if they match, I get both coins; if they differ you get both. Initially, I have m

coins and you n. Let $\mathbb{E}(m, n)$ be the expected length of play until one of us has no coins left. Write down a linear relation between $\mathbb{E}(m, n)$, $\mathbb{E}(m-1, n+1)$ and $E(m+1, n-1)$. Expressing this as a function of $m+n=k$ and n, deduce that $\mathbb{E}(m, n)$ is a quadratic function of m and n, and hence, using appropriate initial conditions, deduce that

$$\mathbb{E}(m, n) = mn.$$

Solution   Proceed as above, getting

$$\mathbb{E}(m, n) = \frac{1}{2}\mathbb{E}(m-1, n+1) + \frac{1}{2}E(m+1, n-1) + 1.$$

The answer is $\mathbb{E}(m, n) = mn$. $\square$

Problem 1.42   In the Aunt Agatha problem (see Problem 1.15), what is the expected time until the clockwork orange falls off the table?

Solution   Continuing the solution of Problem 1.15, let E_ℓ be the (conditional) expected time while starting at distance $10 \times \ell$ cm from the left end. Then $E_0 = E_{20} = 0$ and

$$E_\ell = \frac{3}{5}E_{\ell-1} + \frac{2}{5}E_{\ell+1} + 1,$$

or $E_{\ell+1} = (5/2)E_\ell - (3/2)E_{\ell-1} - 5/2$. For vectors $\boldsymbol{u}_\ell = (E_\ell, E_{\ell+1})$ and $\boldsymbol{v} = (0, -5/2)$ we have $\boldsymbol{u}_\ell = \boldsymbol{u}_{\ell-1} A + \boldsymbol{v}$. Here A is as before:

$$A = \begin{pmatrix} 0 & -3/2 \\ 1 & 5/2 \end{pmatrix},$$

with the eigenvalues $\lambda_1 = 3/2$ and $\lambda_2 = 1$ and eigenvectors $\boldsymbol{e}_1 = (1, 3/2)$ and $\boldsymbol{e}_2 = (1, 1)$. Note that $\boldsymbol{v} = 5(\boldsymbol{e}_2 - \boldsymbol{e}_1)$.

Iterating yields

$$\boldsymbol{u}_\ell = \boldsymbol{u}_0 A^\ell + \boldsymbol{v} \sum_{1 \le j \le l-1} A^j$$

and if vector $\boldsymbol{u}_0 = a_1 \boldsymbol{e}_1 + a_2 \boldsymbol{e}_2$, then

$$\boldsymbol{u}_1 = \left(a_1 \lambda_1^l + a_2 - 5\left(\frac{\lambda_1^\ell - 1}{\lambda_1 - 1} - \ell \right),\ \frac{3}{2}a_1 \lambda_1^\ell + a_2 - 5\left(\frac{3}{2}\frac{\lambda_1^\ell - 1}{\lambda_1 - 1} - \ell \right) \right).$$

Now, substituting into the first component $E_0 = 0$ (for $\ell = 0$) yields

$$a_1 + a_2 = 0,$$

and into the second component $E_{20} = 0$ (for $\ell = 19$):

$$a_1 \left(\frac{3}{2} \right)^{20} + a_2 = 5\left(\frac{(3/2)^{20} - 3/2}{1/2} - 19 \right) = 5\left(\frac{(3/2)^{20} - 1}{1/2} - 20 \right).$$

Thus,

$$a_1 = -a_2 = \frac{10((3/2)^{20} - 1) - 100}{(3/2)^{20} - 1} = 10 - \frac{100}{(3/2)^{20} - 1},$$

and

$$E_{10} = \left[10 - \frac{100}{(3/2)^{20} - 1}\right]\left[\left(\frac{3}{2}\right)^{10} - 1\right] + 50 - 10\left[\left(\frac{3}{2}\right)^{10} - 1\right]$$

$$= 50 + 10\left[\left(\frac{3}{2}\right)^{10} - 1\right] - 10\left[\left(\frac{3}{2}\right)^{10} - 1\right] - \frac{100}{(3/2)^{10} + 1}$$

$$= 50 - \frac{100}{(3/2)^{10} + 1}.$$

A rough estimate yields $E_{10} \geq 48$.

Remark   A much shorter solution (sometimes adopted by students who feel it is correct but cannot justify it) is as follows. Let τ denote the time of the fall-off. As was found in the solution of Problem 1.15, the position S_τ at time τ is

$$S_\tau = \begin{cases} 20, & \text{with probability } \dfrac{1}{(3/2)^{10} + 1}, \\ 0, & \text{with probability } 1 - \dfrac{1}{(3/2)^{10} + 1}. \end{cases}$$

After n jumps, the position is the sum of independent RVs

$$S_n = S_0 + X_1 + \cdots + X_n,$$

where S_0 is the initial position (in our case, equal to 10) and X_i the increment at the jth jump:

$$X_i = \begin{cases} +1, & \text{with probability } \dfrac{2}{5}, \\ -1, & \text{with probability } \dfrac{3}{5}. \end{cases}$$

Writing

$$U_n = S_0 + (X_1 - \mathbb{E}X_1) + \cdots + (X_n - \mathbb{E}X_n) = S_n - n\mathbb{E}X_1$$

yields $\mathbb{E}U_n = \mathbb{E}S_0$. The fact (requiring use of the martingale theory) is that the same is true for the random time τ: $\mathbb{E}U_\tau = \mathbb{E}S_0$. Then

$$10 = 20\frac{1}{(3/2)^{10} + 1} - \mathbb{E}\tau\left(\frac{2}{5} - \frac{3}{5}\right),$$

whence

$$\mathbb{E}\tau = 50 - \frac{100}{(3/2)^{10} + 1} = E_{10},$$

as before.   □

Problem 1.43   Define the covariance Cov (X, Y) of two random variables X and Y. Show that

$$\operatorname{Var}(X + Y) = \operatorname{Var}X + \operatorname{Var}Y + 2\operatorname{Cov}(X, Y).$$

A fair die has two green faces, two red faces and two blue faces, and the die is thrown once. Let

$$X = \begin{cases} 1 & \text{if a green face is uppermost} \\ 0 & \text{otherwise,} \end{cases}$$

$$Y = \begin{cases} 1 & \text{if a blue face is uppermost} \\ 0 & \text{otherwise.} \end{cases}$$

Find Cov (X, Y).

Solution   Set: $\mu_X = \mathbb{E}X$, $\mu_Y = \mathbb{E}Y$, with $\mathbb{E}(X + Y) = \mathbb{E}X + \mathbb{E}Y$. Then

$$\operatorname{Cov}(X, Y) = \mathbb{E}(X - \mu_X)(Y - \mu_Y),$$

and

$$\begin{aligned}
\operatorname{Var}(X + Y) &= \mathbb{E}(X + Y - \mu_X - \mu_Y)^2 \\
&= \mathbb{E}(X - \mu_X)^2 + \mathbb{E}(Y - \mu_Y)^2 + 2\mathbb{E}(X - \mu_X)(Y - \mu_Y) \\
&= \operatorname{Var}X + \operatorname{Var}Y + 2\operatorname{Cov}(X, Y).
\end{aligned}$$

Now

$$\mathbb{P}(\text{green}) = \mathbb{P}(\text{blue}) = \frac{1}{3},$$

so

$$\mu_X = \mathbb{P}(\text{green}) = \frac{1}{3}; \text{ similarly, } \mu_Y = \frac{1}{3}.$$

Finally,

$$\operatorname{Cov}(X, Y) = \mathbb{E}XY - \mu_X\mu_Y - \mu_X\mu_Y + \mu_X\mu_Y = \mathbb{E}XY - \mu_X\mu_Y.$$

But $XY = 0$ with probability 1, thus $\mathbb{E}XY = 0$. Hence, Cov $(X, Y) = -1/9$.   □

Problem 1.44   Let X be an integer-valued RV with distribution

$$\mathbb{P}(X = n) = \frac{n^{-s}}{\zeta(s)},$$

where $s > 1$, and

$$\zeta(s) = \sum_{n \geq 1} n^{-s}.$$

Let $1 < p_1 < p_2 < p_3 < \cdots$ be the primes and let A_k be the event $\{X \text{ is divisible by } p_k\}$. Find $\mathbb{P}(A_k)$ and show that the events $A_1, A_2, \ldots$ are independent. Deduce that

$$\prod_{k=1}^{\infty}(1 - p_k^{-s}) = \frac{1}{\zeta(s)}.$$

Solution Write

$$\mathbb{P}(A_k) = \sum_{n: \text{ divisible by } p_k} \frac{n^{-s}}{\zeta(s)} = \sum_{n: n=p_k l} \frac{n^{-s}}{\zeta(s)} = p_k^{-s} \sum_{l \geq 1} \frac{l^{-s}}{\zeta(s)} = p_k^{-s}.$$

Similarly, $\forall$ collection $A_{k_1}, \ldots, A_{k_i}$ $(1 \leq k_1 < \cdots < k_l)$:

$$\mathbb{P}\left(A_{k_1} \cap \cdots \cap A_{k_i}\right) = p_{k_1}^{-s} \cdots p_{k_i}^{-s} = \prod_{l=1}^{i} \mathbb{P}\left(A_{k_l}\right),$$

i.e. the events $A_1, A_2, \ldots$ are independent.

Finally, $1 - p_k^{-s}$ is the probability that X is not divisible by p_k. Then $\prod_{k=1}^{\infty}(1 - p_k^{-s})$ gives the probability that X is not divisible by any prime, i.e. $X = 1$. The last probability equals $1/\zeta(s)$. $\square$

Remark $\zeta(s)$ is the famous Riemann zeta-function that is the subject of the Riemann hypothesis, one of the few problems posed in the nineteenth century which remain unsolved. The above problem gives a representation of the zeta-function as an infinite product over the prime numbers.

The name of G.F.B. Riemann (1826–1866), the remarkable German mathematician, is also related to Geometry (Riemannian manifolds, Riemannian metric). In the context of this book, we speak of *Riemann integration*, as opposed to a more general Lebesgue integration; see below.

1.5   The binomial, Poisson and geometric distributions. Probability generating, moment generating and characteristic functions

> Life is good for only two things,
> discovering mathematics and teaching mathematics.
> S.-D. Poisson (1781–1840), French mathematician

The binomial distribution appears naturally in the coin-toss setting. Consider the random variable X equal to the number of heads shown in the course of n trials with the same type of coin. A convenient representation is

$$X = Y_1 + \cdots + Y_n$$

where RVs $Y_1, \ldots, Y_n$ are independent and identically distributed

$$Y_j = \begin{cases} 1, & \text{if the } j\text{th trial shows head,} \\ 0, & \text{if the } j\text{th trial shows tail.} \end{cases}$$

Assuming that $\mathbb{P}(Y_j = 1)$, the probability of a head, equals p and $\mathbb{P}(Y_j = 0)$, that of a tail, $q = 1 - p$, we have

$$\mathbb{P}(X = k) = \binom{n}{k} p^k q^{n-k}, \ 0 \le k \le n. \tag{1.37}$$

This probability distribution (and the RV itself) is called binomial (or (n, p)-binomial) because it is related to the binomial expansion

$$\sum_{0 \le k \le n} \binom{n}{k} p^k q^{n-k} = (p + q)^n = 1.$$

Values of binomial probabilities are shown in Figure 1.3.

Because of the above representation $X = Y_1 + \cdots + Y_n$, the sum $X + X'$ of independent (n, p)- and (n', p)-binomial RVs X and X' is $((n + n'), p)$-binomial. This representation also yields that

$$\mathbb{E}X = n\mathbb{E}Y_1 = np, \ \operatorname{Var} X = n \operatorname{Var} Y_1 = npq. \tag{1.38}$$

We write $X \sim \mathrm{Bi}(n, p)$ for a binomial RV X.

Other well-known expansions also give rise to useful probability distributions. For example, if we toss a coin until the first head, the outcomes are numbers $0, 1, \ldots$ (indicating the number of tails before the first head was shown). Let X denote the number of tails before the first head. The probability of outcome k is

$$\mathbb{P}(X = k) = pq^k, \ k = 0, 1, 2, \ldots, \tag{1.39}$$

and equals the probability of sequence $TT \ldots TH$ where first k digits are T and the $(k + 1)$th one H. The sum of all probabilities (proportional to the sum of a geometric progression) is $p/(1 - q) = 1$. The 'tail' probability $\mathbb{P}(X \ge k) = q^k, \ k \ge 0$.

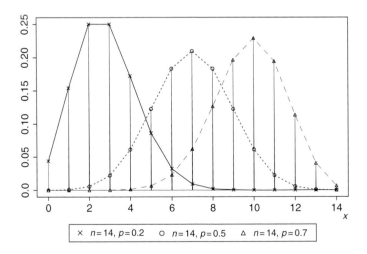

Figure 1.3 The binomial PMFs.

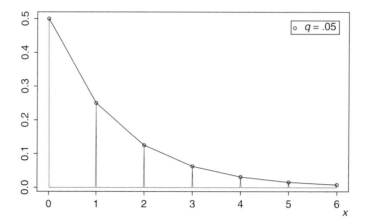

Figure 1.4 The geometric PMF.

Not surprisingly, the RV X giving the number of trials before the first head is said to have a *geometric distribution* (or being geometric), with parameter q. The diagram of values of geometric probabilities is shown in Figure 1.4.

The expectation and the variance of a geometric distribution are calculated below:

$$\mathbb{E}X = \sum_{k \geq 1}(1-q)kq^k = \sum_{k \geq 1}q^k = \frac{q}{p}, \tag{1.40}$$

and

$$\text{Var}\, X = \mathbb{E}X^2 - (\mathbb{E}X)^2 = \sum_{k \geq 1}k^2pq^k - \frac{q^2}{p^2}$$

$$= pq^2 \sum_{k \geq 2}k(k-1)q^{k-2} + pq \sum_{k \geq 1}kq^{k-1} - \frac{q^2}{p^2}$$

$$= pq^2 \frac{d^2}{dq^2}\frac{1}{p} + \frac{q}{p} - \frac{q^2}{p^2} = pq^2\frac{2}{p^3} + \frac{q}{p} - \frac{q^2}{p^2} = \frac{q^2}{p^2} + \frac{q}{p} = \frac{q}{p^2}. \tag{1.41}$$

Note a special property of a geometric RV: $\forall\, m < n$:

$$\mathbb{P}(X \geq n | X \geq m) = \mathbb{P}(X \geq n - m); \tag{1.42}$$

this is called the *memoryless property*. Another property of geometric distributions is that the minimum $\min[X, X']$ of two independent geometric RVs X and X' with parameters q and q' is geometric, with parameter qq':

$$\mathbb{P}(\min[X, X'] \geq k) = (qq')^k, \quad k = 0, 1, \ldots$$

Sometimes, the geometric distribution is defined by probabilities $p_k = pq^{k-1}$, for $k = 1, 2, \ldots$ which counts the number of trials up to (and including) the first head. This leads to a different value of the mean value $1/p$; the variance remains the same. We write $X \sim \text{Geom}\,(q)$ for a geometric RV X (in either definition).

The geometric distribution often arises in various situations. For example, the number of hops made by a bird before flying fits the geometric distribution.

A generalisation of the geometric distribution is a *negative binomial distribution*. Here, the corresponding RV X gives the number of tails before the rth head. In other words, $X = X_1 + \cdots + X_r$, where $X_i \sim \text{Geom}(q)$, independently. A direct calculation shows that

$$\mathbb{P}(X = k) = \binom{k+r-1}{k} p^r q^k, \ k = 0, 1, 2, \ldots \tag{1.43}$$

In fact, $X = k$ means that there were altogether $k + r$ trials, with k tails and r heads, and the last toss was a head.

We write $X \sim \text{NegBin}(q, r)$ for the negative binomial distribution.

Another example is the Poisson distribution, with parameter $\lambda \geq 0$; it emerges from the expansion $e^\lambda = \sum_{k \geq 0} \lambda^k / k!$ and named after S.-D. Poisson (1781–1840), a prominent French mathematician and mathematical physicist. Here we again assign probabilities to non-negative integers, and the probability assigned to k equals $\lambda^k e^{-\lambda} / k!$. An RV X with

$$\mathbb{P}(X = k) = \frac{\lambda^k}{k!} e^{-\lambda}, \ k = 0, 1, 2, \ldots, \tag{1.44}$$

is called Poisson. These probabilities arise from the binomial probabilities in the limit $n \to \infty$, with $p = \lambda/n \to 0$:

$$\frac{n!}{(n-k)!k!} \left(\frac{\lambda}{n}\right)^k \left(1 - \frac{\lambda}{n}\right)^{n-k}$$

$$= \frac{n(n-1)\cdots(n-k+1)}{n^k} \times \frac{\lambda^k}{k!} \times \frac{(1-\lambda/n)^n}{(1-\lambda/n)^k}$$

which approaches $\lambda^k e^{-\lambda} / k!$.

The last observation explains the fact that the sum $X + X'$ of two independent Poisson RVs X and X', with parameters λ and λ', is Poisson, with parameter $\lambda + \lambda'$. This fact can also be established directly.

The graphs of values of Poisson probabilities are in Figure 1.5.

The expectation $\mathbb{E}X$ and the variance $\text{Var } X$ of a Poisson RV equal λ:

$$\sum_{k \geq 0} \frac{\lambda^k}{k!} e^{-\lambda} k = \lambda, \ \sum_{k \geq 0} \frac{\lambda^k}{k!} e^{-\lambda} k^2 - \lambda^2 = \lambda. \tag{1.45}$$

We write $X \sim \text{Po}(\lambda)$ for a Poisson RV X.

The Poisson distribution is widely used in various situations. A famous (albeit chilling) example (with which both authors of this book began their studies in Probability) is the number of Prussian cavalrymen killed by a horse kick in each of the 16 corps in each of the years 1875–1894. This can be perfectly fitted by the Poisson distribution! It is amazing that this example found its way into most textbooks until the end of the twentieth century, without diminishing the enthusiasm of several generations of students.

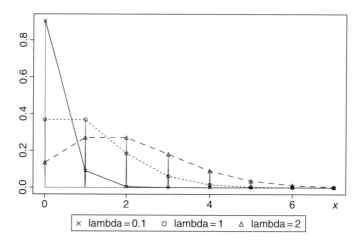

Figure 1.5 The Poisson PMFs.

The *probability generating function* (PGF) $\phi(s)$ $(=\phi_X(s))$ of an RV X taking finitely or countably many non-negative integer values n with probabilities p_n is defined as

$$\phi_X(s) = \mathbb{E}s^X = \sum_n p_n s^n = \sum_{\omega \in \Omega} p(\omega)s^{X(\omega)} = \mathbb{E}s^X; \qquad (1.46)$$

it is usually considered for $-1 \le s \le 1$ to guarantee convergence. A list of PGFs can be found in Appendix 1, Table A1.1; here we give a few often used examples. If X takes values 1 and 0 with probabilities p and $1-p$, then

$$\phi_X(s) = ps + 1 - p; \qquad (1.47)$$

if $X \sim \mathrm{Bin}(n,\ p)$, then

$$\phi_X(s) = \sum_{k=1}^{n} \binom{n}{k} p^k q^{n-k} s^k = [ps + (1-p)]^n; \qquad (1.48)$$

if $X \sim \mathrm{Geom}\ (p)$, then

$$\Phi_X(s) = \sum_{k \ge 0} p(1-p)^k s^k = \frac{p}{1 - s(1-p)}; \qquad (1.49)$$

and if $X \sim \mathrm{Po}\,(\lambda)$, then

$$\phi_X(s) = \sum_{k \ge 0} \frac{\lambda^k}{k!} e^{-\mu} s^k = e^{\lambda(s-1)}. \qquad (1.50)$$

An important fact is that the PGF $\phi_X(s)$ determines the distribution of RV X uniquely: if $\phi_X(s) = \phi_Y(s) \ \forall\ 0 < s < 1$, then $\mathbb{P}(X=n) = \mathbb{P}(Y=n) \ \forall\ n$. In this case we write $X \sim Y$. For X and Y taking finitely many values the uniqueness is obvious as $\phi_X(s)$ and $\phi_Y(s)$ are linear combinations of monomials s^n (i.e. polynomials); if two polynomials coincide, their coefficients also coincide. This is also true for RVs taking countably many values, but here one has to deal with power series (i.e. the Taylor decomposition).

Also,

$$\mathbb{E}X = \frac{d}{ds}\phi_X(s)\Big|_{s=1} \quad \text{and} \quad \mathbb{E}X(X-1) = \frac{d^2}{ds^2}\phi_X(s)\Big|_{s=1}, \tag{1.51}$$

implying that

$$\operatorname{Var}X = \mathbb{E}X(X-1) + \mathbb{E}X - (\mathbb{E}X)^2$$

$$= \left(\frac{d^2}{ds^2}\phi_X(s) + \frac{d}{ds}\phi_X(s) - \left[\frac{d}{ds}\phi_X(s)\right]^2\right)\Big|_{s=1}. \tag{1.52}$$

Another useful observation is that RVs X and Y are independent iff

$$\phi_{X+Y}(s) = \phi_X(s)\phi_Y(s). \tag{1.53}$$

In fact, let X, Y be independent. Then by convolution formula (1.14), $\phi_{X+Y}(s)$ equals

$$\sum_n \mathbb{P}(X+Y=n)s^n = \sum_n \sum_k \mathbb{P}(X=k)s^k \mathbb{P}(Y=n-k)s^{n-k},$$

and by changing $n-k \mapsto l$ is equal to

$$\left[\sum_k \mathbb{P}(X=k)s^k\right]\left[\sum_k \mathbb{P}(N=l)s^l\right] = \phi_X(s)\phi_Y(s).$$

The converse fact is proved by referring to the uniqueness of the coefficients of a power series. A short proof is achieved by observing that if X and Y are independent then s^X and s^Y are independent, and hence by virtue of formula (1.27):

$$\mathbb{E}s^{X+Y} = \mathbb{E}s^X s^Y = \mathbb{E}s^X \mathbb{E}s^Y.$$

The use of PGFs is illustrated in the following example.

Example 1.4   There is a random number N of birds on a rock; each bird is a seagull with probability p and a puffin with probability $1-p$. If S is the number of seagulls then the PGFs $\phi_N(s)$ and $\phi_S(s)$ are related by

$$\phi_S(s) = \phi_N(ps+1-p).$$

In fact, according to the definition, $\phi_S(s) = \mathbb{E}_s^{\ S} = \sum_{n\geq 0} s^n P_S(n)$, with $P_S(n) = \mathbb{P}(S=n)$, and similarly for $\phi_N(s)$. Then

$$\phi_S(s) = \sum_n s^n \sum_{m\geq n} P_N(m) \binom{m}{n} p^n (1-p)^{m-n}$$

$$= \sum_m P_N(m) \sum_{n=0}^m \binom{m}{n}(sp)^n(1-p)^{m-n}$$

$$= \sum_m P_N(m)(ps+(1-p))^m = \phi_N(ps+(1-p)).$$

In short-hand notation:

$$\phi_S(s) = \mathbb{E}\left[\mathbb{E}\left(s^S|N\right)\right] = \sum_m P_N(m)\mathbb{E}\left(s^S|N=m\right)$$

$$= \sum_m P_N(m)\left(ps+(1-p)\right)^m = \phi_N\left[ps+(1-p)\right].$$

Similarly, for the number of puffins,

$$\phi_{N-S}(s) = \phi_N\left[(1-p)s+p\right].$$

Now, suppose N has the Poisson distribution with parameter μ. Then

$$\phi_N(s) = e^{\mu(s-1)}, \quad \text{and} \quad \phi_S(s) = e^{\mu(ps+1-p-1)} = e^{\mu p(s-1)}.$$

By the uniqueness of the probability distribution with a given PGF, $S \sim \text{Po}(\mu p)$. Similarly,

$$\phi_{N-S}(s) = e^{\mu(1-p)(s-1)}, \quad \text{i.e. } N - S \sim \text{Po}[\mu(1-p)].$$

Also,

$$\phi_N(s) = \phi_S(s)\phi_{N-S}(s),$$

i.e. S and $N - S$ are independent.

Of course, one could proceed directly to prove that $S \sim \text{Po}(p\mu)$, $N - S \sim \text{Po}[(1-p)\mu]$ and S and $N - S$ are independent. First,

$$\mathbb{P}(S=k) = \sum_{n \geq k} \mathbb{P}(S=k|N=n)\mathbb{P}(N=n)$$

$$= \sum_{n \geq k} \frac{n!}{k!(n-k)!}p^k(1-p)^{n-k}\frac{\mu^n e^{-\mu}}{n!}$$

$$= \frac{(p\mu)^k e^{-p\mu}}{k!} \sum_{n-k \geq 0} \frac{e^{-\mu(1-p)}[\mu(1-p)]^{n-k}}{(n-k)!} = \frac{(p\mu)^k e^{-p\mu}}{k!},$$

and similarly $\mathbb{P}(N-S=k) = [(1-p)\mu]^k e^{-(1-p)\mu}/k!$. Next,

$$\mathbb{P}(S=k, N-S=l) = \mathbb{P}(S=k|N=k+l)\mathbb{P}(N=k+l)$$

$$= \binom{k+l}{k}p^k(1-p)^l\frac{\mu^{k+l}e^{-\mu}}{(k+l)!}$$

$$= \frac{(p\mu)^k e^{-p\mu}}{k!} \times \frac{((1-p)\mu)^l e^{-\mu(1-p)}}{l!}$$

$$= \mathbb{P}(S=k)\mathbb{P}(N-S=l),$$

i.e. S and $N - S$ are independent.

An elegant representation for S and $N - S$ is:

$$S = \sum_{n=1}^{N} X_i, \quad N - S = \sum_{n-1}^{N} (1 - X_i),$$

where X_1, X_2, ... are IID RVs taking values 1 and 0 with probabilities p and $(1 - P)$, respectively. It is instructive to repeat the above proof that $S \sim \text{Po}(p\mu)$ and $(N - S) \sim \text{Po}[(1 - p)\mu]$ by using this representation (and assuming of course that $N \sim \text{Po}(\mu)$).

An interesting fact is that a converse assertion also holds: here the use of PGFs will be very efficient. Let N be an arbitrary RV, let $p = 1/2$ and suppose that RVs S and $N - S$ are independent, without assuming that either of them is Poisson. By independence, symmetry and the equation $\phi_S(s) = \phi_N(ps + (1 - p))$:

$$\phi_N(s) = \phi_S(s)\phi_{N-S}(s) = \left[\phi_N\left(\frac{1}{2}s + 1 - \frac{1}{2}\right)\right]^2.$$

Then the function $\Phi(s) = \phi_N(1 - s)$ satisfies

$$\Phi(s) = \left[\Phi\left(\frac{s}{2}\right)\right]^2 = \cdots = \left[\Phi\left(\frac{s}{2^n}\right)\right]^{2^n}.$$

Thus, if s is small, the Taylor expansion yields

$$\Phi(s) = \phi_N(1 - s) = \phi_N(1) - \phi'_N(1)s + o(s^2) = 1 - \mu s + O(s^2),$$

where $\mu = \mathbb{E}N$. Hence, as $n \to \infty$

$$\Phi(s) = \left[1 - \frac{\mu s}{2^n} + o\left(\frac{s^2}{4^n}\right)\right]^{2^n} \to e^{-\mu s}.$$

Thus, $\phi_N(1 - s) = e^{\mu(s-1)}$, i.e. $N \sim \text{Po}(\mu)$. Then, by the above argument, S and $N - S$ will be Poisson of parameter $\mu/2$. ■

Function

$$M_X(\theta) = \phi_X(e^{\theta}) = \mathbb{E}e^{\theta X} \tag{1.54}$$

is called the *moment generating function* (MGF). It is considered for real values of the argument θ, but may not exist for some of them. If X takes non-negative values only, $M_X(\theta)$ exists $\forall \; \theta \leq 0$ and possibly for some $\theta > 0$. Then one can also use the function

$$L_X(\theta) = \phi_X(e^{-\theta}) = \mathbb{E}e^{-\theta X}, \tag{1.55}$$

which is called the *Laplace transform*. The name 'moment generating function' comes from the representation $M_X(\theta) = \sum_{n=0}^{\infty} (\mathbb{E}X^n)\theta^n/n!$ as the expected value $\mathbb{E}X^n$ is called the nth *moment* of RV X.

On the other hand, the *characteristic function* (CHF) $\psi(t)$ $(=\psi_X(t))$ can be correctly defined for all real t:

$$\psi(t) = \mathbb{E}e^{itX} = \sum_j p_j e^{itx_j} = \sum_{\omega \in \Omega} p(\omega)e^{itX(\omega)}, \tag{1.56}$$

though it takes complex values, of modulus ≤ 1. The usefulness of these two functions can be seen from the following properties.

(i)   The MGFs and CHFs can be effectively used for RVs taking real values, not necessary non-negative integers. For many important RVs the MGFs and CHFs are written in a convenient and transparent form. The CHFs are particularly important when one analyses the convergence of RVs. Of course, $\psi(t) = M_X(it)$ and $M_X(\theta) = \phi_X(e^\theta)$.

(ii) The MGFs and CHFs multiply when we add independent RVs: if X and Y are independent then, by equation (1.27),

$$M_{X+Y}(\theta) = M_X(\theta)M_Y(\theta), \quad \psi_{X+Y}(\theta) = \psi_X(t)\psi_Y(t). \tag{1.57}$$

In fact, if X and Y are independent then $e^{\theta X}$ and $e^{\theta Y}$ are independent, and

$$M_{X+Y}(\theta) = \mathbb{E}e^{\theta(X+Y)} = \mathbb{E}e^{\theta X}e^{\theta Y} = \mathbb{E}e^{\theta X}\mathbb{E}e^{\theta Y} = M_X(\theta)M_Y(\theta),$$

and similarly for the characteristic functions.

(iii) The expected value $\mathbb{E}X$ and the variances $\text{Var } X$ (and in fact other moments $\mathbb{E}X^j$) are expressed in terms of their derivatives at $t=0$, viz.

$$\mathbb{E}X = \frac{d}{d\theta}M_X(\theta)\Big|_{\theta=0} = \frac{1}{i}\frac{d}{dt}\psi_X(t)\Big|_{t=0}, \tag{1.58}$$

$$\text{Var } X = \left(\frac{d^2}{d\theta^2}M_X(\theta) - \left[\frac{d}{d\theta}M_X(\theta)\right]^2\right)\Big|_{\theta=0}$$

$$= \left(-\frac{d^2}{dt^2}\psi_X(t) + \left[\frac{d}{dt}\psi_X(t)\right]^2\right)\Big|_{t=0}. \tag{1.59}$$

(iv) Each of $M_X(\theta)$ and $\psi_X(t)$ uniquely defines the distribution of the RV: if $M_X(\theta) = M_Y(\theta)$ or $\psi_X(t) = \psi_Y(t)$ in the entire domain of existence then variables X and Y take the same collection of values and with the same probabilities. In this case we write, as before, $X \sim Y$. This property also ensures that if $M_{X+Y}(\theta) = M_X(\theta)M_Y(\theta)$ or $\psi_{X+Y} = \psi_X(t)\psi_Y(t)$ then RVs X and Y are independent.

Property (iv) is helpful for identifying various probability distributions; its full proof is beyond the limits of this course.

Problem 1.45   Let N be a discrete random variable with

$$\mathbb{P}(N=k) = (1-p)^{k-1}p, \quad k = 1, 2, 3, \ldots,$$

where $0 < p < 1$. Show that $\mathbb{E}N = 1/p$.

Solution   A direct calculation:

$$\mathbb{E}N = \sum_{k=1}^{\infty} k(1-p)^{k-1}p = -p\frac{\mathrm{d}}{\mathrm{d}p}\left(\frac{1}{1-(1-p)}\right) = \frac{1}{p}.$$

Alternatively: $\mathbb{E}N = 1 \times p + (1-p) \times (1 + \mathbb{E}N)$, implying that $\mathbb{E}N = 1/p$. $\square$

Problem 1.46   (i) Suppose that the RV X is geometrically distributed, so that $\mathbb{P}(X = n) = pq^{n}$, $n = 0, 1, \ldots$, where $q = 1 - p$ and $0 < p < 1$. Show that

$$\mathbb{E}\left(\frac{1}{(X+1)}\right) = -\frac{p}{q}\ln p.$$

(ii) Two dice are repeatedly thrown until neither of them shows the number 6. Find the probability that at the final throw at least one of the dice shows the number 5.

Suppose that your gain is the total Y shown by the two dice at the last throw, and that the time taken to achieve it is the total number N of throws.

(a)   Find your expected gain $\mathbb{E}Y$.
(b)   Find your expected rate of return $\mathbb{E}(Y/N)$, using the approximation $\ln(36/25) \approx 11/30$.

Solution   (i) Observe that $N = X + 1$ with probability of success $p = 25/36$. Then

$$\mathbb{E}\left(\frac{1}{N}\right) = \sum_{n\geq 1}\frac{1}{n}q^{n-1}p = \frac{p}{q}\sum_{n\geq 1}\frac{1}{n}q^{n}$$

$$= \frac{p}{q}\sum_{n\geq 1}\int_{0}^{q}x^{n-1}\mathrm{d}x = \frac{p}{q}\int_{0}^{q}\frac{1}{1-x}\mathrm{d}x$$

$$= \frac{p}{q}\int_{p}^{1}\frac{1}{u}\mathrm{d}u = \frac{p}{q}\ln\frac{1}{p}.$$

(ii) By symmetry,

$$\mathbb{P}(\text{at least one 5 at the last throw})$$
$$= \frac{\mathbb{P}(\text{at least one 5 and no 6 at the last throw})}{\mathbb{P}(\text{no 6 at the last throw})} = \frac{9}{25}.$$

Next, $Y = Y_{1} + Y_{2}$, where Y_{i} is the number shown by die i. Hence, (a)

$$\mathbb{E}Y = 2\mathbb{E}Y_{1}, \ \ \mathbb{E}Y_{1} = (1+2+3+4+5)/5 = 3, \ \ \mathbb{E}Y = 6.$$

Now, $N \sim$ Geom (q), with parameter $q = 11/36$. Furthermore, Y and N are independent:

$$\mathbb{P}(Y = y, N = n) = q^{n-1}p\mathbb{P}(\text{total } y \text{ by the } n\text{th throw}|\text{no 6})$$
$$= \mathbb{P}(Y = y)\mathbb{P}(N = n).$$

So, (b)

$$\mathbb{E}\left(\frac{Y}{N}\right) = \mathbb{E}Y\mathbb{E}\left(\frac{1}{N}\right) = 6 \times \frac{25}{11} \ln\left(\frac{36}{25}\right) \approx 5. \quad \square$$

Problem 1.47   (i) Suppose that X and Y are discrete random variables with finite mean and variance. Establish the following results:

(a) $\mathbb{E}(X+Y) = \mathbb{E}X + \mathbb{E}Y$.
(b) $\text{Var}(X+Y) = \text{Var}\,X + \text{Var}\,Y + 2\text{Cov}\,(X,Y)$.
(c) X and Y independent implies that $\text{Cov}\,(X,Y) = 0$.

(ii) A coin shows heads with probability $p > 0$ and tails with probability $q = 1 - p$. Let X_n be the number of tosses needed to obtain n heads. Find the PGF for X_1 and compute its mean and variance. What is the mean and variance of X_n?

Solution   (i) (a) Write

$$\mathbb{E}X = \sum_a a\mathbb{P}(X=a), \quad \mathbb{E}Y = \sum_b b\mathbb{P}(Y=b)$$

and observe that

$$\{\omega : X = a\} = \cup_b\{\omega : X = a, Y = b\}$$

where the union is pair-wise disjoint. Then

$$\mathbb{E}X + \mathbb{E}Y = \sum_{a,b} a\mathbb{P}(X=a, \ Y=b) + b\mathbb{P}(X=a, \ Y=b)$$

$$= \sum_c c \sum_{a+b=c} \mathbb{P}(X=a, Y=b)$$

$$= \sum_c c\mathbb{P}(X+Y=c) = \mathbb{E}(X+Y).$$

(b) By definition, $\text{Var}(X) = \mathbb{E}((X - \mathbb{E}X)^2) = \mathbb{E}X^2 - (\mathbb{E}X)^2$. Then

$$\text{Var}(X+Y) = \mathbb{E}(X+Y - (\mathbb{E}X + \mathbb{E}Y))^2$$

$$= \mathbb{E}(X - \mathbb{E}X)^2 + \mathbb{E}(Y - \mathbb{E}Y)^2$$

$$+ 2\mathbb{E}(X - \mathbb{E}X)(Y - \mathbb{E}Y)$$

$$= \text{Var}\,X + \text{Var}\,Y + 2\text{Cov}\,(X,Y).$$

(c) If X, Y are independent,

$$\mathbb{E}XY = \sum_c c\mathbb{P}(XY=c) = \sum_c c \sum_{ab=c} \mathbb{P}(X=a, \ Y=b)$$

$$= \sum_{a,b} ab\mathbb{P}(X=a)\mathbb{P}(Y=b)$$

$$= \left[\sum_a a \mathbb{P}(X=a) \right] \left[\sum_b b \mathbb{P}(Y=b) \right]$$

$$= \mathbb{E}X \mathbb{E}Y.$$

Now if X, Y are independent then so are $X - \mathbb{E}X$, $Y - \mathbb{E}Y$, and so

$$\text{Cov } (X, Y) = \mathbb{E}(X - \mathbb{E}X)(Y - \mathbb{E}Y) = 0.$$

(ii) We have $\mathbb{P}(X_1 = k) = pq^{k-1}$, and so the PGF $\phi(s) = \phi_{X_1}(s) = ps/(1-qs)$ and the derivative $\phi'(s) = p/(1-qs)^2$. Hence,

$$\mathbb{E}X_1 = \phi'(1) = \frac{1}{p}, \ \mathbb{E}X_1^2 = \phi''(1) + \phi'(1) = \frac{1+q}{p^2}, \text{ and Var } X_1 = \frac{q}{p^2}.$$

So $\mathbb{E}X_n = n/p$, Var $X_n = np/q^2$. □

Problem 1.48   Each of the random variables U and V takes the values ± 1. Their joint distribution is given by

$$\mathbb{P}(U=1) = \mathbb{P}(U=-1) = \frac{1}{2},$$

$$\mathbb{P}(V=1|U=1) = \mathbb{P}(V=-1|U=-1) = \frac{1}{3},$$

$$\mathbb{P}(V=-1|U=1) = \mathbb{P}(V=1|U=-1) = \frac{2}{3}.$$

(i) Find the probability that $x^2 + Ux + V = 0$ has at least one real root.
   (ii) Find the expected value of the larger root of $x^2 + Ux + V = 0$ given that there is at least one real root.
   (iii) Find the probability that $x^2 + (U+V)x + U + V = 0$ has at least one real root.

Solution   Write

$$\mathbb{P}(U=1, V=1) = \frac{1}{6}, \ \mathbb{P}(U=-1, V=1) = \frac{1}{3},$$

$$\mathbb{P}(U=1, V=-1) = \frac{1}{3}, \ \mathbb{P}(U=-1, V=-1) = \frac{1}{6}.$$

(i) $x^2 + Ux + V$ has a real root iff $U^2 - 4V \geq 0$ which means $V = -1$. Clearly, if $V = -1$, then $U^2 - 4V = 5$. So the probability of a real root is $1/2$.
   (ii) The expected value of the larger root is

$$\frac{(-1+\sqrt{5})}{2} \mathbb{P}(U=1|V=-1) + \frac{(1+\sqrt{5})}{2} \mathbb{P}(U=-1|V=-1)$$

$$= \frac{1}{\mathbb{P}(V=-1)} \left(\frac{(-1+\sqrt{5})}{2} \frac{1}{3} + \frac{(1+\sqrt{5})}{2} \frac{1}{6} \right) = \frac{\sqrt{5}}{2} - \frac{1}{6}.$$

(iii) $x^2 + Wx + W$ has a real root if $W^2 - 4W \geq 0$. If $W = U + V$, W takes values $2, 0, -2$ and the equation has a real root if $W = 0$ or -2. Then $\mathbb{P}(W = 0) = 2/3$ and $\mathbb{P}(W = 0 \text{ or } - 2) = 5/6$. □

Problem 1.49   In a sequence of independent trials, X is the number of trials up to and including the ath success. Show that

$$\mathbb{P}(X = r) = \binom{r-1}{a-1} p^a q^{r-a}, r = a, a+1, \ldots$$

Verify that the PGF for this distribution is $p^a s^a (1 - qs)^{-a}$. Show that $\mathbb{E}X = a/p$ and $\mathrm{Var}\, X = aq/p^2$. Show how X can be represented as the sum of a independent random variables, all with the same distribution. Use this representation to derive again the mean and variance of X.

Solution   Probability $\mathbb{P}(X = r)$ is represented as

$$\mathbb{P}(\{a - 1 \text{ successes occurred in } r - 1 \text{ trials}\} \cap \{\text{success on the } r\text{th trial}\})$$

$$= \binom{r-1}{a-1} p^{a-1} q^{r-1-a+1} p = \binom{r-1}{a-1} p^a q^{r-a}, \quad r \geq a.$$

The PGF for this distribution is

$$\phi(s) = p^a s^a \sum_{r \geq a} \binom{r-1}{a-1} (qs)^{r-a} = p^a s^a \sum_{k \geq 0} \binom{k+a-1}{a-1} (qs)^k.$$

Observe that

$$\binom{k+a-1}{a-1}$$

coincides with the number of ways to represent k as a sum of a non-negative integers

$$k = k_1 + \cdots + k_a.$$

This fact may be geometrically interpreted as follows. You take k 'stars' and you intersperse them with $a - 1$ 'bits':

$$| * \cdots * | * | * ||| * \cdots * | *.$$

The number of stars between the $(j-1)$th and jth bits give you the value k_j; k_1 is the number of stars to the left of the first bit and k_a is the number of stars to the right of the $(a-1)$th bit. But the number of different diagrams is

$$\binom{k+a-1}{a-1}.$$

Now use the multinomial expansion formula

$$\left(\sum_k b_k\right)^a = \sum_k \sum_{k_1,\dots,k_a:k_1+\cdots+k_a=k} \prod_j b_{k_j}$$

and obtain

$$\sum_{k\geq 0}\binom{k+a-1}{a-1}(qs)^k = \left(\sum_{k\geq 0}(qs)^k\right)^a = (1-qs)^{-a}$$

and $\phi(s)=p^a s^a (1-qs)^{-a}$. Further,

$$\mathbb{E}X = (p^a s^a(1-qs)^{-a})'|_{s=1} = \frac{a}{p}.$$

$$\mathbb{E}(X(X-1)) = (p^a s^a(1-qs)^{-a})''|_{s=1}$$
$$= a(a-1) + \frac{2a^2(1-p)}{p} + \frac{a(a+1)(1-p)^2}{p^2}.$$

Hence,

$$\operatorname{Var}X = \mathbb{E}X(X-1) + \mathbb{E}X - (\mathbb{E}X)^2 = \phi''(1) + \phi'(1) - (\phi'(1))^2$$
$$= \frac{a(a-1)p^2 + 2a^2 p(1-p) + a(a+1)(1-p)^2}{p^2} + \frac{a}{p} - \frac{a^2}{p^2} = \frac{aq}{p^2}.$$

As $\phi(s)=(\psi(s))^a$, where $\psi(s)=ps(1-qs)^{-1}$, you conclude that X may be represented as a sum $\sum_{j=1}^a Y_j$, where $Y_1,\dots,Y_a$ are IID RVs with the PGF $\psi(s)$. In fact, Y_j is the number of trials between the jth and $(j+1)$th successes including the trial of the $(j+1)$th success. So,

$$\mathbb{E}X = a\mathbb{E}Y_j, \quad \operatorname{Var}X = a\operatorname{Var}Y_j.$$

As $Y_j \sim \operatorname{Geom}(q)$, $\mathbb{E}Y_j = 1/p$ and $\operatorname{Var}(Y_j) = q/p^2$. $\square$

Problem 1.50   What is a Poisson distribution? Suppose that X and Y are independent Poisson variables with parameters λ and μ respectively. Show that $X+Y$ is Poisson with parameter $\lambda+\mu$. Are X and $X+Y$ independent? What is the conditional probability $\mathbb{P}(X=r|X+Y=n)$?

Solution   A Poisson distribution assigns probabilities

$$p_r = e^{-\lambda}\frac{\lambda^r}{r!}$$

to non-negative integers $r=0,1,\dots$ Here $\lambda>0$ is the parameter of the distribution. The PGF of a Poisson distribution is

$$\phi_X(s) = \mathbb{E}s^X = \sum_{r=0}^\infty e^{-\lambda} s^r \frac{\lambda^r}{r!} = e^{\lambda(s-1)}.$$

By independence,

$$\phi_{X+Y}(s) = \phi_X(s)\phi_Y(s) = e^{(\lambda+\mu)(s-1)},$$

and by the uniqueness theorem, $X + Y \sim \text{Po}(\lambda + \mu)$. A similar conclusion follows from a direct calculation:

$$\mathbb{P}(X + Y = r) = \sum_{i,j:i+j=r} \mathbb{P}(X = i, Y = j)$$

$$= \sum_{i,j:i+j=r} \mathbb{P}(X = i)\mathbb{P}(Y = j)$$

$$= \sum_{i,j:i+j=r} \frac{e^{-\lambda}\lambda^i}{i!} \frac{e^{-\mu}\mu^j}{j!}$$

$$= \frac{e^{-(\lambda+\mu)}}{r!} \sum_{i,j:i+j=r} \binom{r}{i} \lambda^i \mu^j = \frac{e^{-(\lambda+\mu)}}{r!}(\lambda + \mu)^r.$$

However, X and $X + Y$ are not independent, as $\mathbb{P}(X + Y = 2|X = 4) = 0$. In fact, $\forall\, r = 0, 1, \ldots,\ n$:

$$\mathbb{P}(X = r|X + Y = n) = \frac{\mathbb{P}(X = r, Y = n - r)}{\mathbb{P}(X + Y = n)}$$

$$= \frac{\mathbb{P}(X = r)\mathbb{P}(Y = n - r)}{\mathbb{P}(X + Y = n)}$$

$$= \frac{e^{-\lambda}\lambda^r e^{-\mu}\mu^{n-r} n!}{e^{-(\lambda+\mu)}(\lambda + \mu)^n r!(n - r)!}$$

$$= \binom{n}{r} \left(\frac{\lambda}{\lambda+\mu}\right)^r \left(\frac{\mu}{\lambda+\mu}\right)^{n-r},$$

i.e. $(X|X + Y = n) \sim \text{Bin}(n, \lambda/\lambda + \mu)$. $\square$

Problem 1.51 Let X be a positive integer-valued RV. Define its PGF ϕ_X. Show that if X and Y are independent positive integer-valued random variables, then

$$\phi_{X+Y} = \phi_X \phi_Y.$$

A non-standard pair of dice is a pair of six-sided unbiased dice whose faces are numbered with strictly positive integers in a non-standard way (for example, $(2, 2, 2, 3, 5, 7)$ and $(1, 1, 5, 6, 7, 8)$). Show that there exists a non-standard pair of dice A and B such that when thrown

$$\mathbb{P}(\text{total shown by A and B is } n)$$
$$= \mathbb{P}(\text{total shown by pair of ordinary dice is } n)$$

for all $2 \le n \le 12$.

Hint: $x + x^2 + x^3 + x^4 + x^5 + x^6 = x(x + 1)(1 + x^2 + x^4) = x(1 + x + x^2)(1 + x^3)$.

Solution   Use the PGF: for an RV V with finitely or countably many values v: $\phi_V(x) = \mathbb{E}x^V = \sum_s x^v \mathbb{P}(V = v)$. The PGF determines the distribution uniquely: if $\phi_U(x) \equiv \phi_V(x)$, then $U \sim V$ in the sense that $\mathbb{P}(U = v) \equiv \mathbb{P}(V = v)$. Also, if $V = V_1 + V_2$, where V_1 and V_2 are independent then $\phi_V(x) = \phi_{V_1}(x)\phi_{V_2}(x)$.

Now let S_2 be the total score shown by a pair of standard dice, then

$$\phi_{S_2}(x) = \phi_S^2(x) = \frac{1}{36}\left(\sum_{k=1}^{6} x^k\right)^2 = \frac{1}{36}x(1+x)(1+x^2+x^4)x(1+x+x^2)(1+x^3),$$

where S is the score shown by a single die.

Therefore, if we arrange a pair of dice A and B such that the score T_A for die A has the PGF

$$\phi_A(x) = \frac{1}{6}x(1+x^2+x^4)(1+x^3) = \frac{1}{6}(x+x^3+x^4+x^5+x^6+x^8)$$

and the score T_B for die B the PGF

$$\phi_B(x) = \frac{1}{6}x(1+x+x^2)(1+x) = \frac{1}{6}(x+2x^2+2x^3+x^4)$$

then the total score $T_A + T_B$ will have the same PGF as S_2. Hence, the die A with faces 1, 3, 4, 5, 6 and 8 and die B with faces 1, 2, 2, 3, 3 and 4 will satisfy the requirement.   □

Problem 1.52   A biased coin has probability p, $0 < p < 1$, of showing heads on a single throw. Show that the PGF of the number of heads in n throws is

$$\phi(s) = (ps + 1 - p)^n.$$

Suppose the coin is thrown N times, where N is a random variable with expectation μ_N and variance σ_N^2, and let Y be the number of heads obtained. Show that the PGF $\phi_Y(s)$ of Y satisfies

$$\phi_Y(s) = \phi_N(ps + 1 - p),$$

where $\phi_N(s)$ is the PGF of N. Hence, or otherwise, find $\mathbb{E}Y$ and Var Y.

Suppose $\mathbb{P}(N = k) = e^{-\lambda}\lambda^k/k!$, $k = 0, 1, 2, \ldots$ Show that Y has a Poisson distribution with parameter λp.

Solution   Denote by X the number of heads after n throws. Then $X \sim \text{Bin}(n, p)$:

$$\mathbb{P}(X = k) = \binom{n}{k}p^k(1-p)^{n-k}, \quad k = 0, 1, \ldots, n,$$

and

$$\phi_X(s) = \mathbb{E}s^X = \sum_{k=0}^{n}\binom{n}{k}(sp)^k(1-p)^{n-k} = (ps + 1 - p)^n.$$

Next,

$$\phi_Y(s) = \mathbb{E}s^Y = \mathbb{E}\big[\mathbb{E}(s^Y|N)\big] = \mathbb{E}(ps+1-p)^N = \phi_N(ps+1-p),$$

where $\phi_N(s) = \mathbb{E}s^N$. Then

$$\phi_Y'(s) = p\phi_N'(ps+1-p), \quad \text{so } \phi_Y'(1) = p\phi_N'(1), \quad \text{i.e. } \mathbb{E}Y = p\mu_N.$$

Further,

$$\phi_Y''(s) = p^2\phi_N''(ps+1-p)$$

and

$$\text{Var}\,Y = p^2\phi_N''(1) + p\phi_N'(1) - p^2\mu_N^2 = p^2\sigma_N^2 + \mu_N p(1-p).$$

Thus, if $N \sim \text{Po}(\lambda)$, with

$$\phi_N(s) = \sum_{k=0}^{\infty} s^k e^{-\lambda}\frac{\lambda^k}{k!} = \exp\big[\lambda(s-1)\big],$$

then

$$\phi_Y(s) = \exp\big[\lambda(ps+1-p-1)\big] = \exp\big[\lambda p(s-1)\big],$$

and $Y \sim \text{Po}(\lambda p)$. $\square$

Problem 1.53 If X and Y are independent Poisson RVs with parameters λ and μ, show that $X+Y$ is Poisson with parameter $\lambda+\mu$.

The proofs of my treatise upon the Binomial Theorem which I hope will have a European vogue have come back from the printers. To my horror I discover that the printers have introduced misprints in such a way that the number of misprints on each page is a Poisson RV with parameter λ. If I proofread the book once, then the probability that I will detect any particular misprint is p independent of anything else. Show that after I have proofread the book once the number of remaining misprints on each page is a Poisson random variable and give its parameter.

My book has 256 pages, the number of misprints on each page is independent of the numbers on other pages, the average number of misprints on each page is 2 and $p=3/4$. How many times must I proofread the book to ensure that the probability that no misprint remains in the book is greater than $1/2$?

Solution You can use the PGFs:

$$\phi_X(t) = \mathbb{E}t^X = \sum_{k\geq 0} \frac{e^{-\lambda}(t\lambda)^k}{k!} = e^{\lambda(t-1)},$$

and similarly $\phi_Y(t) = e^{\mu(t-1)}$, with $\phi_{X+Y}(t) = \phi_X(t)\phi_Y(t) = e^{(\lambda+\mu)(t-1)}$ owing to the independence. Then, by the uniqueness, $X+Y \sim \text{Po}(\lambda+\mu)$.

Another way is by direct calculation: by the convolution formula,

$$\mathbb{P}(X+Y=r) = \sum_{k=0}^{r} \mathbb{P}(X=k)\mathbb{P}(Y=r-k)$$

$$= \sum_{k=0}^{r} \frac{e^{-\lambda}\lambda^k}{k!} \times \frac{e^{-\mu}\mu^{r-k}}{(r-k)!}$$

$$= \frac{e^{-\lambda-\mu}}{r!} \sum_{k=0}^{r} \frac{r!}{k!(r-k)!} \lambda^k \mu^{r-k}$$

$$= \frac{e^{-\lambda-\mu}(\lambda+\mu)^r}{r!}.$$

If X is the number of original and Z of remaining misprints then $Z \sim \text{Po}(p\lambda)$. In fact, the PGF $\phi_Z(t) = \mathbb{E}t^Z$ can be written as $\mathbb{E}\big[\mathbb{E}(t^Z|X)\big]$ and equals

$$\sum_{k\geq0} \frac{e^{-\lambda}\lambda^k}{k!} \mathbb{E}(t^Z|X=k) = \sum_{k\geq0} \frac{e^{-\lambda}\lambda^k}{k!} \sum_{r=0}^{k} t^r \frac{k!}{r!(k-r)!} p^r (1-p)^{k-r}$$

$$= \sum_{k\geq0} \frac{e^{-\lambda}\lambda^k}{k!} (tp+1-p)^k$$

$$= e^{\lambda(-1+tp-p+1)} = e^{\lambda p(t-1)}.$$

This implies that $Z \sim \text{Po}(p\lambda)$. A direct calculation also works:

$$\mathbb{P}(Z=r) = \sum_{k\geq r} \mathbb{P}(Z=r|X=k)\mathbb{P}(X=k)$$

$$= \sum_{k\geq r} \frac{e^{-\lambda}\lambda^k}{k!} \frac{k!}{r!(k-r)!} p^r (1-p)^{k-r}$$

$$= \frac{e^{-p\lambda}(p\lambda)^r}{r!} \sum_{k-r\geq0} \frac{e^{-(1-p)\lambda}[\lambda(1-p)]^{k-r}}{(k-r)!} = \frac{e^{-p\lambda}(p\lambda)^r}{r!}.$$

Finally, after n proofreadings, each $Z_i \sim \text{Po}(p^n\lambda)$, $1 \leq i \leq 256$, with $p = 3/4$ and $\lambda = 2$. If $R = \sum_{i=1}^{256} Z_i$, then $R \sim \text{Po}(256 \times 2 \times (3/4)^n)$, and

$$\mathbb{P}(R=0) = e^{-512\cdot(3/4)^n} \quad \text{should be} \quad \geq \frac{1}{2}.$$

Hence, we must have $512\,(3/4)^n \leq \ln 2$, i.e. $(3/4)^n \leq (\ln 2)/2^8$ or

$$n \geq \frac{8\ln 2 - \ln\ln 2}{\ln(4/3)} \approx 20.54939. \quad \square$$

Problem 1.54   State the precise relation between the binomial and Poisson distributions. The lottery in Gambland sells on average 10^7 tickets. To win the first prize one must guess 7 numbers out of 50. Assume that individual bets are independent and random (this is a rather unrealistic assumption). For a given integer $n \geq 0$, write a formula for the Poisson approximation of the probability that there are at least n first prize winners in this week's draw. Give a rough estimate of this value for $n = 5, 10$. The *Stirling formula* $n! \approx \sqrt{2\pi n} n^n e^{-n}$ can be used without proof.

Solution   The limit

$$\lim_{n \to \infty} \binom{n}{k} \left(\frac{\lambda}{n}\right)^k \left(1 - \frac{\lambda}{n}\right)^{n-k} = e^{-\lambda} \frac{\lambda^k}{k!}$$

means that if $X_n \sim \mathrm{Bin}(n, \lambda/n)$ then $X_n \Rightarrow Y \sim \mathrm{Po}(\lambda)$ as $n \to \infty$. This fact is used to approximate various binomial probabilities.

In the example

$$n = 10^7, \quad p = 1 \bigg/ \binom{50}{7}, \quad \text{and } \lambda = 10^7 \bigg/ \binom{50}{7}.$$

Further,

$$\binom{50}{7} \approx 10^8, \quad \lambda \approx 0.1, \quad e^{-\lambda} \approx 0.9.$$

Then

$$\mathbb{P}(\geq n \text{ winners}) = e^{-\lambda} \sum_{k=n}^{\infty} \frac{\lambda^k}{k!} \approx e^{-\lambda} \frac{\lambda^n}{n!}$$

yields:

$$\text{for } n = 5: \approx 0.9 \times \frac{10^{-5}}{120} \approx 7.5 \times 10^{-8},$$

$$\text{for } n = 10: \approx 0.9 \times \frac{10^{-10}}{3 \cdot 10^6} \approx 3 \times 10^{-17}. \quad \square$$

J. Stirling (1692–1770) was a Scottish mathematician and engineer. Apart from his numerous mathematical achievements (and the Stirling formula is one of them), he was interested in such applied problems as the form of the Earth, in which his results were highly praised by his contemporaries. His life was not easy as he was an active Jacobite, a supporter of James, the prominent pretender to the English throne. He had to flee abroad and spent some years in Venice where he continued his academic work.

We conclude this section with a more challenging problem.

Problem 1.55   (i) Let $S_n = X_1 + \cdots + X_n$, $S_0 = 0$, and $X_1, \ldots, X_n$ be IID with

$$X_i = \begin{cases} 1, & \text{probability } p, \\ -1, & \text{probability } (1 - p), \end{cases}$$

where $p \geq 1/2$. Set $\tau_b = \min[n : S_n = b]$, $b > 0$, and $\tau_1 = \tau$. Prove that the Laplace transform $L_\tau(\theta)$ (cf. (1.55)) has the form

$$L_\tau(\theta) = \mathbb{E}e^{-\theta\tau} = \frac{1}{2(1-p)}\left[e^\theta - \sqrt{e^{2\theta} - 4p(1-p)}\right].$$

Verify equations for the mean and the variance:

$$\mathbb{E}\tau_b = \frac{b}{2p-1}, \quad \text{Var}\,(\tau_b) = \frac{4p(1-p)^b}{(2p-1)^3}.$$

(ii) Now assume that $p < 1/2$. Check that

$$\mathbb{P}(\tau < \infty) = \frac{p}{1-p}.$$

Prove the following formulas:

$$\mathbb{E}\left[e^{-\theta\tau}I(\tau < \infty)\right] = \frac{1}{2(1-p)}\left(e^\theta - \sqrt{e^{2\theta} - 4p(1-p)}\right),$$

and

$$\mathbb{E}\left[\tau I(\tau < \infty)\right] = \frac{p}{(1-p)(1-2p)}.$$

Here and below,

$$I(\tau < \infty) = \begin{cases} 1, & \text{if } \tau < \infty, \\ 0, & \text{if } \tau = \infty. \end{cases}$$

(iii) Compute $\text{Var}\,[\tau I(\tau < \infty)]$.

Solution   (i) We have that $\tau_i = \sum_{j=1}^{i} T_j$ where $T_j \sim \tau_1$, independently. Then

$$\mathbb{E}\tau_b = b\mathbb{E}\tau, \quad \text{Var}\,(\tau_b) = b\text{Var}\,(\tau).$$

Next,

$$L_\tau(\theta) = pe^{-\theta} + (1-p)\mathbb{E}\left[e^{-\theta(1+\tau'+\tau'')}\right],$$

where τ' and τ'' are independent with the same distribution as τ. Hence, $\mathbb{E}\left[e^{-\theta(\tau'+\tau'')}\right] = (\mathbb{E}e^{-\theta\tau})^2$ and $x = L_\tau(\theta)$ satisfies the quadratic equation:

$$x = pe^{-\theta} + (1-p)e^{-\theta}x^2.$$

Thus,

$$L_T(\theta) = \frac{1}{2(1-p)}[e^\theta - \sqrt{e^{2\theta} - 4p(1-p)}];$$

the minus sign is chosen since $L_T(\theta) \to 0$ when $\theta \to \infty$. Moreover, omitting subscript τ,

$$L'(\theta) = \frac{1}{2(1-p)} \left[e^{\theta} - \frac{e^{2\theta}}{\sqrt{e^{2\theta} - 4p(1-p)}} \right],$$

implying that

$$L'(0) = -\frac{1}{2p-1}.$$

Alternatively, one can derive an equation $L'(0) = p + (1-p)(1 + 2L'(0))$ that implies the same result.

Using the formula $\mathrm{Var}\,(\tau) = L''(0) - (L'(0))^2$ we get

$$\begin{aligned}
\mathrm{Var}\,(\tau) &= \frac{1}{2(1-p)} \left[1 - \frac{2}{(2p-1)} + \frac{1}{2(2p-1)^3} \right] - \frac{1}{(2p-1)^2} \\
&= \frac{8p^3 - 20p^2 + 14p - 2 + 4p^2 - 6p + 2}{2(1-p)(2p-1)^3} = \frac{4p(1-p)}{(2p-1)^3}.
\end{aligned}$$

Alternatively, we can derive an equation for $y = \mathrm{Var}\,(\tau) = \mathbb{E}(\tau - \mathbb{E}\tau)^2$:

$$y = p(1 - \mathbb{E}\tau)^2 + (1-p)\mathbb{E}(1 + \tau' + \tau'' - \mathbb{E}\tau)^2,$$

where τ' and τ'' are independent RVs with the same distribution as τ. This implies

$$\begin{aligned}
y &= p(1 - \mathbb{E}\tau)^2 + (1-p)\mathbb{E}[1 + \mathbb{E}\tau + (\tau' + \tau'' - 2\mathbb{E}\tau)]^2 \\
&= p(1 - \mathbb{E}\tau)^2 + (1-p)[\mathrm{Var}\,(\tau' + \tau'') + (1 + \mathbb{E}\tau)^2],
\end{aligned}$$

since $\mathbb{E}(\tau' + \tau'') = 2\mathbb{E}\tau$. Finally, observe that $\mathrm{Var}\,(\tau' + \tau'') = 2\mathrm{Var}(\tau)$ to get

$$y = p\left(1 - \frac{1}{2p-1}\right)^2 + (1-p)\left[2y + \left(\frac{1}{2p-1}\right)^2\right], \text{ i.e. } y = \frac{4p(1-p)}{(2p-1)^3}.$$

(ii) Observe that $z = \mathbb{P}\,(\tau < \infty)$ is the minimal solution of the equation $z = p + (1-p)z^2$. The equation for $x = \mathbb{E}[e^{-\theta\tau}I(\tau < \infty)]$ takes the same form as in (i):

$$x = pe^{-\theta} + (1-p)e^{-\theta}x^2.$$

However, the form of the solution is different:

$$\mathbb{E}[\tau | \tau < \infty] = -\left\{ \frac{1}{2(1-p)} \left[1 - \frac{1}{\sqrt{(1-2q)^2}} \right] \right\} = \frac{p}{(1-p)(1-2p)},$$

as the square root is written differently.

(iii) Differentiating $\mathbb{E}[e^{-\theta\tau}I(\tau<\infty)]$ twice with respect to θ we get that Var $[\tau I(\tau < \infty)]$ equals

$$\frac{1}{2(1-p)(1-2p)^3}[(1-2p)^3 - 2(1-2p)^2 + 1]$$

$$-\left[\frac{p}{(1-p)(1-2p)}\right]^2 = \frac{p(1-4p^2+4p^3)}{(1-p)^2(1-2p)^3}. \qquad \square$$

1.6   Chebyshev's and Markov's inequalities. Jensen's inequality. The Law of Large Numbers and the De Moivre–Laplace Theorem

> Probabilists do it with large numbers.
> (From the series *'How they do it'*.)

Chebyshev's inequality is perhaps the most famous in the whole probability theory (and probably the most famous achievement of the prominent Russian mathematician P.L. Chebyshev (1821–1894)). It states that if X is a random variable with finite expectation and variance then $\forall \epsilon > 0$:

$$\mathbb{P}(|X - \mathbb{E}X| \geq \epsilon) \leq \frac{1}{\epsilon^2}\operatorname{Var} X. \tag{1.60}$$

Chebyshev's inequality gave rise to a number of generalisations. One is Markov's inequality (after Chebyshev's pupil A.A. Markov (1856–1922), another prominent Russian mathematician). Markov's inequality is that for any non-negative RV Y with a finite expectation, $\forall \epsilon > 0$:

$$\mathbb{P}(Y \geq \epsilon) \leq \frac{1}{\epsilon}\mathbb{E}Y. \tag{1.61}$$

Chebyshev's inequality is obtained from Markov's by setting $Y = |X - \mathbb{E}X|^2$ and observing that the events $\{|X - \mathbb{E}X| \geq \epsilon\}$ and $\{|X - \mathbb{E}X|^2 \geq \epsilon^2\}$ are the same.

The names of Chebyshev and Markov are associated with the rise of the Russian (more precisely, St Petersburg) school of probability theory. Neither of them could be described as having an ordinary personality. Chebyshev had wide interests in various branches of contemporary science (and also in the political, economical and social life of the period). This included the study of ballistics in response to demands by his brother who was a distinguished artillery general in the Russian Imperial Army. Markov was a well-known liberal opposed to the tsarist regime: in 1913, when Russia celebrated the 300th anniversary of the Imperial House of Romanov, he and some of his colleagues defiantly organised a celebration of the 200th anniversary of the Law of Large Numbers (LLN). See below.

We will now prove Markov's inequality. This is quite straightforward: if the values are $x_1, x_2, \ldots$ and taken with probabilities $p_1, p_j, \ldots$, then

$$\mathbb{E}X = \sum_j x_j p_j \geq \sum_{j:x_j \geq \epsilon} x_j p_j \geq \epsilon \sum_{j:x_j \geq \epsilon} p_j = \epsilon\mathbb{P}(X \geq \epsilon).$$

This can be made shorter using indicator $I(X \geq \epsilon)$:

$$\mathbb{E}X \geq \mathbb{E}\left[XI(X \geq \epsilon)\right] \geq \epsilon\mathbb{E}I(X \geq \epsilon) = \epsilon\mathbb{P}(X \geq \epsilon). \tag{1.62}$$

Here we also see how the argument develops: first, the inequality $X \geq XI(X \geq \epsilon)$ holds because $X \geq 0$ (and of course $1 \geq I(X \geq \epsilon)$). This implies that $\mathbb{E}X \geq \mathbb{E}\left[XI(X \geq \epsilon)\right]$. Similarly, the inequality $XI(X \geq \epsilon) \geq \epsilon I(X \geq \epsilon)$ implies that $\mathbb{E}\left[XI(X \geq \epsilon)\right] \geq \mathbb{E}\epsilon I(X \geq \epsilon)$. The latter is equal to $\epsilon\mathbb{E}I(X \geq \epsilon)$ and finally, to $\epsilon\mathbb{P}(X \geq \epsilon)$.

It has to be noted that in Chebyshev's and Markov's inequalities $\epsilon > 0$ does not have to be small or large: the inequality holds for any positive value.

In general, if $g: \mathbb{R} \to \mathbb{R}$ is a monotone non-decreasing function and X a real-valued RV then, $\forall x \in \mathbb{R}$ with $g(x) > 0$ and a finite mean $\mathbb{E}g(X)$

$$\mathbb{P}(X \geq x) \leq \frac{1}{g(x)}\mathbb{E}g(X); \tag{1.63}$$

a popular case is where $g(x) = e^{ax}$ with $a > 0$:

$$\mathbb{P}(X \geq x) \leq \frac{1}{e^{ax}}\mathbb{E}e^{aX} \tag{1.64}$$

(Chernoff's inequality).

The domain of applications of these inequalities is huge (and not restricted to probability theory); we will discuss one of them here: the LLN.

Another example of a powerful inequality used in more than one area of mathematics is Jensen's inequality. It is named after J.L. Jensen (1859–1925), a Danish analyst who used it in his 1906 paper. Actually, the inequality was discovered in 1889 by O. Hölder, a German analyst, but for some reason is not named after him (maybe because there already was a Hölder inequality proved to be extremely important in analysis and differential equations). Let X be an RV with values in an (open, half-open or closed) interval $J \subseteq \mathbb{R}$ (possibly unbounded, i.e., coinciding with a half-line or the whole line), with a finite expectation $\mathbb{E}X$, and $g: J \to \mathbb{R}$ a convex (concave) real-valued function such that the expectation $\mathbb{E}g(X)$ is finite. Jensen's inequality asserts that

$$\mathbb{E}g(X) \geq g(\mathbb{E}X) \text{ respectively, } \mathbb{E}g(X) \leq g(\mathbb{E}X). \tag{1.65}$$

In other words, $\forall \, x_1, \ldots, x_n \in (a, b)$ and probabilities $p_1, \ldots, p_n$ (with $p_1, \ldots, p_n \geq 0$ and $p_1 + \cdots + p_n = 1$):

$$g\left(\sum_{j=1}^{n} p_j x_j\right) \leq \sum_{j=1}^{n} p_j g(x_j) \quad \text{respectively,} \quad g\left(\sum_{j=1}^{n} p_j x_j\right) \geq \sum_{j=1}^{n} p_j g(x_j). \tag{1.66}$$

Here we adopt the following definition: a function g on $[a, \, b]$ is convex if $\forall x, y \in [a, b]$ and $\lambda \in (0, 1)$,

$$g(\lambda x + (1 - \lambda)y) \leq \lambda g(x) + (1 - \lambda)g(y), \tag{1.67}$$

or, in other words, for a convex function $g: J \to \mathbb{R}$ defined on an interval $J \subseteq \mathbb{R} \; \forall \; x_1, \; x_2 \in J$ and $p_1, \; p_2 \in [0, 1]$ with $p_1 + p_2 = 1$,

$$g(p_1 x_1 + p_2 x_2) \le p_1 g(x_1) + p_2 g(x_2). \tag{1.68}$$

See Figure 1.6.

To prove inequality (1.66), it is natural to use induction in n. For $n = 2$ bound (1.66) is merely bound (1.68). The induction step from n to $n+1$ is as follows. Write

$$g \left(\sum_{i=1}^{n+1} p_i x_i \right) \le p_{n+1} g \; (x_{n+1}) + (1 - p_{n+1}) \, g \left(\sum_{i=1}^{n} \frac{p_i}{1 - p_{n+1}} x_i \right)$$

and use the induction hypothesis for probabilities $p_i / (1 - p_{n+1})$, $1 \le i \le n$. This yields the bound

$$p_{n+1} g(x_{n+1}) + (1 - p_{n+1}) g \left(\sum_{i=1}^{n} \frac{p_i}{1 - p_{n+1}} x_i \right)$$

$$\le p_{n+1} g(x_{n+1}) + \sum_{i=1}^{n} p_i g(x_i) = \sum_{i=1}^{n+1} p_i g(x_i).$$

If X takes infinitely many values, then a further analytic argument is required which we will not perform here. (One would need to use the fact that a convex/concave function g is always continuous in interval J, where it has been defined; g may be not differentiable, but only at an at most countable set of points $x \in J$, and at each point $x \in J$, where g is twice-differentiable, $-g''(x) \overset{\le}{\underset{\ge}{}} 0$.)

Jensen's inequality can be strengthened, by characterising the cases of equality. Call a convex (concave) function g strictly convex (concave) if equality in (1.66) is achieved iff either $x_1 = \cdots = x_n$ or all p_j except for one are equal to zero (and the remaining to one). For such function g, equality in (1.65) is achieved iff RV X is constant.

An immediate corollary of Jensen's inequality, with $g(x) = x^s$, $x \in [0, \infty)$, is that $\forall \; s \ge 1 : (\mathbb{E} X)^s \le \mathbb{E} X^s$, $\forall$ RV $X \ge 0$ with finite expected value $\mathbb{E} X^s$. For $0 < s < 1$, the

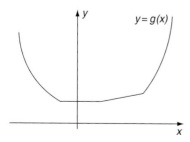

Figure 1.6

inequality is reversed: $(\mathbb{E}X)^s \geq \mathbb{E}X^s$. Another corollary, with $g(x) = \ln x$, $x \in (0, \infty)$, is that $\sum_j x_j p_j \geq \prod_j x_j^{p_j}$ for any positive $x_1, \ldots, x_n$ and probabilities $p_1, \ldots, p_n$. For $p_1 = \cdots = p_n$, this becomes the famous arithmetic–geometric mean inequality:

$$\frac{1}{n}(x_1 + \cdots + x_n) \geq (x_1 \cdots x_n)^{1/n}. \tag{1.69}$$

We now turn to the LLN. In its *weak* form the statement (and its proof) is simple.

> *Let $X_1, \ldots, X_n$ be IID RVs with the (finite) mean value $\mathbb{E}X_j = \mu$ and variance* Var $X = \sigma^2$. *Set*
>
> $$S_n = X_1 + \cdots + X_n.$$
>
> *Then, $\forall \epsilon > 0$,*
>
> $$\mathbb{P}\left(\left|\frac{1}{n}S_n - \mu\right| \geq \epsilon\right) \to 0 \text{ as } n \to \infty. \tag{1.70}$$
>
> *Verbally, the averaged sum S_n/n of IID RVs $X_1, \ldots, X_n$ with mean $\mathbb{E}X_j = \mu$ and* Var $X_j = \sigma^2$ *converges in probability to μ.*

The proof uses Chebyshev's inequality:

$$\mathbb{P}\left(\left|\frac{1}{n}S_n - \mu\right| \geq \epsilon\right) = \mathbb{P}\left(\left|\frac{1}{n}\sum_{i=1}^{n}(X_i - \mu)\right| \geq \epsilon\right)$$

$$\leq \frac{1}{\epsilon^2} \times \frac{1}{n^2}\text{Var}\left(\sum_{i=1}^{n}X_i\right)$$

$$= \frac{1}{\epsilon^2 n^2}\sum_{i=1}^{n}\text{Var } X_i = \frac{1}{\epsilon^2 n^2}n\sigma^2 = \frac{\sigma^2}{n\epsilon^2}$$

which vanishes as $n \to \infty$.

It is instructive to observe that the proof goes through when we simply assume that $X_1, X_2, \ldots$ are such that Var $(\sum_{i=1}^{n} X_i) = o(n^2)$. For indicator IID RVs X_i with

$$X_i = \begin{cases} 1, \text{ probability } p \\ 0, \text{ probability } 1 - p \end{cases} \quad \text{(heads and tails in a biased coin-tossing)}$$

and $\mathbb{E}X_i = p$, the LLN says that after a large number n trials, the proportion of heads will be close to p, the probability of a head at a single toss.

The LLN for IID variables X_i taking values 0 and 1 was known to seventeenth and eighteenth century mathematicians, notably J. Bernoulli (1654–1705). He was a member of the extraordinary Bernoulli family of mathematicians and natural scientists of Flemish origin. Several generations of this family resided in Prussia, Russia, Switzerland and other countries and dominated the scientific development on the European continent.

It turns out that the assumption that the variance σ^2 is finite is not necessary in the LLN. There are also several forms of convergence which emerge in connection with the LLN;

some of them will be discussed in forthcoming chapters. Here we only mention the *strong* form of the LLN:

For IID RVs X_1, X_2, ... with finite mean μ:

$$\mathbb{P}\left(\lim_{n\to\infty}\frac{S_n}{n}=\mu\right)=1.$$

The next step in studying the sum (1.69) is the *Central limit Theorem* (CLT). Consider IID RVs X_1, X_2, ..., with finite mean $\mathbb{E}X_j=a$ and variance Var $X_j=\sigma^2$ and a finite higher moment $\mathbb{E}|X_j-\mathbb{E}X_j|^{2+\delta}=\rho$, for some fixed $\delta>0$. The *integral* CLT asserts that the following convergence holds: $\forall\ x\in\mathbb{R}$,

$$\lim_{n\to\infty}\mathbb{P}\left(\frac{S_n-na}{\sigma\sqrt{n}}<x\right)=\frac{1}{\sqrt{2\pi}}\int_{-\infty}^{x}\mathrm{e}^{-y^2/2}\mathrm{d}y. \tag{1.71}$$

The map

$$\Phi:x\in\mathbb{R}\mapsto\frac{1}{\sqrt{2\pi}}\int_{-\infty}^{x}\mathrm{e}^{-y^2/2}\mathrm{d}y \tag{1.72}$$

defines the so-called standard normal, or $N(0, 1)$, cumulative distribution function $\Phi(x)$, an object of paramount importance in probability theory and statistics. It is also called a Gaussian distribution function, named after K.-F. Gauss (1777–1855), the famous German mathematician, astronomer and physicist, who made a profound impact on a number of areas of mathematics. He identified the distribution while working on the theory of errors in astronomical observations. Gaussian distribution fitted the pattern of errors much better than 'double-exponential' distribution previously used by Laplace.

The values of $\Phi(x)$ (or $\bar{\Phi}(x)=1-\Phi(x)$) have been calculated with a great accuracy for a narrow mesh of values of x and constitute a major part of the probabilistic and statistical tables. See Table 1.1.

This table specifies $1-\Phi(x)$ for $0\le x<3$ with step 0.05.

At the present stage it is useful to memorise four facts.

(i)

$$\lim_{x\to-\infty}\Phi(x)=0,\ \lim_{x\to\infty}\Phi(x)=\frac{1}{\sqrt{2\pi}}\int_{-\infty}^{\infty}\mathrm{e}^{-y^2/2}\mathrm{d}y=1$$

(these are standard properties of a distribution function; see below).

(ii) $\Phi(x)=1-\Phi(-x)\ \forall x\in\mathbb{R}$, implying that

$$\Phi(0)=\frac{1}{\sqrt{2\pi}}\int_{-\infty}^{0}\mathrm{e}^{-y^2/2}\mathrm{d}y=\frac{1}{2}$$

(which means that the median of the standard Gaussian distribution is 0 (see below)). This follows from the previous property and the fact that the integrand $\mathrm{e}^{-y^2/2}$ is an even function.

Table 1.1. *Values of* $1 - \Phi(x)$

x	+0.00	+0.05	x	+0.00	+0.05
0.0	0.5000	0.4801	1.5	0.0668	0.0606
0.1	0.4602	0.4404	1.6	0.0548	0.0495
0.2	0.4207	0.4013	1.7	0.0446	0.0401
0.3	0.3821	0.3632	1.8	0.0359	0.0322
0.4	0.3446	0.3264	1.9	0.0288	0.0256
0.5	0.3085	0.2912	2.0	0.0228	0.0202
0.6	0.2743	0.2578	2.1	0.0179	0.0158
0.7	0.2420	0.2266	2.2	0.0129	0.0122
0.8	0.2119	0.1977	2.3	0.0107	0.0094
0.9	0.1841	0.1711	2.4	0.0082	0.0071
1.0	0.1587	0.1469	2.5	0.0062	0.0054
1.1	0.1357	0.1251	2.6	0.0047	0.0040
1.2	0.1151	0.1056	2.7	0.0035	0.0030
1.3	0.0968	0.0885	2.8	0.0026	0.0022
1.4	0.0808	0.0735	2.9	0.0019	0.0016

(iii)

$$\frac{1}{\sqrt{2\pi}} \int_{-\infty}^{\infty} y e^{-y^2/2} dy = 0$$

(which means that the mean value of the standard Gaussian distribution is 0). This follows from the above observation as $y e^{-y^2/2}$ is an odd function.

(iv)

$$\frac{1}{\sqrt{2\pi}} \int_{-\infty}^{\infty} y^2 e^{-y^2/2} dy = 1$$

(which means that the variance of the standard Gaussian distribution is 1). This again can be deduced by integration by parts; again see below.

(v)   Function $x \mapsto \Phi(x)$ is strictly increasing with x and continuous in x. Hence, the inverse function Φ^{-1} is correctly defined, taking $\gamma \in (0, 1)$ to $x \in \mathbb{R}$ such that $\Phi(x) = \gamma$. The inverse function plays an important rôle in statistics.

The proof of the claim, in (i), that $\Phi(\infty) = 1$, i.e.

$$\int_{-\infty}^{\infty} e^{-x^2/2} dx = \sqrt{2\pi}$$

is quite elegant. For brevity, write $\int_{\mathbb{R}}$ for the integral $\int_{-\infty}^{\infty}$ over the whole line $\mathbb{R}$. If $\int_{\mathbb{R}} e^{-x^2/2} dx = G$, then

$$G^2 = \int_{\mathbb{R}} e^{-x^2/2} dx \int_{\mathbb{R}} e^{-y^2/2} dy = \int_{\mathbb{R}^2} e^{-(x^2+y^2)/2} dy dx$$

which in polar co-ordinates equals

$$\int_0^\infty re^{-r^2/2}\int_0^{2\pi} d\theta dr = 2\pi \int_0^\infty re^{-r^2/2}dr = 2\pi \int_0^\infty e^{-u}du = 2\pi.$$

Hence, $G = \sqrt{2\pi}$. Furthermore, in (iv):

$$\int_{\mathbb{R}} y^2 e^{-y^2/2}dy = \left(-ye^{-y^2/2}\right)\Big|_{-\infty}^\infty + \int_{\mathbb{R}} e^{-y^2/2}dy = \sqrt{2\pi}.$$

There also exists a *local* CLT that deals with a detailed analysis of probabilities

$$\mathbb{P}\left(\frac{S_n - na}{\sigma\sqrt{n}} = x_n\right) = \mathbb{P}(S_n = na + x_n\sigma\sqrt{n})$$

for a suitable range of values $x_n \in \mathbb{R}$. The aim here is to produce asymptotical expressions for these probabilities of the form $e^{-x_n^2/2}/\sqrt{2\pi n}\sigma$ which takes into account only the mean value a and the variance σ^2.

The modern method of proving the CLT is based on the fact that the convergence in (1.71) is equivalent to the convergence of the characteristic functions $\psi_{(S_n - na)/(\sigma\sqrt{n})}(t)$ to the characteristic function of the Gaussian distribution. The basics of this method will be discussed later.

In the rest of this section we will focus on the CLT for the above example of coin-tossing where

$$X_i = \begin{cases} 1, & \text{probability } p \\ 0, & \text{probability } 1-p \end{cases}, \text{independently}.$$

Again, for this example, the CLT goes back to the eighteenth century and is often called *De Moivre–Laplace Theorem* (DMLT), after two French mathematicians. One of them, A. De Moivre (1667–1754), fled to England after the Revocation of the Edict of Nantes, and had a notable career in teaching and research. The other, P.-S. Laplace (1749–1827), made significant contributions in several areas of mathematics. He also served (briefly and not very successfully) as the Interior Minister under Napoleon (Laplace examined the young Napoleon in mathematics in the French Royal Artillery Corps and gave the promising officer the highest mark). Despite his association with Napoleon, Laplace was the first to vote to oust Napoleon from power in 1814 in the French Senate; the returning Bourbons promoted him from the title of Count to Marquis.

An initial form of the theorem was produced by De Moivre in 1733. He was also the first to identify the normal distribution (which was named after Gauss almost a hundred years later).

Informally, DMLT asserts that if $X_1, X_2, \ldots$ is an IID sequence, with $\mathbb{P}(X_j = 1) = p$, $\mathbb{P}(X_j = 0) = 1 - p$ and hence $\mathbb{E}X_j = p$, Var $X_j = p(1-p)$, then the RV $(S_n - np)/\sqrt{np(1-p)}$ has, approximately, the standard $\mathrm{N}(0,1)$ distribution. At the formal level, one again distinguishes a local and an integral DMLT. For a given positive integer $m \leq n$, write

$$\mathbb{P}(S = m) = \mathbb{P}\left(\frac{S_n - np}{\sqrt{np(1-p)}} = z_n(m)\right), \quad \text{where} \quad z_n(m) = \frac{m - np}{\sqrt{np(1-p)}}.$$

The formal statement of the local DMLT is:

As $n, m \to \infty$, the ratio

$$\mathbb{P}(S_n = m) \Big/ \left\{\frac{1}{\sqrt{2\pi np(1-p)}} \exp\left[-\frac{1}{2} z_n(m)^2\right]\right\} \to 1 \qquad (1.73)$$

as long as $m - np = o(n^{2/3})$. More precisely, (1.73) holds uniformly in n, m for which the expression $(m - np)n^{-2/3}$ is confined to a bounded interval and tends to 0.

Recall, $\mathbb{P}(S_n = m)$ is the probability that an event of probability p occurs m times in n independent trials.

As before, the integral DMLT deals with cumulative probabilities and states that $\forall\, x \in \mathbb{R}$:

$$\lim_{n \to \infty} \mathbb{P}\left(\frac{S_n - np}{\sqrt{np(1-p)}} < x\right) = \frac{1}{\sqrt{2\pi}} \int_{-\infty}^{x} e^{-y^2/2} dy. \qquad (1.74)$$

Equivalently, $\forall\, -\infty \leq a < b \leq \infty$:

$$\lim_{n \to \infty} \mathbb{P}\left(a < \frac{S_n - np}{\sqrt{np(1-p)}} < b\right) = \frac{1}{\sqrt{2\pi}} \int_{a}^{b} e^{-x^2/2} dx. \qquad (1.75)$$

Although the integral DMLT looks more amenable, its proof is longer than that of the local DMLT (and uses it as a part). None of these theorems is formally proved in Cambridge IA Probability, although they are widely employed in subsequent courses (particularly, IB Statistics). The proof below is given here for completeness. It is not used in the problems from this volume but gives a useful insight into how the normal distribution arises as an asymptotical distribution for (properly normalised) sums of independent RVs.

The local DMLT can be proved by a direct argument. The main step is the fact that for any sequence of positive integers $\{m_n\}$ such that $m_n \leq n$ and m_n, $n - m_n \to \infty$,

$$\mathbb{P}(S_n = m_n) \Big/ \left\{\left[2\pi n \frac{m_n}{n}\left(1 - \frac{m_n}{n}\right)\right]^{-1/2} \exp\left[-nh\left(\frac{m_n}{n}, p\right)\right]\right\} \to 1. \qquad (1.76)$$

Here,

$$h\left(\frac{m_n}{n}, p\right) = \frac{m_n}{n} \ln \frac{m_n}{np} + \frac{n - m_n}{n} \ln \frac{n - m_n}{n(1-p)}, \qquad (1.77)$$

which is a particular case of a more general definition: for $p^* \in (0, 1)$:

$$h(p^*, p) = p^* \ln \frac{p^*}{p} + (1 - p^*) \ln \frac{1 - p^*}{1 - p}. \qquad (1.78)$$

Convergence (1.73) then follows for $m_n - np = o(n^{2/3})$ as for p^* close to p, the Taylor expansion yields:

$$h(p^*, p) = \frac{1}{2}\left(\frac{1}{p} + \frac{1}{1-p}\right)(p^* - p)^2 + O(|p^* - p|^3), \tag{1.79}$$

as

$$h(p^*, p)\Big|_{p^*=p} = \left(\frac{d}{dp^*}h(p^*, p)\right)\Big|_{p^*=p} = 0.$$

Remark Equation (1.76) (and the particular form $-nh(p^*, p)$ of the exponent, with $p^* = m_n/n$ and $h(p^*, p)$ given by formula (1.79), is important for the asymptotical analysis of various probabilities related to sums of independent RVs. In particular, formulas (1.77)–(1.79) play a significant rôle in information theory and the theory of large deviations. Function $h(p^*, p)$ is called the *relative entropy* of the probability distribution $\{p^*, 1 - p^*\}$ with respect to probability distribution $\{p, 1 - p\}$.

The proof of equation (1.79) is straightforward and uses the Stirling formula:

$$n! \approx \sqrt{2\pi n}n^n e^{-n}. \tag{1.80}$$

Note that this formula admits a more precise form:

$$n! = \sqrt{2\pi n}n^n e^{-n+\theta(n)}, \quad \text{where} \quad \frac{1}{12n+1} < \theta(n) < \frac{1}{12n}. \tag{1.81}$$

However, for our purposes formula (1.80) is enough.

Omitting subscript n in m_n, the probability $\mathbb{P}(S_n = m)$ equals

$$\binom{n}{m}p^m(1-p)^{n-m} \approx \left[\frac{n}{2\pi m(n-m)}\right]^{1/2}\frac{n^n}{m^m(n-m)^{n-m}}p^m(1-p)^{n-m}$$

$$= \left[2\pi n\frac{m}{n}\left(1 - \frac{m}{n}\right)\right]^{-1/2}$$

$$\times \exp\left[-m\ln\frac{m}{n} - (n-m)\ln\frac{n-m}{n} + m\ln p + (n-m)\ln(1-p)\right].$$

But the RHS of the last equation coincides with the denominator in formula (1.76).

To derive the integral DMLT, we take the sum

$$\sum_m \mathbb{P}(S_n = m)I\,(a < z_n(m) < b)$$

and interpret it as the area between the x-axis and the piece-wise horizontal line representing the graph of the function

$$\Pi_n : x \in \mathbb{R} \mapsto \sqrt{np(1-p)}\mathbb{P}(S_n = m), \quad \text{for } z_n(m) \le x < z_n(m+1),$$

on the interval $z_n(\underline{m}) \le x \le z_n(\overline{m})$ where $\underline{m}$ and $\overline{m}$ are uniquely determined from the condition that

$$a \le z_n(\underline{m}) < a + \frac{1}{\sqrt{np(1-p)}}, \quad b \le z_n(\overline{m}) < b + \frac{1}{\sqrt{np(1-p)}}.$$

Of course, this area equals the integral

$$\int_{z_n(\underline{m})}^{z_n(\overline{m})} \Pi_n(x) dx.$$

There are two things here that make the calculation tricky. First, the integrand Π_n, and, second, the limits of integration $z_n(\underline{m})$ and $z_n(\overline{m})$ vary with n. To start with, one considers first the case where a and b are both finite reals. Then, because $z_n(\underline{m}) \to a$ and $z_n(\overline{m}) \to b$, the above integral differs from

$$\int_a^b \Pi_n(x) dx$$

negligibly. Next, in the interval (a, b) we can use the local DMLT which asserts that

$$\Pi_n(x) \to \frac{1}{\sqrt{2\pi}} e^{-x^2/2}.$$

With a bit of analytical work one deduces from this that

$$\int_a^b \Pi_n(x) dx \to \int_a^b \frac{1}{\sqrt{2\pi}} e^{-x^2/2} dx.$$

To finish the proof, we must cover the case where a and/or b are infinite. This is done by exploiting the fact that the convergence in the local DMLT is uniform when $z_n(m_n)$ is confined to a finite interval, and the limiting function $x \mapsto e^{-x^2/2}/\sqrt{2\pi}$ is monotone decreasing with $|x|$ and integrable. The details are omitted.

One important aspect of the CLT is that it provides a (fairly accurate) normal approximation to other distributions.

Problem 1.56 A cubic die is thrown n times, and Y_n is the total number of spots shown. Show that $\mathbb{E} Y_n = 7n/2$ and Var $Y_n = 35n/12$. State Chebyshev's inequality and find an n such that

$$\mathbb{P}\left(\left|\frac{Y_n}{n} - 3.5\right| > 0.1\right) \le 0.1.$$

Solution Let $Y_n = \sum_{i=1}^n X_i$, where X_i is the number of spots on the ith throw. RVs $X_1, \ldots, X_n$ are independent and identically distributed, with $\mathbb{P}(X_i = r) = \frac{1}{6}, r = 1, 2, \ldots, 6$. Hence,

$$\mathbb{E} X_i = \frac{3}{2}, \text{ Var }(X_i) = \frac{11}{6}, \ \mathbb{E} Y_n = \frac{7n}{2}, \text{ Var }(Y_n) = \frac{35n}{12}.$$

Chebyshev's inequality is $\mathbb{P}(|X - \mathbb{E}X| \ge \epsilon) \le (1/\epsilon^2)\text{Var } X$ and is valid $\forall$ RV X with a finite mean and variance.

By Chebyshev's inequality, and independence,

$$\mathbb{P}\left(\left|\frac{Y_n}{n} - 3.5\right| > 0.1\right) = \mathbb{P}\left(|Y_n - \mathbb{E}Y_n| > 0.1n\right)$$

$$\leq \frac{\operatorname{Var} Y_n}{n^2 0.1^2} = \frac{35n/12}{0.01n^2} = \frac{1750}{6n}.$$

We want $1750/(6n) \leq 0.1$, so $n \geq 2920$. $\square$

Problem 1.57   A coin shows heads with probability $p > 0$ and tails with probability $q = 1 - p$. Let X_n be the number of tosses needed to obtain n heads. Find the PGF for X_1 and compute its mean and variance. What is the mean and variance of X_n?

Now suppose that $p = 1/2$. What bound does Chebyshev's inequality give for $\mathbb{P}(X_{100} > 400)$?

Solution   We have $\mathbb{P}(X_1 = k) = pq^{k-1}$, and so PGF $\phi_{X_1}(s) = ps/(1 - qs)$. Then $\phi'_{X_1}(s) = p/(1 - qs)^2$, and so

$$\mathbb{E}X_1 = \phi'_{X_1}(1) = \frac{1}{p}, \operatorname{Var}(X_1) = \frac{q}{p^2}.$$

We conclude that $\mathbb{E}X_n = n/p$, $\operatorname{Var} X = nq/p^2$. Now set $p = q = 1/2$, $\mathbb{E}X_{100} = 200$, $\operatorname{Var}(X_{100}) = 200$, and write

$$\mathbb{P}(X_{100} \geq 400) \leq \mathbb{P}(|X_{100} - 200| > 200) \leq \frac{200}{200^2} = \frac{1}{200}.   \square$$

Remark $X_n \sim \operatorname{NegBin}(n, q)$.

Problem 1.58   In this problem, and in a number of problems below, we use the term 'sample' as a substitute for 'outcome'. This terminology is particularly useful in statistics; see Chapters 3 and 4.

(i) How large a random sample should be taken from a distribution in order for the probability to be at least 0.99 that the sample mean will be within two standard deviations of the mean of the distribution? Use Chebyshev's inequality to determine a sample size that will be sufficient, whatever the distribution.

(ii) How large a random sample should be taken from a normal distribution in order for the probability to be at least 0.99 that the sample mean will be within one standard deviation of the mean of the distribution?

Hint: $\Phi(2.58) = 0.995$

Solution   (i) The sample mean $\overline{X}$ has mean μ and variance σ^2/n. Hence, by Chebyshev's inequality

$$\mathbb{P}(|\overline{X} - \mu| \geq 2\sigma) \leq \frac{\sigma^2}{n(2\sigma)^2} = \frac{1}{4n}.$$

Thus $n = 25$ is sufficient. If more is known about the distribution of X_i, then a smaller sample size may suffice: the case of normally distributed X_i is considered in (ii).

(ii) If $X_i \sim N(\mu, \sigma^2)$, then $\overline{X} - \mu \sim N(0, \sigma^2/n)$, and probability $\mathbb{P}(|\overline{X} - \mu| \geq \sigma)$ equals

$$\mathbb{P}\left(|\overline{X} - \mu| \Big/ \left(\frac{\sigma^2}{n}\right)^{1/2} \geq \sigma \Big/ \left(\frac{\sigma^2}{n}\right)^{1/2}\right) = \mathbb{P}(|Z| \geq n^{1/2})$$

where $Z \sim N(0, 1)$. But $\mathbb{P}(|Z| \geq 2.58) = 0.99$, and so we require that $n^{1/2} \geq 2.58$, i.e. $n \geq 7$. As we see, knowledge that the distribution is normal allows a much smaller sample size, even to meet a tighter condition. $\square$

Problem 1.59   What does it mean to say that a function $g : (0, \infty) \to \mathbb{R}$ is convex? If $f : (0, \infty) \to \mathbb{R}$ is such that $-f$ is convex, show that

$$\int_n^{n+1} f(x)\,dx \geq \frac{f(n) + f(n+1)}{2},$$

and deduce that

$$\int_1^N f(x)\,dx \geq \frac{1}{2}f(1) + f(2) + \cdots + f(N-1) + \frac{1}{2}f(N),$$

for all integers $N \geq 2$. By choosing an appropriate f, show that

$$N^{N+1/2}e^{-(N-1)} \geq N!$$

for all integers $N \geq 2$.

Solution $g : (0, \infty) \to \mathbb{R}$ is convex if $\forall \ x, y \in (0, \infty)$ and $p \in (0, 1)$: $g(px + (1-p)y) \leq pg(x) + (1-p)g(y)$. When g is twice differentiable, the convexity follows from the bound $g'(x) > 0, x > 0$.

If $-f$ is convex, then

$$-f(pn + (1-p)(n+1)) \leq -pf(n) - (1-p)f(n+1).$$

That is for $x = pn + (1-p)(n+1) = n + 1 - p$

$$f(x) \geq pf(n) + (1-p)f(n+1),$$

implying that

$$\int_n^{n+1} f(x)\,dx \geq \int_n^{n+1} [pf(n) + (1-p)f(n+1)]\,dx.$$

As $dx = -dp = d(1-p)$,

$$\int_n^{n+1} f(x)\,dx \geq \int_0^1 [(1-p)f(n) + pf(n+1)]\,dp = \frac{1}{2}f(n) + \frac{1}{2}f(n+1).$$

Finally, summing these inequalities from $n = 1$ to $n = N - 1$ gives the first inequality.

Now choose $f(x) = \ln x$: here, $f''(x) = 1/x$, $f'(x) = -1/x^2$. Hence, $-f$ is convex. By the above:

$$\int_1^N \ln x \, dx \geq \ln 2 + \cdots + \ln (N-1) + \ln (N^{1/2}),$$

i.e.

$$\frac{1}{2} \ln N + \int_1^N \ln x \, dx = \frac{1}{2} \ln N + (x \ln x - x) \big|_1^N$$

$$= \left(N + \frac{1}{2} \right) \ln N - (N-1) \geq \ln (N!)$$

which is the same as $N! \leq N^{N+1/2} e^{-(N-1)}$. □

Problem 1.60   A random sample is taken in order to find the proportion of Labour voters in a population. Find a sample size that the probability of a sample error less than 0.04 will be 0.99 or greater.

Solution $0.04\sqrt{n} \geq 2.58\sqrt{pq}$ where $pq \leq 1/4$. So, $n \approx 1040$.   □

Problem 1.61   State and prove Chebyshev's inequality.
   Show that if $X_1, X_2, \ldots$ are independent identically distributed RVs with finite mean μ and variance σ^2, then

$$\mathbb{P}\left(\left| n^{-1} \sum_{i=1}^n X_i - \mu \right| \geq \epsilon \right) \to 0$$

as $n \to \infty$ for all $\epsilon > 0$.
   Suppose that $Y_1, Y_2, \ldots$ are independent identically distributed random variables such that $\mathbb{P}(Y_j = 4^r) = 2^{-r}$ for all integers $r \geq 1$. Show that

$$\mathbb{P}(\text{at least one of } Y_1, Y_2, \ldots, Y_{2^n} \text{ takes value } 4^n) \to 1 - e^{-1}$$

as $n \to \infty$, and deduce that, whatever the value of K,

$$\mathbb{P}\left(2^{-n} \sum_{i=1}^{2^n} Y_i > K \right) \not\to 0.$$

Solution   Chebychev's inequality:

$$\mathbb{P}\left(|X - \mathbb{E}X| \geq b \right) \leq \frac{1}{b^2} \operatorname{Var} X, \ \forall \, b > 0,$$

follows from the argument below.

$$\operatorname{Var} X = \mathbb{E}\,(X - \mathbb{E}X)^2 \geq \mathbb{E}\left((X - \mathbb{E}X)^2 I\left((X - \mathbb{E}X)^2 \geq b^2 \right) \right)$$

$$\geq b^2 \mathbb{E}I\left((X - \mathbb{E}X)^2 \geq b^2 \right)$$

$$= b^2 \mathbb{P}\left((X - \mathbb{E}X)^2 \geq b^2 \right) = b^2 \mathbb{P}\left(|X - \mathbb{E}X| \geq b \right).$$

We apply it to $n^{-1}\sum_{i=1}^{n} X_i$, obtaining

$$\mathbb{P}\left(n^{-1}\left|\sum_{i=1}^{n}(X_i - \mu)\right| > \epsilon\right) \le \frac{1}{n^2 \epsilon^2}\text{Var}\left(\sum_{i=1}^{n}(X_i - \mu)\right)$$

$$= \frac{1}{n^2 \epsilon^2} n \text{Var}(X_1) \to 0$$

as $n \to \infty$. For RVs $Y_1, Y_2, \ldots$ specified in the question:

$$q_n := \mathbb{P}(\text{at least one of } Y_1, Y_2, \ldots, Y_{2^n} \text{ takes value } 4^n)$$

$$= 1 - \prod_{i=1}^{2^n}\mathbb{P}(Y_i \neq 4^n) = 1 - [\mathbb{P}(Y_1 \neq 4^n)]^{2^n}$$

$$= 1 - (1 - \mathbb{P}(Y_1 = 4^n))^{2^n}$$

$$= 1 - (1 - 2^{-n})^{2^n} \to 1 - e^{-1}.$$

Thus, if $2^n > K$,

$$\mathbb{P}\left(2^{-n}\sum_{i=1}^{2^n} Y_i > K\right) \ge q_n \to 1 - e^{-1} > 0.$$

We see that if the Y_i have no finite mean, the averaged sum does not exhibit convergence to a finite value. □

Problem 1.62   (i) Suppose that X and Y are discrete random variables taking finitely many values. Show that $\mathbb{E}(X+Y) = \mathbb{E}X + \mathbb{E}Y$.

(ii) On a dry road I cycle at 20 mph; when the road is wet at 10 mph. The distance from home to the lecture building is three miles, and the 9.00 am lecture course is 24 lectures. The probability that on a given morning the road is dry is $1/2$, but there is no reason to believe that dry and wet mornings follow independently. Find the expected time to cycle to a single lecture and the expected time for the whole course.

A student friend (not a mathematician) proposes a straightforward answer:

$$\text{average cycling time for the whole course} = \frac{3 \times 24}{\frac{1}{2}10 + \frac{1}{2}20} = 4\,\text{h}\,48\,\text{min}.$$

Explain why his answer gives a shorter time.

Solution   (i) For RVs X, Y, with finitely many values x and y,

$$\mathbb{E}(X+Y) = \sum_{x,y}(x+y)\mathbb{P}(X=x, Y=y)$$

$$= \sum_x x \sum_y \mathbb{P}(X=x, Y=y) + \sum_y y \sum_x \mathbb{P}(X=x, Y=y).$$

The internal sums $\sum_y \mathbb{P}(X=x, Y=y)$ and $\sum_x \mathbb{P}(X=x, Y=y)$ equal, respectively, $\mathbb{P}(X=x)$ and $\mathbb{P}(Y=y)$. Thus,

$$\mathbb{E}(X+Y) = \sum_x x\mathbb{P}(X=x) + \sum_y y\mathbb{P}(Y=y) = \mathbb{E}X + \mathbb{E}Y.$$

(ii) If T is the expected time to a single lecture, then

$$\mathbb{E}(\text{time to lectures}) = 3\mathbb{E}\left(\frac{1}{\text{speed}}\right) = 3\left(\frac{1}{2} \times \frac{1}{10} + \frac{1}{2} \times \frac{1}{20}\right) = 13.5 \,\text{min.}$$

The total time $= 24 \times 13.5 = 5$ h 24 min; the assumption of independence is not needed, as $\mathbb{E}\sum_i T_i = \sum_i \mathbb{E}T_i$ holds in any case. The 'straightforward' answer gives a shorter time:

$$\frac{3 \times 24}{(1/2) \times 10 + (1/2) \times 20} < \frac{1}{2} \times \frac{3 \times 24}{10} + \frac{1}{2} \times \frac{3 \times 24}{20}.$$

However, the average speed $\neq (1/2) \times 10 + (1/2) \times 20 = 15$. This is a particular case of Jensen's inequality with a strictly convex function

$$g(x) = \frac{3 \times 24}{x}, \quad x \in (0, \infty). \quad \square$$

Problem 1.63   What is a convex function? State and prove Jensen's inequality for a convex function of an RV which takes finitely many values.

Deduce that, if X is a non-negative random variable taking finitely many values, then

$$\mathbb{E}[X] \leq (\mathbb{E}[X^2])^{1/2} \leq (\mathbb{E}[X^3])^{1/3} \leq \cdots$$

Solution   (The second part only) Consider $g(x) = x^{n+1/n}$, a (strictly) convex function on $J = [0, \infty)$. By Jensen's inequality:

$$(\mathbb{E}X)^{n+1/n} \leq \mathbb{E}X^{n+1/n}.$$

Finally, let $Y^n = X$ to obtain

$$(\mathbb{E}Y^n)^{1/n} \leq (\mathbb{E}Y^{n+1})^{1/n+1}. \quad \square$$

Problem 1.64   (i) If X is a bounded random variable show that

$$\mathbb{P}(X \geq \lambda) \leq e^{-\lambda}\mathbb{E}(e^X).$$

(ii) By looking at power series expansions, or otherwise, check that

$$\cosh t \leq e^{t^2/2}.$$

If Y is a random variable with $\mathbb{P}(Y=a) = \mathbb{P}(Y=-a) = 1/2$ show that

$$\mathbb{E}(e^{\theta Y}) \leq e^{a^2\theta^2/2}.$$

If Y_1, Y_2, ..., Y_n are independent random variables with $\mathbb{P}(Y_k = a_k) = \mathbb{P}(Y_k = -a_k) = 1/2$ and $Z = Y_1 + Y_2 + \cdots + Y_n$, show, explaining your reasoning carefully, that

$$\mathbb{E}(e^{\theta Z}) \leq e^{A^2 \theta^2 / 2},$$

where A^2 is to be given explicitly in terms of the a_k.

By using (i), or otherwise, show that, if $\lambda > 0$,

$$\mathbb{P}(Z \geq \lambda) \leq e^{(A^2 \theta^2 - 2\lambda\theta)/2}$$

for all $\theta > 0$. Find the θ that minimises $e^{(A^2 \theta^2 - 2\lambda\theta)/2}$ and show that

$$\mathbb{P}(Z \geq \lambda) \leq e^{-\lambda^2/(2A^2)}.$$

Explain why

$$\mathbb{P}(|Z| \geq \lambda) \leq 2e^{-\lambda^2/(2A^2)}.$$

(iii) If $a_1 = a_2 = \cdots = a_n = 1$ in (ii), show that

$$\mathbb{P}(|Y_1 + Y_2 + \cdots + Y_n| \geq (2n \ln \epsilon^{-1})^{1/2}) \leq 2\epsilon$$

whenever $\epsilon > 0$.

Solution   (i) Chernoff's inequality: $\mathbb{P}(X \geq \lambda) \leq e^{-\lambda} \mathbb{E}\left[e^X I(X \geq \lambda)\right] \leq e^{-\lambda} \mathbb{E}e^X$.

(ii) We have

$$\cosh t = \sum_{n=0}^{\infty} \frac{t^{2n}}{(2n)!} \quad \text{and} \quad e^{t^2/2} = \sum_{n=0}^{\infty} \frac{(t^2/2)^n}{(n)!} = \sum_{n=0}^{\infty} \frac{t^{2n}}{2^n n!}.$$

Now, $(2n)(2n-1)\cdots(n+1) > 2^n$ for $n \geq 2$. So,

$$\cosh t \leq e^{t^2/2} \quad \text{and} \quad \mathbb{E}(e^{\theta Y}) = \cosh(\theta a) \leq e^{a^2 \theta^2/2}.$$

Similarly,

$$\mathbb{E}(e^{\theta Y}) = \mathbb{E}(e^{\theta X_1 + \cdots + \theta X_n}) = \prod_{k=1}^{n} \cosh(\theta a_k) \leq \prod_{k=1}^{n} e^{a_k^2 \theta^2/2} = e^{A^2 \theta^2/2},$$

with $A^2 = \sum_{k=1}^{n} a_k^2$. This implies

$$\mathbb{P}(Z \geq \lambda) \leq e^{-\lambda\theta} \mathbb{E}(e^{\theta Z}) \leq e^{(A^2 \theta^2 - 2\lambda\theta)/2}$$

for all $\theta > 0$.

The function $f = e^{(A^2 \theta^2 - 2\lambda\theta)/2}$ achieves its minimum for $\theta = \lambda/A^2$. Thus,

$$\mathbb{P}(Z \geq \lambda) \leq e^{-\lambda^2/2A^2}.$$

Now consider $-Z = -(Y_1 + \cdots + Y_n) = Y_1' + \cdots + Y_n'$, where $Y_k' = -Y_k$, $k = 1, \ldots, n$, has the same distribution as Y_k, $k = 1, \ldots, n$; then

$$\mathbb{P}(-Z \geq \lambda) \leq e^{-\lambda^2/2A^2}.$$

Thus

$$\mathbb{P}(|Z| \geq \lambda) \leq 2e^{-\lambda^2 2A^2}.$$

If $a_1 = a_2 = \cdots = a_n = 1$, then $A^2 = n$, and

$$\mathbb{P}(|Y_1 + Y_2 + \cdots + Y_n| \geq (2n \ln \epsilon^{-1})^{1/2}) \leq 2\exp\left(-\frac{2n \ln \epsilon^{-1}}{2n}\right) = 2\epsilon. \quad \square$$

Problem 1.65 Let (X_k) be a sequence of independent identically distributed random variables with mean μ and variance σ^2. Show that

$$\sum_{k=1}^{n}(X_k - \overline{X})^2 = \sum_{k=1}^{n}(X_k - \mu)^2 - n(\overline{X} - \mu)^2,$$

where $\overline{X} = \frac{1}{n}\sum_{k=1}^{n}X_k$. Prove that, if $\mathbb{E}(X_1 - \mu)^4 < \infty$, then for every $\epsilon > 0$

$$\mathbb{P}\left(\left|\frac{1}{n}\sum_{k=1}^{n}(X_k - \overline{X})^2 - \sigma^2\right| > \epsilon\right) \to 0$$

as $n \to \infty$.

Hint: By Chebyshev's inequality

$$\mathbb{P}\left(\left|\frac{1}{n}\sum_{k=1}^{n}(X_k - \mu)^2 - \sigma^2\right| > \epsilon/2\right) \to 0.$$

Problem 1.66 Let $x_1, x_2, \ldots, x_n$ be positive real numbers. Then geometric mean (GM) lies between the harmonic mean (HM) and arithmetic mean (AM):

$$\left(\frac{1}{n}\sum_{i=1}^{n}\frac{1}{x_i}\right)^{-1} \leq \left(\prod_{i=1}^{n}x_i\right)^{1/n} \leq \frac{1}{n}\sum_{i=1}^{n}x_i.$$

The second inequality is the *AM–GM inequality*; establish the first inequality (called the *HM–GM inequality*).

Solution An AM–GM inequality: induction in n. For $n = 2$ the inequality is equivalent to $4x_1 x_2 \leq (x_1 + x_2)^2$.

The inductive passage: AM–GM inequality is equivalent to

$$\frac{1}{n}\sum_{i=1}^{n}\ln x_i \leq \ln \sum_{i=1}^{n}x_i.$$

Function $\ln y$ is (strictly) concave on $[0, \infty)$ (which means that $\ln' y < 0$). Therefore, for any $\alpha \in [0, 1]$ and any $y_1, y_2 > 0$

$$\ln(\alpha y_1 + (1 - \alpha) y_2) \geq \alpha \ln y_1 + (1 - \alpha) \ln y_2.$$

Take $\alpha = 1/n$, $y_1 = x_1$, $y_2 = \sum_{j=2}^{n} x_j / (n - 1)$ to obtain

$$\ln\left(\frac{1}{n}\sum_{i=1}^{n} x_i\right) \geq \frac{1}{n}\ln x_1 + \frac{n-1}{n}\ln\left(\frac{1}{n-1}\sum_{i=2}^{n} x_i\right).$$

Finally, according to the induction hypothesis

$$\ln\left(\frac{1}{n-1}\sum_{i=2}^{n} x_i\right) \geq \frac{1}{n-1}\sum_{i=2}^{n}\ln x_i.$$

To prove the HM–GM inequality, apply the AM–GM inequality to $1/x_1, \dots, 1/x_n$:

$$\frac{1}{n}\sum_{i=1}^{n}\frac{1}{x_i} \geq \prod_{i=1}^{n}\left(\frac{1}{x_i}\right)^{1/n}.$$

Hence,

$$\left(\frac{1}{n}\sum_{i=1}^{n}\frac{1}{x_i}\right)^{-1} \leq \left(\prod_{i=1}^{n}\left(\frac{1}{x_i}\right)^{1/n}\right)^{-1} = \left(\prod_{i=1}^{n} x_i\right)^{1/n}. \qquad \square$$

Problem 1.67   Let X be a positive random variable taking only finitely many values. Show that

$$\mathbb{E}\frac{1}{X} \geq \frac{1}{\mathbb{E}X},$$

and that the inequality is strict unless $\mathbb{P}(X = \mathbb{E}X) = 1$.

Solution   Let X take values $x_1, x_2, \dots, x_n > 0$ with probabilities p_i. Then this inequality is equivalent to $\mathbb{E}X \geq [\mathbb{E}1/X]^{-1}$, i.e.

$$\sum_{i=1}^{n} p_i x_i \geq \left(\sum_{i=1}^{n} p_i \frac{1}{x_i}\right)^{-1}. \tag{1.82}$$

We shall deduce inequality (1.82) from a bound which is a generalisation of the above AM–GM inequality:

$$\prod_{i=1}^{n} x_i^{p_i} \leq \sum_{i=1}^{n} p_i x_i. \tag{1.83}$$

In fact, applying inequality (1.83) to the values $1/x_i$ yields

$$\left(\sum_{i=1}^{n} p_i \frac{1}{x_i}\right)^{-1} \leq \prod_{i=1}^{n} x_i^{p_i}.$$

Then equation (1.82) follows, again by (1.83).

To prove bound (1.83), assume that all x_j are pair-wise distinct, and proceed by induction in n. For $n = 1$, bound (1.82) becomes an equality. Assume inequality (1.82) holds for $n - 1$ and prove that

$$\sum_{i=1}^{n} p_i \ln x_i \le \ln \left(\sum_{i=1}^{n} p_i x_i \right).$$

We again use the strict concavity of ln:

$$\ln \left[\alpha y_1 + (1 - \alpha) y_2 \right] \ge \alpha \ln y_1 + (1 - \alpha) \ln y_2.$$

Take $\alpha = p_1$, $y_1 = x_1$, $y_2 = \sum_{j=2}^{n} p_j x_j / (1 - p_1)$ to obtain

$$\ln \sum_{i=1}^{n} p_i x_i = \ln \left(p_1 x_1 + (1 - p_1) \sum_{i=2}^{n} p'_i x_i \right)$$

$$\ge p_1 \ln x_1 + (1 - p_1) \ln \left(\sum_{i=2}^{n} p'_i x_i \right)$$

where $p'_i = p_i / (1 - p_1)$, $i = 2, \ldots, n$. We can now use the induction hypothesis

$$\ln \left(\sum_{i=2}^{n} p'_i x_i \right) \ge \sum_{i=2}^{n} p'_i \ln x_i$$

to get the required result. The equality holds iff either $p_i (1 - p_i) = 0$ or $x_1 = \sum_{i=2}^{n} p'_i x_i$. Scanning the situation for $x_2, \ldots, x_n$, we conclude that the equality occurs iff either $p_i (1 - p_i) = 0$ for some (and hence for all) i or $x_1 = \cdots = x_n$. According to our agreement, this means that $n = 1$, i.e.

$$\mathbb{P}(X = \mathbb{E}X) = 1. \quad \square$$

Problem 1.68   Let $b_1, b_2, \ldots, b_n$ be a rearrangement of the positive real numbers $a_1, a_2, \ldots, a_n$. Prove that

$$\sum_{i=1}^{n} \frac{a_i}{b_i} \ge n.$$

Hint: $\prod_{i=1}^{n} (a_i / b_i) = 1$.

Problem 1.69   Let X be an RV for which $\mathbb{E}X = \mu$ and $\mathbb{E}(X - \mu)^4 = \beta_4$. Prove that

$$\mathbb{P}(|X - \mu| \ge t) \le \frac{\beta_4}{t^4}.$$

Hint: Use Markov's inequality for $|X - \mu|^4$.

Problem 1.70   What is a convex function? State and prove Jensen's inequality for convex functions. Use it to prove the arithmetic–geometric mean inequality which states that if $a_1, a_2, \ldots, a_n > 0$, then

$$\frac{a_1 + a_2 + \cdots + a_n}{n} \geq (a_1 a_2 \cdots a_n)^{1/n}.$$

(You may assume that a function with positive second derivative is convex.)

Solution   A function $f : (a, b) \to \mathbb{R}$ is convex if $\forall x, \; x' \in (a, b)$ and $p \in (0, 1)$:

$$f(px + (1 - p)x') \leq pf(x) + (1 - p)f(x').$$

Jensen's inequality for RVs with finitely many values $x_1, \ldots, x_n$ is that $\forall f$ as above,

$$f\left(\sum_{i=1}^n p_i x_i\right) \leq \sum_{i=1}^n p_i f(x_i)$$

$\forall x_1, \ldots, x_n \in (a, b)$ and $p_1, \ldots, p_n \in (0, 1)$ with $p_1 + \cdots + p_n = 1$.

For the proof, use induction in n. For $n = 1$, the inequality is trivially true. (For $n = 2$, it is equivalent to the definition of a convex function.) Suppose it is true for some n. Then, for $n + 1$, let $x_1, \ldots, x_{n+1} \in (a, b)$ and $p_1, \ldots, p_{n+1} \in (0, 1)$ with $p_1 + \cdots + p_{n+1} = 1$. Setting $p'_i = p_i/(1 - p_{n+1})$, we have that $p'_1, \ldots, p'_n \in (0, 1)$ and $p'_1 + \cdots + p'_n = 1$. Then, by the definition of convexity and induction hypothesis,

$$f\left(\sum_{i=1}^{n+1} p_i x_i\right) = f\left((1 - p_{n+1})\sum_{i=1}^n p'_i x_i + p_{n+1} x_{n+1}\right)$$

$$\leq (1 - p_{n+1})f\left(\sum_{i=1}^n p'_i x_i\right) + p_{n+1} f(x_{n+1})$$

$$\leq (1 - p_{n+1})\sum_{i=1}^n p'_i f(x_i) + p_{n+1} f(x_{n+1}) = \sum_{i=1}^{n+1} p_i f(x_i).$$

So, the inequality holds for $n + 1$. Hence, it is true for all n.

For $f : (0, \infty) \to \mathbb{R}$ with $f(x) = -\ln x$: $f'(x) = -1/x$ and $f'(x) = 1/x^2 > 0$. So, f is convex. By Jensen's inequality, with $p_i = 1/n, \; i = 1, \ldots, n$:

$$f\left(\frac{1}{n}\sum_i a_i\right) \leq \frac{1}{n}\sum_i f(a_i),$$

i.e.

$$-\ln\left(\frac{1}{n}\sum_i a_i\right) \leq -\frac{1}{n}\sum_i \ln a_i, \text{ i.e. } \ln\left(\frac{1}{n}\sum_i a_i\right) \geq \ln\left(\prod_i a_i\right)^{1/n}.$$

Thus,

$$\frac{1}{n}\sum_i a_i \geq \left(\prod_i a_i\right)^{1/n}. \quad \square$$

Problem 1.71   A box contains N plastic chips labelled by the numbers $1, \ldots, N$. An experiment consists of drawing n of these chips from the box, where $n \le N$. We assume that each chip is equally likely to be selected and that the drawing is without replacement. Let $X_1, \ldots, X_n$ be random variables, where X_i is the number on the ith chip drawn from the box, $i = 1, \ldots, n$. Set $Y = X_1 + X_2 + \cdots + X_n$.

(i) Check that Var $(X_i) = (N+1)(N-1)/12$.
Hint: $\sum_{i=1}^{N} i^2 = N(N+1)(N+2)/6$.
(ii) Check that Cov $(X_i, X_j) = -(N+1)/12$, $i \ne j$.
(iii) Using the formula

$$\text{Var}(Y) = \sum_{i=1}^{N} \text{Var}(X_i) + \sum_{i \ne j} \text{Cov}(X_i, X_j),$$

or otherwise, prove that

$$\text{Var}(Y) = \frac{n(N+1)(N-n)}{12}.$$

Solution   (i) Clearly, $\mathbb{E}X_i = (N+1)/2$. Then

$$\text{Var}(X_i) = \sum_{k=1}^{N} \frac{k^2}{N} - \left(\frac{N+1}{2}\right)^2$$

$$= \frac{1}{N} \frac{N(N+1)(2N+1)}{6} - \left(\frac{N+1}{2}\right)^2$$

$$= \frac{(N+1)(N-1)}{12}.$$

(ii) As X_i and X_j cannot be equal to each other,

$$\frac{1}{N(N-1)} \sum_{k \ne s} \left(k - \frac{N+1}{2}\right)\left(s - \frac{(N-1)}{2}\right)$$

$$= \frac{1}{N(N-1)} \sum_{k=1}^{N} \left(k - \frac{N+1}{2}\right) \sum_{s=1}^{N} \left(s - \frac{N+1}{2}\right)$$

$$- \frac{1}{N-1} \sum_{k=1}^{N} \left(k - \frac{N+1}{2}\right)^2 \frac{1}{N}.$$

The first sum equals zero, and the second equals $-(\text{Var}\, X_i)/(N-1)$. Hence, Cov $(X_i, X_j) = -(N+1)/12$.
(iii)

$$\text{Var}(Y) = n\frac{(N+1)(N-1)}{12} - n(n-1)\frac{(N+1)}{12} = \frac{n(N+1)(N-n)}{12}. \qquad \square$$

1.7   Branching processes

> Life is a school of probability.
>> W. Bagehot (1826–1877), British economist

Branching processes are a fine chapter of probability theory. Historically, the concept of a *branching process* was conceived to calculate the survival probabilities of noble families. The name of W.R. Hamilton (1805–1865), the famous Irish mathematician, should be mentioned here, as well as F. Galton (1822–1911), the English scientist and explorer, and H.W. Watson (1827–1903), the English mathematician. Since the 1940s branching processes have been used extensively in natural sciences, in particular to calculate products of nuclear fission (physics) and the size of populations (biology). Later they found powerful applications in computer science (algorithms on logical trees) and other disciplines.

The model giving rise to a branching process is simple and elegant. Initially, we have an item (a particle or a biological organism) that produces a random number of 'offspring' each of which produces a random number of offspring and so on. This generates a 'tree-like' structure where a descendant has a link to the parent and a number of links to its own offspring. See Figure 1.7.

Each site of the emerging (random) tree has a path that joins it with the ultimate ancestor (called the origin, or the root of the tree). The length of the path, which is equal to the number of links in it, measures the number of generations behind the given site (and the item it represents). Each site gives rise to a subtree that grows from it (for some sites there may be no continuation, when the number of offspring is zero).

The main assumption is that the process carries on with maximum independence and homogeneity: the number of offspring produced from a given parent is independent of the

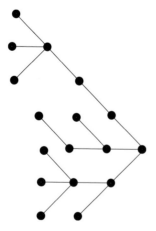

Figure 1.7

numbers related to other sites. More precisely, we consider RVs $X_0, X_1, X_2, \ldots$, where X_n gives the size of the population in the nth generation. That is

$X_0 = 1$,
$X_1 =$ the number of offspring after the 1st fission,
$X_2 =$ the number of offspring after the 2nd fission,
etc.

RVs X_n and X_{n+1} are related by the following recursion:

$$X_{n+1} = \sum_{i=1}^{X_n} Y_i^{(n)}, \tag{1.84}$$

where $Y_i^{(n)}$ is the number of descendants produced by the ith member of the nth generation. RVs $Y_i^{(n)}$ are supposed to be IID, and their common distribution determines the branching process.

The first important exercise is to calculate the mean value $\mathbb{E}X_n$, i.e. the expected size of the nth generation. By using the conditional expectation,

$$\mathbb{E}X_n = \mathbb{E}\left[\mathbb{E}\left(X_n | X_{n-1}\right)\right] = \sum_m \mathbb{P}(X_{n-1} = m)\mathbb{E}\left(X_n | X_{n-1} = m\right)$$

$$= \sum_m \mathbb{P}(X_{n-1} = m)\mathbb{E}\sum_{i=1}^{m} Y_i^{(n-1)} = \sum_m \mathbb{P}(X_{n-1} = m)m\mathbb{E}Y_i^{(n-1)}$$

$$= \mathbb{E}Y_1^{(n-1)} \sum_m \mathbb{P}(X_{n-1} = m)m = \mathbb{E}Y_1^{(n-1)}\mathbb{E}X_{n-1}. \tag{1.85}$$

Value $\mathbb{E}Y_i^{(k)}$ does not depend on k and i, and we denote it by $\mathbb{E}Y$ for short. Then, recurrently,

$$\mathbb{E}X_1 = \mathbb{E}Y, \ \mathbb{E}X_2 = (\mathbb{E}Y)^2, \ \ldots, \ \mathbb{E}X_n = (\mathbb{E}Y)^n, \ldots \tag{1.86}$$

We see that if $\mathbb{E}Y < 1$, $\mathbb{E}X_n \to 0$ with n, i.e. the process eventually dies out. This case is often referred to as *subcritical*. On the contrary, if $\mathbb{E}Y > 1$ (a *supercritical* process), then $\mathbb{E}X_n \to \infty$. The borderline case $\mathbb{E}Y = 1$ is called *critical*.

Remark   In formula (1.86) we did not use the independence assumption.

A convenient characteristic is the common PGF $\phi(s) = \mathbb{E}s^Y$ of RVs $Y_i^{(n)}$ (again it does not depend on n and i). Here, an important fact is that if $\phi_n(s) = \mathbb{E}s^{X_n}$ is the PGF of the size of the nth generation, then $\phi_1(s) = \phi(s)$ and, recursively,

$$\phi_{n+1}(s) = \phi_n(\phi(s)), \ n \geq 1. \tag{1.87}$$

See Problems 1.72 and 1.80. In other words,

$$\phi_n(s) = \phi\left(\phi(\ldots \phi(s)\ldots)\right) = \phi \circ \cdots \circ \phi(s) \ n \ \text{times}, \tag{1.88}$$

where $\phi \circ \cdots \circ \phi$ stands for the iteration of the map $s \mapsto \phi(s)$. In particular, $\phi_{n+1}(s) = \phi(\phi_n(s))$.

This construction leads to an interesting analysis of *extinction probabilities*

$$\pi_n := \mathbb{P}(X_n = 0) = \phi_n(0) = \phi^{\circ n}(0). \tag{1.89}$$

As $\pi_{n+1} = \phi(\pi_n)$, intuitively we would expect the limit $\pi = \lim_{n \to \infty} \pi_n$ to be a fixed point of map $s \mapsto \phi(s)$, i.e. a solution to $z = \phi(z)$. One such point is 1 (as $\phi(1) = 1$), but there may be another solution lying between 0 and 1. An important fact here is that function ϕ is convex, and the value of $\phi(0)$ is between 0 and 1. Then if $\phi'(1) > 1$, there will be a root of equation $z = \phi(z)$ in $(0, 1)$. Otherwise, $z = 1$ will be the smallest positive root. See Figure 1.8.

In fact, it is not difficult to check that the limiting extinction probability q exists and

$$\pi = \text{the least non-negative solution to } z = \phi(z). \tag{1.90}$$

Indeed, if there exists $z \in (0, 1)$ with $z = \phi(z)$, then it is unique, and also $0 < \phi'(z) < 1$ and $0 < \phi(0) = \mathbb{P}(Y = 0) < z$ (as ϕ is convex and $\phi(1) = 1$). Then $\phi(0) < \phi(\phi(0)) < \cdots < z$ because ϕ is monotone (and $\phi(z) = z$). The sequence $\phi^{\circ n}(0)$ must then have a limit which, by continuity, coincides with z.

If the least non-negative fixed point is $z = 1$, then the above analysis can be repeated without changes, yielding that $\pi = 1$. We conclude that if $\mathbb{P}(Y = 0) > 0$, then $\pi > 0$ (actually, $\pi > \mathbb{P}(Y = 0)$). On the other hand, if $\phi(0) = 0$ (i.e. $\mathbb{P}(Y = 0) = 0$), then, trivially, $\pi = 0$. This establishes equation (1.90). We see that even in the supercritical case (with $\phi'(1) > 1$) the limiting extinction probability π can be arbitrarily close to 1.

A slight modification of the above construction arises when we initially have several items (possibly a random number X_0).

Problem 1.72   In a branching process every individual has probability p_k of producing exactly k offspring, $k = 0, 1, \ldots$, and the individuals of each generation produce offspring independently of each other and of individuals in preceding generations. Let X_n represent the size of the nth generation. Assume that $X_0 = 1$ and $p_0 > 0$ and let $\phi_n(s)$ be the PGF of X_n. Thus

$$\phi_1(s) = \mathbb{E}s^{X_1} = \sum_{k=0}^{n} p_k s^k.$$

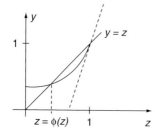
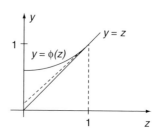

Figure 1.8

(i) Prove that

$$\phi_{n+1}(s) = \phi_n(\phi_1(s)).$$

(ii) Prove that for $n < m$

$$\mathbb{E}[s^{X_n}|X_m = 0] = \frac{\phi_n(s\phi_{n-m}(0))}{\phi_m(0)}.$$

Solution (i) By definition,

$$\phi_{n+1}(s) = \mathbb{E}s^{X_{n+1}} = \sum_{k=0}^{\infty} \mathbb{P}(X_{n+1} = k)s^k,$$

where X_n denotes the size of the nth generation. We write

$$\mathbb{E}s^{X_{n+1}} = \sum_l p_l^{(n)} \mathbb{E}(s^{X_{n+1}}|X_n = l).$$

Here $p_l^{(n)} = \mathbb{P}(X_n = l)$, and the conditional expectation is

$$\mathbb{E}(s^{X_{n+1}}|X_n = l) = \sum_{k=l}^{\infty} \mathbb{P}(X_{n+1} = k|X_n = l)s^k.$$

Now observe that

$$\mathbb{E}(s^{X_{n+1}}|X_n = l) = (\mathbb{E}s^{X_1})^l = (\phi_1(s))^l$$

because (a) under the condition that $X_n = l$,

$$X_{n+1} = \sum_{j=1}^{l} \tilde{X}_j$$

where $\tilde{X}_j$ is the number of offspring produced by the jth individual of the nth generation, (b) all the $\tilde{X}_j$ are IID and $\mathbb{E}(s^{\tilde{X}_j}|X_n = l) = \mathbb{E}s^{X_1} = \phi_1(s)$. This relation yields

$$\mathbb{E}s^{X_{n+1}} = \sum_l p_l^{(n)}(\phi_1(s))^l = \phi_n(\phi_1(s)).$$

(ii) Denote by $I_0^{(m)}$ the indicator $I(X_m = 1)$. Then $\mathbb{P}(I_0^{(m)} = 1) = \mathbb{E}I_0^{(m)} = \mathbb{P}(X_m = 0) = \phi_m(0)$. Furthermore,

$$\mathbb{E}[s^{X_n}|X_m = 0] = \mathbb{E}(s^{X_n}I_0^{(m)})/\phi_m(0).$$

Hence, it suffices to check that

$$\mathbb{E}(s^{X_n}I_0^{(m)}) = \phi_n(s\phi_{n-m}(0)).$$

Indeed,

$$\mathbb{E}\big(s^{X_n} I_0^{(m)}\big) = \sum_k s^k \mathbb{P}(X_n = k, X_m = 0)$$

$$= \sum_k s^k \mathbb{P}(X_n = k)\mathbb{P}(X_m = 0 | X_n = k).$$

Now, since $\mathbb{P}(X_m = 0 | X_n = k) = \phi_{m-n}(0)^k$,

$$\mathbb{E}\big(s^{X_n} I_0^{(m)}\big) = \phi_n(s\phi_{n-m}(0)). \quad \square$$

Problem 1.73   A laboratory keeps a population of aphids. The probability of an aphid passing a day uneventfully is $q < 1$. Given that a day is not uneventful, there is a probability r that the aphid will have one offspring, a probability s that it will have two offspring, and a probability t that it will die of exhaustion, where $r + s + t = 1$. Offspring are ready to reproduce the next day. The fates of different aphids are independent, as are the events of different days. The laboratory starts out with one aphid.

Let X_n be the number of aphids at the end of n days. Show how to obtain an expression for the PGF $\phi_n(z)$ of X_n. What is the expected value of X_n?

Show that the probability of extinction does not depend on q and that if $2r + 3s \le 1$, the aphids will certainly die out. Find the probability of extinction if $r = 1/5$, $s = 2/5$ and $t = 2/5$.

Solution   Denote by $\phi_{X_1}(z)$ the PGF of X_1, i.e. the number of aphids generated, at the end of a single day, by a single aphid (including the initial aphid). Then

$$\phi_{X_1}(z) = (1 - q)t + qz + (1 - q)rz^2 + (1 - q)sz^3, \ z > 0.$$

Write $\mathbb{E}X_n = (\mathbb{E}X_1)^n$ with $\mathbb{E}X_1 = \phi'_{X_1}(1) = q + 2(1 - q)r + 3(1 - q)s$. Indeed, $\phi_{X_n}(z) = \phi_{X_{n-1}}(\phi_{X_1}(z))$ implies $\mathbb{E}X_n = \mathbb{E}X_{n-1}\mathbb{E}X_1$ or $\mathbb{E}X_n = (\mathbb{E}X_1)^n$. The probability of extinction at the end of n days is $\phi_{X_n}(0)$. It is non-decreasing with n and tends to a limit as $n \to \infty$ giving the probability of (eventual) extinction π. As we already know $\pi < 1$ iff $\mathbb{E}X_1 > 1$. The extinction probability π is the minimal positive root of

$$(1 - q)t + qz + (1 - q)rz^2 + (1 - q)sz^3 = z$$

or, after division by $(1 - q)$ (since $q < 1$):

$$t - z + rz^2 + sz^3 = 0.$$

The last equation does not depend on q, hence π also does not depend on q. Condition $\mathbb{E}X_1 \le 1$ is equivalent to $2r + 3s \le 1$. In the case $r = 1/5, s = 2/5, t = 2/5$, the equation takes the form $2z^3 + z^2 - 5z + 2 = 0$. Dividing by $(z - 1)$ (as $z = 1$ is a root) one gets a quadratic equation $2z^2 + 3z - 2 = 0$, with roots $z_\pm = (-3 \pm 5)/4$. The positive root is $1/2$, and it gives the extinction probability. $\square$

Problem 1.74 Let $\phi(s) = 1 - p(1-s)^\beta$, where $0 < p < 1$ and $0 < \beta < 1$. Prove that $\phi(s)$ is a PGF and that its iterates are

$$\phi_n(s) = 1 - p^{1+\beta+\cdots+\beta^{n-1}}(1-s)^{\beta^n}, \quad n = 1, 2, \ldots$$

Find the mean m of the associated distribution and the extinction probability $\pi = \lim_{n\to\infty} \phi_n(0)$ for a branching process with offspring distribution determined by ϕ.

Solution The coefficients in Taylor's expansion of $\phi(s)$ are

$$a_k = \frac{d^k \phi}{ds^k}(s)\big|_{s=0} = p\beta(\beta-1)\cdots(\beta-k+1)(-1)^{k-1} \geq 0,$$

$$k = 1, 2, \ldots, \quad a_0 = 1 - p,$$

and $\phi(1) = \sum_{k\geq 0} a_k/k! = 1$. Thus, $\phi(s)$ is the PGF for the probabilities $p_k = a_k / k!$.
The second iterate, $\phi_2(s) = \phi(\phi(s))$ is of the form

$$\phi(\phi(s)) = 1 - p[1 - \phi(s)]^\beta = 1 - pp^\beta(1-s)^{\beta^2}.$$

Assume inductively that $\phi_k(s) = 1 - p^{1+\beta+\cdots+\beta^{k-1}}(1-s)^{\beta^k}$, $k \leq n-1$. Then

$$\phi_n(s) = 1 - p[1 - \phi_{n-1}(s)]^\beta = 1 - p\left[p^{1+\beta+\cdots+\beta^{n-2}}(1-s)^{\beta^{n-1}}\right]^\beta$$

$$= 1 - p^{1+\beta+\cdots+\beta^{n-1}}(1-s)^{\beta^n},$$

as required.
Finally, the mean value is $\phi'(1) = \lim_{s\to 1-} \phi'(s) = +\infty$ and the extinction probability,

$$\pi = \lim_{n\to\infty} \phi_n(0) = 1 - p^{1/(1-\beta)}. \quad \square$$

Problem 1.75 At time 0, a blood culture starts with one red cell. At the end of 1 min, the red cell dies and is replaced by one of the following combinations with probabilities as indicated

$$\text{two red cells: } \frac{1}{4}; \quad \text{one red, one white: } \frac{2}{3}; \quad \text{two white cells: } \frac{1}{12}.$$

Each red cell lives for 1 min and gives birth to offspring in the same way as the parent cell. Each white cell lives for 1 min and dies without reproducing. Assume that individual cells behave independently.
    (i) At time $n + \frac{1}{2}$ min after the culture began, what is the probability that no white cells have yet appeared?
    (ii) What is the probability that the entire culture dies out eventually?

Solution (i) The event

$$\{\text{by time } n + \frac{1}{2} \text{ no white cells have yet appeared}\}$$

implies that the total number of cells equals the total number of red cells and equals 2^n. Then, denoting the probability of this event by p_n, we find that

$$p_0 = 1, \quad p_1 = \frac{1}{4}, \quad \text{and } p_{n+1} = p_n \left(\frac{1}{4}\right)^{2^n}, \quad n \geq 1,$$

whence

$$p_n = \left(\frac{1}{4}\right)^{2^{n-1}} \left(\frac{1}{4}\right)^{2^{n-2}} \cdots \left(\frac{1}{4}\right)^{2^0} = \left(\frac{1}{4}\right)^{2^n - 1}, \quad n \geq 1.$$

(ii) The extinction probability π obeys

$$\pi = \frac{1}{4}\pi^2 + \frac{2}{3}\pi + \frac{1}{12} \quad \text{or} \quad \frac{1}{4}\pi^2 - \frac{1}{3}\pi + \frac{1}{12} = 0,$$

whence

$$\pi = \frac{2}{3} \pm \frac{1}{3}.$$

We must take $\pi = 1/3$, the smaller root.   □

Problem 1.76 Let $\{X_n\}$ be a branching process such that $X_0 = 1$, $\mathbb{E}X_1 = \mu$. If $Y_n = X_0 + \cdots + X_n$, and for $0 \leq s \leq 1$

$$\psi_n(s) \equiv \mathbb{E}s^{Y_n},$$

prove that

$$\psi_{n+1}(s) = s\phi(\psi_n(s)),$$

where $\phi(s) \equiv \mathbb{E}s^{X_1}$. Deduce that, if $Y = \sum_{n \geq 0} X_n$, then $\psi(s) \equiv \mathbb{E}s^Y$ satisfies

$$\psi(s) = s\phi(\psi(s)), \ 0 \leq s \leq 1.$$

If $\mu < 1$, prove that $\mathbb{E}Y = (1 - \mu)^{-1}$.

Solution Write $Y_{n+1} = 1 + Y_n^1 + \cdots + Y_n^{X_1}$, where Y_n^j is the total number of offspring produced by individual j from the first generation. Then the RVs Y_n^j are IID and have the PGF $\psi_n(s)$. Hence

$$\psi_{n+1}(s) = \mathbb{E}s^{Y_{n+1}} = s\sum_j \mathbb{P}(X_1 = j) \prod_{l=1}^{j} \psi_n(s)$$

$$= s\sum_j \mathbb{P}(X_1 = j)(\psi_n(s))^j = s\phi(\psi_n(s)).$$

The infinite series $Y = \sum_{n \geq 0} X_n$ has the PGF $\psi(s) = \lim_{n \to \infty} \psi_n(s)$. Hence, it obeys

$$\psi(s) = s\phi(\psi(s)).$$

By induction, $\mathbb{E}Y_n = 1 + \mu + \cdots + \mu^n$. In fact, $Y_1 = 1 + X_1$ and $\mathbb{E}Y_1 = 1 + \mu$. Assume that the formula holds for $n \leq k - 1$. Then

$$\mathbb{E}Y_k = (\psi_k(1))' = \phi(\psi_{k-1}(1)) + \phi'(\psi_{k-1}(1))\psi'_{k-1}(1)$$
$$= 1 + \mu(1 + \mu + \cdots + \mu^{k-1}) = 1 + \mu + \mu^2 + \cdots + \mu^k,$$

which completes the induction. Therefore, $\mathbb{E}Y = (1 - \mu)^{-1}$. $\Box$

Problem 1.77 Green's disease turns your hair pink and your toenails blue but has no other symptoms. A sufferer from Green's disease has a probability p_n of infecting n further uninfected individuals ($n = 0, 1, 2, 3$) but will not infect more than 3. (The standard assumptions of elementary branching processes apply.) Write down an expression for e, the expected number of individuals infected by a single sufferer.

Starting from the first principles, find the probability π that the disease will die out if, initially, there is only one case.

Let e_A and π_A be the values of e and π when $p_0 = 2/5$, $p_1 = p_2 = 0$ and $p_3 = 3/5$. Let e_B and π_B be the values of e and π when $p_0 = p_1 = 1/10$, $p_2 = 4/5$ and $p_3 = 0$. Show that $e_A > e_B$ but $\pi_A > \pi_B$.

Solution The expression for e is $e = p_1 + 2p_2 + 3p_3$. Let X_j be the number of individuals in the jth generation of the disease and $X_0 = 1$. Assume that each sufferer dies once passing the disease on to $n \leq 3$ others. Call π the probability that the disease dies out:

$$\pi = \sum_k \mathbb{P}(X_1 = k)\pi^k = p_0 + p_1\pi + p_2\pi^2 + p_3\pi^3.$$

Direct calculation shows that $e_A = 9/5$, $e_B = 17/10$. Value π_A is identified as the smallest positive root of the equation

$$0 = p_0 + (p_1 - 1)\pi + p_2\pi^2 + p_3\pi^3 = (\pi - 1)\left(\frac{3}{5}\pi^2 + \frac{3}{5}\pi - \frac{2}{5}\right).$$

Hence

$$\pi_A = \frac{-3 + \sqrt{33}}{6} \approx 0.46.$$

Similarly, π_B is the smallest positive root of the equation

$$0 = (\pi - 1)\left(\frac{4}{5}\pi - \frac{1}{10}\right),$$

and $\pi_B = 1/8$. So, $e_A > e_B$ and $\pi_A > \pi_B$. $\Box$

Problem 1.78 Suppose that $(X_r, r \geq 0)$ is a branching process with $X_0 = 1$ and that the PGF for X_1 is $\phi(s)$. Establish an iterative formula for the PGF $\phi_r(s)$ for X_r. State a result in terms of $\phi(s)$ about the probability of eventual extinction.

(i) Suppose the probability that an individual leaves k descendants in the next generation is $p_k = 1/2^{k+1}$, for $k \geq 0$. Show from the result you state that extinction is certain. Prove further that

$$\phi_r(s) = \frac{r - (r-1)s}{(r+1) - rs}, \quad r \geq 1,$$

and deduce the probability that the rth generation is empty.

(ii) Suppose that every individual leaves at most three descendants in the next generation, and that the probability of leaving k descendants in the next generation is

$$p_k = \binom{3}{k} \frac{1}{2^3}, \quad k = 0, \ 1, \ 2, \ 3.$$

What is the probability of extinction?

Solution (i) Let Y_i^n be the number of offspring of individual i in generation n. Then

$$X_{n+1} = Y_1^n + \cdots + Y_{X_n}^n,$$

and

$$\mathbb{E}(s^{X_{n+1}}) = \mathbb{E}\big[\mathbb{E}(s^{X_{n+1}} | X_n)\big] = \sum_{k=0}^{\infty} \mathbb{P}(X_n = k) \mathbb{E}(s^{Y_1^n + \cdots + Y_k^n})$$

$$= \sum_{k=0}^{\infty} \mathbb{P}(X_n = k) \mathbb{E}(s^{Y_1^n})^k = \sum_{k=0}^{\infty} \mathbb{P}(X_n = k) \phi(s)^k = \phi_n(\phi(s)),$$

and so $\phi_{n+1}(s) = \phi_n(\phi(s))$. The probability of extinction is the least $s \in (0, 1]$ such that $s = \phi(s)$. Further,

$$\phi(s) = \frac{1}{2} + \frac{1}{2^2}s + \cdots = \frac{1}{2 - s}.$$

Solving $s = 1/(2 - s)$ yields $(s - 1)^2 = 0$, i.e. $s = 1$. Hence the extinction is certain. The formula for $\phi_r(s)$ is established by induction:

$$\phi_{r+1}(s) = \frac{r - (r-1)/(2 - s)}{r + 1 - r/(2 - s)} = \frac{(r+1) - rs}{(r+2) - (r+1)s}.$$

Hence, the probability that the rth generation is empty is

$$\phi_r(0) = \frac{r}{r+1}.$$

(ii) $\phi(s) = \frac{1}{2^3}(1+s)^3$ whence solving $2^3 s = (1+s)^3$ or $(s-1)(s^2 + 4s - 1) = 0$ we get solutions $s = 1$, $s = \pm\sqrt{5} - 2$, and the extinction probability is $\sqrt{5} - 2 \approx 0.24$. □

Problem 1.79 Consider a Galton–Watson process (i.e. the branching process where the number of offspring is random and independent for each division), with $\mathbb{P}(X_1 = 0) = 2/5$ and $\mathbb{P}(X_1 = 2) = 3/5$. Compute the distribution of the random variable X_2. For generation 3 find all probabilities $\mathbb{P}(X_3 = 2k)$, $k = 0, 1, 2, 3, 4$. Find the extinction probability for this model.

Solution   The extinction probability $\pi = 2/3$ and the PGF for X_2

$$\phi_{X_2}(s) = \frac{2}{5} + \frac{12}{125}s + \frac{36}{125}s^2 + \left(\frac{3}{5}\right)^3 s^4.$$

The PGF $\phi_{X_3}(s) = \phi_{X_2}(2/5 + 3s^2/5)$. Then

$$\mathbb{P}(X_2 = 0) = \frac{2}{5} + \frac{12}{125}, \ \mathbb{P}(X_2 = 2) = \frac{36}{125}, \ \mathbb{P}(X_2 = 4) = \left(\frac{3}{5}\right)^3;$$

$$\mathbb{P}(X_3 = 0) = \frac{2}{5} + \frac{12}{125} + \frac{36}{125}\left(\frac{2}{5}\right)^2 + \left(\frac{3}{5}\right)^3\left(\frac{2}{5}\right)^4 = 0.54761;$$

$$\mathbb{P}(X_3 = 2) = \frac{2^4 \times 3^3}{5^5} + \frac{2^5 \times 3^4}{5^7} = 0.17142;$$

$$\mathbb{P}(X_3 = 4) = \frac{4 \times 3^4}{5^5} + \frac{4^2 \times 3^5}{5^7} + \frac{8 \times 3^5}{5^7} = 0.17833;$$

$$\mathbb{P}(X_3 = 6) = \frac{3^6 \times 2^3}{5^7} = 0.07465;$$

$$\mathbb{P}(X_3 = 8) = \left(\frac{3}{5}\right)^7 = 0.02799. \quad \square$$

Problem 1.80   By developing the theory of extinction probabilities, or otherwise, solve the following problem.

No-one in their right mind would wish to be a guest at the Virtual Reality Hotel. The rooms are numbered 0 to $(3^N - 3)/2$, where N is a very large integer. If $0 \leq i \leq (3^{N-1} - 3)/2$ and $j = 1, 2, 3$ there is a door between Room i and Room $3i + j$ through which (if it is unlocked) guests may pass in both directions. In addition, any room with a number higher than $(3^{N-1} - 3)/2$ has an open window through which guests can (and should) escape into the street. So far as the guests are concerned, there are no other doors or windows. Figure 1.9 shows part of the floor plan of the hotel.

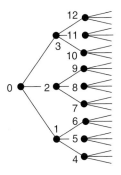

Figure 1.9

Each door in the hotel is locked with probability $1/3$ independently of the others. An arriving guest is placed in Room 0 and can then wander freely (insofar as the locked doors allow). Show that the guest's chance of escape is about $(9 - \sqrt{27})/4$.

Solution Denote by X_r the number of rooms available at level r from 0. Writing $\phi_r(t) = \mathbb{E}t^{X_r}$, with $\phi_1 = \phi$:

$$\phi_{r+1}(t) = \mathbb{E}t^{X_{r+1}} = \sum_i \mathbb{P}(X_r = i)\mathbb{E}\left(t^{X_{r+1}}\big|X_r = i\right)$$

$$= \sum_i \mathbb{P}(X_r = i)\left(\mathbb{E}t^{X_1}\right)^i$$

$$= \sum_i \mathbb{P}(X_r = i)(\phi(t))^i = \phi_r(\phi(t)).$$

Then $\phi_{r+1}(t) = \phi(\phi(\ldots \phi(t) \ldots)) = \phi(\phi_r(t))$, and

$$\mathbb{P}(\text{can't reach level } r) = \phi(\phi_r(0)).$$

Now PGF $\phi(t)$ equals

$$\left(\frac{1}{3}\right)^3 + 3\frac{2}{3}\left(\frac{1}{3}\right)^2 t + 3\left(\frac{2}{3}\right)^2\frac{1}{3}t^2 + \left(\frac{2}{3}\right)^3 t^3 = \frac{1}{27}(1 + 6t + 12t^2 + 8t^3).$$

Hence, the equation $\phi(t) = t$ becomes

$$27t = 1 + 6t + 12t^2 + 8t^3, \text{ i.e. } 1 - 21t + 12t^2 + 8t^3 = 0.$$

By factorising $1 - 21t + 12t^2 + 8t^3 = (t-1)(8t^2 + 20t - 1)$, we find that the roots are

$$t = 1, \quad \frac{-5 \pm \sqrt{27}}{4}.$$

The root between 0 and 1 is $\left(\sqrt{27} - 5\right)/4$. The sequence $\phi_n(0)$ is monotone increasing and bounded above. Hence, it converges as $N \to \infty$:

$$\mathbb{P}(\text{no escape}) \to \frac{\sqrt{27} - 5}{4}.$$

Then

$$\mathbb{P}(\text{escape}) \to 1 - \frac{\sqrt{27} - 5}{4} = \frac{9 - \sqrt{27}}{4} \approx 0.950962. \quad \square$$

Problem 1.81 (i) A mature individual produces offspring according to the PGF $\phi(s)$. Suppose we start with a population of k immature individuals, each of which grows to maturity with probability p and then reproduces, independently of the other individuals. Find the PGF of the number of (immature) individuals in the next generation.

(ii) Find the PGF of the number of mature individuals in the next generation, given that there are k mature individuals in the parent generation.

Hint: (i) Let R be the number of immature descendants of an immature individual. If $X^{(n)}$ is the number of immature individuals in generation n, then, given that $X^{(n)} = k$,

$$X^{(n+1)} = R_1 + \cdots + R_k,$$

where $R_i \sim R$, independently. The conditional PGF is

$$\mathbb{E}\left(s^{X(n+1)}|X^{(n)} = k\right) = (g[\phi(s)])^k,$$

where $g(x) = 1 - p + px$.

(ii) Let U be the number of mature descendants of a mature individual, and $Z^{(n)}$ be the number of mature individuals in generation n. Then, as before, conditional on $Z^{(n)} = k$,

$$Z^{(n+1)} = U_1 + \cdots + U_k,$$

where $U_i \sim U$, independently. The conditional PGF is

$$\mathbb{E}\left(s^{Z(n+1)}|Z^{(n)} = k\right) = (\phi[g(s)])^k.$$

Problem 1.82 Show that the distributions in parts (i) and (ii) of Problem 1.81 have the same mean, but not necessarily the same variance.

Hint:

$$\frac{d^2}{ds^2}\mathbb{E}\left(s^{X^{n+1}}|X^{(n)} = 1\right) = p\phi''(1), \quad \frac{d^2}{ds^2}\mathbb{E}\left(s^{Z(n+1)}|Z^{(n)} = 1\right) = p^2\phi''(1).$$

2   Continuous outcomes

2.1   Uniform distribution. Probability density functions. Random variables. Independence

> Probabilists do it continuously but discreetly.
>> (From the series 'How they do it'.)

> Bye, Bi, Variate
>> (From the series 'Movies that never made it to the Big Screen'.)

After developing a background in probabilistic models with discrete outcomes we can now progress further and do exercises where uncountably many outcomes are explicitly involved. Here, the events are associated with subsets of a continuous space (a real line $\mathbb{R}$, an interval (a, b), a plane $\mathbb{R}^2$, a square, etc.). The simplest case is where the outcome space Ω is represented by a 'nice' bounded set and the probability distribution corresponds to a unit mass uniformly spread over it. Then an event (i.e. a subset) $A \subseteq \Omega$ acquires the probability

$$\mathbb{P}(A) = \frac{v(A)}{v(\Omega)},\tag{2.1}$$

where $v(A)$ is the standard Euclidean volume (or area or length) of A and $v(\Omega)$ that of Ω.

The term 'uniformly spread' is the key here; an example below shows that one has to be careful about what exactly it means in the given context.

Example 2.1   This is known as *Bertrand's paradox*. A chord has been chosen at random in a circle of radius r. What is the probability that it is longer than the side of the equilateral triangle inscribed in the circle? The answer is different in the following three cases:

(i)   the middle point of the chord is distributed uniformly inside the circle;
(ii)   one endpoint is fixed and the second is uniformly distributed over the circumference;
(iii) the distance between the middle point of the chord and the centre of the circle is uniformly distributed over the interval $[0, r]$.

In fact, in case (i), the middle point of the chord must lie inside the circle inscribed in the triangle. Hence,

$$\mathbb{P}(\text{chord longer}) = \frac{\text{area of the inscribed circle}}{\text{area of the original circle}} = \frac{\pi r^2/4}{\pi r^2} = \frac{1}{4}.$$

In case (ii), the second endpoint must then lie on the opposite third of the circumference. Hence,

$$\mathbb{P}(\text{chord longer}) = \frac{1}{3}.$$

Finally, in case (iii), the middle point of the chord must be at distance $\leq r/2$ from the centre. Hence,

$$\mathbb{P}(\text{chord longer}) = \frac{1}{2}. \quad \blacksquare$$

A useful observation is that we can think in terms of a *uniform probability density function* assigning to a point $x \in \Omega$ the value

$$f_\Omega^{\text{uni}}(x) = \frac{1}{v(\Omega)} I_\Omega(x), \tag{2.2}$$

with the probability of event $A \subseteq \Omega$ calculated as the integral

$$\mathbb{P}(A) = \int_A f_\Omega^{\text{uni}}(x) \mathrm{d}x = \frac{1}{v(\Omega)} \int_A \mathrm{d}x \tag{2.3}$$

giving of course the same answer as formula (2.1). Because $f_\Omega^{\text{uni}} \geq 0$ and $\int_\Omega f_\Omega^{\text{uni}}(x)\mathrm{d}x = 1$, the probability of event $A \subseteq \Omega$ is always between 0 and 1. Note that the mass assigned to a single outcome ω represented by a point of Ω is zero. Hence the mass assigned to any finite or countable set of outcomes is zero (as it is the sum of the masses assigned to each outcome); to get a positive mass (and thus a positive probability), an event A must be uncountable.

Problem 2.1   Alice and Bob agree to meet in the Copper Kettle after their Saturday lectures. They arrive at times that are independent and uniformly distributed between 12:00 and 13:00. Each is prepared to wait s minutes before leaving. Find a minimal s such that the probability that they meet is at least $1/2$.

Solution   The set Ω is the unit square $\mathcal{S}$ with co-ordinates $0 \leq x, y \leq 1$ (measuring the time in fractions of an hour between 12:00 and 13:00). Each $\omega = (x, y) \in \Omega$ specifies Alice's arrival time x and Bob's y. Then the event: 'they arrive within s minutes of each other' is a strip around the diagonal $x = y$:

$$A = \left\{ (x, y) \in \mathcal{S} : |x - y| \leq \frac{s}{60} \right\}.$$

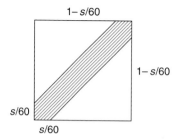

1– s/60

1– s/60

s/60

s/60

Figure 2.1

Its complement is formed by two triangles, of area $(1/2)(1 - s/60)^2$ each. So, the area $v(A)$ is

$$1 - \left(1 - \frac{s}{60}\right)^2$$

and we want it to be $\geq 1/2$. See Figure 2.1.

This gives $s \geq 60(1 - \sqrt{2}/2) \approx 18$ minutes. □

Problem 2.2   A stick is broken into two at random; then the longer half is broken again into two pieces at random. What is the probability that the three pieces will make a triangle?

Solution   Let the stick length be ℓ. If x is the place of the first break, then $0 \leq x \leq \ell$ and x is uniform on $(0, \ell)$. If $x \geq \ell/2$, then the second break point y is uniformly chosen on the interval $(0, x)$. See Figure 2.2. Otherwise y is uniformly chosen on (x, ℓ). Thus

$$\Omega = \{(x, y) : 0 \leq x, y \leq \ell; \ y \leq x \ \text{for} \ x \geq \ell/2 \ \text{and} \ x \leq y \leq \ell \ \text{for} \ x \leq \ell/2\},$$

and the area $v(\Omega) = 3\ell^2/4$. See Figure 2.3.

To make a triangle (x, y) must lie in A, where

$$A = \left\{\max x, y] > \frac{\ell}{2}, \ \min[x, y] < \frac{\ell}{2}, \ |x - y| < \frac{\ell}{2}\right\},$$

which yields the area $\ell^2/4$. The answer then is $1/3$.

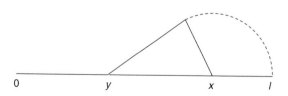

0

y

x

l

Figure 2.2

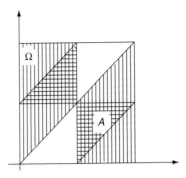

Figure 2.3

It is also possible to reduce Ω to a 'half' of the above set just assuming that x is always the length of the longer stick:

$$\Omega = \{(x, y): \ell/2 \leq x \leq \ell;\ y \leq x\};$$

the probability will still be $1/3$. $\square$

Problem 2.3   A stick is broken in two places chosen beforehand, completely at random along its length. What is the probability that the three pieces will make a triangle?

Answer: $1/4$. The loss in probability occurs as we include into the outcomes the possibility that a shorter stick is broken again while in the previous example it was excluded.

Problem 2.4   The ecu coin is a disc of diameter $4/5$ units. In the traditional game of drop the ecu, an ecu coin is dropped at random onto a grid of squares formed by lines one unit apart. If the coin covers part of a line you lose your ecu. If not, you still lose your ecu but the band plays the national anthem of your choice. What is the probability that you get to choose your national anthem?

Solution   Without loss of generality, assume that the centre of the coin lies in the unit square with corners $(0; 0)$, $(0; 1)$, $(1; 0)$, $(1; 1)$. You will hear an anthem when the centre lies in the inside square $\mathcal{S}$ described by

$$\frac{2}{5} \leq x \leq \frac{3}{5},\ \frac{2}{5} \leq y \leq \frac{3}{5}.$$

Hence,

$$\mathbb{P}\,(\text{anthem}) = \text{area of } S = \frac{1}{25}.\quad \square$$

There are several serious questions arising here which we will address later. One is the so-called measurability: there exist weird sets $A \subset \Omega$ (even when Ω is a unit interval

(0,1)) which do not have a correctly defined volume (or length). In general, how does one measure the volume of a set A in a continuous space? Such sets may not be particularly difficult to describe (for instance, Cantor's continuum $\mathcal{K}$ has a correctly defined length), but calculating their volume, area or length goes beyond standard Riemann integration, let alone elementary formulas. (As a matter of fact, the correct length of $\mathcal{K}$ is zero.) To develop a complete theory, we would need the so-called *Lebesgue integration*, which is called after H.L. Lebesgue (1875–1941), the famous French mathematician. Lebesgue was of a very humble origin, but became a top mathematician. He was renowned for flawless and elegant presentations and written works. In turn, the Lebesgue integration requires the concept of a sigma-algebra and an additive measure which leads to a far-reaching generalisation of the concept of length, area and volume encapsulated in a concept of a *measure*. We will discuss these issues in later volumes.

An issue to discuss now is: what if the distribution of the mass is not uniform? This question is not only of a purely theoretical interest. In many practical models Ω is represented by an unbounded subset of a Euclidean space $\mathbb{R}^d$ whose volume is infinite (e.g. by $\mathbb{R}^d$ itself or by $\mathbb{R}_+ = [0, \infty)$, for $d = 1$). Then the denominator $v(\Omega)$ in equation (2.1) becomes infinite. Here, the recipe is: consider a function $f \geq 0$ with $\int_\Omega f(x)dx = 1$ and set

$$\mathbb{P}(A) = \int_A f(x)dx, \quad A \subseteq \Omega$$

(cf. equation (2.3)). Such a function f is interpreted as a (general) *probability density function* (PDF). The following natural (and important) examples appear in problems below:

A *uniform distribution* on an interval (a, b), $a < b$: here $\Omega = (a, b)$ and

$$f(x) = \frac{1}{b-a}I(a < x < b). \tag{2.4}$$

A *Gaussian*, or *normal*, *distribution*, with $\Omega = \mathbb{R}$ and

$$f(x) = \frac{1}{\sqrt{2\pi}\sigma}\exp\left(-\frac{1}{2\sigma^2}(x-\mu)^2\right), \quad x \in \mathbb{R}. \tag{2.5}$$

Here $\mu \in \mathbb{R}$, $\sigma > 0$ are parameters specifying the distribution.

The graphs of normal PDFs on an interval around the origin and away from it are plotted in Figures 2.4 and 2.5.

This is the famous curve about which the great French mathematician J.-H. Poincaré (1854–1912) said 'Experimentalists think that it is a mathematical theorem while the mathematicians believe it to be an experimental fact.'

An *exponential distribution*: here $\Omega = \mathbb{R}_+$ and

$$f(x) = \lambda e^{-\lambda x}I(x \geq 0), \quad x \in \mathbb{R}. \tag{2.6}$$

Here $\lambda > 0$ is a parameter specifying the distribution.

The graphs of the exponential PDFs are shown in Figure 2.6.

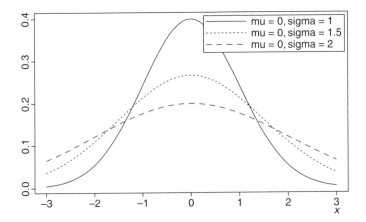

Figure 2.4 The normal PDFs, I.

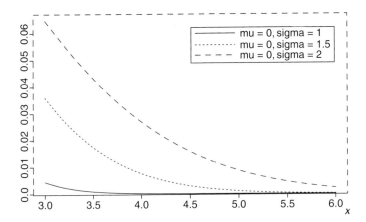

Figure 2.5 The normal PDFs, II.

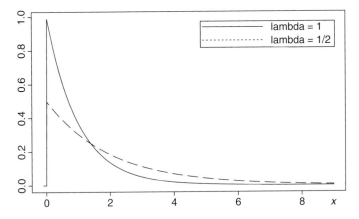

Figure 2.6 The exponential PDFs.

A generalisation of formula (2.6) is the *Gamma distribution*. Here again $\Omega = \mathbb{R}_+$ and the PDF is

$$f(x) = \frac{\lambda^\alpha}{\Gamma(\alpha)} x^{\alpha-1} e^{-\lambda x} I(x \geq 0), \tag{2.7}$$

with parameters $\alpha, \ \lambda > 0$. Here $\Gamma(\alpha) = \int_0^\infty x^{\alpha-1} e^{-x} dx$ (the value of the *Gamma function*) is the normalising constant. Recall, for a positive integer argument, $\Gamma(n) = (n-1)!$; in general, for $\alpha > 1$: $\Gamma(\alpha) = (\alpha-1)\Gamma(\alpha-1)$.

The Gamma distribution plays a prominent rôle in statistics and will repeatedly appear in later chapters. The graphs of the Gamma PDFs are sketched in Figure 2.7.

Another example is a *Cauchy distribution*, with $\Omega = \mathbb{R}$ and

$$f(x) = \frac{1}{\pi} \frac{\tau}{\tau^2 + (x-\alpha)^2}, \ x \in \mathbb{R}, \tag{2.8}$$

with parameters $\alpha \in \mathbb{R}$ and $\tau > 0$. There is a story that the Cauchy distribution was discovered by Poisson in 1824 when he proposed a counterexample to the CLT. See below. The graphs of the Cauchy PDFs are sketched in Figure 2.8.

Cauchy was a staunch royalist and a devoted Catholic and, unlike many other prominent French scientists of the period, he had difficult relations with the Republican regime. In 1830, during one of the nineteenth century French revolutions, he went into voluntary exile to Turin and Prague where he gave private mathematics lessons to the children of the Bourbon Royal family. His admission to the French Academy occurred only in 1838, after he had returned to Paris.

The Gaussian distribution will be discussed in detail below. At this stage we only indicate a generalisation to the multidimensional case where $\Omega = \mathbb{R}^d$ and

$$f(\mathbf{x}) = \frac{1}{(\sqrt{2\pi})^d (\det \Sigma)^{1/2}} \exp\left[-\frac{1}{2} \langle \mathbf{x} - \mu, \Sigma^{-1}(\mathbf{x} - \mu) \rangle \right]. \tag{2.9}$$

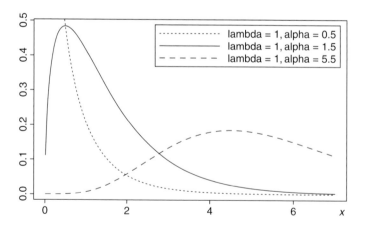

Figure 2.7 The Gamma PDFs.

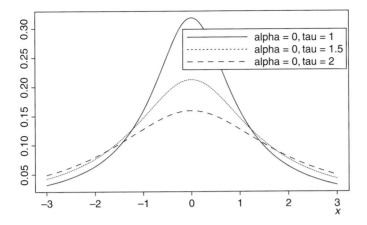

Figure 2.8 The Cauchy PDFs.

Here, **x** and $\boldsymbol{\mu}$ are real d-dimensional vectors:

$$\mathbf{x} = \begin{pmatrix} x_1 \\ \vdots \\ x_d \end{pmatrix}, \; \boldsymbol{\mu} = \begin{pmatrix} \mu_1 \\ \vdots \\ \mu_d \end{pmatrix} \in \mathbb{R}^d,$$

and Σ is an invertible positive-definite $d \times d$ real matrix, with the determinant det $\Sigma > 0$ and the inverse matrix $\Sigma^{-1} = (\Sigma_{ij}^{-1})$. Matrix Σ is called *positive-definite* if it can be represented as the product $\Sigma = AA^*$ and strictly positive-definite if in this representation matrix A is *invertible*, i.e. the *inverse* A^{-1} exists (in which case the inverse matrix $\Sigma^{-1} = A^{*-1}A^{-1}$ also exists). It is easy to see that a positive-definite matrix Σ is always symmetric (or Hermitian), i.e. obeys $\Sigma^* = \Sigma$. Hence, a positive-definite matrix has an orthonormal basis of eigenvectors, and its eigenvalues are non-negative (positive if it is strictly positive-definite). Further, $\langle \, , \, \rangle$ stands for the Euclidean scalar product in $\mathbb{R}^d$:

$$\langle \mathbf{x} - \boldsymbol{\mu}, \Sigma^{-1}(\mathbf{x} - \boldsymbol{\mu}) \rangle = \sum_{i,j=1}^{d} (x_i - \mu_i) \Sigma_{ij}^{-1} (x_j - \mu_j).$$

A PDF of this form is called a *multivariate normal*, or *Gaussian, distribution*.

Remark   We already have seen a number of probability distributions bearing personal names (Gauss, Poisson, Cauchy; more will appear in Chapters 3 and 4). Another example (though not as frequent) is the *Simpson distribution*. Here we take $X, \; Y \sim \mathrm{U}(0,1)$, independently. Then $X + Y$ has a 'triangular' PDF known as Simpson's PDF:

$$f_{X+Y}(u) = \begin{cases} u, & 0 \le u \le 1, \\ 2 - u, & 1 \le u \le 2, \\ 0, & \notin [0,2]. \end{cases}$$

T. Simpson (1700–1761), an English scientist, left a notable mark in interpolation and numerical methods of integration. He was the most distinguished of the group of itinerant lecturers who taught in fashionable London coffee-houses (a popular way of spreading scientific information in eighteenth-century England).

As before, we face a question: what type of function f can serve as PDFs? (The example with $f(x) = I(x \in (0, 1) \setminus \mathcal{K})$, where $\mathcal{K} \subset (0, 1)$ is Cantor's set, is typical. Here $f \geq 0$ by definition but how $\int_0^1 f(x)dx$ should be defined?) And again, the answer lies in the theory of Lebesgue integration. Fortunately, in 'realistic' models, these matters arise rarely and are overshadowed by far more practical issues.

So, from now until the end of this chapter our basic model will be where outcomes ω run over an 'allowed' subset of Ω (such subsets are called measurable and will be introduced later). Quite often Ω will be $\mathbb{R}^d$. The probability $\mathbb{P}(A)$ will be calculated for every such set A (called an *event* in Ω) as

$$\mathbb{P}(A) = \int_A f(\mathbf{x})d\mathbf{x}. \tag{2.10}$$

Here f is a given PDF $f \geq 0$ with $\int_\Omega f(\mathbf{x})d\mathbf{x} = 1$.

As in the discrete case, we have an intuitively plausible property of additivity: if $A_1, A_2, \ldots$ is a (finite or countable) sequence of pair-wise disjoint events then

$$\mathbb{P}\left(\bigcup_j A_j\right) = \sum_j \mathbb{P}(A_j), \tag{2.11}$$

while, in a general case, $\mathbb{P}\left(\bigcup_j A_j\right) \leq \sum_j \mathbb{P}(A_j)$. As $\mathbb{P}(\Omega) = 1$, we obtain that for the complement $A^c = \Omega \setminus A$, $\mathbb{P}(A^c) = 1 - \mathbb{P}(A)$, and for the set-theoretical difference $A \setminus B$, $\mathbb{P}(A \setminus B) = \mathbb{P}(A) - \mathbb{P}(A \cap B)$. Of course, more advanced facts that we learned in the discrete space case remain true, such as the inclusion–exclusion formula.

In this setting, the concept of a *random variable* develops, unsurprisingly, from its discrete-outcome analogue: a RV is a function

$$X: \omega \in \Omega \mapsto X(\omega),$$

with real or complex values $X(\omega)$ (in the complex case we again consider a pair of real RVs representing the real and imaginary parts). Formally, a real RV must have the property that $\forall\, x \in \mathbb{R}$, the set $\{\omega \in \Omega : X(\omega) < x\}$ is an event in Ω to which the probability $\mathbb{P}(X < y)$ can be assigned. Then with each real-valued RV we associate its *cumulative distribution function* (CDF)

$$y \in \mathbb{R} \mapsto F_X(x) = \mathbb{P}(X < x) \tag{2.12}$$

varying monotonically from 0 to 1 as y increases from $-\infty$ to ∞. See Figure 2.9.

The quantity

$$\overline{F}_X(x) = 1 - F_X(x) = \mathbb{P}(X \geq x) \tag{2.13}$$

describing *tail probabilities* is also often used.

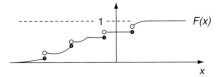

Figure 2.9

Observe that definition (2.12) leads to CDF $F_X(x)$ that is left-continuous (in Figure 2.9 it is presented by black dots). It means that $F_X(x_n) \nearrow F_X(x)$ whenever $x_n \nearrow x$. On the other hand, $\forall\, x$, the right-hand limit $\lim_{x_n \searrow x} F_X(x_n)$ also exists and is $\geq F_X(x)$, but the equality is not guaranteed (in the figure this is represented by circles). Of course the tail probability $\overline{F}_X(x)$ is again left-continuous.

However, if we adopt the definition that $F_X(x) = \mathbb{P}(X \leq x)$ (which is the case in some textbooks) then F_X will become right-continuous (as well as $\overline{F}_X(x)$).

Example 2.2 If $X \equiv b$ is a (real) constant, the CDF F_X is the Heaviside-type function:

$$F_X(y) = I(y > b). \tag{2.14}$$

If $X = I_A$, the indicator function of an event A, then $\mathbb{P}(X < y)$ equals 0 for $y \leq 0$, $1 - \mathbb{P}(A)$ for $0 < y \leq 1$ and 1 for $y > 1$. More generally, if X admits a discrete set of (real) values (i.e. finitely or countably many, without accumulation points on $\mathbb{R}$), say $y_j \in \mathbb{R}$, with $y_j < y_{j+1}$, then $F_X(y)$ is constant on each interval $y_j < x \leq y_{j+1}$ and has jumps at points y_j of size $\mathbb{P}(X = y_j)$.

Observe that all previously discussed discrete-outcome examples of RVs may be fitted into this framework. For instance, if $X \sim \mathrm{Bin}\,(n,\, p)$, then

$$F_X(y) = \sum_{0 \leq m < y,\, m \leq n} \binom{n}{m} p^m (1-p)^{n-m} I(y > 0). \tag{2.15}$$

If $X \sim \mathrm{Po}\,(\lambda)$,

$$F_X(y) = e^{-\lambda} \sum_{0 \leq n < y} \frac{\lambda^n}{n!} I(y > 0); \tag{2.16}$$

Figure 2.10 shows the graphs of $F_X \sim \mathrm{Po}\,(\lambda)$.

If $X \sim \mathrm{Geom}\,(q)$,

$$F_X(y) = I(y > 0)(1-q) \sum_{0 \leq n < y} q^n. \tag{2.17}$$

The graph of the CDF of RV $X \sim \mathrm{Geom}\,(q)$ is plotted in Figure 2.11, together with that of a Poisson RV with $\lambda = 1$ (both RVs have the same mean value 1).

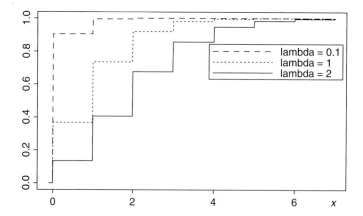

Figure 2.10 The Poisson CDFs.

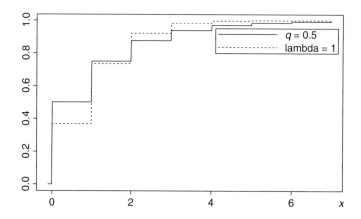

Figure 2.11 The geometric and Poisson CDFs.

We say that an RV X has a uniform, Gaussian, exponential, Gamma or Cauchy distribution (with the corresponding parameters) if CDF $F_X(y)$ is prescribed by the corresponding PDF, i.e. $\mathbb{P}(X < y) = \int f(x)I(x < y)\mathrm{d}x$. For example: for a uniform RV

$$F_X(y) = \begin{cases} 0, y \le a, \\ (y-a)/(b-a), a < y < b, \\ 1, y \ge b; \end{cases} \tag{2.18}$$

for a Gaussian

$$F_X(y) = \frac{1}{\sqrt{2\pi}\sigma} \int_{-\infty}^{y} \exp\left[-\frac{1}{2\sigma^2}(x-\mu)^2\right]\mathrm{d}x = \Phi\left(\frac{y-\mu}{\sigma}\right), \quad y \in \mathbb{R}; \tag{2.19}$$

for an exponential RV

$$F_X(y) = \begin{cases} 0, & y \le 0, \\ 1 - e^{-\lambda y}, & y > 0; \end{cases} \tag{2.20}$$

for a Gamma RV

$$F_X(y) = \frac{\lambda^\alpha}{\Gamma(\alpha)} \int_0^y x^{\alpha-1} e^{-\lambda x} dx I(y > 0); \tag{2.21}$$

and for a Cauchy RV

$$F_X(y) = \frac{1}{\pi} \left[\tan^{-1} \left(\frac{y-\alpha}{\tau} \right) + \frac{\pi}{2} \right], \quad y \in \mathbb{R}. \tag{2.22}$$

Correspondingly, we write $X \sim U(a, b)$, $X \sim N(\mu, \sigma^2)$, $X \sim \text{Exp}(\lambda)$, $X \sim \text{Gam}(\alpha, \lambda)$ and $X \sim \text{Ca}(\alpha, \tau)$.

In Figures 2.12–2.15 we show some graphics for these CDFs.

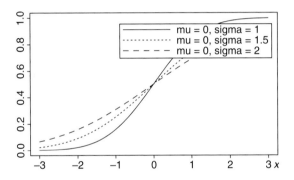

Figure 2.12 The normal CDFs.

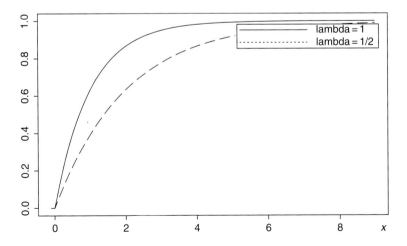

Figure 2.13 The exponential CDFs.

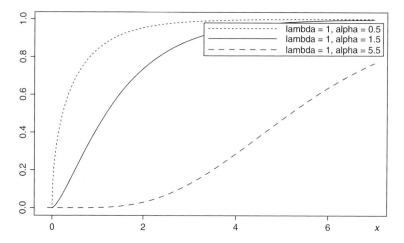

Figure 2.14 The Gamma CDFs.

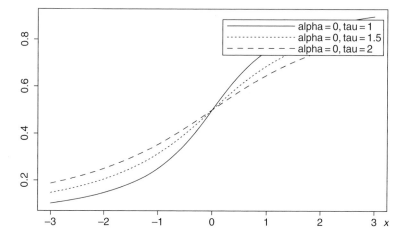

Figure 2.15 The Cauchy CDFs.

In general, we say that X has a PDF f (and write $X \sim f$) if $\forall\ y \in \mathbb{R}$,

$$\mathbb{P}(X < y) = \int_{-\infty}^{y} f(x)\mathrm{d}x. \tag{2.23}$$

Then, $\forall\ a, b \in \mathbb{R}$ with $a < b$:

$$\mathbb{P}(a < X < b) = \int_{a}^{b} f(x)\mathrm{d}x, \tag{2.24}$$

and in general, $\forall$ measurable set $A \subset \mathbb{R}$: $\mathbb{P}(X \in A) = \int_{A} f(x)\mathrm{d}x$.

   Note that, in all calculations involving PDFs, the sets C with $\int_{C} \mathrm{d}x = 0$ (sets of measure 0) can be disregarded. Therefore, probabilities $\mathbb{P}(a \leq X \leq b)$ and $\mathbb{P}(a < X < b)$ coincide. (This is, of course, not true for discrete RVs.)

The *median* $m(X)$ of RV X gives the value that 'divides' the range of X into two pieces of equal mass. In terms of the CDF and PDF:

$$m(X) = \max\left[y : \overline{F}_X(y) \geq \frac{1}{2}\right] = \max\left[y : \int_y^\infty f_X(x)\mathrm{d}x \geq \frac{1}{2}\right]. \tag{2.25}$$

If F_X is strictly monotone and continuous, then, obviously, $m(X)$ equals the unique value y for which $F_X(y) = 1/2$. In other words, $m(X)$ is the unique y for which

$$\int_{-\infty}^y f_X(x)\mathrm{d}x = \int_y^\infty f_X(x)\mathrm{d}x.$$

The *mode* of an RV X with a bounded PDF f_X is the value x where f_X attains its maximum; sometimes one refers to local maxima as local modes.

Problem 2.5 My Mum and I plan to take part in a televised National IQ Test where we will answer a large number of questions, together with selected groups of mathematics professors, fashion hairdressers, brass band players and others, representing various sections of society (not forgetting celebrities of the day of course). The IQ index, we were told, is calculated differently for different ages. For me, it is equal to

$$-80\ln(1-x),$$

where x is the fraction of my correct answers, which can be anything between 0 and 1. In Mum's case, the IQ index is given by a different formula

$$-70\ln\frac{3/4 - y}{3/4} = -70\ln(3/4 - y) + 70\ln 3/4,$$

where y is her fraction of correct answers. (In her age group one does not expect it to exceed 3/4 (sorry, Mum!).)

We each aim to obtain at least 110. What is the probability that we will do this? What is the probability that my IQ will be better than hers?

Solution Again, we employ the uniform distribution assumption. The outcome $\omega = (x_1, x_2)$ is uniformly spread in the set Ω, which is the rectangle $(0, 1) \times (0, 3/4)$, of area $3/4$. We have a pair of RVs:

$$X(\omega) = -80\ln(1 - x_1), \text{ for my IQ,}$$
$$Y(\omega) = -70\ln\frac{3/4 - x_2}{3/4}, \text{ for Mum's IQ,}$$

and $\forall\ y > 0$:

$$\mathbb{P}(X < y) = \frac{1}{3/4}\int_0^1 \int_0^{3/4} I\left[-\ln(1 - x_1) < \frac{1}{80}y\right]\mathrm{d}x_2\mathrm{d}x_1$$

$$= \int_0^{1-e^{-y/80}} \mathrm{d}x_1 = 1 - e^{-y/80}, \text{ i.e. } X \sim \mathrm{Exp}\,(1/80),$$

$$\mathbb{P}(Y < y) = \frac{1}{3/4} \int_0^1 \int_0^{3/4} I\left(-\ln \frac{3/4 - x_2}{3/4} < \frac{1}{70}y\right) dx_2 dx_1$$

$$= \frac{1}{3/4} \int_0^{3(1-e^{-y/70})/4} dx_2 = 1 - e^{-y/70}, \text{ i.e. } Y \sim \text{Exp}(1/70).$$

Next, $\mathbb{P}(\min[X, Y] < y) = 1 - \mathbb{P}(\min[X, Y] \ge y)$, and

$$\mathbb{P}(\min[X, Y] \ge y) = \frac{1}{3/4} \int_0^1 \int_0^{3/4} I\left(-\ln(1 - x_1) \ge \frac{y}{80}, -\ln \frac{3/4 - x_2}{3/4} \ge \frac{y}{70}\right) dx_2 dx_1$$

$$= e^{-y/80} e^{-y/70} = e^{-3y/112}, \text{ i.e. } \min[X, Y] \sim \text{Exp}(3/112).$$

Therefore,

$$\mathbb{P}(\text{both reach } 110) = e^{-3 \times 110/112} \approx e^{-3}$$

(pretty small). To increase the probability, we have to work hard to change the underlying uniform distribution by something more biased towards higher values of x_1 and x_2, the fractions of correct answers.

To calculate $\mathbb{P}(X > Y)$, it is advisable to use the *Jacobian* $\partial(x_1, x_2)/\partial(u_1, u_2)$ of the 'inverse' change of variables $x_1 = 1 - e^{-u_1/80}$, $x_2 = 3(1 - e^{-u_2/70})/4$ (the 'direct' change is $u_1 = -80 \ln(1 - x_1)$, $u_2 = -70 \ln[(3/4 - X_2)/(3/4)]$. Indeed,

$$\frac{\partial(x_1, x_2)}{\partial(u_1, u_2)} = \frac{3}{4} \frac{1}{80} e^{-u_1/80} \frac{1}{70} e^{-u_2/70}, \quad u_1, u_2 > 0,$$

and

$$\mathbb{P}(X > Y) = \frac{1}{3/4} \int_0^1 \int_0^{3/4} I\left(-80 \ln(1 - x_1) > -70 \ln \frac{3/4 - x_2}{3/4}\right) dx_2 dx_1$$

$$= \frac{1}{3/4} \int_0^\infty \int_0^\infty \frac{3}{4} \frac{1}{80} e^{-u_1/80} \frac{1}{70} e^{-u_2/70} I(u_1 > u_2) du_2 du_1$$

$$= \int_0^\infty \frac{1}{70} e^{-u_2/70} \int_{u_2}^\infty \frac{1}{80} e^{-u_1/80} du_1 du_2 = \frac{8}{15}. \quad \square$$

In the above examples, the CDF F either had the form

$$F(y) = \int_{-\infty}^y f(x) dx, \quad y \in \mathbb{R},$$

or was locally constant, with positive jumps at the points of a discrete set $\mathbb{X} \subset \mathbb{R}$. In the first case one says that the corresponding RV has an *absolutely continuous* distribution (with a PDF f), and in the second one says it has a *discrete* distribution concentrated on $\mathbb{X}$. It is not hard to check that absolute continuity implies that F is continuous (but not vice versa), and for the discrete distributions, the CDF is locally constant, i.e. manifestly discontinuous. However, a combination of these two types is also possible.

Furthermore, there are CDFs that do not belong to any of these cases but we will not discuss them here in any detail (a basic example is *Cantor's staircase* function, which is continuous but grows from 0 to 1 on the set $\mathcal{K}$ which has length zero). See Figure 2.16.

Returning to RVs, it has to be said that for many purposes, the detailed information about what exactly the outcome space Ω is where $X(\omega)$ is defined is actually irrelevant. For example, normal RVs arise in a great variety of models in statistics, but what matters is that they are jointly or individually Gaussian, i.e. have a prescribed PDF. Also, an exponential RV arises in many models and may be associated with a lifetime of an item or a time between subsequent changes of a state in a system, or in a purely geometric context. It is essential to be able to think of such RVs without referring to a particular Ω.

On the other hand, a standard way to represent a real RV X with a prescribed PDF $f(x)$, $x \in \mathbb{R}$, is as follows. You choose Ω to be the *support of PDF f*, i.e. the set $\{x \in \mathbb{R} : f(x) > 0\}$, define the probability $\mathbb{P}(A)$ as $\int_A f(x)\mathrm{d}x$ (see equation (2.10)) and set $X(\omega) = \omega$ (or, if you like, $X(x) = x$, $x \in \Omega$). In fact, then, trivially, the event $\{X < y\}$ will coincide with the set $\{x \in \mathbb{R} : f(x) > 0, \ x < y\}$ and its probability will be

$$\mathbb{P}(X < y) = \int f(x) I(x < y) \mathrm{d}x = \int_{-\infty}^{y} f(x)\mathrm{d}x.$$

In the final part of the solution to Problem 2.5 we did exactly this: the change of variables $u_1 = -80 \ln(1 - x_1)$, $u_2 = -70 \ln[(3/4 - x_2)/(3/4)]$ with the inverse Jacobian

$$\frac{3}{4}\frac{1}{80}e^{-u_1/80}\frac{1}{70}e^{-u_2/70}$$

has put us on the half-lines $u_1 > 0$ and $u_2 > 0$, with the PDFs $e^{-u_1/80}/80$ and $e^{-u_1/70}/70$ and the factor $I(u_1 > u_2)$ indicating the event.

So, to visualise a uniform RV on interval (a, b), we take the model with $\Omega = (a, b)$ and define f by equation (2.4); for an exponential or Gamma distribution, $\Omega = \mathbb{R}_+$, the positive half-line, and f is defined by equation (2.6) or (2.7), and for a Gaussian or Cauchy distribution, $\Omega = \mathbb{R}$, the whole line, and f is defined by equation (2.5) or (2.8).

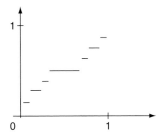

Figure 2.16

In all cases, the standard equation $X(x) = x$ defines an RV X with the corresponding distribution.

Such a representation of an RV X with a given PDF/CDF will be particularly helpful when we have to deal with a function $Y = g(X)$. See below.

So far we have encountered two types of RVs: either (i) with a discrete set of values (finite or countable) or (ii) with a PDF (on a subset of $\mathbb{R}$ or $\mathbb{C}$). These types do not exhaust all occurring situations. In particular, a number of applications require consideration of an RV X that represents a 'mixture' of the two above types where a positive portion of a probability mass is sitting at a point (or points) and another portion is spread out with a PDF over an interval in $\mathbb{R}$. Then the corresponding CDF F_X has jumps at the points x_j where probability $\mathbb{P}(X = x_j) > 0$, of a size equal to $\mathbb{P}(X = x_j) > 0$, and is absolutely continuous outside these points. A typical example is the CDF F_W of the waiting time W in a queue with random arrival and service times (a popular setting is a hairdresser's shop, where the customer waits until the hairdressers finish with the previous customers). You may be lucky in entering the shop when the queue is empty: then your waiting time is 0 (the probability of such event, however small, is > 0). Otherwise you will wait for some positive time; under the simplest assumptions about the arrival and service times:

$$F_W(y) = \mathbb{P}(W < y) = \begin{cases} 0, & y \le 0, \\ 1 - \dfrac{\lambda}{\mu} e^{-(\mu - \lambda)y}, & y > 0. \end{cases} \tag{2.26}$$

Here $\lambda, \mu > 0$ are the rates of two exponential RVs: λ is the rate of the interarrival time and μ that of the service time. Formula (2.26) makes sense when $\lambda < \mu$, i.e. the service rate exceeds the arrival rate which guarantees that the queue does not become overflowed with time. The probability $\mathbb{P}(W = 0)$ that you wait zero time is then equal to $1 - \lambda/\mu > 0$ and the probability that you wait a time $> y$ equals $\lambda e^{-(\mu - \lambda)y}/\mu$.

In this example we can say that the distribution of RV W has a discrete component (concentrated at point 0) and an absolutely continuous component (concentrated on $(0, \infty)$).

Very often we want to find the PDF or CDF of a random variable Y that is a function $g(X)$ of another random variable X, with a given PDF (CDF).

Problem 2.6   The area of a circle is exponentially distributed with parameter λ. Find the PDF of the radius of the circle.

Answer: $f_R(y) = 2\pi\lambda y e^{-\lambda\pi y^2} I(y > 0)$.

Problem 2.7   The radius of a circle is exponentially distributed with parameter λ. Find the PDF of the area of the circle.

Answer: $f_{\text{area}}(s) = \lambda e^{-\lambda\sqrt{s/\pi}}/\sqrt{4\pi s}$.

In these two problems we have dealt with two mutually inverse maps given by the square and the square root. For a general function g, the answer is the result of a straightforward calculation involving the inverse Jacobian. More precisely, if $Y = g(X)$, then the direct change of variables is $y = g(x)$, and

$$f_Y(y) = I(y \in \text{Range}(g)) \sum_{x: g(x)=y} f_X(x) \frac{1}{|g'(x)|}. \qquad (2.27)$$

Here, Range (g) stands for the set $\{y: y = g(x) \text{ for some } x \in \mathbb{R}\}$, and we have assumed that the inverse image of y is a discrete set which allows us to write the summation.

Equation (2.27) holds whenever the RHS is a correctly defined PDF (which allows that $g'(x) = 0$ on a 'thin' set of points y).

Example 2.3   If $b, c \in \mathbb{R}$ are constants, $c \neq 0$, then

$$f_{X+b}(y) = f_X(y - b), \quad \text{and} \quad f_{cX}(y) = \frac{1}{|c|} f_X(y/c).$$

Combining these two formulas, it is easy to see that the normal and Cauchy distributions have the following scaling properties:

$$\text{if } X \sim N(\mu, \sigma^2), \text{ then } \frac{1}{\sigma}(X - \mu) \sim N(0, 1),$$

and

$$\text{if } X \sim \text{Ca}(\alpha, \tau), \text{ then } \frac{1}{\tau}(X - \alpha) \sim \text{Ca}(1, 0).$$

Also,

$$\text{if } X \sim N(\mu, \sigma^2) \text{ then } cX + b \sim N(c\mu + b, c^2\sigma^2),$$

and

$$\text{if } X \sim \text{Ca}(\alpha, \tau) \text{ then } cX + b \sim \text{Ca}(c\alpha + b, c\tau). \quad \blacksquare$$

A formula emerging from Problem 2.6 is

$$f_{\sqrt{X}}(y) = 2\sqrt{y} f_X(y^2) I(y > 0)$$

(assuming that RV X takes non-negative values). Similarly, in Problem 2.7,

$$f_{X^2}(y) = \frac{1}{2\sqrt{y}} (f_X(\sqrt{y}) + f_X(-\sqrt{y})) I(y > 0)$$

(which is equal to $1/(2\sqrt{y}) f_X(\sqrt{y}) I(y > 0)$ if X is non-negative).

Assuming that g is one-to-one, at least on the range of RV X, formula (2.27) is simplified as the summation is reduced to a single inverse image $x(y) = g^{-1}(y)$:

$$f_Y(y) = f_X(x(y)) \frac{1}{|g'(x(y))|} I(y \in \text{Range}(g))$$

$$= f_X[x(y)] |x'(y)| I(y \in \text{Range}(g)). \qquad (2.28)$$

Example 2.4   Let $X \sim \mathrm{U}(0, 1)$. Given $a, b, c, d > 0$, with $d \leq c$, consider RV

$$Y = \frac{a + bX}{c - dX}.$$

Using formula (2.28) one immediately obtains the PDF of Y:

$$f_Y(y) = \frac{bc + ad}{(b + dy)^2}, \quad y \in \left(\frac{a}{c}, \frac{a + b}{c - d} \right). \quad \blacksquare$$

Dealing with a pair X, Y or a general collection of RVs $X_1, \ldots, X_n$, it is convenient to use the concept of the *joint PDF* and *joint CDF* (just as we used joint probabilities in the discrete case). Formally, we set, for real RVs $X_1, \ldots, X_n$:

$$F_{\mathbf{X}}(y_1, \ldots, y_n) = \mathbb{P}(X_1 < y_1, \ldots, X_n < y_1) \tag{2.29}$$

and we say that they have a joint PDF f if, $\forall \, y_1, \ldots, y_n \in \mathbb{R}$,

$$F_{\mathbf{X}}(y_1, \ldots, y_n) = \int_{-\infty}^{y_1} \cdots \int_{-\infty}^{y_n} f(\mathbf{x}) \mathrm{d}\mathbf{x}. \tag{2.30}$$

Here $\mathbf{X}$ is a random vector formed by RVs $X_1, \ldots, X_n$, and $\mathbf{x}$ is a point in $\mathbb{R}^n$, with entries $x_1, \ldots, x_n$:

$$\mathbf{X} = \begin{pmatrix} X_1 \\ \vdots \\ X_n \end{pmatrix}, \quad \mathbf{x} = \begin{pmatrix} x_1 \\ \vdots \\ x_n \end{pmatrix}$$

Next, $\mathrm{d}\mathbf{x} = \mathrm{d}x_1 \times \cdots \times \mathrm{d}x_n$ the Euclidean volume element. Then, given a collection of intervals $[a_j, b_j)$, $j = 1, \ldots, n$:

$$\mathbb{P}(a_1 < X_1 < b_1, \ldots, a_n < X_n < b_n) = \int_{a_1}^{b_1} \cdots \int_{a_n}^{b_n} f(\mathbf{x}) \mathrm{d}\mathbf{x},$$

and, in fact, $\mathbb{P}(\mathbf{X} \in A) = \int_A f(\mathbf{x}) \mathrm{d}\mathbf{x}$ for $\forall$ measurable subsets $A \in \mathbb{R}^n$.

In Chapters 3 and 4 of this book (dealing with IB statistics) we will repeatedly use the notation $f_{\mathbf{X}}$ for the joint PDF of RVs $X_1, \ldots, X_n$ constituting vector $\mathbf{X}$; in the case of two random variables X, Y we will write $f_{X,Y}(x, y)$ for the joint PDF and $F_{X,Y}(x, y)$ for the joint CDF, $x, y \in \mathbb{R}$. A convenient formula for $f_{X,Y}$ is

$$f_{X,Y}(x, y) = \frac{\mathrm{d}^2}{\mathrm{d}x \mathrm{d}y} F_{X,Y}(x, y). \tag{2.31}$$

As before, joint PDF $f(x, y)$ has only to be non-negative and have the integral $\int_{\mathbb{R}^2} f(x, y) \mathrm{d}y \mathrm{d}x = 1$; it may be unbounded and discontinuous. The same is true about $f_{\mathbf{X}}(\mathbf{x})$. We will write $(X, Y) \sim f$ if $f = f_{X,Y}$.

In Problem 2.5, the joint CDF of the two IQs is

$$F_{X,Y}(x, y) = v\left(\left\{(x_1, x_2) : 0 \le x_1, x_2 \le 1,\right.\right.$$

$$\left.\left. -80\ln(1 - x_1) < x, \ -70\ln\frac{3/4 - x_2}{3/4} < y\right\}\right)I(x, y > 0)$$

$$= \left(1 - e^{-x/80}\right)\left(1 - e^{-y/70}\right)I(x, y > 0) = F_X(x)F_Y(y).$$

Naturally, the joint PDF is also the product:

$$f_{X,Y}(x, y) = \frac{1}{80}e^{-x/80}I(x > 0)\frac{1}{70}e^{-y/70}I(y > 0).$$

If we know the joint PDF $f_{X,Y}$ of two RVs X and Y, then their marginal PDFs f_X and f_Y can be found by integration in the complementary variable:

$$f_X(x) = \int_{\mathbb{R}} f_{X,Y}(x, y)\mathrm{d}y, \ \ f_Y(y) = \int_{\mathbb{R}} f_{X,Y}(x, y)\mathrm{d}x, \tag{2.32}$$

This equation is intuitively clear: we apply a continuous analogue of the formula of complete probability, by integrating over all values of Y and thereby 'removing' RV Y from consideration. Formally, $\forall\ y \in R$,

$$F_X(y) = \mathbb{P}(X < y) = \mathbb{P}(X < y, -\infty < Y < \infty)$$

$$= \int_{-\infty}^{y}\int_{-\infty}^{\infty} f_{X,Y}(x, x')\mathrm{d}x'\mathrm{d}x = \int_{-\infty}^{y} g_X(x)\mathrm{d}x, \tag{2.33}$$

where

$$g_X(x) = \int_{-\infty}^{\infty} f_{X,Y}(x, x')\mathrm{d}x'.$$

On the other hand, again $\forall\ y \in R$,

$$F_X(y) = \int_{-\infty}^{y} f_X(x)\mathrm{d}x. \tag{2.34}$$

Comparing the RHSs of equations (2.33) and (2.34), one deduces that $f_X = g_X$, as required.

Example 2.5 Consider a bivariate normal pair of RVs X, Y. Their joint PDF is of the form (2.9) with $d = 2$. Here, positive-definite matrices Σ and Σ^{-1} can be written as

$$\Sigma = \begin{pmatrix} \sigma_1^2 & \sigma_1\sigma_2 r \\ \sigma_1\sigma_2 r & \sigma_2^2 \end{pmatrix}, \ \Sigma^{-1} = \frac{1}{1 - r^2}\begin{pmatrix} 1/\sigma_1^2 & -r/(\sigma_1\sigma_2) \\ -r/(\sigma_1\sigma_2) & 1/\sigma_2^2 \end{pmatrix},$$

where σ_1, σ_2 are non-zero real numbers (with $\sigma_1^2, \sigma_2^2 > 0$) and r is real, with $|r| \lessgtr 1$. Equation (2.9) then takes the form

$$f_{X,Y}(x, y) = \frac{1}{2\pi\sigma_1\sigma_2\sqrt{1 - r^2}}\exp\left\{\frac{-1}{2(1 - r^2)}\right.$$

$$\left. \times \left[\frac{(x - \mu_1)^2}{\sigma_1^2} - 2r\frac{(x - \mu_1)(y - \mu_2)}{\sigma_1\sigma_2} + \frac{(y - \mu_2)^2}{\sigma_2^2}\right]\right\}. \tag{2.35}$$

We want to check that marginally, $X \sim N(\mu_1, \sigma_1^2)$ and $Y \sim N(\mu_2, \sigma_2^2)$, i.e. to calculate the marginal PDFs. For simplicity we omit, whenever possible, limits of integration; integral $\int$ below means $\int_{-\infty}^{\infty}$, or $\int_{\mathbb{R}}$, an integral over the whole line. Write

$$f_X(x) = \frac{1}{2\pi\sigma_1\sigma_2\sqrt{1 - r^2}} \int \exp\left\{-\frac{1}{2(1 - r^2)}\right.$$
$$\times \left.\left[\frac{(x - \mu_1)^2}{\sigma_1^2} - 2r\frac{(x - \mu_1)(y - \mu_2)}{\sigma_1\sigma_2} + \frac{(y - \mu_2)^2}{\sigma_2^2}\right]\right\} dy$$

$$= \frac{1}{2\pi\sigma_1\sigma_2\sqrt{1 - r^2}} \int \exp\left\{-\frac{1}{2(1 - r^2)}\right.$$
$$\times \left.\left[\frac{(1 - r^2)(x - \mu_1)^2}{\sigma_1^2} + \left(\frac{y - \mu_2}{\sigma_2} - r\frac{x - \mu_1}{\sigma_1}\right)^2\right]\right\} dy$$

$$= \frac{e^{-(x - \mu_1)^2/2\sigma_1^2}}{\sqrt{2\pi}\sigma_1} \times \left\{\frac{1}{\sqrt{2\pi}\sigma_2\sqrt{1 - r^2}} \int \exp\left[-\frac{(y_1 - \nu_1)^2}{2\sigma_2^2(1 - r^2)}\right] dy_1\right\}.$$

Here

$$y_1 = y - \mu_2, \quad \nu_1 = r\frac{\sigma_2}{\sigma_1}(x - \mu_1).$$

The last factor in the braces equals 1, and we obtain that

$$f_X(x) = \frac{1}{\sqrt{2\pi}\sigma_1} e^{-(x - \mu_1)^2/2\sigma_1^2},$$

i.e. $X \sim N(\mu_1, \sigma_1^2)$. Similarly, $Y \sim N(\mu_2, \sigma_2^2)$. ∎

The above 'standard representation' principle where a real-valued RV $X \sim f$ was identified as $X(x) = x$ for $x \in \mathbb{R}$ with the probability $\mathbb{P}(A)$ given by formula (2.12) also works for joint distributions. The recipe is similar: if you have a pair $(X, Y) \sim f$, then take $\omega = (x, y) \in \Omega = \mathbb{R}^2$, set

$$\mathbb{P}(A) = \int_A f(x, y)dydx \text{ for } A \subset \mathbb{R}^2$$

and define

$$X(\omega) = x, \quad Y(\omega) = y.$$

A similar recipe works in the case of general collections $X_1, \ldots, X_n$.

A number of problems below are related to transformations of a pair (X, Y) to (U, V), with

$$U = g_1(X, Y), \quad V = g_2(X, Y).$$

Then, to calculate the joint PDF $f_{U,V}$ from $f_{X,Y}$, one uses a formula similar to formula (2.27). Namely, if the change of variables $(x, y) \mapsto (u, v)$, with $u = g_1(x, y)$, $v = g_2(x, y)$, is one-to-one on the range of RVs X and Y, then

$$f_{U,V}(u, v) = I\big((u, v) \in \text{Range } (g_1, g_2)\big) f_{X,Y}\big(x(u, v), y(u, v)\big) \left| \frac{\partial(x, y)}{\partial(u, v)} \right|. \qquad (2.36)$$

Here,

$$\frac{\partial(x, y)}{\partial(u, v)} = \det \begin{pmatrix} \partial x/\partial u & \partial x/\partial v \\ \partial y/\partial u & \partial y/\partial v \end{pmatrix}$$

is the Jacobian of the inverse change $(u, v) \mapsto (x, y)$:

$$\det \begin{pmatrix} \partial x/\partial u & \partial x/\partial v \\ \partial y/\partial u & \partial y/\partial v \end{pmatrix} = \det \begin{pmatrix} \partial u/\partial x & \partial u/\partial y \\ \partial v/\partial x & \partial v/\partial y \end{pmatrix}^{-1}.$$

The presence of the absolute value $|\partial(x, y)/\partial(u, v)|$ guarantees that $f_{U,V}(u, v) \geq 0$.

In particular, the PDFs of the sum $X + Y$ and the product XY are calculated as

$$f_{X+Y}(u) = \int f_{X,Y}(x, u - x) dx = \int f_{X,Y}(u - y, y) dy \qquad (2.37)$$

and

$$f_{XY}(u) = \int f_{X,Y}(x, u/x) \frac{1}{|x|} dx = \int f_{X,Y}(u/y, y) \frac{1}{|y|} dy; \qquad (2.38)$$

cf. equations (1.21) and (1.22). For the ratio X/Y:

$$f_{X/Y}(u) = \int |y| f_{X,Y}(yu, y) dy. \qquad (2.39)$$

The derivation of formula (2.37) is as follows. If $U = X + Y$, then the corresponding change of variables is $u = x + y$ and, say $v = y$, with the inverse $x = u - v$, $y = v$. The Jacobian

$$\frac{\partial(u, v)}{\partial(x, y)} = 1 = \frac{\partial(x, y)}{\partial(u, v)},$$

hence

$$f_{X+Y,Y}(u, v) = f_{X,Y}(u - v, v).$$

Integrating in dv yields

$$f_{X+Y}(u) = \int f_{X,Y}(u - v, v) dv,$$

which is the last integral on the RHS of (2.37). The middle integral is obtained by using the change of variables $u = x + y$, $v = x$ (or simply by observing that X and Y are interchangeable).

The derivation of formulas (2.38) and (2.39) is similar, with $1/|x|$, $1/|y|$ and $|y|$ emerging as the absolute values of the corresponding (inverse) Jacobians.

For completeness, we produce a general formula, assuming that a map $u_1 = g_1(\mathbf{x}),\dots,$ $u_n = g_n(\mathbf{x})$ is one-to-one on the range of RVs $X_1, \dots, X_n$. Here, for the random vectors

$$\mathbf{X} = \begin{pmatrix} X_1 \\ \vdots \\ X_n \end{pmatrix} \text{ and } \mathbf{U} = \begin{pmatrix} U_1 \\ \vdots \\ U_n \end{pmatrix},'$$

with $U_i = g_i(\mathbf{X})$:

$$f_{\mathbf{U}}(u_1,\dots,u_n) = \left| \frac{\partial(x_1,\dots,x_n)}{\partial(u_1,\dots,u_n)} \right| f_{\mathbf{X}}(x_1(\mathbf{u}),\dots,x_n(\mathbf{u}))$$
$$\times I\left(\mathbf{u} \in \text{Range}(g_1,\dots,g_n)\right). \tag{2.40}$$

Example 2.6   Given a bivariate normal pair X, Y, consider the PDF of $X + Y$. From equations (2.35), (2.37):

$$f_{X+Y}(u) = \frac{1}{2\pi\sigma_1\sigma_2\sqrt{1-r^2}} \int \exp\left\{\frac{-1}{2(1-r^2)}\left[\frac{(x-\mu_1)^2}{\sigma_1^2}\right.\right.$$
$$\left.\left. -2r\frac{(x-\mu_1)(u-x-\mu_2)}{\sigma_1\sigma_2} + \frac{(u-x-\mu_2)^2}{\sigma_2^2}\right]\right\} dx$$
$$= \frac{1}{2\pi\sigma_1\sigma_2\sqrt{1-r^2}} \int \exp\left\{\frac{-1}{2(1-r^2)}\right.$$
$$\left. \times \left[\frac{x_1^2}{\sigma_1^2} - 2r\frac{x_1(u_1-x_1)}{\sigma_1\sigma_2} + \frac{(u_1-x_1)^2}{\sigma_2^2}\right]\right\} dx_1,$$

where $x_1 = x - \mu_1$, $u_1 = u - \mu_1 - \mu_2$.

Extracting the complete square yields

$$\frac{x_1^2}{\sigma_1^2} - 2r\frac{x_1(u_1-x_1)}{\sigma_1\sigma_2} + \frac{(u_1-x_1)^2}{\sigma_2^2}$$
$$= \left(x_1\frac{\sqrt{\sigma_1^2+2r\sigma_1\sigma_2+\sigma_2^2}}{\sigma_1\sigma_2} - \frac{u_1}{\sigma_2}\frac{\sigma_1+r\sigma_2}{\sqrt{\sigma_1^2+2r\sigma_1\sigma_2+\sigma_2^2}}\right)^2 + \frac{u_1^2(1-r^2)}{\sigma_1^2+2r\sigma_1\sigma_2+\sigma_2^2}.$$

We then obtain that

$$f_{X+Y}(u) = \frac{\exp\left[-\dfrac{u_1^2}{2(\sigma_1^2+2r\sigma_1\sigma_2+\sigma_2^2)}\right]}{2\pi\sqrt{\sigma_1^2+2r\sigma_1\sigma_2+\sigma_2^2}} \int e^{-v^2/2} dv$$
$$= \frac{\exp\left[-\dfrac{(u-\mu_1-\mu_2)^2}{2(\sigma_1^2+2r\sigma_1\sigma_2+\sigma_2^2)}\right]}{\sqrt{2\pi(\sigma_1^2+2r\sigma_1\sigma_2+\sigma_2^2)}}. \tag{2.41}$$

Here, the integration variable is

$$v = \frac{1}{\sqrt{1-r^2}} \left(x_1 \frac{\sqrt{\sigma_1^2 + 2r\sigma_1\sigma_2 + \sigma_2^2}}{\sigma_1\sigma_2} - \frac{u_1}{\sigma_2} \frac{\sigma_1 + r\sigma_2}{\sqrt{\sigma_1^2 + 2r\sigma_1\sigma_2 + \sigma_2^2}} \right).$$

We see that

$$X + Y \sim \mathrm{N}\left(\mu_1 + \mu_2, \sigma_1^2 + 2r\sigma_1\sigma_2 + \sigma_2^2\right),$$

with the mean value $\mu_1 + \mu_2$ and variance $\sigma_1^2 + 2r\sigma_1\sigma_2 + \sigma_2^2$.   ∎

Another useful example is

Example 2.7   Assume that RVs X_1 and X_2 are independent, and each has an exponential distribution with parameter λ. We want to find the joint PDF of

$$Y_1 = X_1 + X_2, \quad \text{and} \quad Y_2 = X_1/X_2,$$

and check if Y_1 and Y_2 are independent.
  Consider the map

$$T: (x_1, x_2) \mapsto (y_1, y_2), \quad \text{where } y_1 = x_1 + x_2, \ y_2 = \frac{x_1}{x_2},$$

where $x_1, x_2, y_1, y_2 \geq 0$. The inverse map T^{-1} acts by

$$T^{-1}: (y_1, y_2) \mapsto (x_1, x_2), \quad \text{where } x_1 = \frac{y_1 y_2}{1 + y_2}, \ x_2 = \frac{y_1}{1 + y_2},$$

and has the Jacobian

$$J(y_1, y_2) = \det \begin{pmatrix} y_2/(1+y_2) & y_1/(1+y_2) - y_1 y_2/(1+y_2)^2 \\ 1/(1+y_2) & -y_1/(1+y_2)^2 \end{pmatrix}$$

$$= -\frac{y_1 y_2}{(1+y_2)^3} - \frac{y_1}{(1+y_2)^3} = -\frac{y_1}{(1+y_2)^2}.$$

Then the joint PDF

$$f_{Y_1,Y_2}(y_1, y_2) = f_{X_1,X_2}\left(\frac{y_1 y_2}{1 + y_2}, \frac{y_1}{1 + y_2}\right) \left| -\frac{y_1}{(1+y_2)^2} \right|.$$

Substituting $f_{X_1,X_2}(x_1, x_2) = \lambda e^{-\lambda x_1} \lambda e^{-\lambda x_2}$, $x_1, x_2 \geq 0$, yields

$$f_{Y_1,Y_2}(y_1, y_2) = \left(\lambda^2 y_1 e^{-\lambda y_1}\right) \left[\frac{1}{(1+y_2)^2} \right], \quad y_1, y_2 \geq 0.$$

The marginal PDFs are

$$f_{Y_1}(y_1) = \int_0^\infty f_{Y_1,Y_2}(y_1, y_2) dy_2 = \lambda^2 y_1 e^{-\lambda y_1}, \quad y_1 \geq 0,$$

and

$$f_{Y_2}(y_2) = \int_0^\infty f_{Y_1,Y_2}(y_1, y_2) dy_1 = \frac{1}{(1+y_2)^2}, \quad y_2 \ge 0.$$

As $f_{Y_1,Y_2}(y_1, y_2) = f_{Y_1}(y_1) f_{Y_2}(y_2)$, RVs Y_1 and Y_2 are independent. ∎

The definition of the conditional probability $\mathbb{P}(A|B)$ does not differ from the discrete case: $\mathbb{P}(A|B) = \mathbb{P}(A \cap B)/\mathbb{P}(B)$. For example, if X is an exponential RV, then, $\forall \, y, w > 0$

$$\mathbb{P}(X \ge y + w | X \ge w) = \frac{\mathbb{P}(X \ge y + w)}{\mathbb{P}(X \ge w)} = \frac{\int_{y+w}^\infty \lambda e^{-\lambda u} du}{\int_w^\infty \lambda e^{-\lambda v} dv}$$

$$= \frac{e^{-\lambda(y+w)}}{e^{-\lambda w}} = e^{-\lambda y} = \mathbb{P}(X \ge y). \tag{2.42}$$

This is called the *memoryless property* of an exponential distribution which is similar to that of the geometric distribution (see equation (1.42)). It is not surprising that the exponential distribution arises as a limit of geometrics as $p = e^{-\lambda/n} \nearrow 1 \; (n \to \infty)$. Namely, if $X \sim \mathrm{Exp}\,(\lambda)$ and $X^{(n)} \sim \mathrm{Geom}\,(e^{-\lambda/n})$ then, $\forall \, y > 0$,

$$\left(1 - e^{-\lambda y}\right) = \mathbb{P}(X < y) = \lim_{n \to \infty} \mathbb{P}(X^{(n)} < ny). \tag{2.43}$$

Speaking of conditional probabilities (in a discrete or continuous setting), it is instructive to think of a *conditional probability distribution*. Consequently, in the continuous setting, we can speak of a conditional PDF. See below.

Of course, the formula of complete probability and Bayes' Theorem still hold true (not only for finite, but also for countable collections of pair-wise disjoint events B_j with $\mathbb{P}(B_j) > 0$).

Other remarkable facts are two Borel–Cantelli (BC) Lemmas, named after E. Borel (1871–1956), the famous French mathematician (and for 15 years the minister for the Navy), and F.P. Cantelli (1875–1966), an Italian mathematician (the founder of the Italian Institute of Actuaries). The first lemma is that if $B_1, B_2, \ldots$ is a sequence of (not necessarily disjoint) events with $\sum_j \mathbb{P}(B_j) < \infty$, then the probability $\mathbb{P}(A) = 0$, where A is the intersection $\bigcap_{n \ge 1} \bigcup_{j \ge n} B_j$. The proof is straightforward if you are well versed in basic manipulations with probabilities: if $A_n = \bigcup_{j \ge n} B_j$ then $A_{n+1} \subseteq A_n$ and $A = \bigcap_n A_n$. Then $A \subseteq A_n$ and hence $\mathbb{P}(A) \le \mathbb{P}(A_n) \; \forall \, n$. But $\mathbb{P}(A_n) \le \sum_{j \ge n} \mathbb{P}(B_j)$ which tends to 0 as $n \to \infty$ because $\sum_j \mathbb{P}(B_j) < \infty$. So, $\mathbb{P}(A) = 0$.

The first BC Lemma has a rather striking interpretation: if $\sum_j \mathbb{P}(B_j) < \infty$, then with probability 1 only finitely many of events B_j can occur at the same time. This is because the above intersection A has the meaning that 'infinitely many of events B_j occurred'. Formally, if outcome $\omega \in A$, then $\omega \in B_j$ for infinitely many j.

The second BC Lemma says that if events $B_1, B_2, \ldots$, are independent and $\sum_j \mathbb{P}(B_j) = \infty$, then $\mathbb{P}(A) = 1$. The proof is again straightforward: $\mathbb{P}(A^c) = \mathbb{P}(\bigcup_{n \ge 1} A_n^c) \le \sum_{n \ge 1} \mathbb{P}(A_n^c)$. Next, one argues that $\mathbb{P}(A_n^c) = \mathbb{P}\left(\bigcap_{j \ge n} B_j^c\right) = \prod_{j \ge n} \mathbb{P}(B_j^c) = \exp\left\{\sum_{j \ge n} \ln\left[1 - \mathbb{P}(B_j)\right]\right\} \le \exp\left[-\sum_{j \ge n} \mathbb{P}(B_j)\right]$, as $\ln(1 - x) \le -x$ for $x \ge 0$. As $\sum_{j \ge n} \mathbb{P}(B_j) = \infty$, $\mathbb{P}(A_n^c) = 0 \; \forall \, n$. Then $\mathbb{P}(A^c) = 0$ and $\mathbb{P}(A) = 1$.

Thus, if $B_1, B_2, \ldots$ are independent events and $\sum_j \mathbb{P}(B_j) = \infty$, then 'with probability 1 there occur infinitely many of them'.

For example, in Problem 1.35 the events {year k is a record} are independent and have probabilities $1/k$. By the second BC Lemma, with probability 1 there will be infinitely many record years if observations are continued indefinitely.

In the continuous case we can also work with conditional probabilities of the type $\mathbb{P}(A|X=x)$, under the condition that an RV X with PDF f_X takes value x. We know that such a condition has probability 0, and yet all calculations, while being performed correctly, yield right answers. The trick is that we consider not a single value x but all of them, within the range of RV X. Formally speaking, we work with a continuous analogue of the formula of complete probability:

$$\mathbb{P}(A) = \int \mathbb{P}(A|X=x)f_X(x)\mathrm{d}x. \tag{2.44}$$

This formula holds provided that $\mathbb{P}(A|X=x)$ is defined in a 'sensible' way which is usually possible in all naturally occurring situations. For example, if $f_{X,Y}(x, y)$ is the joint PDF of RVs X and Y and A is the event $\{a < Y < b\}$, then

$$\mathbb{P}(A|X=x) = \int_a^b f_{X,Y}(x, y)\mathrm{d}y \Big/ \int f_{X,Y}(x, y)\mathrm{d}y.$$

The next step is then to introduce the *conditional PDF* $f_{Y|X}(y|x)$ of RV Y, conditional on $\{X=x\}$:

$$f_{Y|X}(y|x) = \frac{f_{X,Y}(x, y)}{f_X(x)}, \tag{2.45}$$

and write natural analogues of the formula of complete probability and the Bayes formula:

$$f_Y(y) = \int f_{Y|X}(y|x)f_X(x)\mathrm{d}x, \quad f_{Y|X}(y|x) = f_{X,Y} \Big/ \int f_{X|Y}(x|z)f_Y(z)\mathrm{d}z. \tag{2.46}$$

As in the discrete case, two events, A and B are called independent if

$$\mathbb{P}(A \cap B) = \mathbb{P}(A)\mathbb{P}(B); \tag{2.47}$$

for a general finite collection $A_1, \ldots, A_n$, $n > 2$, we require that $\forall\ k = 2, \ldots, n$ and $1 \le i_1 < \ldots < i_k \le n$:

$$\mathbb{P}\left(\cap_{1 \le j \le k} A_{i_j}\right) = \prod_{1 \le j \le k} \mathbb{P}\left(A_{i_j}\right). \tag{2.48}$$

Similarly, RVs X and Y, on the same outcome space Ω, are called *independent* if

$$\mathbb{P}(X < x, Y < y) = \mathbb{P}(X < x)\mathbb{P}(Y < y)\ \forall\, x, y \in \mathbb{R}. \tag{2.49}$$

In other words, the joint CDF $F_{X,Y}(x, y)$ decomposes into the product $F_X(x)F_Y(y)$. Finally, a collection of RVs $X_1, \ldots, X_n$ (again on the same space Ω), $n > 2$, is called independent if $\forall\ k = 2, \ldots, n$ and $1 \le i_1 < \cdots < i_k \le n$:

$$\mathbb{P}(X_{i_1} < y_1, \ldots, X_{i_k} < y_k) = \prod_{1 \le j \le k} \mathbb{P}(X_{i_j} < y_j),\ y_1, \ldots, y_k \in \mathbb{R}. \tag{2.50}$$

Remark   We can require equation (2.50) for the whole collection $X_1, \ldots, X_n$ only, but allowing y_i to take value $+\infty$:

$$\mathbb{P}(X_1 < y_1, \ldots, X_n < y_n) = \prod_{1 \le j \le n} \mathbb{P}(X_j < y_j), \quad y_1, \ldots, y_n \in \overline{\mathbb{R}} = \mathbb{R} \cup \{\infty\}. \quad (2.51)$$

Indeed, if some values y_i in equation (2.51) is equal to ∞, it means that the corresponding condition $y_i < \infty$ is trivially fulfilled and can be omitted.

If RVs under consideration have a joint PDF $f_{X,Y}$ or $f_{\mathbf{X}}$, then equations (2.50) and (2.51) are equivalent to its product-decomposition:

$$f_{X,Y}(x, y) = f_X(x) f_Y(y), \quad f_{\mathbf{X}}(\mathbf{x}) = f_{X_1}(x_1) \ldots f_{X_n}(x_n), \quad (2.52)$$

The formal proof of this fact is as follows. The decomposition $F_{X,Y}(y_1, y_2) = F_X(y_1) F_Y(y_2)$ means that $\forall \; y_1, y_2 \in \mathbb{R}$:

$$\int_{-\infty}^{y_1} \int_{-\infty}^{y_2} f_{X,Y}(x, y) \, dy \, dx = \int_{-\infty}^{y_1} \int_{-\infty}^{y_2} f_X(x) f_Y(y) \, dy \, dx.$$

One deduces then that the integrands must coincide, i.e. $f_{X,Y}(x, y) = f_X(x) f_Y(y)$. The inverse implication is straightforward. The argument for a general collection $X_1, \ldots, X_n$ is analogous.

For independent RVs X, Y, equations (2.37)–(2.39) become

$$f_{X+Y}(u) = \int f_X(x) f_Y(u - x) \, dx = \int f_X(u - y) f_Y(y) \, dy, \quad (2.53)$$

$$f_{XY}(u) = \int f_X(x) f_Y(u/x) \frac{1}{|x|} \, dx = \int f_X(u/y) f_Y(y) \frac{1}{|y|} \, dy. \quad (2.54)$$

and

$$f_{X/Y}(u) = \int |y| f_X(yu) f_Y(y) \, dy. \quad (2.55)$$

Cf. Problem 2.13. As in the discrete case, equation (2.53) is called the convolution formula (for densities).

The concept of IID RVs employed in the previous chapter will continue to play a prominent rôle, particularly in Section 2.3.

Problem 2.8   Random variables X and Y are independent and exponentially distributed with parameters λ and μ. Set

$$U = \max [X, Y], \quad V = \min [X, Y].$$

Are U and V independent? Is RV U independent of the event $\{X > Y\}$ (i.e. of the RV $I(X > Y)$)?

Hint: Check that $\mathbb{P}(V > y_1, \; U < y_2) \ne \mathbb{P}(V > y_1) \mathbb{P}(U < y_2)$ and $\mathbb{P}(X > Y, U < y) \ne \mathbb{P}(X > Y) P(U < y)$.

Problem 2.9   Random variables X and Y are independent and exponentially distributed, each with parameter λ. Show that the random variables $X + Y$ and $X/(X + Y)$ are independent and find their distributions.

   Hint: Check that

$$f_{U,V}(u, v) = \lambda^2 u e^{-\lambda u} I(u > 0) I(0 < v < 1) = f_U(u) f_V(v).$$

Problem 2.10   A shot is fired at a circular target. The vertical and horizontal co-ordinates of the point of impact (taking the centre of the target as origin) are independent random variables, each distributed N(0, 1).

   Show that the distance of the point of impact from the centre has the PDF $re^{-r^2/2}$ for $r > 0$. Find the median of this distribution.

Solution   In fact, $X \sim N(0, 1)$, $Y \sim N(0, 1)$, and $f_{X,Y}(x, y) = (1/2\pi)e^{-(x^2+y^2)/2}$, $x, y \in \mathbb{R}$. Set $R = \sqrt{X^2 + Y^2}$ and $\Theta = \tan^{-1}(Y/X)$. The range of the map

$$\begin{pmatrix} x \\ y \end{pmatrix} \mapsto \begin{pmatrix} r \\ \theta \end{pmatrix}$$

is $\{r > 0, \ \theta \in (-\pi, \pi)\}$ and the Jacobian

$$\frac{\partial(r, \theta)}{\partial(x, y)} = \det \begin{pmatrix} \partial r/\partial x & \partial r/\partial y \\ \partial \theta/\partial x & \partial \theta/\partial y \end{pmatrix}$$

equals

$$\det \begin{pmatrix} \dfrac{x}{r} & \dfrac{y}{r} \\[2mm] \dfrac{-y/x^2}{1 + (y/x)^2} & \dfrac{1/x}{1 + (y/x)^2} \end{pmatrix} = \frac{1}{r}, \ x, y \in \mathbb{R}, \ r > 0, \ \theta \in (-\pi, \pi).$$

Then the inverse map

$$\begin{pmatrix} r \\ \theta \end{pmatrix} \mapsto \begin{pmatrix} x \\ y \end{pmatrix}$$

has the Jacobian $\partial(x, y)/\partial(r, \theta) = r$. Hence,

$$f_{R,\Theta}(r, \theta) = re^{-r^2/2} I(r > 0) \frac{1}{2\pi} I(-\pi, \pi).$$

Integrating in $d\theta$ yields

$$f_R(r) = re^{-r^2/2} I(r > 0),$$

as required. To find the median $m(R)$, consider the equation

$$\int_0^y re^{-r^2/2} dr = \int_y^\infty re^{-r^2/2} dr,$$

giving $e^{-y^2/2} = 1/2$. So, $m(R) = \sqrt{\ln 4}$.   □

We conclude this section's theoretical considerations with the following remark. Let X be an RV with CDF F and $0 < y < 1$ be a value taken by F, with $F(x^*) = y$ at the left point $x^* = \inf[x \in \mathbb{R} : F(x) \geq y]$. Then

$$\mathbb{P}(F(X) < y) = y. \tag{2.56}$$

In fact, in this case the event $\{F(X) < y\} = \{X < x^*\}$, implying $\mathbb{P}(F(X) < y) = F(x^*)$. In general, if $g : \mathbb{R} \to \mathbb{R}$ is another function and there is a unique point $a \in \mathbb{R}$ such that $F(a) = g(a)$, $F(x) < g(x)$ for $x < a$ and $F(x) \geq g(x)$ for $x \geq a$, then

$$\mathbb{P}(F(X) < g(X)) = g(a). \tag{2.57}$$

Problem 2.11   A random sample $X_1, \ldots, X_{2n+1}$ is taken from a distribution with PDF f. Let $Y_1, \ldots, Y_{2n+1}$ be values of $X_1, \ldots, X_{2n+1}$ arranged in increasing order. Find the distribution of each of Y_k, $k = 1, \ldots, 2n+1$.

Solution   If $X_j \sim f$, then

$$P_{Y_k}(y) = P(Y_k < y) = \sum_{j=k}^{2n+1} \binom{2n+1}{j} F(y)^j [1 - F(y)]^{2n+1-j}, \quad y \in \mathbf{R}.$$

The PDF f_{Y_k}, $k = 1, \ldots, 2n+1$, can be obtained as follows. Y_k takes value x, iff $k-1$ values among $X_1, \ldots, X_{2n+1}$ are less than x, $2n+1-k$ are greater than x, and the remaining one is equal to x. Hence

$$f_{Y_k}(x) = \frac{(2n+1)!}{(k-1)!(2n+1-k)!} [F(x)]^{k-1} [1 - F(x)]^{2n+1-k} f(x).$$

In particular, if $X_i \sim U[0, 1]$, the PDF of the *the sample median* Y_{n+1} is

$$f_{Y_{n+1}}(x) = \frac{(2n+1)!}{n!n!} [x(1-x)]^n. \quad \square$$

Problem 2.12   (i) X and Y are independent RVs, with continuous symmetric distributions, with PDFs f_X and f_Y respectively. Show that the PDF of $Z = X/Y$ is

$$h(a) = 2 \int_0^\infty y f_X(ay) f_Y(y) dy.$$

(ii) X and Y are independent normal random variables distributed $N(0, \sigma^2)$ and $N(0, \tau^2)$. Show that $Z = X/Y$ has PDF $h(a) = d/[\pi(d^2 + a^2)]$, where $d = \sigma/\tau$ (the PDF of the Cauchy distribution).

Hint:

$$F_Z(a) = 2 \int_0^\infty \int_{-\infty}^{ay} f_X(x) f_Y(y) dx dy.$$

Problem 2.13 Let X and Y be independent RVs with respective PDFs f_X and f_Y. Show that $Z = Y/X$ has the PDF

$$h(z) = \int f_X(x) f_Y(zx) |x| dx.$$

Deduce that $T = \tan^{-1}(Y/X)$ is uniformly distributed on $(-\pi/2, \pi/2)$ if and only if

$$\int f_X(x) f_Y(xz) |x| dx = \frac{1}{\pi(1 + z^2)}$$

for $z \in \mathbb{R}$. Verify that this holds if X and Y both have the normal distribution with mean 0 and non-zero variance σ^2.

Solution The distribution function of RV Z is

$$F_Z(u) = \mathbb{P}(Z < u) = \mathbb{P}(Y/X < u, X > 0) + \mathbb{P}(Y/X < u, X < 0)$$

$$= \int_0^\infty f_X(x) \int_{-\infty}^{ux} f_Y(y) dy dx + \int_{-\infty}^0 f_X(x) \int_{ux}^\infty f_Y(y) dy dx.$$

Then the PDF

$$f_Z(u) = \frac{d}{du} F_Z(u) = \int_0^\infty f_X(x) f_Y(ux) x dx - \int_{-\infty}^0 f_X(x) f_Y(ux) x dx$$

$$= \int f_X(x) f_Y(ux) |x| dx,$$

which agrees with the formula obtained via the Jacobian.

If T is uniformly distributed, then

$$F_Z(u) = \mathbb{P}(\tan^{-1} Z \le \tan^{-1} u) = \frac{\tan^{-1} u + \pi/2}{\pi},$$

$$f_Z(u) = \frac{d}{du} F_Z(u) = \frac{1}{\pi(u^2 + 1)}.$$

Conversely,

$$f_Z(u) = \frac{1}{\pi(1 + u^2)} \text{ implying } F_Z(u) = \frac{1}{\pi} \tan^{-1} u + \frac{1}{2}.$$

We deduce that

$$\mathbb{P}(T \le u) = \frac{1}{\pi} u + \frac{1}{2}$$

or $f_T(u) = 1/\pi$ on $(-\pi/2, \pi/2)$.
Finally, take

$$f_X(x) = \frac{1}{\sqrt{2\pi}\sigma} e^{-x^2/2\sigma^2}, \quad f_Y(y) = \frac{1}{\sqrt{2\pi}\sigma} e^{-y^2/2\sigma^2}.$$

Then

$$
\int f_X(x)f_Y(ux)|x|dx = \frac{1}{\pi\sigma^2}\int\limits_0^\infty e^{-x^2/(2\sigma^2)-x^2u^2/(2\sigma^2)}xdx
$$

$$
= \frac{1}{\pi\sigma^2}\left(\frac{-\sigma^2}{u^2+1}e^{-x^2(1+u^2)/2\sigma^2}\right)\bigg|_0^\infty
$$

$$
= \frac{1}{\pi(1+u^2)}. \quad \square
$$

Problem 2.14 Let $X_1, X_2, \ldots$ be independent Cauchy random variables, each with PDF

$$
f(x) = \frac{d}{\pi(d^2+x^2)}.
$$

Show that $A_n = (X_1 + X_2 + \cdots + X_n)/n$ has the same distribution as X_1.

Solution For $d=1$ and $n=2$, $f(x) = 1/[\pi(1+x^2)]$. Then $A_2 = (X_1+X_2)/2$ has the CDF $F_{A_2}(x) = \mathbb{P}(X_1+X_2 < 2x)$, with the PDF

$$
g(x) = 2\tilde{g}(2x), \quad \text{where } \tilde{g}(y) = \frac{1}{\pi^2}\int \frac{1}{1+u^2}\frac{du}{1+(y-u)^2}.
$$

Now we use the identity

$$
m\int \frac{1}{(1+y^2)(m^2+(x-y)^2)}dy = \frac{\pi(m+1)}{(m+1)^2+x^2},
$$

which is a simple but tedious exercise (it becomes straightforward if you use complex integration). This yields

$$
\tilde{g}(y) = \frac{2}{\pi}\frac{1}{4+y^2}
$$

which implies

$$
g(x) = 2\tilde{g}(2x) = \frac{1}{\pi}\frac{1}{1+x^2} = f(x).
$$

In general, if $Y = qX_1 + (1-q)X_2$, then its PDF is

$$
\frac{d}{dx}\int f(u)\int_{-\infty}^{(x-qu)/(1-q)} f(y)dydu = \frac{1}{1-q}\int_{\mathbb{R}^1} f\left(\frac{x-qu}{1-q}\right)f(u)du,
$$

which, by the same identity (with $m=(1-q)/q$), is

$$
= \frac{1}{q}\frac{1}{\pi}\left(\frac{1-q}{q}+1\right)\frac{1}{1/q^2+x^2/q^2} = f(x).
$$

Now, for $d = 1$ and a general n we can write $S_n = [(n-1)S_n + X_n]/n$. Make the induction hypothesis: $S_{n-1} \sim f(x)$. Then by the above argument (with two summands, S_{n-1} and X_n), $S_n \sim f(x)$.

Finally, for a general d, we set $\widetilde{X}_i = X_i/d$. Then $\widetilde{S}_n \sim \widetilde{f}(x) = 1/[\pi(1+x^2)]$. Hence, for the original variables, $S_n \sim f(x) = d/[\pi(d^2 + x^2)]$.   □

Remark   A shorter solution uses characteristic functions, see equation (2.91). For the proof, see Problem 2.52.

Problem 2.15   If X, Y and Z are independent RVs each uniformly distributed on $(0, 1)$, show that $(XY)^Z$ is also uniformly distributed on $[0, 1]$.

Solution   Take

$$\ln[(XY)^Z] = Z(\ln X + \ln Y).$$

To prove that $(XY)^Z$ is uniformly distributed on $[0, 1]$ is the same as to prove that $-Z(\ln X + \ln Y)$ is exponentially distributed on $[0, \infty)$. Now $W = -\ln X - \ln Y$ has the PDF

$$f_W(y) = \begin{cases} ye^{-y}, & 0 < y < \infty, \\ 0, & y \le 0. \end{cases}$$

The joint PDF $f_{Z,W}(x, y)$ is of the form

$$f_{Z,W}(x, y) = \begin{cases} ye^{-y}, & \text{if } 0 < x < 1,\ 0 < y < \infty, \\ 0, & \text{otherwise.} \end{cases}$$

We are interested in the product ZW. It is convenient to pass from x, y to variables $u = xy$, $v = y/x$ with the inverse Jacobian $1/(2v)$. In the new variables the joint PDF $f_{U,V}$ of RVs $U = ZW$ and $V = W/Z$ reads

$$f_{U,V}(u, v) = \begin{cases} \frac{1}{2v}(uv)^{1/2}e^{-(uv)^{1/2}}, & u, v > 0,\ 0 < u/v < 1, \\ 0, \text{otherwise.} \end{cases}$$

The PDF of U then is

$$f_U(u) = \int f_{U,V}(u, v)\mathrm{d}v = \int_u^\infty \frac{1}{2v}(uv)^{1/2}e^{-(uv)^{1/2}}\mathrm{d}v$$

$$= -\int_u^\infty \mathrm{d}\left(e^{-(uv)^{1/2}}\right) = -\left[e^{-(uv)^{1/2}}\right]_u^\infty = e^{-u}, \quad u > 0,$$

with $f_U(u) = 0$ for $u < 0$.   □

Problem 2.16 Let RVs $X_1, \ldots, X_n$ be independent and exponentially distributed, with the same parameter λ. By using induction on n or otherwise, show that the RVs

$$\max\,[X_1, \ldots, X_n] \text{ and } X_1 + \frac{1}{2}X_2 + \cdots + \frac{1}{n}X_n$$

have the same distribution.

Solution Write

$$Y_n = \max[X_1, \ldots, X_n], \ \ Z_n = X_1 + \frac{1}{2}X_2 + \cdots + \frac{1}{n}X_n.$$

Now write

$$\mathbb{P}(Y_n < y) = (\mathbb{P}(X_1 < y))^n = (1 - e^{-\lambda y})^n.$$

For $n = 1$, $Y_1 = Z_1$. Now use induction in n:

$$\mathbb{P}(Z_n < y) = \mathbb{P}\left(Z_{n-1} + \frac{1}{n}X_n < y\right).$$

As $X_n/n \sim \mathrm{Exp}\,(n\lambda)$, the last probability equals

$$\int\limits_0^y (1 - e^{-\lambda z})^{n-1} n\lambda e^{-n\lambda(y-z)}\mathrm{d}z$$

$$= n\lambda e^{-n\lambda y} \int\limits_0^y e^{\lambda z}(1 - e^{-\lambda z})^{n-1} e^{(n-1)\lambda z}\mathrm{d}z = ne^{-n\lambda y} \int\limits_1^{e^{\lambda y}} (u - 1)^{n-1}\mathrm{d}u$$

$$= ne^{-n\lambda y} \int\limits_0^{e^{\lambda y}-1} v^{n-1}\mathrm{d}v = e^{-n\lambda y}(e^{\lambda y} - 1)^n = (1 - e^{-\lambda y})^n. \quad \square$$

Problem 2.17 Suppose $\alpha \geq 1$ and X_α is a positive real-valued RV with PDF

$$f_\alpha(t) = A_\alpha t^{\alpha-1} \exp\,(-t^\alpha)$$

for $t > 0$, where A_α is a constant. Find A_α and show that, if $\alpha > 1$ and $s, t > 0$,

$$\mathbb{P}(X_\alpha \geq s + t | X_\alpha \geq t) < \mathbb{P}(X_\alpha \geq s).$$

What is the corresponding relation for $\alpha = 1$?

Solution We must have $\int_0^\infty f_\alpha(t)\mathrm{d}t = 1$, so

$$A_\alpha^{-1} = \int\limits_0^\infty t^{\alpha-1} \exp\,(-t^\alpha)\mathrm{d}t$$

$$= \alpha^{-1} \int\limits_0^\infty \exp\,(-t^\alpha)\mathrm{d}(t^\alpha) = \alpha^{-1}\left[e^{-t^\alpha}\right]_0^\infty = \alpha^{-1},$$

and $A_\alpha = \alpha$. If $\alpha > 1$, then, $\forall\, s, t > 0$,

$$\mathbb{P}(X_\alpha \geq s + t | X_\alpha \geq t) = \frac{\mathbb{P}(X_\alpha \geq s + t)}{\mathbb{P}(X_\alpha \geq t)} = \frac{\int_{s+t}^\infty \exp(-u^\alpha)\mathrm{d}(u^\alpha)}{\int_t^\infty \exp(-u^\alpha)\mathrm{d}(u^\alpha)}$$

$$= \frac{\exp(-(s+t)^\alpha)}{\exp(-t^\alpha)} = \exp(t^\alpha - (s+t)^\alpha)$$

$$= \exp(-s^\alpha + \text{negative terms})$$

$$< \exp(-s^\alpha) = \mathbb{P}(X_\alpha \geq s).$$

If $\alpha = 1$, $t^\alpha - (s + t)^\alpha = -s$, and the above inequality becomes an equality as $\mathbb{P}(X_\alpha \geq t) = \exp(-t)$. (This is the memoryless property of the exponential distribution.) $\quad\square$

Remark   If you interpret X_α as the lifetime of a certain device (e.g. a bulb), then the inequality $\mathbb{P}(X_\alpha \geq s + t | X_\alpha \geq t) < \mathbb{P}(X_\alpha \geq s)$ emphasises an 'aging' phenomenon, where an old device that had been in use for time t is less likely to serve for duration s than a new one. There are examples where the inequality is reversed: the quality of a device (or an individual) improves in the course of service.

Problem 2.18   Let a point in the plane have Cartesian co-ordinates (X, Y) and polar co-ordinates (R, Θ). If X and Y are independent identically distributed RVs each having a normal distribution with mean zero, show that R^2 has an exponential distribution and is independent of Θ.

Solution   The joint PDF $f_{X,Y}$ is $(1/2\pi\sigma^2)\mathrm{e}^{-(x^2+y^2)/2\sigma^2}$, and R and Θ are defined by

$$R^2 = T = X^2 + Y^2, \quad \Theta = \tan^{-1}\frac{Y}{X}.$$

Then

$$f_{R^2,\Theta}(t, \theta) = f_{X,Y}\big(x(t, \theta), y(t, \theta)\big) I(t > 0) I(0 < \theta < 2\pi) \left| \frac{\partial(x, y)}{\partial(t, \theta)} \right|,$$

where the inverse Jacobian

$$\frac{\partial(x, y)}{\partial(t, \theta)} = \left(\frac{\partial(t, \theta)}{\partial(x, y)}\right)^{-1} = \det\left[\begin{pmatrix} 2x & 2y \\ -y/(x^2+y^2) & x/(x^2+y^2) \end{pmatrix}\right]^{-1}$$

$$= \left(\frac{2x^2 + 2y^2}{x^2 + y^2}\right)^{-1} = \frac{1}{2}.$$

Hence,

$$f_{R^2,\Theta}(t, \theta) = \frac{1}{4\pi\sigma^2}\mathrm{e}^{-t/2\sigma^2} I(t > 0) I(0 < \theta < 2\pi) = f_{R^2}(t) f_\Theta(\theta),$$

with

$$f_{R^2}(t) = \frac{1}{2\sigma^2}\mathrm{e}^{-t/2\sigma^2} I(t > 0), \quad f_\Theta(\theta) = \frac{1}{2\pi}I(0 < \theta < 2\pi).$$

Thus, $R^2 \sim \mathrm{Exp}(1/2\sigma^2)$, $\Theta \sim \mathrm{U}(0, 2\pi)$ and R^2 and Θ are independent. $\quad\square$

2.2   Expectation, conditional expectation, variance, generating function, characteristic function

Tales of the Expected Value
(From the series *'Movies that never made it to the Big Screen'*.)

They usually have difficult or threatening names such as
Bernoulli, De Moivre–Laplace, Chebyshev, Poisson.
Where are the probabilists with names such as Smith,
Brown, or Johnson?
(From the series *'Why they are misunderstood'*.)

The *mean* and the *variance* of an RV X with PDF f_X are calculated similarly to the discrete-value case. Namely,

$$\mathbb{E}X = \int x f_X(x) dx \quad \text{and} \quad \operatorname{Var} X = \int (x - \mathbb{E}X)^2 f_X(x) dx. \tag{2.58}$$

Clearly,

$$\operatorname{Var} X = \int (x^2 - 2x\mathbb{E}X + (\mathbb{E}X)^2) f_X(x) dx$$

$$= \int x^2 f_X(x) dx - 2\mathbb{E}X \int x f_X(x) dx + (\mathbb{E}X)^2 \int f_X(x) dx$$

$$= \mathbb{E}X^2 - 2(\mathbb{E}X)^2 + (\mathbb{E}X)^2 = \mathbb{E}X^2 - (\mathbb{E}X)^2. \tag{2.59}$$

Of course, these formulas make sense when the integrals exist; in the case of the expectation we require that

$$\int |x| f_X(x) dx < \infty.$$

Otherwise, i.e. if $\int |x| f_X(x) dx = \infty$, it is said that X does not have a finite expectation (or $\mathbb{E}X$ does not exist). Similarly, with $\operatorname{Var} X$.

Remark   Sometimes one allows a further classification, regarding the contributions to integral $\int x f_X(x) dx$ from $\int_0^\infty$ and $\int_{-\infty}^0$. Indeed, $\int_0^\infty x f_X(x) dx$ corresponds to $\mathbb{E}X_+$ while $\int_{-\infty}^0 (-x) f_X(x) dx$ to $\mathbb{E}X_-$, where $X_+ = \max[0, X]$ and $X_- = -\min[0, X]$. Dealing with integrals $\int_0^\infty$ and $\int_{-\infty}^0$ is simpler as value x keeps its sign on each of the intervals $(0, \infty)$ and $(-\infty, 0)$. Then one says that $\mathbb{E}X = \infty$ if $\int_0^\infty x f_X(x) dx = \infty$ and $\int_{-\infty}^0 (-x) f_X(x) dx < \infty$. Similarly, $\mathbb{E}X = -\infty$ if $\int_0^\infty x f_X(x) dx < \infty$ but $\int_{-\infty}^0 (-x) f_X(x) dx = \infty$.

Formulas (2.58) and (2.59) are in agreement with the standard representation of RVs, where $X(x) = x$ (or $X(\omega) = \omega$). Indeed, we can say that in $\mathbb{E}X$ we integrate values $X(x)$ and in $\operatorname{Var} X$ values $(X(x) - \mathbb{E}X)^2$ against the PDF $f_X(x)$, which makes strong analogies with the discrete case formulas $\mathbb{E}X = \sum_\omega X(\omega) p_X(\omega)$ and $\operatorname{Var} X = \sum_\omega (X(\omega) - \mathbb{E}X)^2 p(\omega)$.

Example 2.8 For $X \sim U(a, b)$, the mean and the variance are

$$\mathbb{E}X = \frac{b+a}{2}, \quad \text{Var } X = \frac{(b-a)^2}{12}. \tag{2.60}$$

In fact,

$$\mathbb{E}X = \frac{1}{b-a} \int_a^b x \, dx = \frac{1}{b-a} \left. \left(\frac{x^2}{2} \right) \right|_a^b = \frac{1}{2} \frac{b^2 - a^2}{b-a} = \frac{b+a}{2},$$

the middle point of (a, b). Further,

$$\frac{1}{b-a} \int_a^b x^2 \, dx = \frac{b^3 - a^3}{3(b-a)} = \frac{1}{3}(b^2 + ab + a^2).$$

Hence, Var $X = \frac{1}{3}(b^2 + ab + a^2) - \frac{1}{4}(b+a)^2 = \frac{1}{12}(b-a)^2.$ ∎

Example 2.9 For $X \sim N(\mu, \sigma^2)$:

$$\mathbb{E}X = \mu, \quad \text{Var } X = \sigma^2. \tag{2.61}$$

In fact,

$$\mathbb{E}X = \frac{1}{\sqrt{2\pi}\sigma} \int x \exp\left[-\frac{(x-\mu)^2}{2\sigma^2} \right] dx$$

$$= \frac{1}{\sqrt{2\pi}\sigma} \int (x - \mu + \mu) \exp\left[-\frac{(x-\mu)^2}{2\sigma^2} \right] dx$$

$$= \frac{1}{\sqrt{2\pi}\sigma} \left[\int x \exp\left(-\frac{x^2}{2\sigma^2} \right) dx + \mu \int \exp\left(-\frac{x^2}{2\sigma^2} \right) dx \right]$$

$$= 0 + \frac{\mu}{\sqrt{2\pi}\sigma} \int \exp\left(-\frac{x^2}{2} \right) dx = \mu,$$

and

$$\text{Var } X = \frac{1}{\sqrt{2\pi}\sigma} \int (x - \mu)^2 \exp\left[-\frac{(x-\mu)^2}{2\sigma^2} \right] dx$$

$$= \frac{1}{\sqrt{2\pi}\sigma} \int x^2 \exp\left(-\frac{x^2}{2\sigma^2} \right) dx$$

$$= \sigma^2 \frac{1}{\sqrt{2\pi}} \int x^2 \exp\left(-\frac{x^2}{2} \right) dx = \sigma^2. \quad ∎$$

Example 2.10 For $X \sim \text{Exp}(\lambda)$,

$$\mathbb{E}X = \frac{1}{\lambda}, \quad \text{Var } X = \frac{1}{\lambda^2}. \tag{2.62}$$

In fact,

$$\mathbb{E}X = \lambda \int_0^\infty x e^{-\lambda x} \, dx = \frac{1}{\lambda} \int_0^\infty x e^{-x} \, dx = \frac{1}{\lambda},$$

and

$$\lambda \int_0^\infty x^2 e^{-\lambda x} dx = \frac{1}{\lambda^2} \int_0^\infty x^2 e^{-x} dx = \frac{2}{\lambda^2},$$

implying that $\mathrm{Var}\, X = 2/\lambda^2 - 1/\lambda^2 = 1/\lambda^2$. ∎

Example 2.11   For $X \sim \mathrm{Gam}(\alpha, \lambda)$,

$$\mathbb{E}X = \frac{\alpha}{\lambda}, \quad \mathrm{Var}\, X = \frac{\alpha}{\lambda^2}. \tag{2.63}$$

In fact,

$$\mathbb{E}X = \frac{\lambda^\alpha}{\Gamma(\alpha)} \int_0^\infty x x^{\alpha-1} e^{-\lambda x} dx = \frac{1}{\lambda \Gamma(\alpha)} \int_0^\infty x^\alpha e^{-x} dx$$

$$= \frac{\Gamma(\alpha+1)}{\lambda \Gamma(\alpha)} = \frac{\alpha}{\lambda}.$$

Next,

$$\frac{\lambda^\alpha}{\Gamma(\alpha)} \int_0^\infty x^2 x^{\alpha-1} e^{-\lambda x} dx = \frac{1}{\lambda^2 \Gamma(\alpha)} \int_0^\infty x^{\alpha+1} e^{-x} dx$$

$$= \frac{\Gamma(\alpha+2)}{\lambda^2 \Gamma(\alpha)} = \frac{(\alpha+1)\alpha}{\lambda^2}.$$

This gives $\mathrm{Var}\, X = (\alpha+1)\alpha/\lambda^2 - \alpha^2/\lambda^2 = \alpha/\lambda^2$. ∎

Example 2.12   Finally, for $X \sim \mathrm{Ca}(\alpha, \tau)$, the integral

$$\int \frac{\tau|x|}{\tau^2 + (x-\alpha)^2} dx = \infty, \tag{2.64}$$

which means that $\mathbb{E}X$ does not exist (let alone $\mathrm{Var}\, X$). ∎

A table of several probability distributions is given in Appendix 1.

In a general situation where the distribution of X has a discrete and an absolutely continuous component, formulas (2.58), (2.59) have to be modified. For example, for the RV W from equation (2.26):

$$\mathbb{E}W = [0 \cdot \mathbb{P}(W=0)] + \frac{\lambda}{\mu} \int_0^\infty x e^{-(\mu-\lambda)x} dx = \frac{\lambda}{\mu(\mu-\lambda)^2}.$$

An important property of expectation is additivity:

$$\mathbb{E}(X+Y) = \mathbb{E}X + \mathbb{E}Y. \tag{2.65}$$

To check this fact, we use the joint PDF $f_{X,Y}$:

$$\mathbb{E}X + \mathbb{E}Y = \int x f_X(x)dx + \int y f_Y(y)dy$$

$$= \int x \int f_{X,Y}(x, y)dy dx + \int y \int f_{X,Y}(x, y)dx dy$$

$$= \int \int (x + y) f_{X,Y}(x, y)dy dx = \mathbb{E}(X + Y).$$

As in the discrete-value case, we want to stress that formula (2.65) is a completely general property that holds for any RVs X, Y. It has been derived here when X and Y have a joint PDF (and in Section 1.4 for discrete RVs), but the additivity holds in the general situation, regardless of whether X and Y are discrete or have marginal or joint PDFs. The proof in its full generality requires Lebesgue integration and will be addressed in a later chapter.

Another property is that if c is a constant then

$$\mathbb{E}(cX) = c\mathbb{E}X. \tag{2.66}$$

Again, in the case of an RV with a PDF, the proof is easy: for $c \neq 0$:

$$\mathbb{E}(cX) = \int x f_{cX}(x)dx = \int x \frac{1}{c} f_X(x/c)dx = c \int x f_X(x)dx;$$

when $c > 0$ the last equality is straightforward and when $c < 0$ one has to change the limits of integration which leads to the result. Combining equations (2.65) and (2.66), we obtain the linearity of expectation:

$$\mathbb{E}(c_1 X + c_2 Y) = c_1 \mathbb{E}X + c_2 \mathbb{E}Y. \tag{2.67}$$

It also holds for any finite or countable collection of RVs:

$$\mathbb{E}\left(\sum_i c_i X_i\right) = \sum_i c_i \mathbb{E}X_i, \tag{2.68}$$

provided that each mean value $\mathbb{E}X_i$ exists and the series $\sum_i c_i \mathbb{E}X_i$ is absolutely convergent. In particular, for RVs X_1, X_2, $\ldots$ with $\mathbb{E}X_i = \mu$, $\mathbb{E}(\sum_{i=1}^n X_i) = n\mu$.

A convenient (and often used) formula is the Law of the Unconscious Statistician:

$$\mathbb{E}g(X) = \int g(x) f_X(x)dx, \tag{2.69}$$

relating the mean value of $Y = g(X)$, a function of an RV X, with the PDF of X. It holds whenever the integral on the RHS exists, i.e. $\int |g(x)| f_X(x)dx < \infty$.

The full proof of formula (2.69) again requires the use of Lebesgue's integration, but its basics are straightforward. To start with, assume that function g is differentiable and $g' > 0$ (so g is monotone increasing and hence invertible: $\forall y$ in the range of g there exists a unique $x = x(y) (= g^{-1}(y)) \in \mathbb{R}$ with $g(x) = y$). Owing to equation (2.28),

$$\mathbb{E}g(X) = \int y f_{g(X)}(y)dy = \int_{\text{Range } (g)} y f_X(x(y)) \frac{1}{g'(x(y))}dy.$$

The change of variables $y = g(x)$, with $x(g(x)) = x$, yields that the last integral is equal to

$$\int g'(x)g(x)f_X(x)\frac{1}{g'(x)}dx = \int g(x)f_X(x)dx,$$

i.e. the RHS of formula (2.69). A similar idea works when $g' < 0$: the minus sign is compensated by the inversion of the integration limits $-\infty$ and ∞.

The spirit of this argument is preserved in the case where g is continuously differentiable and the derivative g' has a discrete set of zeroes (i.e. finite or countable, but without accumulation points on $\mathbb{R}$). This condition includes, for example, all polynomials. Then we can divide the line $\mathbb{R}$ into intervals where g' has the same sign and repeat the argument locally, on each such interval. Summing over these intervals would again give formula (2.68).

Another instructive (and simple) observation is that formula (2.69) holds for any locally constant function g, taking finitely or countably many values y_j: here

$$\mathbb{E}g(X) = \sum_j y_j \mathbb{P}(g(X) = y_j)$$

$$= \sum_j y_j \int f_X(x)I(x: g(x) = y_j)dx = \int g(x)f_X(x)dx.$$

By using linearity, it is possible to extend formula (2.69) to the class of functions g continuously differentiable on each of (finitely or countably many) intervals partitioning $\mathbb{R}$, viz. $g(x) = \begin{cases} 0, & x \le 0, \\ x, & x \ge 0. \end{cases}$ This class will cover all applications considered in this volume.

Examples of using formula (2.69) are the equations

$$\mathbb{E}I(a < X < b) = \int_a^b f_X(x)dx = \mathbb{P}(a < X < b)$$

and

$$\mathbb{E}XI(a < X < b) = \int_a^b xf_X(x)dx.$$

A similar formula holds for the expectation $\mathbb{E}g(X, Y)$, expressing it in terms of a two-dimensional integral against joint PDF $f_{X,Y}$:

$$\mathbb{E}g(X, Y) = \int_{\mathbb{R}^2} g(x, y)f_{X,Y}(x, y)dxdy. \qquad (2.70)$$

Here, important examples of function g are: (i) the sum $x + y$, with

$$\mathbb{E}(X + Y) = \int \int (x + y)f_{X,Y}(x, y)dxdy$$

$$= \int xf_X(x)dx + \int yf_Y(y)dy = \mathbb{E}X + \mathbb{E}Y$$

(cf. equation (2.65)), and (ii) the product $g(x, y) = xy$, where

$$\mathbb{E}(XY) = \int \int xyf_{X,Y}(x, y)dxdy. \qquad (2.71)$$

Note that the CS inequality (1.30) holds:

$$|\mathbb{E}(XY)| \le \left(\mathbb{E}X^2 \mathbb{E}Y^2\right)^{1/2},$$

as its proof in Section 1.4 uses only linearity of the expectation.

The *covariance* Cov (X, Y) of RVs X and Y is defined by

$$\mathrm{Cov}\,(X, Y) = \int\int (x - \mathbb{E}X)(y - \mathbb{E}Y) f_{X,Y}(x, y) \mathrm{d}x \mathrm{d}y$$

$$= \mathbb{E}(X - \mathbb{E}X)(Y - \mathbb{E}Y) \tag{2.72}$$

and coincides with

$$\int\int f_{X,Y}(x, y)(xy - \mathbb{E}X\mathbb{E}Y) \mathrm{d}y \mathrm{d}x = \mathbb{E}(XY) - \mathbb{E}X\mathbb{E}Y. \tag{2.73}$$

By the CS inequality, $|\mathrm{Cov}\,(X, Y)| \le (\mathrm{Var}\,X)^{1/2}(\mathrm{Var}\,Y)^{1/2}$, with equality iff X and Y are proportional, i.e. $X = cY$ where c is a (real) scalar.

As in the discrete case,

$$\mathrm{Var}\,(X + Y) = \mathrm{Var}\,X + \mathrm{Var}\,Y + 2\mathrm{Cov}\,(X, Y) \tag{2.74}$$

and

$$\mathrm{Var}\,(cX) = c^2 \mathrm{Var}\,X. \tag{2.75}$$

If RVs X and Y are independent and with finite mean values, then

$$\mathbb{E}(XY) = \mathbb{E}X\mathbb{E}Y. \tag{2.76}$$

The proof here resembles that given for RVs with discrete values: by formula (2.69),

$$\mathbb{E}(XY) = \int\int yx f_X(x) f_Y(y) \mathrm{d}x \mathrm{d}y$$

$$= \int x f_X(x) \mathrm{d}x \int y f_Y(y) \mathrm{d}y = \mathbb{E}X\mathbb{E}Y.$$

An immediate consequence is that for independent RVs

$$\mathrm{Cov}\,(X, Y) = 0, \tag{2.77}$$

and hence

$$\mathrm{Var}\,(X + Y) = \mathrm{Var}\,X + \mathrm{Var}\,Y. \tag{2.78}$$

However, as before, neither equation (2.76) nor equation (2.77) implies independence. An instructive example is as follows.

Example 2.13   Consider a random point inside a unit circle: let X and Y be its co-ordinates. Assume that the point is distributed so that the probability that it falls within a subset A is proportional to the area of A. In the polar representation:

$$X = R\cos\Theta, \; Y = R\sin\Theta,$$

where $R \in (0, 1)$ is the distance between the point and the centre of the circle and $\Theta \in [0, 2\pi)$ is the angle formed by the radius through the random point and the horizontal line. The area in the polar co-ordinates ρ, θ is

$$\frac{1}{\pi}\rho\mathbf{1}(0 \le \rho \le 1)\mathbf{1}(0 \le \theta \le 2\pi)d\rho d\theta = f_R(\rho)d\rho f_\Theta(\theta)d\theta,$$

whence R and Θ are independent, and

$$f_R(\rho) = 2\rho\mathbf{1}(0 \le \rho \le 1), \quad f_\Theta(\theta) = \frac{1}{2\pi}\mathbf{1}(0 \le \theta \le 2\pi).$$

Then

$$\mathbb{E}XY = \mathbb{E}(R^2 \cos\Theta \sin\Theta) = \mathbb{E}(R^2)\mathbb{E}\left(\frac{1}{2}\sin(2\Theta)\right) = 0,$$

as

$$\mathbb{E}\sin(2\Theta) = \frac{1}{2\pi}\int_0^{2\pi}\sin(2\theta)d\theta = 0.$$

Similarly, $\mathbb{E}X = \mathbb{E}R\mathbb{E}\cos\Theta = 0$, and $\mathbb{E}Y = \mathbb{E}R\mathbb{E}\sin\Theta = 0$. So, Cov $(X, Y) = 0$. But X and Y are not independent:

$$\mathbb{P}\left(X > \frac{\sqrt{2}}{2}, Y > \frac{\sqrt{2}}{2}\right) = 0, \quad \text{but } \mathbb{P}\left(X > \frac{\sqrt{2}}{2}\right) = \mathbb{P}\left(Y > \frac{\sqrt{2}}{2}\right) > 0. \quad \blacksquare$$

Generalising formula (2.78), for any finite or countable collection of independent RVs and real numbers

$$\text{Var}\left(\sum_i c_i X_i\right) = \sum_i c_i^2 \text{Var}\, X_i, \tag{2.79}$$

provided that each variance Var X_i exists and $\sum_i c_i^2$ Var $X_i < \infty$. In particular, for IID RVs $X_1, X_2, \ldots$ with Var $X_i = \sigma^2$, Var $(\sum_{i=1}^n X_i) = n\sigma^2$.

The *correlation coefficient* of two RVs X, Y is defined by

$$\text{Corr}\,(X, Y) = \frac{\text{Cov}\,(X, Y)}{\sqrt{\text{Var}\,X}\sqrt{\text{Var}\,Y}}. \tag{2.80}$$

As $|\text{Cov}\,(X, Y)| \le (\text{Var}\,X)^{1/2}(\text{Var}\,Y)^{1/2}$, $-1 \le \text{Corr}(X, Y) \le 1$. Furthermore, Corr $(X, Y) = 0$ if X and Y are independent (but not only if), Corr $(X, Y) = 1$ iff $X = cY$ with $c > 0$, and -1 iff $X = cY$ with $c < 0$.

Example 2.14   For a bivariate normal pair X, Y, with the joint PDF $f_{X,Y}$ of the form (2.35), parameter $r \in [-1, 1]$ can be identified with the correlation coefficient Corr (X, Y). More precisely,

$$\text{Var}\,X = \sigma_1^2, \ \text{Var}\,Y = \sigma_2^2, \ \text{and Cov}\,(X, Y) = r\sigma_1\sigma_2. \tag{2.81}$$

In fact, the claim about the variances is straightforward as $X \sim N(\mu_1, \sigma_1^2)$ and $Y \sim N(\mu_2, \sigma_2^2)$.

Next, the covariance is

$$\text{Cov}(X, Y) = \frac{1}{2\pi\sigma_1\sigma_2\sqrt{1 - r^2}} \int \int (x - \mu_1)(y - \mu_2)$$

$$\times \exp\left\{-\frac{1}{2(1 - r^2)}\left[\frac{(x - \mu_1)^2}{\sigma_1^2} - 2r\frac{(x - \mu_1)(y - \mu_2)}{\sigma_1\sigma_2} + \frac{(y - \mu_2)^2}{\sigma_2^2}\right]\right\} dydx$$

$$= \frac{1}{2\pi\sigma_1\sigma_2\sqrt{1 - r^2}} \int \int xy\exp\left\{-\frac{1}{2(1 - r^2)}\left[\frac{x^2}{\sigma_1^2} - 2r\frac{xy}{\sigma_1\sigma_2} + \frac{y^2}{\sigma_2^2}\right]\right\} dydx$$

$$= \frac{1}{2\pi\sigma_1\sigma_2\sqrt{1 - r^2}} \int x \int y\exp\left\{-\frac{1}{2(1 - r^2)}\left[\frac{x^2(1 - r^2)}{\sigma_1^2} + \frac{y_1^2}{\sigma_2^2}\right]\right\} dydx.$$

where now

$$y_1 = y - rx\frac{\sigma_2}{\sigma_1}.$$

Therefore,

$$\text{Cov}(X, Y) = \frac{1}{2\pi\sigma_1\sigma_2\sqrt{1 - r^2}} \int xe^{-x^2/2\sigma_1^2} \int \left(y_1 + rx\frac{\sigma_2}{\sigma_1}\right)$$

$$\times \exp\left[-\frac{y_1^2}{2\sigma_2^2(1 - r^2)}\right] dy_1 dx$$

$$= \frac{r\sigma_2}{\sqrt{2\pi\sigma_1^2}} \int x^2 e^{-x^2/2\sigma_1^2} dx = \frac{r\sigma_1\sigma_2}{\sqrt{2\pi}} \int x^2 e^{-x^2/2} dx = r\sigma_1\sigma_2.$$

In other words, the off-diagonal entries of the matrices Σ and Σ^{-1} have been identified with the covariance. Hence,

$$\text{Corr}(X, Y) = r.$$

A simplified version of the above calculation, where $\mu_1 = \mu_2 = 0$ and $\sigma_1^2 = \sigma_2^2 = 1$ is repeated in Problem 2.23.

We see that if X and Y are independent, then $r = 0$ and both Σ and Σ^{-1} are diagonal matrices. The joint PDF becomes the product of the marginal PDFs:

$$f_{X,Y}(x, y) = \frac{1}{2\pi\sigma_1\sigma_2}\exp\left[-\frac{(x - \mu_1)^2}{2\sigma_1^2} - \frac{(y - \mu_2)^2}{2\sigma_2^2}\right]$$

$$= \frac{e^{-(x-\mu_1)^2/2\sigma_1^2}}{\sqrt{2\pi}\sigma_1}\frac{e^{-(y-\mu_2)^2/2\sigma_2^2}}{\sqrt{2\pi}\sigma_2}.$$

An important observation is that the inverse is also true: if $r = 0$, then Σ and Σ^{-1} are diagonal and $f_{X,Y}(x, y)$ factorises into a product. Hence, for jointly normal RVs, $\text{Cov}(X, Y) = 0$ iff X, Y are independent. ∎

The notion of the PGF (and MGF) also emerges in the continuous case:

$$\phi_X(s) = \mathbb{E}s^X = \int s^x f_X(x)dx, \quad s > 0,$$

$$M_X(\theta) = \mathbb{E}e^{\theta X} = \int e^{\theta x} f_X(x)dx = \phi_X(e^\theta), \quad \theta \in \mathbb{R}. \tag{2.82}$$

The PGFs and MGFs are used for non-negative RVs. But even then $\phi_X(s)$ and $M_X(\theta)$ may not exist for some positive s or θ. For instance, for $X \sim \text{Exp}(\lambda)$:

$$\phi_X(s) = \lambda \int_0^\infty s^x e^{-\lambda x}dx = \frac{\lambda}{\lambda - \ln s}, \quad M_X(\theta) = \frac{\lambda}{\lambda - \theta}$$

Here, ϕ_X exists only when $\ln s < \lambda$, i.e. $s < e^\lambda$, and $M_X(\theta)$ when $\theta < \lambda$. For $X \sim \text{Ca}(\alpha, \tau)$, the MGF does not exist: $\mathbb{E}e^{\theta X} \equiv \infty$, $\theta \in \mathbb{R}$.

In some applications (especially when the RV takes non-negative values), one uses the argument $e^{-\theta}$ instead of e^θ; the function

$$L_X(\theta) = \mathbb{E}e^{-\theta X} = M_X(-\theta) = \int e^{-\theta x} f_X(x)dx, \quad \theta \in \mathbb{R}, \tag{2.83}$$

which as in the discrete case, is called the Laplace transform (LTF) of PDF f_X. Cf. equation (1.55).

On the other hand, the characteristic function (CHF) $\psi_X(t)$ defined by

$$\psi_X(t) = \mathbb{E}e^{itX} = \int e^{itx} f_X(x)dx, \quad t \in \mathbb{R}, \tag{2.84}$$

exists $\forall\, t \in \mathbb{R}$ and PDF f_X. Cf. (1.56). Moreover: $\psi_X(0) = \mathbb{E}1 = 1$ and

$$|\psi_X(t)| \leq \int |e^{itx}| f_X(x)dx = \int f_X(x)dx = 1.$$

Furthermore, as a function of t, ψ_X is (uniformly) continuous on the whole line. In fact:

$$|\psi_X(t+\delta) - \psi_X(t)| \leq \int |e^{i(t+\delta)x} - e^{itx}| f_X(x)dx$$

$$= \int |e^{itx}| \, |e^{i\delta x} - 1| f_X(x)dx$$

$$= \int |e^{i\delta x} - 1| f_X(x)dx.$$

The RHS does not depend on the choice of $t \in \mathbb{R}$ (uniformity) and goes to 0 as $\delta \to 0$.

The last fact holds because $|e^{i\delta x} - 1| \to 0$, and the whole integrand $|e^{i\delta x} - 1| f_X(x)$ is $\leq 2f_X(x)$, an integrable function. A formal argument is as follows: given $\epsilon > 0$, take $A > 0$ so large that

$$2\int_{-\infty}^{-A} f_X(x)dx + 2\int_A^\infty f_X(x)dx < \epsilon/2.$$

Next, take δ so small that

$$|e^{i\delta x} - 1| < \epsilon/4A, \quad \forall\, -A < x < A;$$

we can do this because $e^{i\delta x} \to 1$ as $\delta \to 0$ uniformly on $(-A, A)$. Then split the entire integral and estimate each summand separately:

$$\int \left|e^{i\delta x} - 1\right| f_X(x) \mathrm{d}x = \left(\int_{-\infty}^{-A} + \int_{-A}^{A} + \int_{A}^{\infty} \right) \left|e^{i\delta x} - 1\right| f_X(x) \mathrm{d}x$$

$$\leq 2 \left(\int_{-\infty}^{-A} + \int_{A}^{\infty} \right) f_X(x) \mathrm{d}x + \int_{-A}^{A} \left|e^{i\delta x} - 1\right| f_X(x) \mathrm{d}x \leq \frac{\epsilon}{2} + \frac{2A\epsilon}{4A} = \epsilon.$$

We want to emphasise a few facts that follow from the definition:

(1) If $Y = cX + b$, then $\psi_Y(t) = e^{itb}\psi_X(ct)$.
(2) As in the discrete case, the mean $\mathbb{E}X$, the variance Var X and higher moments $\mathbb{E}X^k$ are expressed in terms of derivatives of ψ_X at $t = 0$:

$$\mathbb{E}X = \frac{1}{i}\frac{\mathrm{d}}{\mathrm{d}t}\psi_X(t)\bigg|_{t=0}, \tag{2.85}$$

$$\text{Var } X = \left\{ -\frac{\mathrm{d}^2}{\mathrm{d}t^2}\psi_X(t) + \left[\frac{\mathrm{d}}{\mathrm{d}t}\psi_X(t)\right]^2 \right\}\bigg|_{t=0}. \tag{2.86}$$

(3) CHF $\psi_X(t)$ uniquely defines PDF $f_X(x)$: if $\psi_X(t) \equiv \psi_Y(t)$, then $f_X(x) \equiv f_Y(x)$. (Formally, the last equality should be understood to mean that f_X and f_Y can differ on a set of measure 0.)
(4) For independent RVs X and Y: $\psi_{X+Y}(t) = \psi_X(t)\psi_Y(t)$. The proof is similar to that in the discrete case.

Example 2.15 In this example we calculate the CHF for: (i) uniform, (ii) normal, (iii) exponential, (iv) Gamma and (v) Cauchy. (i) If $X \sim U(a, b)$, then

$$\psi_X(t) = \frac{1}{b-a}\int_a^b e^{itx}\mathrm{d}x = \frac{e^{itb} - e^{ita}}{it(b-a)}. \tag{2.87}$$

(ii) If $X \sim N(\mu, \sigma^2)$, then

$$\psi_X(t) = \exp\left(it\mu - \frac{1}{2}t^2\sigma^2\right). \tag{2.88}$$

To prove this fact, set $\mu = 0$, $\sigma^2 = 1$ and take the derivative in t and integrate by parts:

$$\frac{\mathrm{d}}{\mathrm{d}t}\psi_X(t) = \frac{\mathrm{d}}{\mathrm{d}t}\left(\frac{1}{\sqrt{2\pi}}\int e^{itx}e^{-x^2/2}\mathrm{d}x\right) = \frac{1}{\sqrt{2\pi}}\int (ix)e^{itx}e^{-x^2/2}\mathrm{d}x$$

$$= \frac{1}{\sqrt{2\pi}}\left(-ie^{-x^2/2}e^{itx}\right)\bigg|_{-\infty}^{\infty} - \frac{1}{\sqrt{2\pi}}\int te^{itx}e^{-x^2/2}\mathrm{d}x = -t\psi_X(t).$$

That is

$$(\ln \psi_X(t))' = -t, \text{ whence } \ln \psi_X(t) = -\frac{t^2}{2} + c, \text{ i.e. } \psi_X(t) = e^{-t^2/2}.$$

As $\psi_X(0) = 1$, $c = 0$. The case of general μ and σ^2 is recovered from the above property (1) as $(X - \mu)/\sigma \sim N(0, 1)$.

The MGF can also be calculated in the same fashion:

$$\mathbb{E}e^{\theta X} = \exp\left(\theta\mu + \frac{1}{2}\theta^2\sigma^2\right).$$

Now, by the uniqueness of the RV with a given CHF or MGF, we can confirm the previously established fact that if $X \sim N(\mu_1, \sigma_1^2)$ and $Y \sim N(\mu_2, \sigma_2^2)$, independently, then $X + Y \sim N(\mu_1 + \mu_2, \sigma_1^2 + \sigma_2^2)$.

(iii) If $X \sim \text{Exp}(\lambda)$, then

$$\psi_X(t) = \lambda \int_0^\infty e^{itx} e^{-\lambda x} dx = \frac{\lambda}{\lambda - it}. \tag{2.89}$$

(iv) If $X \sim \text{Gam}(\alpha, \lambda)$, then

$$\psi_X(t) = \left(1 - \frac{it}{\lambda}\right)^{-\alpha}. \tag{2.90}$$

To prove this fact, we again differentiate with respect to t and integrate by parts:

$$\frac{d}{dt}\psi_X(t) = \frac{d}{dt}\frac{\lambda^\alpha}{\Gamma(\alpha)} \int_0^\infty e^{itx} x^{\alpha-1} e^{-\lambda x} dx = \frac{\lambda^\alpha}{\Gamma(\alpha)} \int_0^\infty ix^\alpha e^{(it-\lambda)x} dx$$

$$= \frac{\lambda^\alpha}{\Gamma(\alpha)} \frac{i\alpha}{\lambda - it} \int_0^\infty x^{\alpha-1} e^{(it-\lambda)x} dx = \frac{i\alpha}{\lambda - it}\psi_X(t).$$

That is

$$(\ln \psi_X(t))' = (-\alpha \ln(\lambda - it))', \text{ whence } \psi_X(t) = c(\lambda - it)^{-\alpha}.$$

As $\psi_X(0) = 1$, $c = \lambda^\alpha$. This yields the result.

Again by using the uniqueness of the RV with a given CHF, we can see that if $X \sim \text{Gam}(\alpha, \lambda)$ and $Y \sim \text{Gam}(\alpha', \lambda)$, independently, then $X + Y \sim \text{Gam}(\alpha + \alpha', \lambda)$. Also, if $X_1, \ldots, X_n$ are IID RVs, with $X_i \sim \text{Exp}(\lambda)$, then $X_1 + \cdots + X_n \sim \text{Gam}(n, \lambda)$.

(v) To find the CHF of a Cauchy distribution, we make a digression. In analysis, the function

$$t \in \mathbb{R} \mapsto \int e^{itx} f(x) dx, \quad t \in \mathbb{R},$$

is called the *Fourier transform* of function f and denoted by $\widehat{f}$. The inverse Fourier transform is the function

$$x \in \mathbb{R} \mapsto \frac{1}{2\pi} \int e^{-itx} \widehat{f}(t) dt, \quad x \in \mathbb{R}.$$

If f_X is a PDF, with $f_X \geq 0$ and $\int f_X(x)dx = 1$, and its Fourier transform $\widehat{f} = \psi_X$ has $\int \left|\widehat{f}(t)\right| dt < \infty$, then the inverse Fourier transform of ψ_X coincides with f_X:

$$f_X(x) = \frac{1}{2\pi} \int e^{-itx} \psi_X(t)dt.$$

Furthermore, if we know that a PDF f_X is the inverse Fourier transform of some function $\psi(t)$:

$$f_X(x) = \frac{1}{2\pi} \int e^{-itx} \psi(t)dx,$$

then $\psi(t) \equiv \psi_X(t)$. (The last fact is merely a re-phrasing of the uniqueness of the PDF with a given CHF.)

The Fourier transform is named after J.B.J. Fourier (1768–1830), a prolific scientist who also played an active and prominent rôle in French political life and administration. He was the Prefect of several French Departements, notably Grenoble where he was extremely popular and is remembered to the present day. In 1798, Fourier went with Napoleon's expedition to Egypt, where he took an active part in the scientific identification of numerous ancient treasures. He was unfortunate enough to be captured by the British forces and spent some time as a prisoner of war. Until the end of his life he was a staunch Bonapartist. However, when Napoleon crossed the Grenoble area on his way to Paris from the Isle of Elba in 1814, Fourier launched an active campaign against the Emperor as he was convinced that France needed peace, not another military adventure.

In mathematics, Fourier created what is now called the Fourier analysis, by studying problems of heat transfer. Fourier analysis is extremely important and is used in literally every theoretical and applied domain. Fourier is also recognised as the scientific creator of modern social statistics.

Now consider the PDF

$$f(x) = \frac{1}{2} e^{-|x|}, \quad x \in \mathbb{R}.$$

Its CHF equals the half-sum of the CHFs of the 'positive' and 'negative' exponential RVs with parameter $\lambda = 1$:

$$\frac{1}{2} \int e^{itx} e^{-|x|} dx = \frac{1}{2} \left(\frac{1}{1-it} + \frac{1}{1+it} \right) = \frac{1}{1+t^2},$$

which is, up to a scalar factor, the PDF of the Cauchy distribution Ca $(0, 1)$. Hence, the inverse Fourier transform of function $e^{-|t|}$

$$\frac{1}{2\pi} \int e^{-itx} e^{-|t|} dt = \frac{1}{\pi} \frac{1}{1+x^2}.$$

Then, by the above observation, for $X \sim \text{Ca}(1, 0)$, $\psi_X(t) = e^{-|t|}$. For $X \sim \text{Ca}(\alpha, \tau)$:

$$\psi_X(t) = e^{i\alpha t - \tau|t|}, \tag{2.91}$$

as $(X - \alpha)/\tau \sim \text{Ca}(0, 1)$.

Note that $\psi_X(t)$ for $X \sim \mathrm{Ca}\,(\alpha,\tau)$ has no derivative at $t=0$. This reflects the fact that X has no finite expectation. On the other hand, if $X \sim \mathrm{Ca}\,(\alpha_1,\tau_1)$ and $Y \sim \mathrm{Ca}\,(\alpha_2,\tau_2)$, independently, then $X+Y \sim \mathrm{Ca}\,(\alpha_1+\alpha_2,\tau_1+\tau_2)$. ∎

Problem 2.19 A radioactive source emits particles in a random direction (with all directions being equally likely). The source is held at a distance d from a photographic plate which is an infinite vertical plane.

(i)   Show that, given the particle hits the plate, the horizontal coordinate of the point of impact (with the point nearest the source as origin) has PDF $d/\pi(d^2+x^2)$.

(ii) Can you compute the mean of this distribution?

Hint: (i) Use spherical co-ordinates r, ψ, θ. The particle hits the photographic plate if the ray along the direction of emission crosses the half-sphere touching the plate; the probability of this equals $1/2$. The conditional distribution of the direction of emission is uniform on this half-sphere. Consequently, the angle ψ such that $x/d=\tan\psi$ and $x/r=\sin\psi$ is uniformly distributed over $(0,\pi)$ and θ uniformly distributed over $(0,2\pi)$. See Figure 2.17.

In fact, the Cauchy distribution emerges as the ratio of jointly normal RVs. Namely, let the joint PDF $f_{X,Y}$ be of the form (2.35). Then, by equation (2.39), the PDF $f_{X/Y}(u)$ equals

$$\frac{1}{2\pi\sigma_1\sigma_2\sqrt{1-r^2}}\int |y|\exp\left[\frac{-y^2}{2\sigma_1^2\sigma_2^2(1-r^2)}(\sigma_2^2u^2-2r\sigma_1\sigma_2u+\sigma_1^2)\right]dy$$

$$=\frac{1}{\pi\sigma_1\sigma_2\sqrt{1-r^2}}\int_0^\infty y\exp\left[\frac{-y^2}{2(1-r^2)}\frac{(\sigma_2^2u^2-2r\sigma_1\sigma_2u+\sigma_1^2)}{\sigma_1^2\sigma_2^2}\right]dy$$

$$=\frac{\sigma_1\sigma_2\sqrt{1-r^2}}{\pi(\sigma_2^2u^2-2r\sigma_1\sigma_2u+\sigma_1^2)}\int_0^\infty e^{-y_1}dy_1,$$

where

$$y_1=\frac{y^2}{2(1-r^2)}\frac{\sigma_2^2u^2-2ru\sigma_1\sigma_2+\sigma_1^2}{\sigma_1^2\sigma_2^2}.$$

We see that

$$f_{X/Y}(u)=\frac{\sqrt{1-r^2}\,\sigma_1/\sigma_2}{\pi((u-r\sigma_1/\sigma_2)^2+(1-r^2)\sigma_1^2/\sigma_2^2)}, \tag{2.92}$$

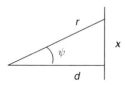

Figure 2.17

i.e. $X/Y \sim \mathrm{Ca}\left(r\sigma_1/\sigma_2, \sqrt{1-r^2}\sigma_1/\sigma_2\right)$. For independent RVs,

$$f_{X/Y}(u) = \frac{\sigma_1/\sigma_2}{\pi(u^2 + \sigma_1^2/\sigma_2^2)}, \tag{2.93}$$

i.e. $X/Y \sim \mathrm{Ca}(0, \sigma_1/\sigma_2)$. Cf. Problem 2.13.

Problem 2.20   Suppose that n items are being tested simultaneously and that the items have independent lifetimes, each exponentially distributed with parameter λ. Determine the mean and variance of the length of time until r items have failed.

Answer:

$$\mathbb{E}T_r = \sum_{i=1}^{r} \frac{1}{(n-i+1)\lambda}, \quad \mathrm{Var}(T_r) = \sum_{i=1}^{r} \frac{1}{(n-i+1)^2\lambda^2}.$$

Problem 2.21   Let $U_1, U_2, \ldots, U_n$ be IID random variables. The *ordered statistic* is an arrangement of their values in the increasing order: $U_{(1)} \leq \cdots \leq U_{(n)}$.

(i)   Let $Z_1, Z_2, \ldots, Z_n$ be IID exponential random variables with PDF $f(x) = e^{-x}$, $x \geq 0$. Show that the distribution of $Z_{(1)}$ is exponential and identify its mean.

(ii)   Let $X_1, \ldots, X_n$ be IID random variables uniformly distributed on the interval $[0, 1]$, with PDF $f(x) = 1$, $0 \leq x \leq 1$. Check that the joint PDF of $X_{(1)}, X_{(2)}, \ldots, X_{(n)}$ has the form

$$f_{X_{(1)}, X_{(2)}, \ldots, X_{(n)}}(x_1, \ldots, x_n) = n! I(0 \leq x_1 \leq \cdots \leq x_n \leq 1).$$

(iii)   For random variables $Z_1, Z_2, \ldots, Z_n$ and $X_1, X_2, \ldots, X_n$ as above, prove that the joint distribution of $X_{(1)}, X_{(2)}, \ldots, X_{(n)}$ is the same as that of

$$\frac{S_1}{S_{n+1}}, \ldots, \frac{S_n}{S_{n+1}}.$$

Here, for $1 \leq i \leq n$, S_i is the sum $\sum_{j=1}^{i} Z_j$, and $S_{n+1} = S_n + Z_{n+1}$, where $Z_{n+1} \sim \mathrm{Exp}(1)$, independently of $Z_1, \ldots, Z_n$.

(iv)   Prove that the joint distribution of the above random variables $X_{(1)}, X_{(2)}, \ldots, X_{(n)}$ is the same as the joint conditional distribution of $S_1, S_2, \ldots, S_n$ given that $S_{n+1} = 1$.

Solution   (i) $\mathbb{P}(Z_{(1)} > x) = \mathbb{P}(Z_1 > x, \ldots, Z_n > x) = e^{-nx}$. Hence, $Z_{(1)}$ is exponential, with mean $1/n$.

(ii) By the definition, we need to take the values of $X_1, \ldots, X_n$ and order them. A fixed collection of non-decreasing values $x_1, \ldots, x_n$ can be produced from $n!$ unordered samples. Hence, the joint PDF $f_{X_{(1)}, X_{(2)}, \ldots, X_{(n)}}(x_1, \ldots, x_n)$ equals

$$n! f(x_1) \ldots f(x_n) I(0 \leq x_1 \leq \cdots \leq x_n \leq 1) = n! I(0 \leq x_1 \leq \cdots \leq x_n \leq 1)$$

in the case of a uniform distribution.

(iii) The joint PDF $f_{S_1/S_{n+1},\ldots,S_n/S_{n+1},S_{n+1}}$ of random variables $S_1/S_{n+1}, \ldots, S_n/S_{n+1}$ and S_{n+1} is calculated as follows:

$$f_{S_1/S_{n+1},\ldots,S_n/S_{n+1},S_{n+1}}(x_1, x_2, \ldots, x_n, t)$$

$$= f_{S_{n+1}}(t)\frac{\partial^n}{\partial x_1 \ldots \partial x_n}\mathbb{P}\left(\frac{S_1}{S_{n+1}} < x_1, \ldots, \frac{S_n}{S_{n+1}} < x_n \middle| S_{n+1} = t\right)$$

$$= f_{S_{n+1}}(t)\frac{\partial^n}{\partial x_1 \ldots \partial x_n}\mathbb{P}\left(S_1 < x_1 t, \ldots, S_n < x_n t \middle| S_{n+1} = t\right)$$

$$= t^n f_{S_{n+1}}(t) f_{S_1,\ldots,S_n}\left(x_1 t, \ldots, x_n t \middle| S_{n+1} = t\right)$$

$$= t^n f_{S_1,\ldots,S_n,S_{n+1}}(tx_1, \ldots, tx_n, t)$$

$$= t^n e^{-tx_1} e^{-t(x_2-x_1)} \ldots e^{-t(1-x_n)} I(0 \le x_1 \le \cdots \le x_n \le 1)$$

$$= t^n e^{-t} I(0 \le x_1 \le \cdots \le x_n \le 1).$$

Hence,

$$f_{S_1/S_{n+1},\ldots,S_n/S_{n+1}}(x_1, x_2, \ldots, x_n) = \int_0^\infty t^n e^{-t} dt\, I(0 \le x_1 \le \cdots \le x_n \le 1).$$

This equals $n! I(0 \le x_1 \le \cdots \le x_n \le 1)$, which is the joint PDF of $X_{(1)}, X_{(2)}, \ldots, X_{(n)}$, by (ii).

(iv) The joint PDF $f_{S_1,\ldots,S_{n+1}}(x_1, \ldots, x_{n+1})$ equals

$$f_{Z_1,\ldots,Z_{n+1}}(x_1, x_2 - x_1, \ldots, x_{n+1} - x_n) I(0 \le x_1 \le \cdots \le x_n \le x_{n+1})$$

$$= e^{-x_1} e^{-(x_2-x_1)} \ldots e^{-(x_{n+1}-x_n)} I(0 \le x_1 \le \cdots \le x_n \le x_{n+1})$$

$$= e^{-(x_{n+1})} I(0 \le x_1 \le \cdots \le x_n \le x_{n+1}).$$

The sum S_{n+1} has PDF $f_{S_{n+1}}(x) = (1/n!) x^n e^{-x} I(x \ge 0)$. Hence, the conditional PDF could be expressed as the ratio

$$\frac{f_{S_1,\ldots,S_{n+1}}(x_1, \ldots, x_n, 1)}{f_{S_{n+1}}(1)} = n! I(0 \le x_1 \le \cdots \le x_n \le 1). \quad \square$$

Problem 2.22 Let $X_1, X_2, \ldots$ be independent RVs each of which is uniformly distributed on $[0, 1]$. Let

$$U_n = \max_{1 \le j \le n} X_j, \quad V_n = \min_{1 \le j \le n} X_j.$$

By considering

$$\mathbb{P}(v \le X_1 \le v + \delta v, v \le X_2, X_3, \ldots, X_{n-1} \le u, u \le X_n \le u + \delta u)$$

with $0 < v < u < 1$, or otherwise, show that (U_n, V_n) has a joint PDF given by

$$f(u, v) = \begin{cases} n(n-1)(u-v)^{n-2}, & \text{if } 0 \le v \le u \le 1, \\ 0, & \text{otherwise.} \end{cases}$$

Find the PDFs f_{U_n} and f_{V_n}. Show that $\mathbb{P}(1 - 1/n \le U_n \le 1)$ tends to a non-zero limit as $n \to \infty$ and find it. What can you say about $\mathbb{P}(0 \le V_n \le 1/n)$?

Find the correlation coefficient τ_n of U_n and V_n. Show that $\tau_n \to 0$ as $n \to \infty$. Why should you expect this?

Solution   The joint CDF of U_n and V_n is given by

$$\mathbb{P}(U_n < u, V_n < v) = \mathbb{P}(U_n < u) - \mathbb{P}(U_n < u, V_n \geq v)$$
$$= \mathbb{P}(U_n < u) - \mathbb{P}(v \leq X_1 < u, \ldots, v \leq X_n < u)$$
$$= (F(u))^n - (F(u) - F(v))^n.$$

Hence, the joint PDF

$$f_{U_n,V_n}(u,v) = \frac{\partial^2}{\partial u \partial v}\left[(F(u))^n - (F(u) - F(v))^n\right]$$
$$= \begin{cases} n(n-1)(u-v)^{n-2}, & \text{if } 0 \leq v \leq u \leq 1 \\ 0, & \text{otherwise.} \end{cases}$$

The marginal PDFs are

$$f_{U_n}(u) = \int_0^u n(n-1)(u-v)^{n-2}dv = nu^{n-1}$$

and

$$f_{V_n}(v) = \int_v^1 n(n-1)(u-v)^{n-2}du = n(1-v)^{n-1}.$$

Then the probability $\mathbb{P}(1 - 1/n \leq U_n \leq 1)$ equals

$$\int_{1-1/n}^1 f_{U_n}(u)du = n\int_{1-1/n}^1 u^{n-1}du = 1 - \left(1 - \frac{1}{n}\right)^n \to 1 - e^{-1} \text{ as } n \to \infty.$$

Similarly, $\mathbb{P}(0 \leq V_n \leq 1/n) \to 1 - e^{-1}$.
    Next,

$$\mathbb{E}U_n = \frac{n}{n+1}, \quad \mathbb{E}V_n = \frac{1}{n+1},$$

and

$$\mathrm{Var}\, U_n = \mathrm{Var}\, V_n = \frac{n}{(n+1)^2(n+2)}.$$

Furthermore, the covariance $\mathrm{Cov}\,(U_n, V_n)$ is calculated as the integral

$$\int_0^1 \int_0^u f(u,v)\left(u - \frac{n}{n+1}\right)\left(v - \frac{1}{n+1}\right)dvdu$$
$$= \int_0^1 \int_0^u n(n-1)(u-v)^{n-2}\left(u - \frac{n}{n+1}\right)\left(v - \frac{1}{n+1}\right)dvdu$$

$$= n(n-1) \int_0^1 \left(u - \frac{n}{n+1} \right) du \int_0^u (u-v)^{n-2} \left(v - \frac{1}{n+1} \right) dv$$

$$= n(n-1) \int_0^1 \left(u - \frac{n}{n+1} \right) \left[\frac{u^n}{n(n-1)} - \frac{u^{n-1}}{(n-1)(n+1)} \right] du$$

$$= \left[\frac{u^{n+2}}{(n+2)} - 2 \frac{nu^{n+1}}{(n+1)^2} + \frac{nu^n}{(n+1)^2} \right]_0^1 = \frac{1}{(n+1)^2(n+2)}.$$

Hence,

$$T_n = \operatorname{Corr}(U_n, V_n) = \frac{\operatorname{Cov}(U_n, V_n)}{\sqrt{\operatorname{Var} U_n} \sqrt{\operatorname{Var} V_n}} = \frac{1}{n} \to 0$$

as $n \to \infty$. □

Problem 2.23   Two real-valued RVs, X and Y, have joint PDF

$$p(x_1, x_2) = \frac{1}{2\pi\sqrt{1-r^2}} \exp \left[-\frac{1}{2(1-r^2)} (x_1^2 - 2rx_1 x_2 + x_2^2) \right],$$

where $-1 < r < 1$. Prove that each of X and Y is normally distributed with mean 0 and variance 1.

Prove that the number r is the correlation coefficient of X and Y.

Solution   The CDF of X is given by

$$\mathbb{P}(X < t) = \int_{-\infty}^t \int p(x_1, x_2) dx_2 dx_1.$$

The internal integral $\int p(x_1, x_2) dx_2$ is equal to

$$\frac{1}{2\pi\sqrt{1-r^2}} \int \exp \left\{ -\frac{1}{2(1-r^2)} \left[(x_2 - rx_1)^2 + (1 - r^2)x_1^2 \right] \right\} dx_2$$

$$= \frac{e^{-x_1^2/2}}{2\pi} \int \exp \left(-\frac{y^2}{2} \right) dy, \quad \left(\text{with } y = \frac{x_2 - rx_1}{\sqrt{1-r^2}} \right)$$

$$= \frac{1}{\sqrt{2\pi}} e^{-x_1^2/2},$$

which specifies the distribution of X:

$$\mathbb{P}(X < t) = \int_{-\infty}^t \frac{1}{\sqrt{2\pi}} e^{-x_1^2/2} dx_1,$$

i.e. $X \sim N(0, 1)$. Similarly, $Y \sim N(0, 1)$.

Hence,

$$\operatorname{Corr}(X, Y) = \operatorname{Cov}(X, Y) = \mathbb{E}(XY) = \int \int x_1 x_2 p(x_1, x_2) dx_2 dx_1.$$

Now

$$\int x_2 p(x_1, x_2) dx_2$$

$$= \frac{1}{2\pi\sqrt{1-r^2}} \int x_2 \exp\left\{-\frac{1}{2(1-r^2)}\left[(x_2 - rx_1)^2 + (1-r^2)x_1^2\right]\right\} dx_2$$

$$= \frac{rx_1 e^{-x_1^2/2}}{\sqrt{2\pi}}.$$

Then

$$\mathbb{E}(XY) = \frac{r}{\sqrt{2\pi}} \int_{-\infty}^{\infty} x_1^2 \exp\left(-\frac{x_1^2}{2}\right) dx_1 = r.$$

Hence, Corr $(X, Y) = r$, as required.   □

Problem 2.24   Let X, Y be independent random variables with values in $[0, \infty)$ and the same PDF $2e^{-x^2}/\sqrt{\pi}$. Let $U = X^2 + Y^2$, $V = Y/X$. Compute the joint PDF $f_{U,V}$ and prove that U, V are independent.

Solution   The joint PDF of X and Y is

$$f_{X,Y}(x, y) = \frac{4}{\pi} e^{-(x^2+y^2)}.$$

The change of variables $u = x^2 + y^2$, $v = y/x$ produces the Jacobian

$$\frac{\partial(u, v)}{\partial(x, y)} = \det\begin{pmatrix} 2x & 2y \\ -y/x^2 & 1/x \end{pmatrix} = 2 + 2\frac{y^2}{x^2} = 2(1 + v^2),$$

with the inverse Jacobian

$$\frac{\partial(x, y)}{\partial(u, v)} = \frac{1}{2(1 + v^2)}.$$

Hence, the joint PDF

$$f_{U,V}(u, v) = f_{X,Y}(x(u, v), y(u, v)) \frac{\partial(x, y)}{\partial(u, v)} = e^{-u} \frac{2}{\pi(1 + v^2)}, \quad u > 0, \ v > 0.$$

It means that U and V are independent, $U \sim \text{Exp}(1)$ and $V \sim \text{Ca}(0, 1)$ restricted to $(0, \infty)$.   □

Problem 2.25   (i) Continuous RVs X and Y have a joint PDF

$$f(x, y) = \frac{(m + n + 2)!}{m!n!}(1 - x)^m y^n,$$

for $0 < y \leq x < 1$, where m, n are given positive integers. Check that f is a proper PDF (i.e. its integral equals 1). Find the marginal distributions of X and Y. Hence calculate

$$\mathbb{P}\left(Y \leq \frac{1}{3}\middle| X = \frac{2}{3}\right).$$

(ii) Let X and Y be random variables. Check that

$$\text{Cov}(X, Y) = \mathbb{E}[1 - X] \times \mathbb{E}Y - \mathbb{E}[(1 - X)Y].$$

(iii) Let X, Y be as in (i). Use the form of $f(x, y)$ to express the expectations $\mathbb{E}(1 - X)$, $\mathbb{E}Y$ and $\mathbb{E}[(1 - X)Y]$ in terms of factorials. Using (ii), or otherwise, show that the covariance $\text{Cov}(X, Y)$ equals

$$\frac{(m + 1)(n + 1)}{(m + n + 3)^2 (m + n + 4)}.$$

Solution   (i) Using the notation $B(m, n) = \Gamma(m)\Gamma(n) \big/ \Gamma(m + n)$,

$$\frac{(m + n + 2)!}{m!n!} \int_0^1 (1 - x)^m dx \int_0^x y^n dy = \frac{(m + n + 2)!}{m!n!} \int_0^1 (1 - x)^m \frac{x^{n+1}}{n + 1} dx$$

$$= \frac{(m + n + 2)!}{m!n!} \frac{1}{n + 1} B(m + 1, n + 2)$$

$$= \frac{(m + n + 2)!}{m!n!} \frac{1}{n + 1} \frac{m!(n + 1)!}{(m + n + 2)!} = 1.$$

Hence,

$$\mathbb{P}\left(Y \leq \frac{1}{3} \Big| X = \frac{2}{3}\right) = \left(\frac{1}{3}\right)^m \int_0^{1/3} y^n dy \Big/ \left(\frac{1}{3}\right)^m \int_0^{2/3} y^n dy = \left(\frac{1}{2}\right)^{n+1}.$$

(ii) Straightforward rearrangement shows that

$$\mathbb{E}(1 - X)\mathbb{E}Y - \mathbb{E}[(1 - X)Y] = \mathbb{E}(XY) - \mathbb{E}X\mathbb{E}Y.$$

(iii) By the definition of Beta function (see Example 3.5),

$$\mathbb{E}(1 - X) = \frac{(m + n + 2)!}{m!n!} \frac{1}{n + 1} B(m + 1, n + 2) = \frac{m + 1}{m + n + 3},$$

$$\mathbb{E}Y = \frac{(m + n + 2)!}{m!n!} \frac{1}{n + 2} B(m + 1, n + 3) = \frac{n + 1}{m + n + 3}$$

and

$$\mathbb{E}[(1 - X)Y] = \frac{(m + n + 2)!}{m!n!} \frac{1}{n + 2} B(m + 2, n + 3)$$

$$= \frac{(m + 1)(n + 1)}{(m + n + 3)(m + n + 4)}.$$

Hence,

$$\text{Cov}(X, Y) = \frac{(m + 1)(n + 1)}{(m + n + 3)^2} - \frac{(m + 1)(n + 1)}{(m + n + 3)(m + n + 4)}$$

$$= \frac{(m + 1)(n + 1)}{(m + n + 3)^2 (m + n + 4)}. \quad \square$$

In Problem 2.26 a *mode* (of a PDF f) is the point of maximum; correspondingly, one speaks of *unimodal, bimodal* or *multimodal PDFs*.

Problem 2.26 A shot is fired at a circular target. The vertical and horizontal co-ordinates of the point of impact (taking the centre of the target as origin) are independent random variables, each distributed normally $N(0, 1)$.

(i) Show that the distance of the point of impact from the centre has PDF $re^{-r^2/2}$ for $r > 0$.

(ii) Show that the mean of this distance is $\sqrt{\pi/2}$, the median is $\sqrt{\ln 4}$, and the mode is 1.

Hint: For part (i), see Problem 2.10. For part (ii): recall, for the median $\hat{x}$

$$\int_0^{\hat{x}} re^{-r^2/2}dr = \int_{\hat{x}}^{\infty} re^{-r^2/2}dr.$$

Problem 2.27 Assume $X_1, X_2, \ldots$ form a sequence of independent RVs, each uniformly distributed on $(0, 1)$. Let

$$N = \min\{n : X_1 + X_2 + \cdots + X_n \geq 1\}.$$

Show that $\mathbb{E}N = e$.

Solution Clearly, N takes the values $2, 3, \ldots$ Then

$$\mathbb{E}N = \sum_{l \geq 2} l\mathbb{P}(N = l) = 1 + \sum_{l \geq 2} \mathbb{P}(N \geq l).$$

Now $\mathbb{P}(N \geq l) = \mathbb{P}(X_1 + X_2 + \cdots + X_{l-1} < 1)$. Finally, setting

$$\mathbb{P}(X_1 + X_2 + \cdots + X_l < y) = q_l(y), 0 < y < 1,$$

write $q_1(y) = y$, and

$$q_l(y) = \int_0^y p_{X_l}(u)q_{l-1}(y - u)du = \int_0^y q_{l-1}(y - u)du,$$

yielding $q_2(y) = y^2/2!, q_3(y) = y^3/3!, \ldots$ The induction hypothesis $q_{l-1}(y) = y^{l-1}/(l-1)!$ now gives

$$q_l(y) = \int_0^y \frac{u^{l-1}}{(l-1)!}du = \frac{y^l}{l!}$$

and we get that

$$\mathbb{E}N = \sum_{l \geq 1} q_l(1) = 1 + \frac{1}{2!} + \frac{1}{3!} + \cdots = e.$$

Alternatively, let $N(x) = \min\{n : X_1 + X_2 + \cdots + X_n \geq x\}$ and $m(x) = \mathbb{E}N(x)$. Then

$$m(x) = 1 + \int_0^x m(u)du, \text{ whence } m'(x) = m(x).$$

Integrating this ordinary differential equation with initial condition $m(0) = 1$ one gets $m(1) = e$. $\square$

Problem 2.28   The RV X has a *log-normal distribution* if $Y = \ln X$ is normally distributed. If $Y \sim N(\mu, \sigma^2)$, calculate the mean and variance of X. The log-normal distribution is sometimes used to represent the size of small particles after a crushing process, or as a model for future commodity prices. When a particle splits, a daughter might be some proportion of the size of the parent particle; when a price moves, it may move by a percentage.

Hint: $\mathbb{E}(X^r) = M_Y(r) = \exp(r\mu + r^2\sigma^2/2)$. For $r = 1, 2$ we immediately get $\mathbb{E}X$ and $\mathbb{E}X^2$.

Problem 2.29   What does it mean to say that the real-valued RVs X and Y are independent, and how, in terms of the joint PDF $f_{X,Y}$ could you recognise whether X and Y are independent? The non-negative RVs X and Y have the joint PDF

$$f_{X,Y}(x, y) = \frac{1}{2}(x + y)e^{-(x+y)}I(x, y > 0), \quad x, y \in \mathbb{R}.$$

Find the PDF f_X and hence deduce that X and Y are not independent. Find the joint PDF of (tX, tY), where t is a positive real number. Suppose now that T is an RV independent of X and Y with PDF

$$p(t) = \begin{cases} 2t, & \text{if } 0 < t < 1 \\ 0, & \text{otherwise.} \end{cases}$$

Prove that TX and TY are independent.

Solution   Clearly,

$$f_X(x) = \int_0^\infty f_{X,Y}(x, y)dy = \frac{1}{2}(1 + x)e^{-x}, \quad x > 0.$$

Similarly, $f_Y(y) = (1 + y)e^{-y}/2$, $y > 0$. Hence, $f_{X,Y} \neq f_X f_Y$, and X and Y are dependent. Next, for $t, x, y \in \mathbb{R}$,

$$f_{T,X,Y}(t, x, y) = t(x + y)e^{-(x+y)}I(0 < t < 1)I(x, y > 0).$$

To find the joint PDF of the RVs TX and TY we must pass to the variables t, u and v, where $u = tx$, $v = ty$. The Jacobian

$$\frac{\partial(t, u, v)}{\partial(t, x, y)} = \det \begin{pmatrix} 1 & x & y \\ 0 & t & 0 \\ 0 & 0 & t \end{pmatrix} = t^2.$$

Hence,

$$f_{TX,TY}(u, v) = \int_0^1 t^{-2} f_{T,X,Y}(t, u/t, v/t) dt$$

$$= \int_0^1 t^{-2} t \frac{(u+v)}{t} \exp\left(-\frac{u+v}{t}\right) dt.$$

Changing the variable $t \to \tau = t^{-1}$ yields

$$f_{TX,TY}(u, v) = \int_1^\infty d\tau (u+v) \exp\left(-\tau(u+v)\right) = e^{-(u+v)}, \ u, v > 0.$$

Hence, TX and TY are independent (and exponentially distributed with mean 1). □

Problem 2.30 A pharmaceutical company produces a drug based on a chemical Amethanol. The strength of a unit of a drug is taken to be $-\ln(1-x)$ where $0 < x < 1$ is the portion of Amethanol in the unit and $1-x$ that of an added placebo substance. You test a sample of three units taken from a large container filled with the Amethanol powder and the added substance in an unknown proportion. The container is thoroughly shaken up before each sampling.

Find the CDF of the strength of each unit and the CDF of the minimal strength.

Solution It is convenient to set $\omega = (x, y, z)$, where $0 \le x, y, z \le 1$ represent the portions of Amethanol in the units. Then Ω is the unit cube $\{\omega = (x, y, z): 0 \le x, y, z \le 1\}$. If X_1, X_2, X_3 are the strengths of the units, then

$$X_1(\omega) = -\ln(1-x), \ X_2(\omega) = -\ln(1-y), \ X_3(\omega) = -\ln(1-z).$$

We assume that the probability mass is spread on Ω uniformly, i.e. the portions x, y, z are uniformly distributed on $(0, 1)$. Then $\forall \ j = 1, 2, 3$, the CDF $F_{X_j}(x) = \mathbb{P}(X_j < y)$ is calculated as

$$\int_0^1 I(-\ln(1-x) < y) dx = \int_0^{1-e^{-y}} dx I(y > 0) = (1 - e^{-y}) I(y > 0),$$

i.e. $X_j \sim \text{Exp}(1)$. Further, $\mathbb{P}(\min_j X_j < y) = 1 - \mathbb{P}(\min_j X_j \ge y)$ and

$$\mathbb{P}(\min_j X_j \ge y) = \mathbb{P}(X_1 \ge y, X_2 \ge y, X_3 \ge y) = \prod_j \mathbb{P}(X_j \ge y) = e^{-3y}.$$

That is $\min_j X_j \sim \text{Exp}(3)$.

In this problem, the joint CDF $F_{X_1,X_2,X_3}(y_1, y_2, y_3)$ of the three units of drug equals the volume of the set

$$\{(x, y, z): 0 \le x, y, z \le 1,$$
$$-\ln(1-x) < y_1, \ -\ln(1-y) < y_2, \ -\ln(1-z) < y_3\}$$

and coincides with the product

$$(1-e^{-y_1})(1-e^{-y_2})(1-e^{-y_3}),$$

i.e. with $F_{X_1}(y_1)F_{X_2}(y_2)F_{X_3}(y_3)$. The joint PDF is also the product:

$$f_{X_1,X_2,X_3}(x_1,x_2,x_3) = e^{-x_1}e^{-x_2}e^{-x_3}. \quad \square$$

Problem 2.31 Let X_1, X_2,... be a sequence of independent identically distributed RVs having common MGF $M(\theta)$, and let N be an RV taking non-negative integer values with PGF $\phi(s)$; assume that N is independent of the sequence (X_i). Show that $Z = X_1 + X_2 + \cdots + X_n$ has MGF $\phi(M(\theta))$.

The sizes of claims made against an insurance company form an independent identically distributed sequence having common PDF $f(x) = e^{-x}$, $x \geq 0$. The number of claims during a given year had the Poisson distribution with parameter λ. Show that the MGF of the total amount T of claims during the year is

$$\psi(\theta) = \exp\{\lambda\theta/(1-\theta)\} \quad \text{for } \theta < 1.$$

Deduce that T has mean λ and variance 2λ.

Solution The MGF

$$M_Z(\theta) = \mathbb{E}e^{\theta Z} = \mathbb{E}\left[\mathbb{E}\left(e^{\theta(X_1 + \cdots + X_n)} \mid N = n\right)\right]$$

$$= \sum_{n=0}^{\infty} \mathbb{P}(N = n)\left(M_{X_1}(\theta)\right)^n = \phi\left(M_{X_1}(\theta)\right),$$

as required.
Similarly, for T:

$$T = X_1 + \cdots + X_N,$$

where $X_1, \ldots, X_n$ represent the sizes of the claims, with the PDF $f_{X_i}(x) = e^{-x}I(x > 0)$, independently, and N stands for the number of claims, with $\mathbb{P}(N = n) = \lambda^n e^{-\lambda}/n!$.

Now, the PGF of N is $\phi(s) = e^{\lambda(s-1)}$, and the MGF of X_i is $M_{X_i}(\theta) = 1/(1-\theta)$. Then the MGF of T is

$$\psi(\theta) = \exp\left[\lambda\left(\frac{1}{1-\theta} - 1\right)\right] = \exp\left[\frac{\lambda\theta}{1-\theta}\right],$$

as required. Finally,

$$\psi'(0) = \mathbb{E}T = \lambda, \quad \psi''(0) = \mathbb{E}T^2 = 2\lambda + \lambda^2,$$

and Var $T = 2\lambda$. $\square$

Problem 2.32 Let X be an exponentially distributed random variable with PDF

$$f(x) = \frac{1}{\mu} e^{-x/\mu}, \quad x > 0,$$

where $\mu > 0$. Show that $\mathbb{E}X = \mu$ and Var $X = \mu^2$.

In an experiment, n independent observations $X_1, \ldots, X_n$ are generated from an exponential distribution with expectation μ, and an estimate $\overline{X} = \sum_{i=1}^n X_i/n$ of μ is obtained. A second independent experiment yields m independent observations $Y_1, \ldots, Y_m$ from the same exponential distribution as in the first experiment, and the second estimate $\overline{Y} = \sum_{j=1}^m Y_j/m$ of μ is obtained. The two estimates are combined into

$$T_p = p\overline{X} + (1-p)\overline{Y},$$

where $0 < p < 1$.

Find $\mathbb{E}T_p$ and Var T_p and show that, for all $\epsilon > 0$,

$$\mathbb{P}\big(|T_p - \mu| > \epsilon\big) \to 0 \quad \text{as } m, n \to \infty.$$

Find the value $\tilde{p}$ of p that minimises Var T_p and interpret $T_{\tilde{p}}$. Show that the ratio of the inverses of the variances of $\overline{X}$ and $\overline{Y}$ is $\tilde{p}/(1-\tilde{p})$.

Solution Integrating by parts:

$$\mathbb{E}X = \frac{1}{\mu} \int_0^\infty x e^{-x/\mu} dx = \left(-x e^{-x/\mu}\right)\big|_0^\infty + \int_0^\infty e^{-x/\mu} dx = \mu$$

and

$$\mathbb{E}X^2 = \int_0^\infty x^2 \frac{1}{\mu} e^{-x/\mu} dx = \left(-x^2 e^{-x/\mu}\right)\big|_0^\infty + 2\mu\mathbb{E}X = 2\mu^2,$$

which yields

$$\text{Var } X = 2\mu^2 - \mu^2 = \mu^2.$$

Now,

$$\mathbb{E}\overline{X} = \frac{1}{n}\sum_{i=1}^n \mathbb{E}X_i = \mu, \quad \mathbb{E}\overline{Y} = \frac{1}{m}\sum_{j=1}^m \mathbb{E}Y_j = \mu,$$

and

$$\mathbb{E}T_p = p\mathbb{E}\overline{X} + (1-p)\mathbb{E}\overline{Y} = \mu.$$

Similarly,

$$\text{Var } \overline{X} = \frac{1}{n^2}\sum_{i=1}^n \text{Var } X_i = \frac{\mu^2}{n}, \quad \text{Var } \overline{Y} = \frac{1}{m^2}\sum_{j=1}^m \text{Var } Y_j = \frac{\mu^2}{m},$$

and

$$\text{Var } T_p = p^2 \text{Var } \overline{X} + (1-p)^2 \text{Var } \overline{Y} = \mu^2 \left[\frac{p^2}{n} + \frac{(1-p)^2}{m} \right].$$

Chebyshev's inequality $\mathbb{P}(|Z| \ge \epsilon) \le (1/\epsilon^2)\mathbb{E}Z^2$ gives

$$\mathbb{P}(|T_p - \mu| > \epsilon) \le \frac{1}{\epsilon^2}\mathbb{E}(T_p - \mu)^2 = \frac{1}{\epsilon^2}\text{Var } T_p$$

$$= \frac{\mu^2}{\epsilon^2}\left[\frac{p^2}{n} + \frac{(1-p)^2}{m} \right] \to 0 \text{ as } n, m \to \infty.$$

To minimise, take

$$\frac{d}{dp}\text{Var } T_p = 2\mu^2 \left(\frac{p}{n} - \frac{1-p}{m} \right) = 0, \text{ with } \widetilde{p} = \frac{n}{n+m}.$$

As

$$\frac{d^2}{dp^2}\text{Var } T_p \bigg|_{p=\widetilde{p}} = 2\mu^2 \left(\frac{1}{n} + \frac{1}{m} \right) > 0,$$

$\widetilde{p}$ is the (global) minimum. Then

$$T_{\widetilde{p}} = \frac{n\overline{X} + m\overline{Y}}{n+m}$$

is the average of the total of $n+m$ observations. Finally,

$$\frac{\widetilde{p}}{1-\widetilde{p}} = \frac{n}{m}, \text{ and } \frac{1/\text{Var } \overline{X}}{1/\text{Var } \overline{Y}} = \frac{\mu^2/m}{\mu^2/n} = \frac{n}{m}. \quad \square$$

Problem 2.33 Let $X_1, X_2, \ldots$ be independent, identically distributed RVs with PDF

$$f(x) = \begin{cases} \dfrac{\alpha}{x^{\alpha+1}}, & x \ge 1, \\ 0, & x < 1, \end{cases}$$

where $\alpha > 0$.

There is an *exceedance* of u by $\{X_i\}$ at j if $X_j > u$. Let $L(u) = \min\{i \ge 1 : X_i > u\}$, where $u > 1$, be the time of the first exceedance of u by $\{X_i\}$. Find $\mathbb{P}(L(u) = k)$ for $k = 1, 2, \ldots$ in terms of α and u. Hence find the expected time $\mathbb{E}L(u)$ to the first exceedance. Show that $\mathbb{E}L(u) \to \infty$ as $u \to \infty$.

Find the limit as $u \to \infty$ of the probability that there is an exceedance of u before time $\mathbb{E}L(u)$.

Solution Observe that

$$\mathbb{P}(X_i > u) = \int_u^\infty \alpha x^{-\alpha-1}dx = (-x^{-\alpha})\big|_u^\infty = \frac{1}{u^\alpha}.$$

Then

$$P(L(u)=k)=\left(1-\frac{1}{u^{\alpha}}\right)^{k-1}\frac{1}{u^{\alpha}}, \quad k=1,2,\ldots,$$

i.e. $L(u) \sim$ Geom $(1-1/u^{\alpha})$, and $\mathbb{E}L(u)=u^{\alpha}$, which tends to ∞ as $u\to\infty$.
  Now, with $[\cdot]$ standing for the integer part:

$$\mathbb{P}\left(L(u)<\mathbb{E}L(u)\right)=\mathbb{P}\left(L(u)\le[u^{\alpha}]\right)=1-\mathbb{P}\left(L(u)>[u^{\alpha}]\right)$$

$$=1-\sum_{k\ge[u^{\alpha}]+1}\left(1-\frac{1}{u^{\alpha}}\right)^{k-1}\frac{1}{u^{\alpha}}$$

$$=1-\frac{1}{u^{\alpha}}\left(1-\frac{1}{u^{\alpha}}\right)^{[u^{\alpha}]}\frac{1}{1-(1-1/u^{\alpha})}$$

$$=1-\left(1-\frac{1}{u^{\alpha}}\right)^{[u^{\alpha}]} \to 1-e^{-1}, \quad \text{as } u\to\infty. \quad \square$$

Problem 2.34   Let A and B be independent RVs each having the uniform distribution on $[0, 1]$. Let $U=\min\{A, B\}$ and $V=\max\{A, B\}$. Find the mean values of U and hence find the covariance of U and V.

Solution   The tail probability, for $0\le x\le1$:

$$1-F_U(x)=\mathbb{P}(U\ge x)=\mathbb{P}(A\ge x,\ B\ge x)=(1-x)^2.$$

Hence, still for $0\le x\le1$: $F_U(x)=2x-x^2$, and

$$f_U(x)=F'_U(x)I(0<x<1)=(2-2x)I(0<x<1).$$

Therefore,

$$\mathbb{E}U=\int_0^1 x(2-2x)\mathrm{d}x=\left(x^2-\frac{2}{3}x^3\right)\Big|_0^1=\frac{1}{3}.$$

As $U+V=A+B$ and $\mathbb{E}(A+B)=\mathbb{E}A+\mathbb{E}B=1/2+1/2=1$,

$$\mathbb{E}V=1-1/3=2/3.$$

Next, as $UV=AB$, Cov $(U,V)=\mathbb{E}(UV)-\mathbb{E}U\mathbb{E}V$ which in turn equals

$$\mathbb{E}(AB)-\frac{2}{9}=\mathbb{E}A\mathbb{E}B-\frac{2}{9}=\frac{1}{4}-\frac{2}{9}=\frac{1}{36}. \quad \square$$

2.3   Normal distributions. Convergence of random variables and distributions. The Central Limit Theorem

> Probabilists do it. After all, it's only normal.
> (From the series *'How they do it'*.)

We have already learned a number of properties of a normal distribution. Its importance was realised at an early stage by, among others, Laplace, Poisson and of course Gauss. However, progress in understanding the special nature of normal distributions was steady and required facts and methods from other fields of mathematics, including analysis and mathematical physics (notably, complex analysis and partial differential equations). Nowadays, the emphasis is on multivariate (multidimensional) normal distributions which play a fundamental rôle wherever probabilistic concepts are in use. Despite an emerging variety of other exciting examples, they remain a firm basis from which further development takes off. In particular, normal distributions form the basis of statistics and financial mathematics.

Recall the properties of Gaussian distributions which we have established so far:

(i)   The PDF of an $N(\mu, \sigma^2)$ RV X is

$$\frac{1}{\sqrt{2\pi}\sigma} \exp\left(-\frac{1}{2\sigma^2}(x-\mu)^2\right), \quad x \in \mathbb{R}, \tag{2.94}$$

with the mean and variance

$$\mathbb{E}X = \mu, \quad \operatorname{Var}X = \sigma^2 \tag{2.95}$$

and the MGF and CHF

$$\mathbb{E}e^{\theta X} = e^{\theta\mu + \frac{1}{2}\theta^2\sigma^2}, \quad \mathbb{E}e^{itX} = e^{it\mu - \frac{1}{2}t^2\sigma^2}, \quad \theta, t \in \mathbb{R}. \tag{2.96}$$

If $X \sim N(\mu, \sigma^2)$, then $(X-\mu)/\sigma \sim N(0,1)$ and $\forall\, b, c \in \mathbb{R}: cX + b \sim N(c\mu + b, c^2\sigma^2)$.

(ii)   Two jointly normal RVs X and Y are independent iff $\operatorname{Cov}(X,Y) = \operatorname{Corr}(X,Y) = 0$.

(iii)   The sum $X + Y$ of two jointly normal RVs $X \sim N(\mu_1, \sigma_1^2)$ and $Y \sim N(\mu_2, \sigma_2^2)$ with $\operatorname{Corr}(X,Y) = r$ is normal, with mean $\mu_1 + \mu_2$ and variance $\sigma_1^2 + 2r\sigma_1\sigma_2 + \sigma_2^2$. See equation (2.41). In particular, if X, Y are independent, $X + Y \sim N(\mu_1 + \mu_2, \sigma_1^2 + \sigma_2^2)$. In general, for independent RVs $X_1, X_2, \ldots$, where $X_i \sim N(\mu_i, \sigma_i^2)$, the linear combination $\sum_i c_i X_i \sim N\left(\sum_i c_i \mu_i, \sum_i c_i^2 \sigma_i^2\right)$.

Problem 2.35   How large a random sample should be taken from a normal distribution in order for the probability to be at least 0.99 that the sample mean will be within one standard deviation of the mean of the distribution?

Hint: $\Phi(2.58) = 0.995$.

Remark   (cf. Problem 1.58). Observe that knowing that the distribution is normal allows a much smaller sample size.

As has been already said, the main fact justifying our interest in Gaussian distributions is that they appear in the celebrated Central Limit Theorem (CLT). The early version of this, the De Moivre–Laplace Theorem (DMLT), was established in Section 1.6. The statement of the DMLT can be extended to a general case of independent and identically distributed RVs. The following theorem was proved in 1900–1901 by a Russian mathematician A.M. Lyapunov (1857–1918).

> *Suppose* X_1, X_2, ... *are IID RVs, with finite mean* $\mathbb{E}X_j = a$ *and variance* Var $X_j = \sigma^2$. *If* $S_n = X_1 + \cdots + X_n$, *with* $\mathbb{E}S_n = na$, $VarS_n = n\sigma^2$, *then* $\forall\ y \in \mathbb{R}$:

$$\lim_{n\to\infty} \mathbb{P}\left(\frac{S_n - \mathbb{E}S_n}{\sqrt{\text{Var } S_n}} < y\right) = \Phi(y). \tag{2.97}$$

> *In fact, the convergence in equation (2.97) is uniform in* y:

$$\lim_{n\to\infty} \sup_{y\in\mathbb{R}} \left| \mathbb{P}\left(\frac{S_n - \mathbb{E}S_n}{\sqrt{\text{Var } S_n}} < y\right) - \Phi(y) \right| = 0.$$

Lyapunov and Markov were contemporaries and close friends. Lyapunov considered himself as Markov's follower (although he was only a year younger). He made his name through Lyapunov's functions, a concept that proved to be very useful in analysis of convergence to equilibrium in various random and deterministic systems. Lyapunov died tragically, committing suicide after the death of his beloved wife, amidst deprivation and terror during the civil war in Russia.

As in Section1.6, limiting relation (2.97) is commonly called the *integral* CLT and often written as $(S_n - \mathbb{E}S_n)/\sqrt{\text{Var } S_n} \sim N(0, 1)$. Here, the CLT was stated for IID RVs, but modern methods can extend it to a much wider situation and provide an accurate bound on the speed of convergence.

The proof of the integral CLT for general IID RVs requires special techniques. A popular method is based on characteristic functions and uses the following result that we will give here without proof. (The proof will be supplied in Volume 3.)

> *Let* $Y, Y_1, Y_2, \ldots$ *be a sequence of RVs with distribution functions* $F_Y, F_{Y_1}, F_{Y_2}, \ldots$ *and characteristic functions* $\psi_Y, \psi_{Y_1}, \psi_{Y_2}, \ldots$. *Suppose that, as* $n \to \infty$, *CHF* $\psi_{Y_n}(t) \to \psi_Y(t) \forall t \in \mathbb{R}$. *Then* $F_{Y_n}(y) \to F_Y(y)$ *at every point* $y \in \mathbb{R}$ *where CDF* F_Y *is continuous.*

In our case, $Y \sim N(0, 1)$, with $F_Y = \Phi$ which is continuous everywhere on $\mathbb{R}$. Setting

$$Y_n = \frac{S_n - an}{\sqrt{n}\sigma},$$

we will have to check that the CHF $\psi_{Y_n}(t) \to e^{-t^2/2} \forall t \in \mathbb{R}$. This follows from a direct calculation which we perform below. Write

$$\frac{S_n - an}{\sigma} = \sum_{1 \le j \le n} \frac{X_j - a}{\sigma}$$

and note that the RV $(X_j - a)/\sigma$ is IID. Then

$$\psi_{(S_n - an)/(\sqrt{n}\sigma)}(t) = \left[\psi_{(X_j - a)/\sigma} \left(\frac{t}{\sqrt{n}} \right) \right]^n.$$

RV $(X_j - a)/\sigma$ has mean 0 and variance 1. Hence its CHF admits the following Taylor expansion near 0:

$$\psi(u) = \psi(0) + u\psi'(0) + \frac{u^2}{2}\psi''(0) + o(u^2) = 1 - \frac{u^2}{2} + o(u^2)$$

(Here and below we omit the subscript $(X_j - a)/\sigma$.) This yields

$$\psi\left(\frac{t}{\sqrt{n}} \right) = 1 - \frac{t^2}{2n} + o\left(\frac{t^2}{n} \right).$$

But then

$$\left[\psi\left(\frac{t}{\sqrt{n}} \right) \right]^n = \left[1 - \frac{t^2}{2n} + o\left(\frac{t^2}{n} \right) \right]^n \to e^{-t^2/2}. \tag{2.98}$$

A short proof of equation (2.98) is to set

$$A(= A_n) = 1 - \frac{t^2}{2n} + o\left(\frac{t^2}{n} \right), \quad B(= B_n) = 1 - \frac{t^2}{2n},$$

and observe that, clearly, $B^n \to e^{-t^2/2}$. Next,

$$A^n - B^n = A^{n-1}(A - B) + A^{n-2}(A - B)B + \cdots + (A - B)B^{n-1},$$

whence

$$|A^n - B^n| \le (\max[1, A, B])^{n-1} |A - B|$$

which goes to 0 as $n \to \infty$.

Problem 2.36 Every year, a major university assigns Class A to $\sim$16 per cent of its mathematics graduates, Class B and Class C each to $\sim$34 per cent and Class D or failure to the remaining 16 per cent. The figures are repeated regardless of the variation in the actual performance in a given year.

A graduating student tries to make sense of such a practice. She assumes that the individual candidate's scores $X_1, \ldots, X_n$ are independent variables that differ only in mean values $\mathbb{E}X_j$, so that 'centred' scores $X_j - \mathbb{E}X_j$ have the same distribution. Next, she considers the average sample total score distribution as approximately $N(\mu, \sigma^2)$. Her guess is that the above practice is related to a standard partition of students' total score values into four categories. Class A is awarded when the score exceeds a certain limit, say a, Class B when it is between b and a, Class C when between c and b and Class D or failure when it is lower that c. Obviously, the thresholds c, b and a may depend on μ and σ.

After a while (and using tables), she convinces herself that it is indeed the case and manages to find simple formulas giving reasonable approximations for a, b and c. Can you reproduce her answer?

Solution   Let X_j be the score of candidate j. Set

$$S_n = \sum_{1 \le j \le n} X_j \quad \text{(the total score)},$$

$$S_n - \mathbb{E}S_n = \sum_{1 \le j \le n} (X_j - \mathbb{E}X_j) \quad \text{(the 'centred' total score)}.$$

Assume that $X_j - \mathbb{E}X_j$ are IID RVs (which, in particular, means that Var $X_j = \sigma^2$ does not depend on j and Var $S_n = n\sigma^2$). Then, owing to the CLT, for n large,

$$\frac{S_n - \mathbb{E}S_n}{\sqrt{n}\sigma} \sim N(0, 1).$$

Thus total average score S_n/n must obey, for large n:

$$\frac{S_n}{n} \sim N\left(\mu, \frac{\sigma^2}{n}\right) \sim \frac{\sigma}{\sqrt{n}}Y + \mu,$$

where $Y \sim N(0, 1)$ and

$$\mu = \frac{1}{n}\mathbb{E}S_n = \frac{1}{n}\sum_{1 \le j \le n} \mathbb{E}X_j.$$

We look for the value α such that $\mathbb{P}(Y > \alpha) = 1 - \Phi(\alpha) \approx 0.16$ which gives $\alpha = 1$. Clearly, $\mathbb{P}(Y > 1) = \mathbb{P}(S_n/n > \sigma/\sqrt{n} + \mu)$. Similarly, the equation $\mathbb{P}(\beta < Y < 1) = \Phi(1) - \Phi(\beta) \approx 0.34$ yields $\beta = 0$. A natural conjecture is that $a = \mu + \sigma$, $b = \mu$ and $c = \mu - \sigma$.

To give a justification for this guess, write $X_j \sim \mathbb{E}X_j + \sigma Y$. This implies

$$\frac{S_n}{n} \sim \frac{1}{\sqrt{n}}(X_j - \mathbb{E}X_j) + \mu$$

and

$$\mathbb{P}\left(\frac{1}{\sqrt{n}}(X_j - \mathbb{E}X_j) + \mu > \frac{\sigma}{\sqrt{n}} + \mu\right) = \mathbb{P}(X_j > \mathbb{E}X_j + \sigma) = 0.16.$$

We do not know μ or σ^2 and use their *estimates*:

$$\hat{\mu} = \frac{S_n}{n} \quad \text{and} \quad \widehat{\sigma^2} = \frac{1}{n-1}\sum_{1 \le j \le n}\left(X_j - \frac{Sn}{n}\right)^2.$$

(see Problem 1.32). Then the categories are defined in the following way: Class A is given when candidate's score exceeds $\hat{\mu} + \sqrt{\widehat{\sigma^2}}$, Class B when it is between $\hat{\mu}$ and $\hat{\mu} + \sqrt{\widehat{\sigma^2}}$, Class C when it is between $\hat{\mu} - \sqrt{\widehat{\sigma^2}}$ and $\hat{\mu}$ and Class D or failure when it is less than $\hat{\mu} - \sqrt{\widehat{\sigma^2}}$. □

Problem 2.37 (continuation of Problem 2.36) Now suppose one wants to assess how accurate is the approximation of the average expected score μ by S_n/n. Assuming that σ, the standard deviation of the individual score X_j, is ≤ 10 mark units, how large should n be to guarantee that the probability of deviation of S_n/n from μ is at most 5 marks does not exceed 0.1?

Solution   We want

$$\mathbb{P}\left(\left|\frac{S_n}{n} - \frac{\mathbb{E}S_n}{n}\right| \geq 5\right) \leq 0.1.$$

Letting, as before, $Y \sim \mathrm{N}(0, 1)$, the CLT yields that the last probability is

$$\approx \mathbb{P}\left(|Y| \geq 5\frac{\sqrt{n}}{\sigma}\right).$$

Thus, we want

$$5\frac{\sqrt{n}}{\sigma} \geq \Phi^{-1}(0.995), \quad \text{i.e. } n \geq \sigma^2\left[\frac{\Phi^{-1}(0.995)}{5}\right]^2,$$

with $\sigma^2 \leq 100$. Here, and below, Φ^{-1} stands for the inverse of Φ. As $\Phi^{-1}(0.995) = 2.58$, we have, in the worst case, that

$$n = \frac{100}{25}\left[\Phi^{-1}(0.995)\right]^2 = 26.63$$

will suffice. □

In the calculations below it will be convenient to use an alternative notation for the scalar product:

$$(\mathbf{x} - \mu)^{\mathrm{T}}\Sigma^{-1}(\mathbf{x} - \mu) = \langle \mathbf{x} - \mu, \Sigma^{-1}(\mathbf{x} - \mu)\rangle = \sum_{i,j=1}^{n}(x_i - \mu_i)\Sigma_{ij}^{-1}(x_j - \mu_j).$$

Problem 2.38   Suppose $X_1, \ldots, X_n$ are independent RVs, $X_j \sim \mathrm{N}(\mu_j, \sigma^2)$, with the same variance. Consider variables $Y_1, \ldots, Y_n$ given by

$$Y_j = \sum_{i=1}^{n} a_{ij}X_i, \quad j = 1, \ldots, n,$$

where $\mathbf{A} = (a_{ij})$ is an $n \times n$ real orthogonal matrix. Prove that $Y_1, \ldots, Y_n$ are independent and determine their distributions.

Comment on the case where variances $\sigma_1^2, \ldots, \sigma_n^2$ are different.

(An $n \times n$ real matrix $\mathbf{A}$ is called *orthogonal* if $\mathbf{A} = (\mathbf{A}^{-1})^{\mathrm{T}}$.)

Solution   In vector notation,

$$\mathbf{Y} := \begin{pmatrix} Y_1 \\ \vdots \\ Y_n \end{pmatrix} = \mathbf{A}^{\mathrm{T}}\mathbf{X}, \quad \text{where } \mathbf{X} = \begin{pmatrix} X_1 \\ \vdots \\ X_n \end{pmatrix}.$$

We have $\mathbf{A}^{\mathrm{T}} = \mathbf{A}^{-1}$, $(\mathbf{A}^{\mathrm{T}})^{-1} = \mathbf{A}$. Moreover, the Jacobians of the mutually inverse linear maps

$$\mathbf{x} = \begin{pmatrix} x_1 \\ \vdots \\ x_n \end{pmatrix} \mapsto \mathbf{y} = \mathbf{A}^{\mathrm{T}}\mathbf{x}, \ \mathbf{y} = \begin{pmatrix} y_1 \\ \vdots \\ y_n \end{pmatrix} \mapsto \mathbf{x} = \mathbf{A}\mathbf{y}, \ \mathbf{x}, \mathbf{y} \in \mathbb{R}^n,$$

are equal to ± 1 (and equal each other). In fact:

$$\frac{\partial(y_1, \dots, y_n)}{\partial(x_1, \dots, x_n)} = \left| \det \mathbf{A}^{\mathrm{T}} \right|, \quad \frac{\partial(x_1, \dots, x_n)}{\partial(y_1, \dots, y_n)} = \left| \det \mathbf{A} \right|,$$

and $\det \mathbf{A}^{\mathrm{T}} = \det \mathbf{A} = \pm 1$.

The PDF $f_{Y_1, \dots, Y_n}(\mathbf{y})$ equals $f_{X_1, \dots, X_n}(\mathbf{A}\mathbf{y})$, which in turn is equal to

$$\left(\frac{1}{(2\pi)^{1/2}\sigma} \right)^n \prod_{1 \le j \le n} \exp\left[-\frac{1}{2\sigma^2} \left(\sum_{1 \le i \le n} a_{ji} y_i - \mu_j \right)^2 \right].$$

Set

$$\mu = \begin{pmatrix} \mu_1 \\ \vdots \\ \mu_n \end{pmatrix} \quad \text{and } \sigma^{-2}\mathbf{I} = \begin{pmatrix} \sigma^{-2} & 0 & \dots & 0 \\ 0 & \sigma^{-2} & \dots & 0 \\ \vdots & \vdots & \dots & \vdots \\ 0 & 0 & \dots & \sigma^{-2} \end{pmatrix},$$

where $\mathbf{I}$ is the $n \times n$ unit matrix. Writing $(\mathbf{A}\mathbf{y})_j$ for the jth entry $(\mathbf{A}^{\mathrm{T}}\mathbf{y})_j = \sum_{1 \le i \le n} a_{ij} y_i$ of vector $\mathbf{A}\mathbf{y}$, the above product of exponentials becomes

$$\exp\left\{ -\frac{1}{2\sigma^2} \sum_{1 \le j \le n} \left[(\mathbf{A}\mathbf{y})_j - \mu_j \right]^2 \right\} = \exp\left[-\frac{1}{2} (\mathbf{A}\mathbf{y} - \mu)^{\mathrm{T}} (\sigma^{-2}\mathbf{I})(\mathbf{A}\mathbf{y} - \mu) \right]$$

$$= \exp\left[-\frac{1}{2} (\mathbf{y} - \mathbf{A}^{\mathrm{T}}\mu)^{\mathrm{T}} \mathbf{A}^{\mathrm{T}} (\sigma^{-2}\mathbf{I}) \mathbf{A} (\mathbf{y} - \mathbf{A}^{\mathrm{T}}\mu) \right].$$

The triple matrix product $\mathbf{A}^{\mathrm{T}}(\sigma^{-2}\mathbf{I})\mathbf{A} = \sigma^{-2}\mathbf{A}^{\mathrm{T}}\mathbf{A} = \sigma^{-2}\mathbf{I}$. Hence, the last expression is equal to

$$\exp\left[-\frac{1}{2} (\mathbf{y} - \mathbf{A}^{\mathrm{T}}\mu)^{\mathrm{T}} \sigma^{-2}\mathbf{I} (\mathbf{y} - \mathbf{A}^{\mathrm{T}}\mu) \right]$$

$$= \exp\left\{ -\frac{1}{2\sigma^2} \sum_{1 \le j \le n} \left[y_j - (\mathbf{A}^{\mathrm{T}}\mu)_j \right]^2 \right\}$$

$$= \prod_{1 \le j \le n} \exp\left\{ -\frac{1}{2\sigma^2} \left[y_j - (\mathbf{A}^{\mathrm{T}}\mu)_j \right]^2 \right\}.$$

Here, as before, $(A^T\mu)_j = \sum_{1 \le i \le n} a_{ij}\mu_i$ stands for the jth entry of vector $A^T\mu$. So, $Y_1, \ldots, Y_n$ are independent, and $Y_j \sim N((A^T\mu)_j, \sigma^2)$.

In the general case where variances σ_i^2 are different, matrix $\sigma^{-2}I$ must be replaced by the diagonal matrix

$$\Sigma^{-1} = \begin{pmatrix} \sigma_1^{-2} & 0 & \cdots & 0 \\ 0 & \sigma_2^{-2} & \cdots & 0 \\ \vdots & \vdots & \cdots & \vdots \\ 0 & 0 & \cdots & \sigma_n^{-2} \end{pmatrix}. \tag{2.99}$$

Random variables $Y_1, \ldots, Y_n$ will be independent iff matrix

$$A^T\Sigma^{-1}A$$

is diagonal. For instance, if A commutes with Σ^{-1}, i.e. $\Sigma^{-1}A = A\Sigma^{-1}$, then

$$A^T\Sigma^{-1}A = A^TA\Sigma^{-1} = \Sigma^{-1},$$

in which case $Y_j \sim N((A\mu)_j, \sigma_j^2)$, with the same variance. $\square$

Problem 2.38 leads us to the general statement that multivariate normal variables $X_1, \ldots, X_n$ are independent iff $\operatorname{Cov}(X_i, X_j) = 0 \ \forall \ 1 \le i < j \le n$. This is a part of properties (IV) below. So far we have established the fact that bivariate normal variables X, Y are independent iff $\operatorname{Cov}(X, Y) = 0$ (see equation (2.35) and Example 2.5).

Recall (see equation (2.9)) that a general multivariate normal vector

$$X = \begin{pmatrix} X_1 \\ \vdots \\ X_n \end{pmatrix}$$

has the PDF $f_X(x)$ of the form

$$f_X(x) = \frac{1}{(\sqrt{2\pi})^n (\det \Sigma)^{1/2}} \exp\left(-\frac{1}{2}\langle x - \mu, \Sigma^{-1}(x - \mu)\rangle\right), \tag{2.100}$$

where Σ is an invertible positive-definite (and hence symmetric) $n \times n$ real matrix, and

$$\det \Sigma^{-1} = (\det \Sigma)^{-1}.$$

For a multivariate normal vector X we will write $X \sim N(\mu, \Sigma)$.

Following the properties (I)–(III) at the beginning of the current section, the next properties of the Gaussian distribution we are going to establish are (IVa) and (IVb)

(IVa) If $X \sim N(\mu, \Sigma)$, with PDF f_X as in formula (2.100),
  (i) Then each $X_i \sim N(\mu_i, \Sigma_{ii})$, $i = 1, \ldots, n$, with mean μ_i and variance Σ_{ii}, the diagonal element of matrix Σ.

(IVb) The off-diagonal element Σ_{ij} equals the covariance $\operatorname{Cov}(X_i, X_j)$, $\forall \ 1 \le i < j \le n$.
  So, the matrices Σ and Σ^{-1} are diagonal and therefore the PDF $f_X(x)$ decomposes

into the product iff Cov $(X_i, X_j) = 0$, $1 \leq i < j \leq n$. In other words, jointly normal RVs $X_1, \ldots, X_n$ are independent iff Cov $(X_i, X_j) = 0$, $1 \leq i < j \leq n$.

Naturally, vector μ is called the mean-value vector and matrix Σ the covariance matrix of a multivariate random vector $\mathbf{X}$.

The proof of assertions (IVa) and (IVb) uses directly the form (2.100) of the joint multivariate normal PDF $f_{\mathbf{X}}(\mathbf{x})$. First, we discuss some algebraic preliminaries. (This actually will provide us with more properties of a multivariate normal distribution.)

It was stated in Section 2.1 that as a positive-definite matrix, Σ has a diagonal form (in the basis formed by its eigenvectors). That is $\exists$ an orthogonal matrix $\mathbf{B} = (b_{ij})$, with $\mathbf{B} = (\mathbf{B}^{-1})^{\mathrm{T}}$ such that $\mathbf{BDB}^{-1} = \Sigma$ and $\mathbf{BD}^{-1}\mathbf{B}^{-1} = \Sigma^{-1}$, where

$$
\mathbf{D} = \begin{pmatrix} \lambda_1^2 & 0 & 0 & \ldots & 0 \\ 0 & \lambda_2^2 & 0 & \ldots & 0 \\ 0 & 0 & 0 & \ldots & \lambda_n^2 \end{pmatrix}, \quad \mathbf{D}^{-1} = \begin{pmatrix} \lambda_1^{-2} & 0 & 0 & \ldots & 0 \\ 0 & \lambda_2^{-2} & 0 & \ldots & 0 \\ 0 & 0 & 0 & \ldots & \lambda_n^{-2} \end{pmatrix},
$$

and $\lambda_1^2, \lambda_2^2, \ldots, \lambda_n^2$ are (positive) eigenvalues of Σ. Note that det $\Sigma = \prod_{i=1}^n \lambda_i^2$ and $(\det \Sigma)^{1/2} = \prod_{i=1}^n \lambda_i$.

If we make the orthogonal change of variables $\mathbf{x} \mapsto \mathbf{y} = \mathbf{B}^{\mathrm{T}}\mathbf{x}$ with the inverse map $\mathbf{y} \mapsto \mathbf{x} = \mathbf{By}$ and the Jacobian $\partial(x_1, \ldots, x_n)/\partial(y_1, \ldots, y_n) = \det \mathbf{B} = \pm 1$, the joint PDF of the new RVs

$$
\mathbf{Y} = \begin{pmatrix} Y_1 \\ \vdots \\ Y_n \end{pmatrix} = \mathbf{B}^{\mathrm{T}} \begin{pmatrix} X_1 \\ \vdots \\ X_n \end{pmatrix}
$$

is $f_{\mathbf{Y}}(\mathbf{y}) = f_{\mathbf{X}}(\mathbf{By})$. More precisely,

$$
f_{\mathbf{Y}}(\mathbf{y}) = \left(\prod_{i=1}^n \frac{1}{\sqrt{2\pi}\lambda_i} \right) \exp\left[-\frac{1}{2}(\mathbf{By} - \mu)^{\mathrm{T}}\Sigma^{-1}(\mathbf{By} - \mu) \right]
$$

$$
= \left(\prod_{i=1}^n \frac{1}{\sqrt{2\pi}\lambda_i} \right) \exp\left[-\frac{1}{2}(\mathbf{y} - \mathbf{B}^{\mathrm{T}}\mu)^{\mathrm{T}} \mathbf{B}^{\mathrm{T}}\mathbf{BD}^{-1}\mathbf{B}^{-1}\mathbf{B}(\mathbf{y} - \mathbf{B}^{\mathrm{T}}\mu) \right]
$$

$$
= \left(\prod_{i=1}^n \frac{1}{\sqrt{2\pi}\lambda_i} \right) \exp\left[-\frac{1}{2}(\mathbf{y} - \mathbf{B}^{\mathrm{T}}\mu)^{\mathrm{T}} \mathbf{D}^{-1}(\mathbf{y} - \mathbf{B}^{\mathrm{T}}\mu) \right]
$$

$$
= \left(\prod_{i=1}^n \frac{1}{\sqrt{2\pi}\lambda_i} \right) \exp\left\{ -\frac{1}{2\lambda_i^2}[y_i - (\mathbf{B}^{\mathrm{T}}\mu)_i]^2 \right\}.
$$

Here, $(\mathbf{B}^{\mathrm{T}}\mu)_i$ is the ith component of the vector $\mathbf{B}^{\mathrm{T}}\mu$. We see that the RVs $Y_1, \ldots, Y_n$ are independent, and $Y_i \sim \mathrm{N}\left((\mathbf{B}^{\mathrm{T}}\mu)_i, \lambda_i^2 \right)$. That is the covariance:

$$
\mathrm{Cov}(Y_i, Y_j) = \mathbb{E}\left[Y_i - (\mathbf{B}^{\mathrm{T}}\mu)_i \right]\left[Y_j - (\mathbf{B}^{\mathrm{T}}\mu)_j \right] = \begin{cases} 0, & i \neq j, \\ \lambda_j^2, & i = j. \end{cases}
$$

Actually, variables Y_i will help us to do calculations with RVs X_j. For example, for the mean value of X_j:

$$\mathbb{E}X_j = \int_{\mathbb{R}^n} \frac{x_j}{(2\pi)^{n/2}\sqrt{\det \Sigma}} \exp\left[-\frac{1}{2}\langle(\mathbf{x} - \mu), \Sigma^{-1}(\mathbf{x} - \mu)\rangle\right] d\mathbf{x}$$

$$= \int_{\mathbb{R}^n} \frac{(\mathbf{By})_j}{(2\pi)^{n/2}\sqrt{\det \Sigma}} \exp\left[-\frac{1}{2}\langle(\mathbf{y} - \mathbf{B}^{\mathsf{T}}\mu), \mathbf{D}^{-1}(\mathbf{y} - \mathbf{B}^{\mathsf{T}}\mu)\rangle\right] d\mathbf{y}$$

$$= \sum_{1 \le i \le n} \int_{\mathbb{R}^n} \frac{y_i b_{ji}}{(2\pi)^{n/2}\sqrt{\det \Sigma}} \exp\left[-\frac{1}{2}\langle(\mathbf{y} - \mathbf{B}^{\mathsf{T}}\mu), \mathbf{D}^{-1}(\mathbf{y} - \mathbf{B}^{\mathsf{T}}\mu)\rangle\right] d\mathbf{y}$$

$$= \sum_{1 \le i \le n} \frac{b_{ji}}{(2\pi)^{1/2}\lambda_i} \int y\exp\left\{-\frac{1}{2\lambda_i^2}[y - (\mathbf{B}^{\mathsf{T}}\mu)_i]^2\right\} d\mathbf{y}$$

$$= \sum_{1 \le i \le n} (\mathbf{B}^{\mathsf{T}}\mu)_i b_{ji} = \mu_j.$$

Similarly, for the covariance $\text{Cov}(X_i, X_j)$:

$$\int_{\mathbb{R}^n} \frac{(x_i - \mu_i)(x_j - \mu_j)}{(2\pi)^{n/2}\sqrt{\det \Sigma}} \exp\left[-\frac{1}{2}\langle(\mathbf{x} - \mu), \Sigma^{-1}(\mathbf{x} - \mu)\rangle\right] d\mathbf{x}$$

$$= \int_{\mathbb{R}^n} \frac{[\mathbf{B}(\mathbf{y} - \mathbf{B}^{\mathsf{T}}\mu)]_i [\mathbf{B}(\mathbf{y} - \mathbf{B}^{\mathsf{T}}\mu)]_j}{(2\pi)^{n/2}\sqrt{\det \Sigma}}$$

$$\times \exp\left[-\frac{1}{2}\langle(\mathbf{y} - \mathbf{B}^{\mathsf{T}}\mu), \mathbf{D}^{-1}(\mathbf{y} - \mathbf{B}^{\mathsf{T}}\mu)\rangle\right] d\mathbf{y}$$

$$= \sum_{1 \le l \le n} \sum_{1 \le m \le n} \text{Cov}(Y_l, Y_m)\, b_{il} b_{jm}$$

$$= \sum_{1 \le m \le n} \lambda_m^2 b_{im} b_{jm} = (\mathbf{BDB}^{-1})_{ij} = \Sigma_{ij}.$$

This proves assertion (IVb). For $i = j$ it gives the variance $\text{Var}\, X_i$.

In fact, an even more powerful tool is the *joint CHF* $\psi_{\mathbf{X}}(\mathbf{t})$ defined by

$$\psi_{\mathbf{X}}(\mathbf{t}) = \mathbb{E}e^{i\langle \mathbf{t}, \mathbf{X}\rangle} = \mathbb{E}\exp\left(i\sum_{j=1}^{n} t_j X_j\right), \quad \mathbf{t}^{\mathsf{T}} = (t_1, \ldots, t_n) \in \mathbb{R}^n. \qquad (2.101)$$

The joint CHF has many features of a marginal CHF. In particular, it determines the joint distribution of a random vector $\mathbf{X}$ uniquely. For multivariate normal vector $\mathbf{X}$ the joint CHF can be calculated explicitly. Indeed,

$$\mathbb{E}e^{i\mathbf{t}^{\mathsf{T}}\mathbf{X}} = \mathbb{E}e^{i\mathbf{t}^{\mathsf{T}}\mathbf{BY}} = \mathbb{E}e^{i(\mathbf{B}^{\mathsf{T}}\mathbf{t})^{\mathsf{T}}\mathbf{Y}} = \prod_{j=1}^{n}\mathbb{E}\exp\left[i\left(\mathbf{B}^{\mathsf{T}}\mathbf{t}\right)_j Y_j\right].$$

Here each factor is a marginal CHF:

$$\mathbb{E}\exp\left[i\left(\mathbf{B}^{\mathsf{T}}\mathbf{t}\right)_j Y_j\right] = \exp\left[i(\mathbf{B}^{\mathsf{T}}\mathbf{t})_j\left(\mathbf{B}^{-1}\mu\right)_j - (\mathbf{B}^{\mathsf{T}}\mathbf{t})_j^2\lambda_j^2/2\right].$$

As $\mathbf{BDB}^{-1} = \Sigma$, the whole product equals

$$\exp\left(\mathrm{i}\mathbf{t}^{\mathrm{T}}\mathbf{BB}^{-1}\mu - \frac{1}{2}\mathbf{t}^{\mathrm{T}}\mathbf{BDB}^{-1}\mathbf{t}\right) = \exp\left(\mathrm{i}\mathbf{t}^{\mathrm{T}}\mu - \frac{1}{2}\mathbf{t}^{\mathrm{T}}\Sigma\mathbf{t}\right).$$

Hence,

$$\psi_{\mathbf{X}}(\mathbf{t}) = e^{\mathrm{i}\mathbf{t}^{\mathrm{T}}\mu - \mathbf{t}^{\mathrm{T}}\Sigma\mathbf{t}/2}. \tag{2.102}$$

Note a distinctive difference in the matrices in the expressions for the multivariate normal PDF and CHF: formula (2.102) has Σ, the covariance matrix, and equation (2.100) has Σ^{-1}.

Now, to obtain the marginal CHF $\psi_{X_j}(t)$, we substitute vector $\mathbf{t} = (0, \ldots, 0, t, 0, \ldots, 0)$ (t in position j) into the RHS of equation (2.102):

$$\psi_{X_j}(t) = \exp\left(\mathrm{i}\mu_j t - t^2 \Sigma_{jj}/2\right).$$

Owing to the uniqueness of a PDF with a given CHF, $X_j \sim \mathrm{N}(\mu_j, \Sigma_{jj})$, as claimed in assertion (IVa).

As a by-product of the above argument, we immediately establish that:

(IVc) If $\mathbf{X} \sim \mathrm{N}(\mu, \Sigma)$, then any subcollection (X_{j_i}) is also jointly normal, with the mean vector (μ_{j_i}) and covariance matrix $(\Sigma_{j_i j_{i'}})$.

For characteristic functions we obtained the following property:

(V) The joint CHF $\psi_{\mathbf{X}}(\mathbf{t})$ of a random vector $\mathbf{X} \sim \mathrm{N}(\mu, \Sigma)$ is of the form

$$\psi_{\mathbf{X}_j}(\mathbf{t}) = \exp\left(\mathrm{i}\langle \mathbf{t}, \mu \rangle - \frac{1}{2}\langle \mathbf{t}, \Sigma\mathbf{t} \rangle\right).$$

Finally, the tools and concepts developed so far also allow us to check that

(VI) A linear combination $\sum_{i=1}^{n} c_i X_i$ of jointly normal RVs, with $\mathbf{X} \sim \mathrm{N}(\mu, \Sigma)$ is normal, with mean $\langle \mathbf{c}, \mu \rangle = \sum_{1 \le i \le n} c_i \mu_i$ and variance $\langle \mathbf{c}, \Sigma\mathbf{c} \rangle = \sum_{1 \le i,j \le n} c_i \Sigma_{ij} c_j$. More generally, if $\mathbf{Y}$ is the random vector of the form $\mathbf{Y} = \mathbf{A}^{\mathrm{T}}\mathbf{X}$ obtained from $\mathbf{X}$ by a linear change of variables with invertible matrix $\mathbf{A}$, then $\mathbf{Y} \sim \mathrm{N}\left(\mathbf{A}^{\mathrm{T}}\mu, \mathbf{A}^{\mathrm{T}}\Sigma\mathbf{A}\right)$. See Example 3.1. (The last fact can also be extended to the case of a non-invertible $\mathbf{A}$ but we will leave this subject to later volumes.)

Problem 2.39 Derive the distribution of the sum of n independent random variables each having the Poisson distribution with parameter λ. Use the CLT to prove that

$$e^{-n}\left(1 + \frac{n}{1!} + \frac{n^2}{2!} + \cdots + \frac{n^n}{n!}\right) \to \frac{1}{2}$$

as $n \to \infty$.

Solution Let $X_1, \ldots, X_n$ be IID Po (1). The PGF

$$\psi_{X_i}(s) = \mathbb{E}s^{X_i} = \sum_{l \ge 0} s^l \frac{1}{l!} e^{-1} = e^{s-1}, \ s \in \mathbb{R}^1.$$

In general, if $Y \sim \mathrm{Po}(\lambda)$, then

$$\psi_Y(s) = \sum_{l \geq 0} \frac{s^l \lambda^l}{l!} e^{-\lambda} = e^{\lambda(s-1)}.$$

Now if $S_n = X_1 + \cdots + X_n$, then

$$\psi_{S_n}(s) = \psi_{X_1}(s) \cdots \psi_{X_n}(s),$$

yielding that $S_n \sim \mathrm{Po}(n)$, with $\mathbb{E} S_n = \mathrm{Var}(S_n) = n$. By the CLT,

$$T_n = \frac{S_n - n}{\sqrt{n}} \sim \mathrm{N}(0, 1) \text{ for } n \text{ large.}$$

But

$$e^{-n}\left(1 + \frac{n}{1!} + \frac{n^2}{2!} + \cdots + \frac{n^n}{n!}\right) = \mathbb{P}(S_n \leq n) = \mathbb{P}(T_n \leq 0)$$

$$\to \frac{1}{\sqrt{2\pi}} \int_{-\infty}^{0} e^{-y^2/2} dy = \frac{1}{2} \text{ as } n \to \infty. \quad \square$$

Problem 2.40 *An algebraic question. If $\mathbf{x}_1, \ldots, \mathbf{x}_n \in \mathbb{R}^n$ are linearly independent column vectors, show that the matrix $n \times n$*

$$\sum_{i=1}^{n} \mathbf{x}_i \mathbf{x}_i^{\mathrm{T}}$$

is invertible.

Solution It is sufficient to show that matrix $\sum_{i=1}^{n} \mathbf{x}_i \mathbf{x}_i^T$ does not send any non-zero vector to zero. Hence, assume that

$$\left(\sum_{i=1}^{n} \mathbf{x}_i \mathbf{x}_i^T\right) \mathbf{c} = 0, \text{ i.e. } \sum_{k=1}^{n}\sum_{i=1}^{n} x_{ij} x_{ik} c_k = \sum_{i=1}^{n} x_{ij} \langle \mathbf{x}_i, \mathbf{c} \rangle = 0, \ 1 \leq j \leq n,$$

where

$$\mathbf{x}_i = \begin{pmatrix} x_{i1} \\ \vdots \\ x_{in} \end{pmatrix}, \ 1 \leq i \leq n, \quad \text{and } \mathbf{c} = \begin{pmatrix} c_1 \\ \vdots \\ c_n \end{pmatrix}.$$

The last equation means that the linear combination

$$\sum_{i=1}^{n} \mathbf{x}_i \langle \mathbf{x}_i, \mathbf{c} \rangle = 0.$$

Since the $\mathbf{x}_i$ are linearly independent, the coefficients $\langle \mathbf{x}_i, \mathbf{c} \rangle = 0$, $1 \leq i \leq n$. But this means that $\mathbf{c} = 0$. $\quad \square$

A couple of problems below have been borrowed from advanced statistical courses; they may be omitted at the first reading, but are useful for those readers who aim to achieve better understanding at this stage.

Problem 2.41   Let $X_1, \ldots, X_n$ be an IID sample from $N(\mu_1, \sigma_1^2)$ and $Y_1, \ldots, Y_m$ be an IID sample from $N(\mu_2, \sigma_2^2)$ and assume the two samples are independent of each other. What is the joint distribution of

the difference $\overline{X} - \overline{Y}$ and the sum $\dfrac{1}{\sigma_1^2} \sum_{i=1}^{n} X_i + \dfrac{1}{\sigma_2^2} \sum_{j=1}^{m} Y_j$?

Solution   We have $\overline{X} \sim N(\mu_1, \sigma_1^2/n)$ and $\overline{Y} \sim N(\mu_2, \sigma_2^2/n)$. Further,

$$f_{\overline{X}, \overline{Y}}(x, y) = \left(\frac{n}{2\pi\sigma_1^2} \right)^{1/2} \left(\frac{m}{2\pi\sigma_2^2} \right)^{1/2} \exp\left[-n \frac{(x - \mu)^2}{2\sigma_1^2} - m \frac{(y - \mu_2)^2}{2\sigma_2^2} \right].$$

We see that both

$$U = \overline{X} - \overline{Y} \text{ and } S = \frac{1}{\sigma_1^2} \sum_{i=1}^{n} X_i + \frac{1}{\sigma_2^2} \sum_{j=1}^{m} Y_j = \frac{n}{\sigma_1^2} \overline{X} + \frac{m}{\sigma_2^2} \overline{Y}$$

are linear combinations of (independent) normal RVs and hence are normal. A straight-forward calculation shows that

$$\mathbb{E}U = \mu_1 - \mu_2, \quad \mathbb{E}S = \frac{n}{\sigma_1^2} \mu_1 + \frac{m}{\sigma_2^2} \mu_2,$$

and

$$\text{Var } U = \frac{m\sigma_1^2 + n\sigma_2^2}{mn}, \quad \text{Var } S = \frac{n}{\sigma_1^2} + \frac{m}{\sigma_2^2}.$$

So

$$f_U(u) = \frac{(mn)^{1/2}}{[2\pi(m\sigma_1^2 + n\sigma_2^2)]^{1/2}} \exp\left\{ \frac{[u - (\mu_1 - \mu_2)]^2}{2(m\sigma_1^2 + n\sigma_2^2)/(mn)} \right\}, \quad u \in \mathbb{R},$$

and

$$f_S(s) = \frac{\sigma_1\sigma_2}{[2\pi(m\sigma_1^2 + n\sigma_2^2)]^{1/2}}$$

$$\times \exp\left\{ -\left[s - \left(\frac{n}{\sigma_1^2} \mu_1 + \frac{m}{\sigma_2^2} \mu_2 \right) \right]^2 \middle/ \frac{2(m\sigma_1^2 + n\sigma_2^2)}{\sigma_1^2 \sigma_2^2} \right\}, \quad s \in \mathbb{R}.$$

Finally, U and S are independent as they have Cov $(U, S) = 0$. Therefore, the joint PDF $f_{U,S}(u, s) = f_U(u) f_S(s)$.   □

Remark   The formulas for $f_U(u)$ and $f_S(s)$ imply that the pair (U, S) forms the so-called 'sufficient statistic' for the pair of unknown parameters (μ_1, μ_2); see Section 3.2.

Problem 2.42   Let $X_1, \ldots, X_n$ be a random sample from the $N(0, \sigma^2)$ distribution, and suppose that the prior distribution for θ is the $N(\mu, \tau^2)$ distribution, where σ^2, μ and τ^2 are known. Determine the posterior distribution for θ, given $X_1, \ldots, X_n$.

Solution The prior PDF is Gaussian:

$$\pi(\theta) = \frac{1}{\sqrt{2\pi}\tau} e^{-(\theta-\mu)^2/(2\tau^2)},$$

and so is the (joint) PDF of $X_1, \ldots, X_n$ for the given value of θ:

$$f_{X_1,\ldots,X_n}(x_1,\ldots,x_n;\theta) = \prod_{i=1}^n \frac{1}{\sqrt{2\pi}\sigma} e^{-(x_i-\theta)^2/(2\sigma^2)}.$$

Thus

$$\pi(\theta)f_{X_1,\ldots,X_n}(x_1,\ldots,x_n;\theta) \propto \exp\left[-\frac{(\theta-\mu)^2}{2\tau^2} - \sum_i \frac{(x_i-\theta)^2}{2\sigma^2}\right]$$

$$= \exp\left[-\frac{1}{2\tau^2}(\theta^2 - 2\mu\theta + \mu^2) - \frac{1}{2\sigma^2}\sum_i\left(x_i^2 - 2\theta\sum_i x_j + \theta^2\right)\right]$$

$$= \exp\left\{-\frac{1}{2}\left[\theta^2\left(\frac{1}{\tau^2} + \frac{n}{\sigma^2}\right) - 2\theta\left(\frac{\mu}{\tau^2} + \frac{n\bar{x}}{\sigma^2}\right)\right]\right\},$$

leaving out terms not involving θ. Here and below

$$\bar{x} = \frac{1}{n}\sum_{i=1}^n x_i.$$

Then the posterior

$$\pi(\theta|x_1,\ldots,x_n) = \frac{\pi(\theta)f_{X_1,\ldots,X_n}(x_1,\ldots,x_n;\theta)}{\int \pi(\theta')f_{X_1,\ldots,X_n}(x_1,\ldots,x_n;\theta')d\theta'}$$

$$= \frac{1}{\sqrt{2\pi}\tau_n}\exp\left[-\frac{(\theta-\mu_n)^2}{2\tau_n^2}\right],$$

where

$$\frac{1}{\tau_n^2} = \frac{1}{\tau^2} + \frac{n}{\sigma^2}, \quad \mu_n = \frac{\mu/\tau^2 + n\bar{x}/\sigma^2}{1/\tau^2 + n/\sigma^2}. \quad \square$$

Problem 2.43 Let $\Theta, X_1, X_2, \ldots$ be RVs. Suppose that, conditional on $\Theta = \theta$, $X_1, X_2, \ldots$ are independent and X_k is normally distributed with mean θ and variance σ_k^2. Suppose that the marginal PDF of Θ is

$$\pi(\theta) = \frac{1}{\sqrt{2\pi}}e^{-\frac{\theta^2}{2}}, \quad \theta \in \mathbb{R}.$$

Calculate the mean and variance of Θ conditional on $X_1 = x_1, \ldots, X_n = x_n$.

Solution A direct calculation shows that the conditional PDF $f_{\Theta|X_1,\ldots,X_n}(\theta) = f(\theta|x_1,\ldots,x_n)$ is a multiple of

$$\exp\left[-\frac{1}{2}\left(1 + \sum_i\frac{1}{\sigma_i^2}\right)\left(\theta - \frac{\sum_i x_i/\sigma_i^2}{1 + \sum_i 1/\sigma_i^2}\right)^2\right];$$

with a coefficient depending on values $x_1, \ldots, x_n$ of $X_1, \ldots, X_n$. This implies that the conditional mean

$$\mathbb{E}(\Theta|X_1, \ldots, X_n) = \frac{1}{1 + \sum_i 1/\sigma_i^2} \sum_i \frac{X_i}{\sigma_i^2}$$

and the conditional variance

$$\text{Var}(\Theta|X_1, \ldots, X_n) = \frac{1}{1 + \sum_i 1/\sigma_i^2}, \quad \text{independently of } X_1, \ldots, X_n. \quad \square$$

Problem 2.44   Let X and Y be independent, identically distributed RVs with the standard normal PDF

$$f(x) = \frac{1}{\sqrt{2\pi}} e^{-x^2/2}, \, x \in \mathbb{R}.$$

Find the joint PDF of $U = X + Y$ and $V = X - Y$. Show that U and V are independent and write down the marginal distribution for U and V. Let

$$Z = \begin{cases} |Y|, & \text{if } X > 0, \\ -|Y|, & \text{if } X < 0. \end{cases}$$

By finding $\mathbb{P}(Z \leq z)$ for $z < 0$ and $z > 0$, show that Z has a standard normal distribution. Explain briefly why the joint distribution of X and Z is not bivariate normal.

Solution   Write $(U, V) = T(X, Y)$. Then for the PDF:

$$f_{U,V}(u, v) = f_{X,Y}(T^{-1}(u, v)) \left| \frac{\partial(x, y)}{\partial(u, v)} \right|.$$

The inverse map is

$$T^{-1}(u, v) = \left(\frac{u+v}{2}, \frac{u-v}{2} \right), \quad \text{with } \frac{\partial(x, y)}{\partial(u, v)} = -\frac{1}{2}.$$

Hence,

$$f_{U,V}(u, v) = \frac{1}{2\pi} e^{-\frac{1}{2}((u+v)^2/4 + (u-v)^2/4)} \frac{1}{2} = \frac{1}{4\pi} e^{-\frac{1}{4}(u^2+v^2)},$$

i.e. U, V are independent. Next, if $z \geq 0$ then

$$\mathbb{P}(Z < z) = \frac{1}{2}(1 + \mathbb{P}(|Y| < z)) = \mathbb{P}(Y < z),$$

and if $z < 0$, then

$$\mathbb{P}(Z < z) = \frac{1}{2}\mathbb{P}(-|Y| < z) = \mathbb{P}(Y > |z|) = \mathbb{P}(Y < z).$$

So Z has the same standard normal distribution as Y. But the joint distribution of (X, Z) gives zero mass to the second and fourth quadrants; hence Z is not independent of X. $\square$

Problem 2.45   Check that the standard normal PDF $p(x) = e^{-x^2/2}/\sqrt{2\pi}$ satisfies the equation

$$\int_y^\infty xp(x)dx = p(y), \quad y > 0.$$

By using this equation and $\sin x = \int_0^x \cos y \, dy$, or otherwise, prove that if X is an $N(0, 1)$ random variable, then

$$(\mathbb{E}\cos X)^2 \le \mathrm{Var}(\sin X) \le \mathbb{E}(\cos X)^2.$$

Solution   Write

$$\frac{1}{\sqrt{2\pi}} \int_y^\infty x e^{-x^2/2} dx = \frac{1}{\sqrt{2\pi}} \int_y^\infty e^{-x^2/2} d\left(\frac{x^2}{2}\right) = \frac{1}{\sqrt{2\pi}} e^{-y^2/2}.$$

Now

$$\mathbb{E}\sin X = \frac{1}{\sqrt{2\pi}} \int e^{-x^2/2} \sin x \, dx = 0$$

as $e^{-x^2/2}\sin x$ is an odd function. Thus,

$$\mathrm{Var}(\sin X) = \mathbb{E}(\sin X)^2 = \int p(x)(\sin x)^2 dx = \int p(x)\left(\int_0^x \cos y \, dy\right)^2 dx.$$

Owing to the CS inequality, the last integral is

$$\le \int p(x)|x| \int_0^x (\cos y)^2 dy dx$$

$$= -\int_{-\infty}^0 (\cos y)^2 \int_{-\infty}^y xp(x) dx dy + \int_0^\infty (\cos y)^2 \int_y^\infty xp(x) dx dy$$

$$= \int_{-\infty}^0 p(y)(\cos y)^2 dy + \int_0^\infty p(y)(\cos y)^2 dy$$

$$= \int p(y)(\cos y)^2 dy = \mathbb{E}(\cos X)^2.$$

On the other hand, as $\mathbb{E}X^2 = 1$,

$$\mathrm{Var}(\sin X) = \mathbb{E}(\sin X)^2 = \mathbb{E}X^2 \mathbb{E}(\sin X)^2$$

$$\ge [\mathbb{E}(X\sin X)]^2 = \left[\int xp(x)dx \int_0^x \cos y \, dy\right]^2$$

$$= \left[-\int_{-\infty}^0 \cos y \int_{-\infty}^y xp(x)dxdy + \int_0^\infty \cos y \int_y^\infty xp(x)dxdy\right]^2$$

$$= \left(\int p(y)\cos y \, dy\right)^2 = (\mathbb{E}\cos X)^2. \quad \square$$

Problem 2.46   In Problem 2.23, prove that X and Y are independent if and only if $r = 0$.

Solution In general, if X, Y are independent then $r = \mathbb{E}XY = \mathbb{E}X\mathbb{E}Y = 0$. In the Gaussian case, the inverse is also true: if $r = 0$, then the joint PDF

$$f_{X,Y}(x, y) = \frac{1}{2\pi}e^{-(x^2+y^2)/2} = \frac{1}{\sqrt{2\pi}}e^{-x^2/2}\frac{1}{\sqrt{2\pi}}e^{-y^2/2} = f_X(x)f_Y(y),$$

i.e. X and Y are independent. □

Problem 2.47 State the CLT for independent identically distributed real RVs with mean μ and variance σ^2.

Suppose that X_1, X_2, ... are independent identically distributed random variables each uniformly distributed over the interval $[0, 1]$. Calculate the mean and variance of $\ln X_1$.

Suppose that $0 \le a < b$. Show that

$$\mathbb{P}\left((X_1 X_2 \ldots X_n)^{n^{-1/2}}e^{n^{1/2}} \in [a, b]\right)$$

tends to a limit and find an expression for it.

Solution Let X_1, X_2, ..., be IID RVs with $\mathbb{E}X_i = \mu$ and Var $X_i = \sigma^2$. The CLT states that $\forall -\infty \le a < b \le \infty$:

$$\lim_{n \to \infty} \mathbb{P}\left(a < \frac{X_1 + \cdots + X_n - n\mu}{\sqrt{n}\sigma} < b\right) = \frac{1}{\sqrt{2\pi}}\int_a^b e^{-x^2/2}dx.$$

Moreover, if $X \sim \mathrm{U}[0, 1]$, then the mean value $\mathbb{E}(\ln X_i)$ equals

$$\int_0^1 \ln x\,dx = -\int_\infty^0 y\,(de^{-y}) = (-ye^{-y})|_\infty^0 + \int_\infty^0 e^{-y}dy = -1.$$

Similarly, the mean value $\mathbb{E}(\ln X_i)^2$ is equal to

$$\int_0^1 (\ln X)^2 dx = \int_\infty^0 y^2 d(e^{-y}) = (y^2 e^{-y})|_\infty^0 - 2\int_\infty^0 e^{-y}y\,dy = 2,$$

and

$$\text{Var}\,(\ln X_i) = 2 - (-1)^2 = 1.$$

Finally,

$$\mathbb{P}\left((X_1 X_2 \ldots X_n)^{n^{-1/2}}e^{n^{1/2}} \in [a, b]\right)$$

$$= \mathbb{P}\left(\frac{1}{\sqrt{n}}\sum_{i=1}^n \ln X_i + \sqrt{n} \in [\ln a, \ln b]\right)$$

$$= \mathbb{P}\left(\frac{1}{\sqrt{n}}\left(\sum_{i=1}^n \ln X_i + n\right) \in [\ln a, \ln b]\right)$$

$$\to \frac{1}{\sqrt{2\pi}}\int_{\ln a}^{\ln b} e^{-x^2/2}dx, \text{ by the CLT, as } n \to \infty. \quad \square$$

Problem 2.48   The RV X_i is normally distributed with mean μ_i and variance σ_i^2, for $i = 1, 2$, and X_1 and X_2 are independent. Find the distribution of $Z = a_1 X_1 + a_2 X_2$, where $a_1, a_2 \in \mathbb{R}$.

(You may assume that $\mathbb{E} e^{\theta X_i} = \exp\left(\theta \mu_i + \theta^2 \sigma_i^2/2\right)$.)

Solution   The MGF

$$\phi_{a_1 X_1 + a_2 X_2}(\theta) = \mathbb{E}\left(e^{\theta(a_1 X_1 + a_2 X_2)}\right) = \mathbb{E}\left(e^{\theta a_1 X_1} e^{\theta a_2 X_2}\right)$$
$$= \mathbb{E}\left(e^{\theta a_1 X_1}\right) \mathbb{E}\left(e^{\theta a_2 X_2}\right) = \phi_{X_1}(a_1 \theta) \phi_{X_2}(a_2 \theta),$$

by independence. Next,

$$\phi_{X_1}(a_1 \theta) \phi_{X_2}(a_2 \theta) = \exp\left(a_1 \theta \mu_1 + \frac{a_1^2 \theta^2 \sigma_1^2}{2} + a_2 \theta \mu_2 + \frac{a_2^2 \theta^2 \sigma_2^2}{2}\right)$$
$$= \exp\left[\theta(a_1 \mu_1 + a_2 \mu_1) + \frac{\theta^2(a_1^2 \sigma_1^2 + a_2^2 \sigma_2^2)}{2}\right] = M_Z(\theta),$$

where $Z \sim N\left((a_1 \mu_1 + a_2 \mu_1), (a_1^2 \sigma_1^2 + a_2^2 \sigma_2^2)\right)$. In view of the uniqueness of a PDF with a given MGF,

$$\left(a_1 X_1 + a_2 X_2\right) \sim N\left((a_1 \mu_1 + a_2 \mu_1), (a_1^2 \sigma_1^2 + a_2^2 \sigma_2^2)\right). \quad \square$$

Problem 2.49   Let X be a normally distributed RV with mean 0 and variance 1. Compute $\mathbb{E} X^r$ for $r = 0, 1, 2, 3, 4$. Let Y be a normally distributed RV with mean μ and variance σ^2. Compute $\mathbb{E} Y^r$ for $r = 0, 1, 2, 3, 4$. State, without proof, what can be said about the sum of two independent RVs.

The President of Statistica relaxes by fishing in the clear waters of Lake Tchebyshev. The number of fish that she catches is a Poisson variable with parameter λ. The weight of each fish in Lake Tchebyshev is an independent normally distributed RV with mean μ and variance σ^2. (Since μ is much larger than σ, fish of negative weight are rare and much prized by gourmets.) Let Z be the total weight of her catch. Compute $\mathbb{E} Z$ and $\mathbb{E} Z^2$.

Show, quoting any results you need, that the probability that the President's catch weighs less than $\lambda \mu/2$ is less than $4(\mu^2 + \sigma^2)\lambda^{-1}\mu^{-2}$.

Solution $\mathbb{E} X^0 = \mathbb{E} 1 = 1$. $\mathbb{E} X^1 = \mathbb{E} X^3 = 0$, by symmetry. Next, $\mathbb{E} X^2$ and $\mathbb{E} X^4$ are found by integration by parts:

$$\frac{1}{\sqrt{2\pi}} \int x^2 e^{-x^2/2} dx = \frac{1}{\sqrt{2\pi}} \left[\left(-x e^{-x^2/2}\right)\Big|_{-\infty}^{\infty} + \int e^{-x^2/2} dx\right] = 1,$$

$$\frac{1}{\sqrt{2\pi}} \int x^4 e^{-x^2/2} dx = \frac{1}{\sqrt{2\pi}} \left[\left(-x^3 e^{-x^2/2}\right)\Big|_{-\infty}^{\infty} + 3 \int x^2 e^{-x^2/2} dx\right] = 3.$$

Further,

$$\mathbb{E}Y^0 = \mathbb{E}1 = 1,$$
$$\mathbb{E}Y^1 = \mathbb{E}(\sigma X + \mu) = \sigma\mathbb{E}X + \mu\mathbb{E}1 = \mu,$$
$$\mathbb{E}Y^2 = \mathbb{E}\left(\sigma^2 X^2 + 2\mu\sigma X + \mu^2\right) = \sigma^2 + \mu^2,$$
$$EY^3 = \mathbb{E}\left(\sigma^3 X^3 + 3\mu\sigma^2 X^2 + 3\mu^2\sigma X + \mu^3\right) = 3\mu\sigma^2 + \mu^3$$

and

$$\mathbb{E}Y^4 = \mathbb{E}\left(\sigma^4 X^4 + 4\mu\sigma^3 X^3 + 6\mu^2\sigma^2 X^2 + 4\mu^3\sigma X + \mu^4\right) = 3\sigma^4 + 6\mu^2\sigma^2 + \mu^4.$$

Now, if X_1, X_2 are independent normal RVs of means μ_1, μ_2 and variances σ_1^2, σ_2^2, then $X_1 + X_2$ is $\mathrm{N}\left(\mu_1 + \mu_2, \sigma_1^2 + \sigma_2^2\right)$.

Thus, if Y_r is the weight of r fish, then $Y_r \sim \mathrm{N}(r\mu, r\sigma^2)$.

Finally,

$$\mathbb{E}Z = \sum_{r\geq 0} \mathbb{P}\,(\text{catch } r)\,\mathbb{E}Y_r = \sum_{r\geq 0} \frac{\lambda^r e^{-\lambda}}{r!} r\mu = \lambda\mu,$$

and similarly

$$\mathbb{E}Z^2 = \sum_{r\geq 0} \frac{\lambda^r e^{-\lambda}}{r!} \mathbb{E}Y_r^2 = \sum_{r\geq 0} \frac{\lambda^r e^{-\lambda}}{r!}\left(r\sigma^2 + r^2\mu^2\right)$$

$$= \lambda\sigma^2 + \mu^2 \sum_{r>1} \frac{\lambda^r e^{-\lambda}}{r!} r(r-1) + \mu^2 \sum_{r\geq 1} \frac{\lambda^r e^{-\lambda}}{r!} r = \lambda(\sigma^2 + \mu^2) + \lambda^2\mu^2.$$

This yields

$$\mathrm{Var}\, Z = \mathbb{E}Z^2 - \left(\mathbb{E}Z\right)^2 = \lambda(\sigma^2 + \mu^2).$$

Then by Chebyshev's inequality:

$$\mathbb{P}\left(Z < \frac{\lambda\mu}{2}\right) \leq \mathbb{P}\left(|Z - \lambda\mu| > \frac{\lambda\mu}{2}\right) \leq \frac{\mathrm{Var}\, Z}{\left(\lambda\mu/2\right)^2} = \frac{4(\sigma^2 + \mu^2)}{\lambda\mu^2}. \quad \square$$

Problem 2.50 Let X and Y be independent and normally distributed RVs with the same PDF

$$\frac{1}{\sqrt{2\pi}} e^{-x^2/2}.$$

Find the PDFs of:

(i) $X + Y$;
(ii) X^2;
(iii) $X^2 + Y^2$.

Solution   (i) For the CDF, we have

$$F_{X+Y}(t) = \frac{1}{2\pi} \int\int e^{-(x^2+y^2)/2} I(x+y \leq t)\,dy\,dx.$$

As PDF $e^{-(x^2+y^2)/2}/2\pi$ is symmetric relative to rotations, the last expression equals

$$\frac{1}{2\pi}\int e^{-(x^2+y^2)/2}I\left(x\leq\frac{t}{\sqrt{2}}\right)dydx=\frac{1}{2\sqrt{\pi}}\int_{-\infty}^{t}e^{-u^2/4}du,$$

whence the PDF

$$f_{X+Y}(x)=\frac{1}{2\sqrt{\pi}}e^{-x^2/4}.$$

(ii) Similarly, for $t\geq 0$:

$$F_{X^2}(t)=\frac{1}{2\pi}\int\int e^{-(x^2+y^2)/2}I(x^2\leq t)dydx$$

$$=\frac{1}{\sqrt{2\pi}}\int_{-\sqrt{t}}^{\sqrt{t}}e^{-x^2/2}dx=\frac{1}{\sqrt{2\pi}}\int_{0}^{t}e^{-u/2}\frac{du}{\sqrt{u}},$$

which yields

$$f_{X^2}(x)=\frac{1}{\sqrt{2\pi x}}e^{-x/2}I(x\geq 0).$$

(iii) Finally:

$$F_{X^2+Y^2}(t)=\frac{1}{2\pi}\int\int e^{-(x^2+y^2)/2}I(x^2+y^2\leq t)dydx$$

$$=\frac{1}{2\pi}\int_{0}^{\infty}\int_{0}^{2\pi}re^{-r^2/2}I(r^2\leq t)d\theta dr=\frac{1}{2}\int_{0}^{t}e^{-u/2}du,$$

and the PDF

$$f_{X^2+Y^2}(x)=\frac{1}{2}e^{-x/2}I(x\geq 0). \quad \square$$

Problem 2.51   The PDF for the t distribution with q degrees of freedom is

$$f(x;q)=\frac{\Gamma((q+1)/2)}{\Gamma(q/2)\sqrt{\pi q}}\left(1+\frac{x^2}{q}\right)^{-(q+1)/2}, \quad -\infty < t < \infty.$$

Cf. equation (3.6). Using properties of the exponential function, and the result that

$$\Gamma\left(\frac{q}{2}+b\right)\rightarrow\sqrt{2\pi}\left(\frac{q}{2}\right)^{b+(q-1)/2}\exp\left(-\frac{q}{2}\right)$$

as $q\rightarrow\infty$, prove that $f(x;q)$ tends to the PDF of an $N(0, 1)$ RV in this limit.
Hint: Write

$$\left(1+\frac{t^2}{q}\right)^{-(q+1)/2}=\left[\left(1+\frac{t^2}{q}\right)^{q}\right]^{-1/2-1/2q}.$$

Derive the PDF of variable $Y = Z^2$, where Z is N(0, 1). The PDF for the F-distribution with $(1, q)$ degrees of freedom is

$$g\,(x;\,q) = \frac{\Gamma\,((q+1)/2)}{\Gamma\,(q/2)\sqrt{\pi q}}\,x^{-1/2}\left(1 + \frac{x}{q}\right)^{-(q+1)/2}, \quad 0 < x < \infty.$$

Using the above limiting results, show that $f\,(x;\,q)$ tends to the PDF of Y as $q \to \infty$.

Solution   (The second part only.) The PDF of Y equals $I(x > 0)\mathrm{e}^{-x/2}/\sqrt{2\pi x}$. The analysis of the ratio of gamma functions shows that

$$\left(1 + \frac{v}{q}\right)^{-(q+1)/2} \to \exp\left(-\frac{v}{2}\right).$$

Therefore, the PDF for the F-distribution tends to

$$\frac{\sqrt{q/2}}{\sqrt{\pi q}}\,x^{-1/2}\exp\left(-\frac{x}{2}\right) = \frac{1}{\sqrt{2\pi x}}\exp\left(-\frac{x}{2}\right),$$

as required.

This result is natural. In fact, by Example 3.4, the $F_{1,q}$-distribution is related to the ratio

$$\frac{X_1^2}{\sum\limits_{j=1}^{q} Y_j^2/q}.$$

where $X_1, Y_1, \ldots, Y_q$ are IID N(0,1). The denominator $\sum_{j=1}^{q} Y_j^2/q$ tends to 1 as $q \to \infty$ by the LLN.   □

Problem 2.52   Let $X_1, X_2, \ldots$ be independent Cauchy RVs, each with PDF

$$f(x) = \frac{d}{\pi(d^2 + x^2)}.$$

Show that $(X_1 + X_2 + \cdots + X_n)/n$ has the same distribution as X_1.

Does this contradict the weak LLN or the CLT?

Hint: The CHF of X_1 is $\mathrm{e}^{-|t|}$, and so is the CHF of $(X_1 + X_2 + \cdots + X_n)/n$. The result follows by the uniqueness of the PDF with a given CHF.

The LLN and the CLT require the existence of the mean value and the variance.

Problem 2.53   Let $X \sim N(\mu, \sigma^2)$ and suppose $h(x)$ is a smooth bounded function, $x \in \mathbb{R}$. Prove *Stein's formula*

$$\mathbb{E}[(X - \mu)h(X)] = \sigma^2\mathbb{E}[h'(X)].$$

Solution   Without loss of generality we assume $\mu = 0$. Then

$$
\begin{aligned}
\mathbb{E}[Xh(X)] &= \frac{1}{\sqrt{2\pi\sigma^2}} \int xh(x)e^{-x^2/2\sigma^2} dx \\
&= \frac{1}{\sqrt{2\pi\sigma^2}} \int h(x)d\left(-\sigma^2 e^{-x^2/2\sigma^2}\right) \\
&= \frac{1}{\sqrt{2\pi\sigma^2}} \int h'(x)\sigma^2 e^{-x^2/2\sigma^2} dx,
\end{aligned}
$$

which holds because the integrals converge absolutely. $\square$

Problem 2.54   Let $X \sim N(\mu, \sigma^2)$ and let Φ be the CDF of $N(0, 1)$. Suppose that $h(x)$ is a smooth bounded function, $x \in \mathbb{R}$. Prove for any real numbers θ, α, β the following equations hold:

$$
\mathbb{E}\left[e^{\theta X} h(X)\right] = e^{\mu\theta + \sigma^2\theta^2/2} \mathbb{E}[h(X + \theta\sigma^2)]
$$

and

$$
\mathbb{E}[\Phi(\alpha X + \beta)] = \Phi\left(\frac{\alpha\mu + \beta}{\sqrt{1 + \alpha^2\sigma^2}}\right).
$$

Solution   Again, assume without loss of generality that $\mu = 0$. Then

$$
\begin{aligned}
\mathbb{E}\left[e^{\theta X} h(X)\right] &= \frac{1}{\sqrt{2\pi\sigma^2}} \int e^{\theta x - x^2/2\sigma^2} h(x)dx \\
&= \frac{1}{\sqrt{2\pi\sigma^2}} \int e^{-(x-\sigma^2\theta)^2/2\sigma^2} e^{\sigma^2\theta^2/2} h(x)dx \\
&= \frac{e^{\sigma^2\theta^2/2}}{\sqrt{2\pi\sigma^2}} \int e^{-(x-\sigma^2\theta)^2/(2\sigma^2)} h(x)dx = e^{\sigma^2\theta^2/2} \mathbb{E}[h(X + \theta\sigma^2)].
\end{aligned}
$$

All the integrals here are absolutely converging.

In the proof of the second formula we keep a general value of μ. If $Z \sim N(\mu, 1)$, independently of X, then

$$
\begin{aligned}
\mathbb{E}[\Phi(\alpha X + \beta)] &= P(Z \le \alpha X + \beta) = \mathbb{P}(Z - \alpha(X - \mu) \le \alpha\mu + \beta) \\
&= \mathbb{P}\left(\frac{Z - \alpha(X - \mu)}{\sqrt{1 + \alpha^2\sigma^2}} \le \frac{\alpha\mu + \beta}{\sqrt{1 + \alpha^2\sigma^2}}\right) = \Phi\left(\frac{\alpha\mu + \beta}{\sqrt{1 + \alpha^2\sigma^2}}\right). \quad \square
\end{aligned}
$$

Problem 2.55   Suppose that (X, Y) has a jointly normal distribution, and $h(x)$ is a smooth bounded function. Prove the following relations:

$$
\mathbb{E}[(Y - \mathbb{E}Y)X] = \frac{\text{Cov}[X, Y]}{\text{Var } X}(X - \mathbb{E}X)
$$

and

$$
\text{Cov}[h(X), Y] = \mathbb{E}[h'(X)] \, \text{Cov}[X, Y].
$$

Solution   Again, assume that both X and Y have mean zero. The joint PDF $f_{X,Y}(x, y)$ is

$$\propto \exp\left[-\frac{1}{2}(x, y)\Sigma^{-1}\begin{pmatrix}x\\y\end{pmatrix}\right]$$

where $\Sigma = (\Sigma_{ij})$ is the 2×2 covariance matrix. Then, conditional on $X = x$, the PDF $f_Y(y|x)$ is

$$\propto \exp\left\{-\frac{1}{2}\left[(\Sigma^{-1})_{22}\,y^2 + 2xy\,(\Sigma^{-1})_{12}\right]\right\}.$$

Recall, Σ^{-1} is the inverse of the covariance matrix Σ. This indicates that the conditional PDF $f_Y(y|x)$ is a Gaussian whose mean is linear in x. That is

$$\mathbb{E}(Y|X) = \gamma X.$$

To find γ, multiply by X and take the expectation. The LHS gives $\mathbb{E}(XY) = \text{Cov}\,[X, Y]$ and the RHS $\gamma\,\text{Var}\,X$. The second equality follows from this result by Stein's formula

$$\text{Cov}\,[h(X), Y] = \mathbb{E}\,[h(X)Y] = \mathbb{E}\left[h(X)\mathbb{E}\,(Y|X)\right]$$

$$= \frac{\text{Cov}\,[X, Y]}{\text{Var}\,X}\,\mathbb{E}\,[Xh(X)] = \mathbb{E}\,[h'(X)]\,\text{Cov}\,[X, Y]. \quad \square$$

Part II
Basic statistics

3 Parameter estimation

3.1 Preliminaries. Some important probability distributions

> All models are wrong but some are useful.
> G.P.E. Box (1919–), American statistician

> Model without a Cause
> (From the series 'Movies that never made it to the Big Screen'.)

In the second half of this volume we discuss the material from the second year (Part IB) Statistics. This material will be treated as a natural continuation of the IA probability course. Statistics, which is called an 'applicable' subject by the Faculty of Mathematics of Cambridge University, occupies a place somewhere between 'pure' and 'applied' disciplines in the current Cambridge University course landscape. One modern definition is that statistics is *a collection of procedures and principles for gaining and processing information in order to make decisions when faced with uncertainty*. It is interesting to compare this with earlier interpretations of the term 'statistics' and related terms. Traditionally, the words 'statistic' and 'statistics' stem from 'state', meaning a political form of government. In fact, the words 'statist' appears in *Hamlet*, Act 5, Scene 2:

> Hamlet: Being thus benetted round with villainies,-
> Ere I could make a prologue to my brains,
> They had begun the play,- I sat me down,
> Devis'd a new commision; wrote it fair:
> I once did hold it, as our's **statists** do,
> A baseness to write fair, and labour'd much
> How to forget that learning; but, sir, now
> It did me yeoman's service. Wilt thou know
> Th' effect of what I wrote?

and then in *Cymbeline*, Act 2, Scene 4:

> Posthumus: I do believe,
> **Statist** though I am none, nor like to be,
> That this will prove a war; . . .

The meaning of the word 'statist' seems to be a person performing a state function. (The glossary to *The Complete Works by William Shakespeare. The Alexander Text* (London and Glasgow: Collins, 1990) simply defines it as 'statesman'.) In a similar sense, the same word is used in Milton's *Paradise Regained*, The Fourth Book, Line 355.

Many of the definitions of statistics that appeared before 1935 can be found in [Wil]; their meaning is essentially 'a description of the past or present political and financial situation of a given realm'. Characteristically, Napoleon described statistics as 'a budget of things'.

The definition of statistics remained a popular occupation well after 1935 [NiFY], with a wide variety of opinions expressed by different authors (and sometimes by a single author over an interval of time). Political and ideological factors added to the confusion: Soviet-era authors concertedly attacked Western writers for portraying statistics as a methodological, rather than a material, science. The limit of absurdity was to proclaim the existence of 'proletarian statistics', as opposed to 'bourgeois statistics'. The former was helping in the 'struggle of the working class against its exploitators', while the latter was 'a servant of the monopolistic capital'.

A particularly divisive issue became the place and rôle of mathematical statistics. For example, G.E.P. Box (1919–), a British-born American statistician who began his career as a chemistry student and then served as a practising statistician in the British Army during World War II, wrote that it was a "mistake to invent the term 'mathematical statistics'. This grave blunder has led to a great number of difficulties."

It is interesting to compare this with two rather different sentences by J.W. Tukey (1915–2000), one of the greatest figures of all time in statistics and many areas of applied mathematics, credited, among many other things, with the invention of the terms 'bit' (short for binary digit) and 'software.' Tukey, who trained as a pure mathematician (his Ph.D. was in topology), said that 'Statistics is a part of a perplexed and confusing network connecting mathematics, scientific philosophy and other branches of science, including experimental sampling, with what one does in analysing and sometimes in collecting data'. On the other hand, Tukey expressed the opinion: 'Statistics is a part of applied mathematics which deals with (although not exclusively) stochastic processes'. The latter point of view was endorsed in a substantial number of universities (Cambridge included) where many of the members of Statistics Departments or units (including holders of Chairs of Mathematical Statistics) are in fact specialists in stochastic processes.

In the statistics part of this book one learns various ways to process observed data and draw inferences from it: point estimation, interval estimation, hypothesis testing and regression modelling. Some of the methods are based on a clear logical foundation, but some appear *ad hoc* and are adopted simply because they provide answers to (important) practical questions.

It may appear that, after decades of painstaking effort (especially in the 1940s–1970s), attempts to provide a unified rigorous foundation for modern statistics have nowadays been all but abandoned by the majority of the academic community. (This is perhaps an overstatement, but it is how it often seems to non-specialists.) However, such an authority as Rao (of the Rao–Blackwell Theorem and the Cramér–Rao inequality; see below) stresses that ties between statistics and mathematics have only become stronger and more diverse.

On the other hand, during the last 30 years there has been a spectacular proliferation of statistical methods in literally every area of scientific analysis, the main justification of their usefulness being that these methods work and work successfully. The advent of modern computational techniques (including the packages SPSS, MINITAB and SPLUS) has made it possible to analyse huge arrays of data and display results in accessible forms. One can say that *computers have freed statisticians from the grip of mathematical tractability* [We].

It has to be stressed that even (or perhaps especially) at the level of an initial statistics course, accurate manual calculations are extremely important for successful examination performance, and candidates are advised to pay serious attention to their numerical work.

The prerequisite for IB Statistics includes brushing up on knowledge of some key facts from IA Probability. This includes basic concepts: probability distributions, PDFs, RVs, expectation, variance, joint distributions, covariance, independence. It is convenient to speak of a probability mass function (PMF) in the case of discrete random variables and a PDF in the case of continuous ones. Traditionally, statistical courses begin with studying some important families of PMFs/PDFs depending on a parameter (or several parameters forming a vector). For instance, Poisson PMFs, Po(λ), are parametrised by $\lambda > 0$, and so are exponential PDFs, Exp (λ). Normal PDFs are parametrised by pairs (μ, σ^2), where $\mu \in \mathbb{R}$ is the mean and $\sigma^2 > 0$ the variance. The 'true' value of a parameter (or several parameters) is considered unknown and we will have to develop the means to make a judgement about what it is.

A significant part of the course is concerned with IID $N(0, 1)$ RVs $X_1, X_2, \ldots$ and their functions. The simplest functions are linear combinations $\sum_{i=1}^{n} a_i X_i$.

Example 3.1 *Linear combinations of independent normal RVs.* We have already discussed linearity properties of normal RVs in Section 2.3; here we recall them with minor modifications. Suppose that $X_1, \ldots, X_n$ are independent, and $X_i \sim N(\mu_i, \sigma_i^2)$. Their joint PDF is

$$f_{\mathbf{X}}(\mathbf{x}) = \prod_{i=1}^{n} \frac{1}{\sqrt{2\pi}\sigma_i} \exp\left[-\frac{1}{2}(x_i - \mu_i)^2/\sigma_i^2\right], \quad \mathbf{x} = \begin{pmatrix} x_1 \\ \vdots \\ x_n \end{pmatrix} \in \mathbb{R}^n. \tag{3.1}$$

Then $\forall$ real $a_1, \ldots, a_n$,

$$\sum_{i} a_i X_i \sim N\left(\sum_{i} a_i \mu_i, \sum_{i} a_i^2 \sigma_i^2\right). \tag{3.2}$$

In particular, if $a_1 = \cdots = a_n = 1/n$, $\mu_1 = \cdots = \mu_n = \mu$ and $\sigma_1 = \cdots = \sigma_n = \sigma$, then

$$\frac{1}{n}\sum_{i=1}^{n} X_i \sim N\left(\mu, \frac{\sigma^2}{n}\right). \tag{3.3}$$

On the other hand,

$$\sum_{i=1}^{n}(X_i - \mu_i) \Bigg/ \left(\sum_{i=1}^{n} \sigma_i^2\right)^{1/2} \sim N(0, 1).$$

Next, suppose $A = (A_{ij})$ is a real invertible $n \times n$ matrix, with det $A \neq 0$, and the inverse matrix $A^{-1} = (A'_{ij})$. Write

$$\mathbf{X} = \begin{pmatrix} X_1 \\ \vdots \\ X_n \end{pmatrix} \quad \text{and} \quad \mathbf{Y} = \begin{pmatrix} Y_1 \\ \vdots \\ Y_n \end{pmatrix}$$

and consider the mutually inverse linear transformations $\mathbf{Y} = A^{\mathrm{T}} \mathbf{X}$ and $\mathbf{X} = \left(A^{-1}\right)^{\mathrm{T}} \mathbf{Y}$, with

$$Y_j = \left(A^{\mathrm{T}} \mathbf{X}\right)_j = \sum_{i=1}^{n} X_i A_{ij}, \quad X_i = \left(\left(A^{-1}\right)^{\mathrm{T}} \mathbf{Y}\right)_i = \sum_{j=1}^{n} Y_j A'_{ji}.$$

Then the RVs $Y_1, \ldots, Y_n$ are jointly normal. More precisely, the joint PDF $f_{\mathbf{Y}}(\mathbf{y})$ is calculated as

$$f_{\mathbf{Y}}(\mathbf{y}) = \frac{1}{|\det A|} f_{\mathbf{X}} \left[\left(A^{-1}\right)^{\mathrm{T}} \mathbf{y} \right]$$

$$= \frac{1}{|\det A|} \prod_{j=1}^{n} \frac{1}{\sqrt{2\pi} \sigma_j} \exp\left[-\frac{1}{2\sigma_j^2} \left(\sum_{i=1}^{n} y_i A'_{ij} - \mu_j \right)^2 \right]$$

$$= \frac{1}{(2\pi)^{n/2}} \frac{1}{[\det (A^{\mathrm{T}} \Sigma A)]^{1/2}}$$

$$\times \exp\left[-\frac{1}{2} \left\langle (\mathbf{y} - A^{\mathrm{T}} \mu), \left(A^{\mathrm{T}} \Sigma A\right)^{-1} (\mathbf{y} - A^{\mathrm{T}} \mu) \right\rangle \right].$$

Here, as before, $\langle , \rangle$ stands for the scalar product in $\mathbb{R}^n$, and

$$\mu = \begin{pmatrix} \mu_1 \\ \vdots \\ \mu_n \end{pmatrix}, \ \mathbf{y} = \begin{pmatrix} y_1 \\ \vdots \\ y_n \end{pmatrix} \in \mathbb{R}^n \text{ and } \Sigma = \begin{pmatrix} \sigma_1^2 & 0 & \cdots & 0 \\ 0 & \sigma_2^2 & \cdots & 0 \\ & & \cdots & \\ 0 & 0 & \cdots & \sigma_n^2 \end{pmatrix}.$$

Recall, μ and Σ are the mean vector and the covariance matrix of $\mathbf{X}$.

We recognise that the mean vector of $\mathbf{Y}$ is $A^{\mathrm{T}} \mu$ and the covariance matrix is $A^{\mathrm{T}} \Sigma A$:

$$\mathbb{E} Y_j = \sum_{i=1}^{n} A_{ij} \mu_i, \ \mathrm{Cov}\,(Y_i, Y_j) = \sum_{k=1}^{n} A_{ki} \sigma_{kk} A_{kj}.$$

Now suppose A is a real orthogonal $n \times n$ matrix, with $\sum_k A_{ki} A_{kj} = \delta_{ij}$, i.e. $A^{\mathrm{T}} A$ equal to the unit $n \times n$ matrix. Then det $A = \pm 1$. Assume that the above RVs X_i have the same variance: $\sigma_1^2 = \cdots = \sigma_n^2 = \sigma^2$. Then

$$\mathrm{Cov}\,(Y_i, Y_j) = \mathrm{Cov}\,\left[\left(A^{\mathrm{T}} \mathbf{X}\right)_i, \left(A^{\mathrm{T}} \mathbf{X}\right)_j \right] = \sum_{k,l=1}^{n} A_{ki} A_{lj} \mathrm{Cov}\,(X_k, X_l)$$

$$= \sigma^2 \sum_{k,l} A_{ki} A_{lj} \delta_{k,l} = \sigma^2 \sum_{k} A_{ki} A_{kj} = \sigma^2 \delta_{i,j}.$$

That is, random vector $\mathbf{X}^{\mathsf{T}}A = \mathbf{Y}^{\mathsf{T}}$ has independent components $Y_1, \ldots, Y_n$, with $Y_j \sim$ $N\left[(A^{\mathsf{T}}\mu)_j, \sigma^2\right]$. ■

Example 3.2   *Sums of squares: the χ^2 distribution.* Another example repeatedly used in what follows is the sum of squares. Let $X_1, X_2, \ldots$ be IID $N(0,1)$ RVs. The distribution of the sum

$$\sum_{i=1}^{n} X_i^2$$

is called the *chi-square, or χ^2 distribution*, with n degrees of freedom, or shortly the χ_n^2 distribution. As we will check below, it has the PDF $f_{\chi_n^2}$ concentrated on the positive half-axis $(0, \infty)$:

$$f_{\chi_n^2}(x) \propto x^{n/2-1} e^{-x/2} I(x > 0),$$

with the constant of proportionality $1/\left[\Gamma(n/2)2^{n/2}\right]$. Here

$$\Gamma(n/2) = \int_0^\infty \frac{1}{2^{n/2}} x^{n/2-1} e^{-x/2} dx.$$

One can recognise the χ_n^2 distribution as $\mathrm{Gam}(\alpha, \lambda)$ with $\alpha = n/2$, $\lambda = 1/2$. On the other hand, if $X_1, \ldots, X_n$ are IID $N(\mu, \sigma^2)$, then

$$\sum_{i=1}^{n}(X_i - \mu)^2 \sim \mathrm{Gam}\left(\frac{n}{2}, \frac{1}{2\sigma^2}\right) \quad \text{and} \quad \frac{1}{\sigma^2}\sum_{i=1}^{n}(X_i - \mu)^2 \sim \chi_n^2. \tag{3.4}$$

The mean value of the χ_n^2 distribution equals n and the variance $2n$. All χ^2 PDFs are unimodal. A sample of graphs of PDF $f_{\chi_n^2}$ is shown in Figure 3.1.

A useful property of the family of χ^2 distributions is that it is closed under independent summation. That is if $Z \sim \chi_n^2$ and $Z' \sim \chi_{n'}^2$, independently, then $Z + Z' \sim \chi_{n+n'}^2$. Of course, χ^2 distributions inherit this property from Gamma distributions.

A quick way to check that

$$f_{\chi_n^2}(x) = \frac{1}{\Gamma(n/2)} \frac{1}{2^{n/2}} x^{n/2-1} e^{-x/2} I(x > 0) \tag{3.5}$$

is to use the MGF or CHF. The MGF $M_{\chi_i^2}(\theta) = \mathbb{E}e^{\theta X_i^2}$ equals

$$\frac{1}{\sqrt{2\pi}} \int e^{\theta x^2} e^{-x^2/2} dx = \frac{1}{\sqrt{2\pi}} \int e^{-(1-2\theta)x^2/2} dx$$

$$= \frac{1}{\sqrt{1-2\theta}} \frac{1}{\sqrt{2\pi}} \int e^{-y^2/2} dy = \frac{1}{\sqrt{1-2\theta}}, \quad \theta < \frac{1}{2},$$

which is the MGF of $\mathrm{Gam}(1/2, 1/2)$. Next, the MGF $M_{Y_n}(t)$ of $Y_n = \sum_{i=1}^{n} X_i^2$ is the power $\left(M_{\chi_i^2}(\theta)\right)^n = (1 - 2\theta)^{-n/2}$. This is the MGF of the $\mathrm{Gam}(n/2, 1/2)$ distribution. Hence $f_{Y_n} \sim \mathrm{Gam}(n/2, 1/2)$, as claimed. ■

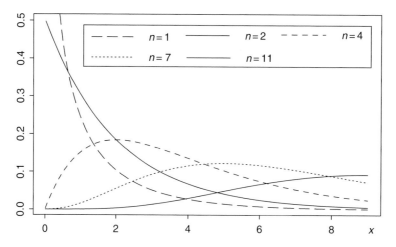

Figure 3.1 The chi-square PDFs.

Example 3.3   *The Student* t *distribution.* If, as above, $X_1, X_2, \ldots$ are IID $N(0,1)$ RVs, then the distribution of the ratio

$$\frac{X_{n+1}}{\left(\sum_{i=1}^{n} X_i^2 / n \right)^{1/2}}$$

is called the *Student distribution*, with n degrees of freedom, or the t_n distribution for short. It has the PDF f_{t_n} spread over the whole axis $\mathbb{R}$ and is symmetric (even) with respect to the inversion $x \mapsto -x$:

$$f_{t_n}(t) \propto \left(1 + \frac{t^2}{n} \right)^{-(n+1)/2},$$

with the proportionality constant

$$\frac{1}{\sqrt{\pi n}} \frac{\Gamma((n+1)/2)}{\Gamma(n/2)}.$$

For $n > 1$ it has, obviously, the mean value 0. For $n > 2$, the variance is $n/(n-2)$. All Student PDFs are unimodal. A sample of graphs of PDF f_{t_n} is shown in Figure 3.2. These PDFs resemble normal PDFs (and, as explained in Problem 2.51, $f_{t_n}(t)$ approaches $e^{-t^2/2}/\sqrt{2\pi}$ as $n \to \infty$). However, for finite n, the 'tails' of f_{t_n} are 'thicker' than those of the normal PDF. In particular, the MGF of a t distribution does not exist (except at $\theta = 0$): if $X \sim t_n$, then $\mathbb{E}e^{\theta X} = \infty \; \forall \; \theta \in \mathbb{R} \setminus \{0\}$.

Note that for $n = 1$, the t_1 distribution coincides with the Cauchy distribution.

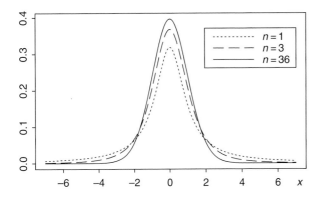

Figure 3.2 The Student PDFs.

To derive the formula for the PDF f_{t_n} of the t_n distribution, observe that it coincides with the PDF of the ratio $T = X/\sqrt{Y/n}$, where RVs X and Y are independent, and $X \sim N(0,1)$, $Y \sim \chi_n^2$. Then

$$f_{X,Y}(x, y) = \frac{1}{\sqrt{2\pi}} \frac{2^{-n/2}}{\Gamma(n/2)} e^{-x^2/2} y^{n/2-1} e^{-y/2} I(y > 0).$$

The Jacobian $\partial(t, u)/\partial(x, y)$ of the change of variables

$$t = \frac{x}{\sqrt{y/n}}, \quad u = y$$

equals $(n/y)^{1/2}$ and the inverse Jacobian $\partial(x, y)/\partial(t, u) = (u/n)^{1/2}$. Then

$$f_{t_n}(t) = f_T(t) = \int_0^\infty f_{X,Y}\left(t(u/n)^{1/2}, u\right)\left(\frac{u}{n}\right)^{1/2} du$$

$$= \frac{1}{\sqrt{2\pi}} \frac{2^{-n/2}}{\Gamma(n/2)} \int_0^\infty e^{-t^2 u/2n} u^{n/2-1} e^{-u/2}\left(\frac{u}{n}\right)^{1/2} du$$

$$= \frac{1}{\sqrt{2\pi}} \frac{2^{-n/2}}{\Gamma(n/2)n^{1/2}} \int_0^\infty e^{-(1+t^2/n)u/2} u^{(n+1)/2-1} du.$$

The last integrand comes from the PDF of $\mathrm{Gam}\left((n+1)/2, 1/2 + t^2/(2n)\right)$. Hence,

$$f_{t_n}(t) = \frac{1}{\sqrt{\pi n}} \frac{\Gamma((n+1)/2)}{\Gamma(n/2)}\left(\frac{1}{1+t^2/n}\right)^{(n+1)/2}, \tag{3.6}$$

which gives the above formula. ∎

Example 3.4 *The Fisher F-distribution.* Now let $X_1, \ldots, X_m$ and $Y_1, \ldots, Y_n$ be IID $N(0,1)$ RVs. The ratio

$$\frac{\sum_{i=1}^m X_i^2/m}{\sum_{j=1}^n Y_j^2/n}$$

has the distribution called the *Fisher*, or F *distribution* with parameters (degrees of free-dom) m, n, or the $F_{m,n}$ distribution for short. The corresponding PDF $f_{F_{m,n}}$ is concentrated on the positive half-axis:

$$f_{F_{m,n}}(x) \propto x^{m/2-1} \left(1 + \frac{m}{n}x\right)^{-(m+n)/2} I(x > 0), \tag{3.7}$$

with the proportionality coefficient

$$\frac{\Gamma((m+n)/2)}{\Gamma(m/2)\Gamma(n/2)} \left(\frac{m}{n}\right)^{m/2}.$$

The F distribution has the mean value $n/(n-2)$ (for $n > 2$, independently of m) and variance

$$\frac{2n^2(m+n-2)}{m(n-2)^2(n-4)} \quad \text{(for } n > 4\text{)}.$$

Observe that

$$\text{if } Z \sim t_n, \text{ then } Z^2 \sim F_{1,n}, \text{ and if } Z \sim F_{m,n}, \text{ then } Z^{-1} \sim F_{n,m}.$$

A sample of graphs of PDF $f_{F_{m,n}}$ is plotted in Figure 3.3.

The Fisher distribution is often called the Snedecor–Fisher distribution. The above formula for the PDF $f_{F_{m,n}}$ can be verified similarly to that for f_{t_n}; we omit the corresponding calculation. ∎

Example 3.5   *The Beta distribution.* The *Beta distribution* is a probability distribution on $(0,1)$ with the PDF

$$f(x) \propto x^{\alpha-1}(1-x)^{\beta-1} I(0 < x < 1), \tag{3.8}$$

where α, $\beta > 0$ are parameters. The proportionality constant equals

$$\frac{\Gamma(\alpha+\beta)}{\Gamma(\alpha)\Gamma(\beta)} := \frac{1}{B(\alpha,\beta)},$$

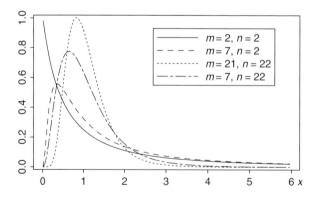

Figure 3.3 The Fisher PDFs.

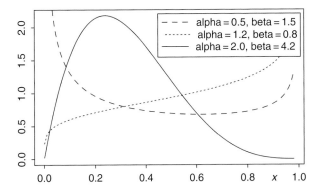

Figure 3.4 The Beta PDFs.

where $B(\alpha, \beta)$ is the *Beta function*. We write $X \sim \mathrm{Bet}(\alpha, \beta)$ if RV X has the PDF as above. A Beta distributed is used to describe various random fractions. It has

$$\mathbb{E}X = \frac{\alpha}{\alpha + \beta}, \quad \mathrm{Var}\, X = \frac{\alpha\beta}{(\alpha + \beta)(\alpha + \beta + 1)}.$$

Beta PDF plots are shown in Figure 3.4.

It is interesting to note that

$$\text{if } X \sim F_{m,n} \text{ then } \frac{(m/n)X}{1 + (m/n)X} = \frac{mX}{n + mX} \sim \mathrm{Bet}\left(\frac{m}{2}, \frac{n}{2}\right). \quad \blacksquare$$

For further examples we refer the reader to the tables of probability distributions in Appendix 1.

It will be important to work with *quantiles* of these (and other) distributions. Given $\gamma \in (0, 1)$, the *upper γ-quantile*, or *upper γ-point*, $a_+(\gamma)$, of a PMF/PDF f is determined from the equation

$$\sum_{x \geq a_+(\gamma)} f(x) = \gamma, \quad \text{or} \quad \int_{a_+(\gamma)}^{\infty} f(x)\mathrm{d}x = \gamma.$$

Similarly, the *lower γ-quantile* (*lower γ-point*) $a_-(\gamma)$ is determined from the equation

$$\sum_{x \leq a_-(\gamma)} f(x) = \gamma, \quad \text{or} \quad \int_{-\infty}^{a_-(\gamma)} f(x)\mathrm{d}x = \gamma.$$

Of course in the case of a PDF

$$a_-(\gamma) = a_+(1 - \gamma), \quad 0 < \gamma < 1. \tag{3.9}$$

In the case of a PMF, equation (3.9) should be modified, taking into account wheteher the value $a_-(\gamma)$ is attained or not (i.e. whether $f(a_-(\gamma)) > 0$ or $f(a_-(\gamma)) = 0$).

If we measure γ as a percentage, we speak of *percentiles* of a given distribution. Quantiles and percentiles of a normal, of a χ^2, of a t- and of an F-distribution can be found in standard statistical tables. Modern packages allow one to calculate them with a high accuracy for practically any given distribution.

Some basic lower percentiles are given in Tables 3.1–3.4 (courtesy of R. Weber). For points of the normal distribution, see also Table 1.1 in Section 1.6.

These tables give values of x such that a certain percentage of the distribution lies less than x. For example, if $X \sim t_3$, then $\mathbb{P}(X \le 5.84) = 0.995$, and $\mathbb{P}(-5.84 \le X \le 5.84) = 0.99$. If $X \sim F_{8,5}$, then $\mathbb{P}(X \le 4.82) = 0.95$.

Table 3.1. *Percentage points of* t_n

n	0.995	0.99	0.975	0.95
1	63.66	31.82	12.71	6.31
2	9.92	6.96	4.30	2.92
3	5.84	4.54	3.18	2.35
4	4.60	3.75	2.78	2.13
5	4.03	3.36	2.57	2.02
6	3.71	3.14	2.45	1.94
7	3.50	3.00	2.36	1.89
8	3.36	2.90	2.31	1.86
9	3.25	2.82	2.26	1.83
10	3.17	2.76	2.23	1.81
11	3.11	2.72	2.20	1.80
12	3.05	2.68	2.18	1.78
13	3.01	2.65	2.16	1.77
14	2.98	2.62	2.14	1.76
15	2.95	2.60	2.13	1.75
16	2.92	2.58	2.12	1.75
17	2.90	2.57	2.11	1.74
18	2.88	2.55	2.10	1.73
19	2.86	2.54	2.09	1.73
20	2.85	2.53	2.09	1.72
21	2.83	2.52	2.08	1.72
22	2.82	2.51	2.07	1.72
23	2.81	2.50	2.07	1.71
24	2.80	2.49	2.06	1.71
25	2.79	2.49	2.06	1.71
26	2.78	2.48	2.06	1.71
27	2.77	2.47	2.05	1.70
28	2.76	2.47	2.05	1.70
29	2.76	2.46	2.05	1.70
30	2.75	2.46	2.04	1.70
40	2.70	2.42	2.02	1.68
60	2.66	2.39	2.00	1.67
120	2.62	2.36	1.98	1.66

Table 3.2. *Percentage points of* N(0,1)

0.995	0.99	0.975	0.95	0.90
2.58	2.33	1.96	1.645	1.282

Table 3.3. *Percentage points of* χ^2_n

n	0.99	0.975	0.95	0.9
1	6.63	5.02	3.84	2.71
2	9.21	7.38	5.99	4.61
3	11.34	9.35	7.81	6.25
4	13.28	11.14	9.49	7.78
5	15.09	12.83	11.07	9.24
6	16.81	14.45	12.59	10.64
7	18.48	16.01	14.07	12.02
8	20.09	17.53	15.51	13.36
9	21.67	19.02	16.92	14.68
10	23.21	20.48	18.31	15.99
11	24.73	21.92	19.68	17.28
12	26.22	23.34	21.03	18.55
13	27.69	24.74	22.36	19.81
14	29.14	26.12	23.68	21.06
15	30.58	27.49	25.00	22.31
16	32.00	28.85	26.30	23.54
17	33.41	30.19	27.59	24.77
18	34.81	31.53	28.87	25.99
19	36.19	32.85	30.14	27.20
20	37.57	34.17	31.41	28.41
30	50.89	46.98	43.77	40.26
40	63.69	59.34	55.76	51.81
50	76.15	71.42	67.50	63.17
60	88.38	83.30	79.08	74.40
70	100.4	95.02	90.53	85.53
80	112.3	106.6	101.8	96.58
90	124.1	118.1	113.1	107.5
100	135.8	129.5	124.3	118.5

Tables 3.1–3.4 can be used to conduct various hypothesis tests with sizes 0.01, 0.05 and 0.10. For the F distribution, only the 95% point is shown; this is what is needed to conduct a one-sided test of size 0.05. Tables for other percentage points can be found in any statistics book or can be calculated using computer software. (These tables were constructed using functions available in Microsoft Excel.)

Note that the percentage points for t_n tend to those for N(0,1) as $n \to \infty$.

Table 3.4. 95% *points of* $F_{m,n}$

n	1	2	3	4	5	6	8	12	16	20	30	40	50
							m						
1	161.4	199.5	215.7	224.5	230.1	233.9	238.8	243.9	246.4	248.0	250.1	251.1	251.7
2	18.51	19.00	19.16	19.25	19.30	19.33	19.37	19.41	19.43	19.45	19.46	19.47	19.48
3	10.13	9.55	9.28	9.12	9.01	8.94	8.85	8.74	8.69	8.66	8.62	8.59	8.58
4	7.71	6.94	6.59	6.39	6.26	6.16	6.04	5.91	5.84	5.80	5.75	5.72	5.70
5	6.61	5.79	5.41	5.19	5.05	4.95	4.82	4.68	4.60	4.56	4.50	4.46	4.44
6	5.99	5.14	4.76	4.53	4.39	4.28	4.15	4.00	3.92	3.87	3.81	3.77	3.75
7	5.59	4.74	4.35	4.12	3.97	3.87	3.73	3.57	3.49	3.44	3.38	3.34	3.32
8	5.32	4.46	4.07	3.84	3.69	3.58	3.44	3.28	3.20	3.15	3.08	3.04	3.02
9	5.12	4.26	3.86	3.63	3.48	3.37	3.23	3.07	2.99	2.94	2.86	2.83	2.80
10	4.96	4.10	3.71	3.48	3.33	3.22	3.07	2.91	2.83	2.77	2.70	2.66	2.64
11	4.84	3.98	3.59	3.36	3.20	3.09	2.95	2.79	2.70	2.65	2.57	2.53	2.51
12	4.75	3.89	3.49	3.26	3.11	3.00	2.85	2.69	2.60	2.54	2.47	2.43	2.40
13	4.67	3.81	3.41	3.18	3.03	2.92	2.77	2.60	2.51	2.46	2.38	2.34	2.31
14	4.60	3.74	3.34	3.11	2.96	2.85	2.70	2.53	2.44	2.39	2.31	2.27	2.24
15	4.54	3.68	3.29	3.06	2.90	2.79	2.64	2.48	2.38	2.33	2.25	2.20	2.18
16	4.49	3.63	3.24	3.01	2.85	2.74	2.59	2.42	2.33	2.28	2.19	2.15	2.12
17	4.45	3.59	3.20	2.96	2.81	2.70	2.55	2.38	2.29	2.23	2.15	2.10	2.08
18	4.41	3.55	3.16	2.93	2.77	2.66	2.51	2.34	2.25	2.19	2.11	2.06	2.04
19	4.38	3.52	3.13	2.90	2.74	2.63	2.48	2.31	2.21	2.16	2.07	2.03	2.00
20	4.35	3.49	3.10	2.87	2.71	2.60	2.45	2.28	2.18	2.12	2.04	1.99	1.97
22	4.30	3.44	3.05	2.82	2.66	2.55	2.40	2.23	2.13	2.07	1.98	1.94	1.91
24	4.26	3.40	3.01	2.78	2.62	2.51	2.36	2.18	2.09	2.03	1.94	1.89	1.86
26	4.23	3.37	2.98	2.74	2.59	2.47	2.32	2.15	2.05	1.99	1.90	1.85	1.82
28	4.20	3.34	2.95	2.71	2.56	2.45	2.29	2.12	2.02	1.96	1.87	1.82	1.79
30	4.17	3.32	2.92	2.69	2.53	2.42	2.27	2.09	1.99	1.93	1.84	1.79	1.76
40	4.08	3.23	2.84	2.61	2.45	2.34	2.18	2.00	1.90	1.84	1.74	1.69	1.66
50	4.03	3.18	2.79	2.56	2.40	2.29	2.13	1.95	1.85	1.78	1.69	1.63	1.60
60	4.00	3.15	2.76	2.53	2.37	2.25	2.10	1.92	1.82	1.75	1.65	1.59	1.56
70	3.98	3.13	2.74	2.50	2.35	2.23	2.07	1.89	1.79	1.72	1.62	1.57	1.53
80	3.96	3.11	2.72	2.49	2.33	2.21	2.06	1.88	1.77	1.70	1.60	1.54	1.51
100	3.94	3.09	2.70	2.46	2.31	2.19	2.03	1.85	1.75	1.68	1.57	1.52	1.48

3.2   Estimators. Unbiasedness

License to Sample
You Only Estimate Twice
The Estimator
(From the series *'Movies that never made it to the Big Screen'*.)

We begin this section with the concepts of unbiasedness and sufficiency. The main model in Chapters 3 and 4 is one in which we observe a *sample* of values of a given number n of IID real RVs $X_1, \ldots, X_n$, with a common PMF/PDF $f(x; \theta)$. The notation $f(x; \theta)$ aims to stress that the PMF/PDF under consideration depends on a parameter θ varying within a given range Θ. The joint PDF/PMF of the random vector $\mathbf{X}$ is denoted by $f_{\mathbf{X}}(\mathbf{x}; \theta)$ or $f(\mathbf{x}; \theta)$ and is given by the product

$$f_{\mathbf{X}}(\mathbf{x}; \theta) = f(x_1; \theta) \dots f(x_n; \theta), \quad \mathbf{X} = \begin{pmatrix} X_1 \\ \vdots \\ X_n \end{pmatrix}, \ \mathbf{x} = \begin{pmatrix} x_1 \\ \vdots \\ x_n \end{pmatrix}. \tag{3.10}$$

Here, and below vector $\mathbf{x}$ is a sample value of $\mathbf{X}$. (It follows the tradition where capital letters refer to RVs and small letters to their sample values.) The probability distribution generated by $f_{\mathbf{X}}(\cdot; \theta)$ is denoted by $\mathbb{P}_\theta$ and the expectation and variance relative to $\mathbb{P}_\theta$ by $\mathbb{E}_\theta$ and Var_θ.

In a continuous model where we are dealing with a PDF, the argument x is allowed to vary in $\mathbb{R}$; more precisely, within a range where $f(x; \theta) > 0$ for at least one $\theta \in \Theta$. Similarly, $\mathbf{x} \in \mathbb{R}^n$ is a vector from the set where $f_{\mathbf{X}}(\mathbf{x}; \theta) > 0$ for at least one value of $\theta \in \Theta$. In a discrete model where $f(x; \theta)$ is a PMF, x varies within a specified discrete set $\mathbb{V} \subset \mathbb{R}$ (say, $\mathbb{Z}_+ = \{0, 1, 2, \dots\}$, the set of non-negative integers in the case of a Poisson distribution). Then $\mathbf{x} \in \mathbb{V}^n$ is a vector with components from $\mathbb{V}$.

The subscript $\mathbf{X}$ in notation $f_{\mathbf{X}}(\mathbf{x}; \theta)$ will often be omitted in the rest of the book.

The precise value of parameter θ is unknown; our aim is to 'estimate' it from sample $\mathbf{x} = (x_1, \dots, x_n)$. This means that we want to determine a function $\theta^*(\mathbf{x})$ depending on sample $\mathbf{x}$ but not on θ which we could take as a projected value of θ. Such a function will be called an *estimator* of θ; its particular value is often called an *estimate*. (Some authors use the term 'an estimate' instead of 'an estimator'; others use 'a point estimator' or even 'a point estimate'.) For example, in the simple case in which θ admits just two values, θ_0 and θ_1, an estimator would assign a value θ_0 or θ_1 to each observed sample $\mathbf{x}$. This would create a partition of the sample space (the set of outcomes) into two domains, one where the estimator takes value θ_0 and another where it is equal to θ_1. In general, as was said, we suppose that $\theta \in \Theta$, a given set of values. (For instance, values of θ and θ^* may be vectors.)

For example, it is well known that the number of hops by a bird before it takes off is described by a geometric distribution. Similarly, emission of alpha-particles by radioactive material is described by a Poisson distribution (this follows immediately if one assumes that the emission mechanism works independently as time progresses). However, the parameter of the distribution may vary with the type of bird or the material used in the emission experiment (and also other factors). It is important to assess the unknown value of the parameter (q or λ) from an observed sample $x_1, \dots, x_n$, where x_i is the number of emitted particles within the ith period of observation. In the 1930s when the experimental techniques were very basic, one simply counted the emitted particles visually. At Cambridge University, people still remember that E. Rutherford (1871–1937), the famous physicist and Director of the Cavendish Laboratory, when recruiting a new member of the staff, asked two straightforward questions: 'Have you got a First?' and 'Can you count?' Answering 'Yes' to both questions was a necessary condition for being hired.

In principle, any function of **x** can be considered as an estimator, but in practice we want it to be 'reasonable'. We therefore need to develop criteria for which estimator is good and which bad. The domain of statistics that emerges is called *parametric estimation*.

Example 3.6   Let $X_1, \ldots, X_n$ be IID and $X_i \sim \mathrm{Po}\,(\lambda)$. An estimator of $\lambda = \mathbb{E}X_i$ is the *sample mean* $\overline{X}$, where

$$\overline{X} = \frac{1}{n}\sum_{i=1}^{n} X_i. \tag{3.11}$$

Observe that $n\overline{X} \sim \mathrm{Po}\,(n\lambda)$. We immediately see that the sample mean has the following useful properties:

(i)   The random value $\overline{X} = \sum_{i=1}^{n} X_i/n$ is grouped around the true value of the parameter:

$$\mathbb{E}\overline{X} = \frac{1}{n}\sum_i \mathbb{E}X_i = \mathbb{E}X_1 = \lambda. \tag{3.12}$$

This property is called *unbiasedness* and will be discussed below in detail.

(ii) $\overline{X}$ approaches the true value as $n \to \infty$:

$$\mathbb{P}\left(\lim_{n\to\infty} \overline{X} = \lambda\right) = 1 \quad \text{(the strong LLN)}. \tag{3.13}$$

Property (ii) is called *consistency*.

> An unbiased and consistent statistician?
> This is the complement to an event of probability 1.
>> (From the series *'Why they are misunderstood'*.)

(iii)   For large n,

$$\sqrt{\frac{n}{\lambda}}(\overline{X} - \lambda) \sim \mathrm{N}(0, 1) \text{ (the CLT)}. \tag{3.14}$$

This property is often called *asymptotic normality*.

> Even when statisticians are normal,
> in most cases they are only asymptotically normal.
>> (From the series *'Why they are misunderstood'*.)

We are also able to see that $\overline{X}$ has another important property:

(iv) $\overline{X}$ has the minimal *mean square error* in a wide class of estimators λ^*:

$$\mathbb{E}_\lambda\left(\overline{X} - \lambda\right)^2 \le \mathbb{E}_\lambda(\lambda^*(\mathbf{X}) - \lambda)^2. \quad \blacksquare \tag{3.15}$$

Example 3.7 Let $X_1, \ldots, X_n$ be IID and $X_i \sim$ Bin (k, p). Recall, $\mathbb{E}X_i = kp$, Var $X_i = kp(1-p)$. Suppose that k is known but value $p = \mathbb{E}X_i/k \in (0, 1)$ is unknown. An estimator of p is $\overline{X}/k$, with $n\overline{X} \sim$ Bin (kn, p). Here, as before:

(i) $\mathbb{E}\overline{X}/k = p$ (unbiasedness).
(ii) $\mathbb{P}\left(\lim_{n\to\infty} \overline{X}/k = p\right) = 1$ (consistency).
(iii) $\sqrt{kn/[p(1-p)]}\,(\overline{X}/k - p) \sim \mathrm{N}(0, 1)$ for n large (asymptotic normality).
(iv) $\overline{X}/k$ has the minimal mean square error in a wide class of estimators.

Now what if we know p but value $k = 1, 2, \ldots$ is unknown? In a similar fashion, $\overline{X}/p$ can be considered as an estimator of k (never mind that it takes non-integer values!). Again, one can check that properties (i)–(iii) hold. ∎

Example 3.8 A frequent example is where $X_1, \ldots, X_n$ are IID and $X_i \sim \mathrm{N}(\mu, \sigma^2)$. When speaking of normal samples, one usually distinguishes three situations:

(I)   the mean $\mu \in \mathbb{R}$ is unknown and variance σ^2 known (say, $\sigma^2 = 1$);
(II) μ is known (say, equal to 0) and $\sigma^2 > 0$ unknown;
(III)   neither μ nor σ is known.

In cases (I) and (III), an estimator for μ is the sample mean

$$\overline{X} = \frac{1}{n}\sum_{i=1}^{n} X_i, \text{ with } \mathbb{E}\overline{X} = \frac{1}{n}\sum_i \mathbb{E}X_i = \mathbb{E}X_1 = \mu. \tag{3.16}$$

From Example 3.1 we know that $\overline{X} \sim \mathrm{N}(\mu, \sigma^2/n)$; see equation (3.3). In case (II), an estimator for σ^2 is $\overline{\Sigma}^2/n$, where

$$\overline{\Sigma}^2 = \sum_i (X_i - \mu)^2, \text{ with } \mathbb{E}\overline{\Sigma}^2 = \sum_i \mathbb{E}(X_i - \mu)^2 = n\mathrm{Var}\,X_1 = n\sigma^2, \tag{3.17}$$

and $\mathbb{E}(\overline{\Sigma}^2/n) = \sigma^2$. From Example 3.2 we deduce that $\overline{\Sigma}^2/\sigma^2 \sim \chi_n^2$.

With regard to an estimator of σ^2 in case (III), it was established that setting

$$S_{XX} = \sum_{i=1}^{n}(X_i - \overline{X})^2 \tag{3.18}$$

yields

$$\mathbb{E}\left(\frac{1}{n-1}S_{XX}\right) = \frac{1}{n-1}\mathbb{E}S_{XX} = \sigma^2. \tag{3.19}$$

See Problem 1.32 where this fact was verified for IID RVs with an arbitrary distribution. Hence, an estimator of σ^2 is provided by $S_{XX}/(n-1)$.

We are now able to specify the distribution of S_{XX}/σ^2 as χ_{n-1}^2; this is a part of the Fisher Theorem (see below).

So, in case (III), the pair

$$\left(\overline{X}, \frac{S_{XX}}{n-1}\right)$$

can be taken as an estimator for vector (μ, σ^2), and we obtain an analogue of property (i) (joint unbiasedness):

$$\left(\mathbb{E}\overline{X}, \mathbb{E}\frac{S_{XX}}{n-1}\right) = (\mu, \sigma^2).$$

Also, as $n \to \infty$, both $\overline{X}$ and $S_{XX}/(n-1)$ approach the estimated values μ and σ^2:

$$\mathbb{P}\left(\lim_{n\to\infty}\overline{X}=\mu, \lim_{n\to\infty}\frac{S_{XX}}{n-1}=\sigma^2\right)=1 \text{ (again the strong LLN)}.$$

This gives an analogue of property (ii) (joint consistency). For $\overline{X}$ this property can be deduced in a straightforward way from the fact that $\overline{X} \sim N(\mu, \sigma^2/n)$ and for $S_{XX}/(n-1)$ from the fact that $S_{XX}/\sigma^2 \sim \chi^2_{n-1}$. The latter remarks also help to check the analogue of property (iii) (joint asymptotic normality): as $n \to \infty$

$$\frac{\sqrt{n}}{\sigma}(\overline{X}-\mu) \sim N(0,1), \quad \frac{\sqrt{n-1}}{\sigma^2}\left(\frac{S_{XX}}{n-1}-\sigma^2\right) \sim N(0,2), \text{ independently}.$$

In other words, the pair

$$\left(\frac{\sqrt{n}}{\sigma}(\overline{X}-\mu), \frac{\sqrt{n-1}}{\sigma^2}\left(\frac{S_{XX}}{n-1}-\sigma^2\right)\right)$$

is asymptotically bivariate normal, with

$$\text{mean vector } \begin{pmatrix}0\\0\end{pmatrix} \text{ and the covariance matrix} \begin{pmatrix}1 & 0\\0 & 2\end{pmatrix}.$$

When checking this fact, you should verify that the variance of S_{XX} equals $2(n-1)\sigma^4$.
    An analogue of property (iv) also holds in this example, although we should be careful about how we define minimal mean square error for a vector estimator. ■

It is clear that the fact that X_i has a specific distribution plays an insignificant rôle here: properties (i)–(iv) are expected to hold in a wide range of situations. In fact, each of them develops into a recognised direction of statistical theory. Here and below, we first of all focus on property (i) and call an estimator θ^* $(=\theta^*(\mathbf{x}))$ of a parameter θ *unbiased* if

$$\mathbb{E}_\theta \theta^*(\mathbf{X}) = \theta, \quad \forall \theta \in \Theta. \tag{3.20}$$

We will also discuss properties of mean square errors.
    So, concluding this section, we summarise that for a vector $\mathbf{X}$ of IID real RVs $X_1, \ldots, X_n$, (I) the sample mean

$$\overline{X} = \frac{1}{n}\sum_{i=1}^{n} X_i \tag{3.21}$$

is always an unbiased estimator of the mean $\mathbb{E}X_1$:

$$\mathbb{E}\overline{X} = \frac{1}{n}\sum_{i=1}^{n}\mathbb{E}X_i = \mathbb{E}X_1, \tag{3.22}$$

(II) in the case of a known mean $\mathbb{E}X_1$,

$$\frac{1}{n}\overline{\Sigma}^2 = \frac{1}{n}\sum_{i=1}^{n}(X_i - \mathbb{E}X_i)^2 \tag{3.23}$$

is an unbiased estimator of the variance Var X_1:

$$\mathbb{E}\left(\frac{1}{n}\overline{\Sigma}^2\right) = \frac{1}{n}\sum_{i=1}^{n}\mathbb{E}(X_i - \mathbb{E}X_1)^2 = \mathbb{E}(X_1 - \mathbb{E}X_1)^2 = \text{Var } X_1, \tag{3.24}$$

and (III) in the case of an unknown mean,

$$\frac{1}{n-1}S_{XX} = \frac{1}{n-1}\sum_{i=1}^{n}(X_i - \overline{X})^2 \tag{3.25}$$

is an unbiased estimator of the variance Var X_1:

$$\mathbb{E}\left(\frac{1}{n-1}S_{XX}\right) = \text{Var } X_1, \tag{3.26}$$

as was shown in Problem 1.32.

Estimators $\overline{\Sigma}^2/n$ and $S_{XX}/(n-1)$ are sometimes called the *sample variances*.

> Statisticians stubbornly insist that the *n* justifies the means.
> (From the series *'Why they are misunderstood'*.)

3.3   Sufficient statistics. The factorisation criterion

> There are two kinds of statistics, the kind you look up,
> and the kind you make up.
> R.T. Stout (1886–1975), American detective-story writer

In general, a *statistic* (or a sample statistic) is an arbitrary function of sample vector **x** or its random counterpart **X**. In the parametric setting that we have adopted, we call a function T of **x** (possibly, with vector values) a *sufficient statistic* for parameter $\theta \in \Theta$ if the conditional distribution of random sample **X** given $T(\mathbf{X})$ does not depend on θ. That is, $\forall\, D \subset \mathbb{R}^n$

$$\mathbb{P}_\theta(\mathbf{X} \in D | T(\mathbf{X})) = \mathbb{E}(I(\mathbf{X} \in D)|T(\mathbf{X})) \text{ is the same } \forall\, \theta \in \Theta. \tag{3.27}$$

The significance of this concept is that the sufficient statistic encapsulates all knowledge about sample **x** needed to produce a 'good' estimator for θ.

In Example 3.6, the sample mean $\overline{X}$ is a sufficient statistic for λ. In fact, $\forall$ non-negative integer-valued vector $\mathbf{x} = (x_1, \ldots, x_n) \in \mathbb{Z}_+^n$ with $\Sigma_i x_i = nt$, the conditional probability $\mathbb{P}_\lambda(\mathbf{X} = \mathbf{x} | \overline{X} = t)$ equals

$$\frac{\mathbb{P}_\lambda(\mathbf{X} = \mathbf{x}, \overline{X} = t)}{\mathbb{P}_\lambda(\overline{X} = t)} = \frac{\mathbb{P}_\lambda(\mathbf{X} = \mathbf{x})}{\mathbb{P}_\lambda(\overline{X} = t)} = \frac{e^{-n\lambda} \prod_i (\lambda^{x_i}/x_i!)}{e^{-n\lambda}(n\lambda)^{nt}/(nt)!} = \frac{(nt)!}{n^{nt}} \prod_i \frac{1}{x_i!},$$

which does not depend on $\lambda > 0$. We used here the fact that the events

$$\{\mathbf{X} = \mathbf{x}, \ \overline{X} = t\} \quad \text{and} \quad \{\mathbf{X} = \mathbf{x}\}$$

coincide (as the equation $\overline{X} = t$ holds trivially) and the fact that $n\overline{X} \sim \text{Po}(n\lambda)$.

So, in general we can write

$$\mathbb{P}_\lambda(\mathbf{X} = \mathbf{x} | \overline{X} = t) = \frac{(nt)!}{n^{nt}} \prod_i \frac{1}{x_i!} I\left(\sum_{i=1}^n x_i = nt\right).$$

Of course, $n\bar{x} = \Sigma_i x_i$ is another sufficient statistic, and $\bar{x}$ and $n\bar{x}$ (or their random counterparts $\overline{X}$ and $n\overline{X}$) are, in a sense, equivalent (as one-to-one images of each other).

Similarly, in Example 3.7, $\bar{x}$ is sufficient for p with a known k. Here, $\forall \, \mathbf{x} \in \mathbb{Z}^n$ with entries $x_i = 0, 1, \ldots, k$ and the sum $\Sigma_i x_i = nt$, the conditional probability $\mathbb{P}_p(\mathbf{X} = \mathbf{x} | \overline{X} = t)$ equals

$$\frac{\mathbb{P}_p(\mathbf{X} = \mathbf{x})}{\mathbb{P}_p(\overline{X} = t)} = \frac{\prod_i \dfrac{k!}{x_i!(k - x_i)!} p^{x_i}(1 - p)^{k - x_i}}{\dfrac{(nk)!}{(nt)!(nk - nt)!} p^{nt}(1 - p)^{nk - nt}} = \frac{(k!)^n}{(nk)!} \frac{(nt)!(nk - nt)!}{\prod_i x_i!(k - x_i)!}.$$

which does not depend on $p \in (0, 1)$. As before, if $\Sigma_i x_i \neq nt$, $\mathbb{P}_p(\mathbf{X} = \mathbf{x} | \overline{X} = t) = 0$ which again does not depend on p.

Consider now Example 3.8 where $X_1, \ldots, X_n$ are IID and $X_i \sim \text{N}(\mu, \sigma^2)$. Here,

(I)   with σ^2 known, the sufficient statistic for μ is

$$\overline{X} = \frac{1}{n} \sum_{i=1}^n X_i,$$

(II)   with μ known, a sufficient statistic for σ^2 is

$$\overline{\Sigma}^2 = \sum_i (X_i - \mu)^2,$$

(III)   with both μ and σ^2 unknown, a sufficient statistic for (μ, σ^2) is

$$\left(\overline{X}, \sum_{i=1}^n X_i^2\right).$$

The most efficient way to check these facts is to use the factorisation criterion.

The *factorisation criterion* is a general statement about sufficient statistics. It says:

T is sufficient for θ iff the PMF/PGF $f_{\mathbf{X}}(\mathbf{x}; \theta)$ can be written as a product $g(T(\mathbf{x}), \theta)h(\mathbf{x})$ for some functions g and h.

The proof in the *discrete* case, with PMF $f_{\mathbf{X}}(\mathbf{x}; \theta) = \mathbb{P}_{\theta}(\mathbf{X} = \mathbf{x})$, is straightforward. In fact, for the 'if' part, assume that the above factorisation holds. Then for sample vector $\mathbf{x} \in \mathbb{V}^n$ with $T(\mathbf{x}) = t$, the conditional probability $\mathbb{P}_{\theta}(\mathbf{X} = \mathbf{x}|T = t)$ equals

$$\frac{\mathbb{P}_{\theta}(\mathbf{X} = \mathbf{x}, T = t)}{\mathbb{P}_{\theta}(T = t)} = \frac{\mathbb{P}_{\theta}(\mathbf{X} = \mathbf{x})}{\mathbb{P}_{\theta}(T = t)} = \frac{g(T(\mathbf{x}), \theta)h(\mathbf{x})}{\sum\limits_{\widetilde{\mathbf{x}} \in \mathbb{V}^n : T(\widetilde{\mathbf{x}}) = t} g(T(\widetilde{\mathbf{x}}), \theta)h(\widetilde{\mathbf{x}})}$$

$$= \frac{g(t, \theta)h(\mathbf{x})}{g(t, \theta) \sum\limits_{\widetilde{\mathbf{x}} \in \mathbb{V}^n : T(\widetilde{\mathbf{x}}) = t} h(\widetilde{\mathbf{x}})} = \frac{h(\mathbf{x})}{\sum\limits_{\widetilde{\mathbf{x}} \in \mathbb{V}^n : T(\widetilde{\mathbf{x}}) = t} h(\widetilde{\mathbf{x}})}.$$

This does not depend on θ.

If, on the other hand, the compatibility condition $T(\mathbf{x}) = t$ fails (i.e. $T(\mathbf{x}) \neq t$) then $\mathbb{P}_{\theta}(\mathbf{X} = \mathbf{x}|T = t) = 0$ which again does not depend on θ. A general formula is

$$\mathbb{P}_{\theta}(\mathbf{X} = \mathbf{x}|T = t) = \frac{h(\mathbf{x})}{\sum\limits_{\widetilde{\mathbf{x}} \in \mathbb{V}^n : T(\widetilde{\mathbf{x}}) = t} h(\widetilde{\mathbf{x}})} I(T(\mathbf{x}) = t).$$

As the RHS does not depend on θ, T is sufficient.

For the 'only if' part of the criterion we assume that $\mathbb{P}_{\theta}(\mathbf{X} = \mathbf{x}|T = t)$ does not depend on θ. Then, again for $\mathbf{x} \in \mathbb{V}^n$ with $T(\mathbf{x}) = t$

$$f_{\mathbf{X}}(\mathbf{x}) = \mathbb{P}_{\theta}(\mathbf{X} = \mathbf{x}) = \mathbb{P}_{\theta}(\mathbf{X} = \mathbf{x}|T = t)\mathbb{P}_{\theta}(T = t).$$

The factor $\mathbb{P}_{\theta}(\mathbf{X} = \mathbf{x}|T = t)$ does not depend on θ; we denote it by $h(\mathbf{x})$. The factor $\mathbb{P}_{\theta}(T = t)$ is then denoted by $g(t, \theta)$, and we obtain the factorisation.

In the *continuous* case the proof goes along similar lines (although to make it formally impeccable one needs some elements of measure theory). Namely, we write the conditional PDF $f_{\mathbf{X}|T}(\mathbf{x}|t)$ as the ratio

$$\frac{f(\mathbf{x}; \theta)}{f_T(t; \theta)} I(T(\mathbf{x}) = t)$$

and represent the PDF $f_T(t; \theta)$ as the integral

$$f_T(t; \theta) = \int_{\{\widetilde{\mathbf{x}} \in \mathbb{R}^n : T(\widetilde{\mathbf{x}}) = t\}} f(\widetilde{\mathbf{x}}; \theta) d\widetilde{\mathbf{x}}$$

over the level surface $\{\widetilde{\mathbf{x}} \in \mathbb{R}^n : T(\widetilde{\mathbf{x}}) = t\}$, against the area element $d(\widetilde{\mathbf{x}}|t)$ on this surface. Then for the 'if' part we again use the representation $f(\mathbf{x}) = g(T(\mathbf{x}), \theta)h(\mathbf{x})$ and arrive at the equation

$$f_{\mathbf{X}|T}(\mathbf{x}|t) = \frac{h(\mathbf{x})}{\int\limits_{\{\widetilde{\mathbf{x}} \in \mathbb{R}^n : T(\widetilde{\mathbf{x}}) = t\}} d\widetilde{\mathbf{x}} h(\widetilde{\mathbf{x}})} I(T(\mathbf{x}) = t),$$

with the RHS independent of θ. For the 'only if' part we simply re-write $f(\mathbf{x}; \theta)$ as $f_{\mathbf{X}|T}(\mathbf{x}|t)f_T(t; \theta)$, where $t = T(\mathbf{x})$, and set, as before,

$$h(\mathbf{x}) = f_{\mathbf{X}|T}(\mathbf{x}|t) \text{ and } g(T(\mathbf{x}), \theta) = f_T(t; \theta).$$

The factorisation criterion means that T is a sufficient statistic when $T(\mathbf{x}) = T(\mathbf{x}')$ implies that the ratio $f_{\mathbf{x}}(\mathbf{x}; \theta)/f_{\mathbf{x}}(\mathbf{x}'; \theta)$ is the same $\forall\, \theta \in \Theta$. The next step is to consider a *minimal* sufficient statistic for which $T(\mathbf{x}) = T(\mathbf{x}')$ is equivalent to the fact that the ratio $f_{\mathbf{x}}(\mathbf{x}; \theta)/f_{\mathbf{x}}(\mathbf{x}'; \theta)$ is the same $\forall\, \theta \in \Theta$. This concept is convenient because any sufficient statistic is a function of the minimal one. In other words, the minimal sufficient statistic has the 'largest' level sets where it takes a constant value, which represents the least amount of detail we should know about sample $\mathbf{x}$. Any further suppression of information about the sample would result in the loss of sufficiency.

In all examples below, sufficient statistics are minimal.

The statement and the proof of the factorisation criterion (in the discrete case) has been extremely popular among the Cambridge University examination questions. See MT-IB 1999-112D (i), 1998-212E (ii), 1997-203G, 1994-403F (ii), 1993-403J (i), 1992-106D. See also SP-IB 1992-103H (i). Here and below, we use the following convention: MT-IB 1997-203G stands for question 3G from the 1997 IB Math Tripos Paper 2, and SP-IB 1992-103H stands for question 3H from the 1992 IB Specimen Paper 1.

The idea behind the factorisation criterion goes back to a 1925 paper by R.A. Fisher (1890–1962), the outstanding UK applied mathematician, statistician and genetist, whose name will be often quoted in this part of the book. (Some authors trace the factorisation criterion to his 1912 paper.) The concept was further developed in Fisher's 1934 work. An important rôle was also played by a 1935 paper by J. Neyman (1894–1981), a Polish–American statistician (born in Moldova, educated in the Ukraine, worked in Poland and the UK and later in the USA). Neyman's name will also appear in this part of the book on many occasions, mainly in connections with the Neyman–Pearson Lemma. See below.

Example 3.9   (i) Let $X_i \sim U(0, \theta)$ where $\theta > 0$ is unknown. Then $T(\mathbf{x}) = \max x_i$, $\mathbf{x} = (x_1, \ldots, x_n) \in \mathbb{R}^n_+$, is a sufficient statistic for θ.

(ii) Now consider $X_i \sim U(\theta, \theta + 1)$, $\theta \in \mathbb{R}$. Here the sample PDF

$$f_{\mathbf{X}}(\mathbf{x}; \theta) = \prod_i I(\theta \le x_i \le \theta + 1)$$

$$= I(\min x_i \ge \theta)I(\max x_i \le \theta + 1), \ \mathbf{x} \in \mathbb{R}^n.$$

We see that the factorisation holds with $T(\mathbf{x})$ being a two-vector $(\min x_i, \max x_i)$, function $g((y_1, y_2), \theta) = I(y_1 \ge \theta)I(y_2 \le \theta + 1)$ and $h(\mathbf{x}) \equiv 1$. Hence, the pair $(\min x_i, \max x_i)$ is sufficient for θ.

(iii) Let $X_1, \ldots, X_n$ form a random sample from a Poisson distribution for which the value of mean θ is unknown. Find a one-dimensional sufficient statistic for θ.

(*Answer*: $T(\mathbf{X}) = \sum_i X_i$.)

(iv) Assume that $X_i \sim N(\mu, \sigma^2)$, where both $\mu \in \mathbb{R}$ and $\sigma^2 > 0$ are unknown. Then $T(\mathbf{x}) = \left(\sum_i x_i, \sum_i x_i^2 \right)$ is a sufficient statistic for $\theta = (\mu, \sigma^2)$. ∎

3.4   Maximum likelihood estimators

> Robin Likelihood – Prince of Liars
>
> (From the series '*Movies that never made it to the Big Screen'*.)

The concept of a *maximum likelihood estimator* (MLE) forms the basis of a powerful (and beautiful) method of constructing good estimators which is now universally adopted (and called the *method of maximum likelihood*). Here, we treat the PMF/PDF $f_{\mathbf{x}}(\mathbf{x}; \theta)$ as a function of $\theta \in \Theta$ depending on the observed sample $\mathbf{x}$ as a parameter. We then take the value of θ that maximises this function on set Θ:

$$\widehat{\theta}(=\widehat{\theta}(\mathbf{x})) = \arg \, \max\left[f(\mathbf{x}; \theta) : \theta \in \Theta\right]. \qquad (3.28)$$

In this context, $f(\mathbf{x}; \theta)$ is often called the *likelihood function* (the likelihood for short) for sample $\mathbf{x}$. Instead of maximising $f(\mathbf{x}; \theta)$, one often prefers to maximise its logarithm $\ell(\mathbf{x}; \theta) = \ln f(\mathbf{x}; \theta)$, which is called the *log-likelihood function*, or log-likelihood (LL) for short. The MLE is then defined as

$$\widehat{\theta} = \arg \, \max\left[\ell(\mathbf{x}; \theta) : \theta \in \Theta\right].$$

The idea of an MLE was conceived in 1921 by Fisher.

Often, the maximiser is unique (although it may lie on the boundary of allowed set Θ). If $\ell(\mathbf{x}; \theta)$ is a smooth function of $\theta \in \Theta$, one could consider stationary points, where the first derivative vanishes

$$\frac{d}{d\theta}\ell(\mathbf{x}; \theta) = 0, \; \left(\text{or } \frac{\partial}{\partial \theta_j}\ell(\mathbf{x}; \theta) = 0, \; j = 1, \ldots, d, \text{ if } \theta = (\theta_1, \ldots, \theta_d)\right). \quad (3.29)$$

Of course, one has to select from the roots of equation (3.29) the local maximisers (by checking the signs of the second derivatives or otherwise) and establish which of these maximisers is global. Luckily, in the examples that follow, the stationary point (when it exists) is always unique. When parameter set Θ is unbounded (say a real line) then to check that a (unique) stationary point gives a global maximiser, it is enough to verify that $\ell(\,\cdot\,; \theta) \to -\infty$ for large values of $|\theta|$.

Finding the MLE (sometimes together with a sufficient statistic) is another hugely popular examination topic. See MT-IB 1998-103E (ii), 1998-212E (iii), 1997-403G, 1996-203G (i), 1995-203G (i), 1994-203F (ii,b), 1994-403F, 1992D-106 (i). See also SP-IB 1992-103H. Moreover, MLEs play an important rôle in (and appear in Tripos examples related to) hypotheses testing and linear regression. See Chapter 4.

Example 3.10   In this example we identify MLEs for some concrete models. (i) The MLE for λ when $X_i \sim \text{Po}(\lambda)$. Here, $\forall \; \mathbf{x} = (x_1, \ldots, x_n) \in \mathbb{Z}_+^n$ with non-negative integer entries $x_i \in \mathbb{Z}_+$, the LL

$$\ell(\mathbf{x}; \lambda) = -n\lambda + \sum_i x_i \ln \lambda - \sum_i \ln (x_i!), \; \lambda > 0.$$

Differentiating with respect to λ yields

$$\frac{\partial}{\partial \lambda}\ell(\mathbf{x}; \lambda) = -n + \frac{1}{\lambda}\sum_i x_i = 0 \text{ and } \widehat{\lambda} = \frac{1}{n}\sum_i x_i = \bar{x}.$$

Furthermore,

$$\frac{\partial^2}{\partial \lambda^2}\ell(\mathbf{x}; \lambda) = -\frac{1}{\lambda^2}\sum_i x_i < 0.$$

So, $\bar{x}$ gives the (global) maximum. We see that the MLE $\widehat{\lambda}$ of λ coincides with the sample mean. In particular, it is unbiased.

(ii) $X_i \sim U\,(0, \theta)$, $\theta > 0$, then $\widehat{\theta}(\mathbf{x}) = \max x_i$, is the MLE for θ.

(iii) Let $X_i \sim U\,(0, \theta + 1)$ (cf. Example 3.9 (ii)). To find the MLE for θ, again look at the likelihood:

$$f(\mathbf{x}; \theta) = I(\max x_i - 1 \leq \theta \leq \min x_i), \quad \theta \in \mathbb{R}.$$

We see that if $\max x_i - 1 < \min x_i$ (which is consistent with the assumption that sample $\mathbf{x}$ is generated by IID $U(0, \theta + 1)$ RVs), we can take any value between $\max x_i - 1$ and $\min x_i$ as the MLE for θ.

It is not hard to guess that the unbiased MLE estimator for θ is the middle point

$$\widehat{\theta} = \frac{1}{2}(\max\ x_i - 1 + \min\ x_i) = \frac{1}{2}(\max\ x_i + \min\ x_i) - \frac{1}{2}.$$

Indeed:

$$\mathbb{E}\widehat{\theta} = \frac{1}{2}(\mathbb{E}\max X_i + \mathbb{E}\min X_i) - \frac{1}{2}$$

$$= \frac{1}{2}\left[\int_\theta^{\theta+1} \mathrm{d}x\, x p_{\min X_i}(x) + \int_\theta^{\theta+1} \mathrm{d}x\, x p_{\max X_i}(x)\right] - \frac{1}{2},$$

with

$$p_{\min X_i}(x) = -\frac{\mathrm{d}}{\mathrm{d}x}\mathbb{P}(\min X_i > x) = -\frac{\mathrm{d}}{\mathrm{d}x}(\mathbb{P}(X_i > x))^n$$

$$= -\frac{\mathrm{d}}{\mathrm{d}x}\left(\int_x^{\theta+1} \mathrm{d}y\right)^n = n(\theta + 1 - x)^{n-1}, \quad \theta < x < \theta + 1,$$

and similarly,

$$p_{\max X_i}(x) = n(x - \theta)^{n-1}, \quad \theta < x < \theta + 1.$$

Then

$$\mathbb{E}\min X_i = n\int_{\theta}^{\theta+1}(\theta+1-x)^{n-1}x\mathrm{d}x = -n\int_{\theta}^{\theta+1}(\theta+1-x)^n\mathrm{d}x$$

$$+(\theta+1)n\int_{\theta}^{\theta+1}(\theta+1-x)^{n-1}\mathrm{d}x = -n\int_0^1 x^n\mathrm{d}x$$

$$+(\theta+1)n\int_0^1 x^{n-1}\mathrm{d}x = -\frac{n}{n+1}+\theta+1 = \theta+\frac{1}{n+1}, \qquad (3.30)$$

and similarly,

$$\mathbb{E}\max X_i = \frac{n}{n+1}+\theta, \qquad (3.31)$$

giving that $\mathbb{E}\widehat{\theta}=\theta$. ∎

The MLEs have a number of handy properties:

(i)   If T is a sufficient statistic for θ, then $\ell(\mathbf{x};\theta) = \ln g(T(\mathbf{x}),\theta) + \ln h(\mathbf{x})$. Maximising the likelihood in θ is then reduced to maximising function $g(T(\mathbf{x}),\theta)$ or its logarithm. That is the MLE $\widehat{\theta}$ will be a function of $T(\mathbf{x})$.

(ii)   Under mild conditions on the distribution of X_i, the MLEs are (or can be chosen to be) asymptotically unbiased, as $n\to\infty$. Furthermore, an MLE $\widehat{\theta}$ is often asymptotically normal. In the scalar case, this means that $\sqrt{n}(\widehat{\theta}-\theta)\sim N(0,v)$, where the variance v is minimal amongst attainable variances of unbiased estimators for θ.

(iii)   The invariance principle for MLEs: If $\widehat{\theta}$ is an MLE for θ and we pass from parameter θ to $\eta = u(\theta)$, where function u is one-to-one, then $u(\widehat{\theta})$ is an MLE for η.

3.5   Normal samples. The Fisher Theorem

> Did you hear about the statistician who was put in jail? He now has zero
> degrees of freedom.
>   (From the series *'Why they are misunderstood'*.)

Example 3.11   In this example we consider the MLE for the pair (μ,σ^3) in IID normal samples. Given $X_i\sim N(\mu,\sigma^2)$, the LL is

$$\ell(\mathbf{x};\mu,\sigma^2) = -\frac{n}{2}\ln(2\pi) - \frac{n}{2}\ln(\sigma^2) - \frac{1}{2\sigma^2}\sum_{i=1}^n(x_i-\mu)^2,$$

$$\mathbf{x} = \begin{pmatrix} x_1 \\ \vdots \\ x_n \end{pmatrix} \in \mathbb{R}^n,\ \mu\in\mathbb{R},\ \sigma^2>0,$$

with

$$\frac{\partial}{\partial\mu}\ell(\mathbf{x};\mu,\sigma^2) = \frac{1}{\sigma^2}\sum_i(x_i-\mu),$$

$$\frac{\partial}{\partial\sigma^2}\ell(\mathbf{x};\mu,\sigma^2) = -\frac{n}{2\sigma^2}+\frac{1}{2\sigma^4}\sum_i(x_i-\mu)^2.$$

The stationary point where $\partial\ell/\partial\mu = \partial\ell/\partial\sigma^2 = 0$ is unique:

$$\widehat{\mu} = \bar{x}, \quad \widehat{\sigma}^2 = \frac{1}{n}S_{xx},$$

where, as in equation (3.18):

$$S_{xx}(=S_{xx}(\mathbf{x})) = \sum_i(x_i-\bar{x})^2. \qquad (3.32)$$

(In some texts, the notation S_{xx}^2 or even $\bar{S}_{xx}^2$ is used, instead of S_{xx}.) The point $(\widehat{\mu},\widehat{\sigma}^2) = (\bar{x}, S_{xx}/n)$ is the global maximiser. This can be seen, for example, because $\ell(\mathbf{x};\mu,\sigma^2)$ goes to $-\infty$ when $|\mu| \to \infty$ and $\sigma^2 \to \infty$, and also $\ell(\mathbf{x};\mu,\sigma^2) \to -\infty$ as $\sigma^2 \to 0$ for every μ. Then $(\bar{x}, S_{xx}/n)$ cannot be a minimum (or a saddle point). Hence, it is the global maximum. Here, $\overline{X}$ is unbiased, but S_{XX}/n has a bias: $S_{XX}/n = (n-1)\sigma^2/n < \sigma^2$. See equation (3.17) and Problem 1.32. However, as $n \to \infty$, the bias disappears: $\mathbb{E}S_{XX}/n \to \sigma^2$. (The unbiased estimator for σ^2 is of course $S_{XX}/(n-1)$.) ∎

An important fact is the following statement, often called the *Fisher Theorem*.

> For IID normal samples, the MLE $(\widehat{\mu},\widehat{\sigma}^2) = (\overline{X}, S_{XX}/n)$ is formed by independent RVs $\overline{X}$ and S_{XX}/n, with

$$\overline{X} \sim \mathrm{N}\left(\mu, \frac{\sigma^2}{n}\right) \quad \left(i.e.\ \sqrt{n}(\overline{X}-\mu) \sim \mathrm{N}(0,\sigma^2)\right)$$

> and

$$\frac{S_{XX}}{\sigma^2} \sim \chi^2_{n-1} \quad \left(i.e.\ \frac{S_{XX}}{\sigma^2} \sim \sum_{i=1}^{n-1} Y_i^2 \text{ where } Y_i \sim \mathrm{N}(0,1), \text{ independently}\right).$$

See (in a slightly different form) MT-IB 1994-203F. The question MT-IB 1998-212E(iii) refers to the Fisher Theorem as a 'standard distributional result'.

The Fisher Theorem implies that the RV $\left[S_{XX}-(n-1)\sigma^2\right]/(\sigma^2\sqrt{2n})$ is asymptotically N(0,1). Then of course,

$$\sqrt{n}\left(\frac{1}{n}S_{XX}-\sigma^2\right) \sim \mathrm{N}(0,2\sigma^4).$$

To prove the Fisher Theorem, first write

$$\sum_{i=1}^n(X_i-\mu)^2 = \sum_{i=1}^n(X_i-\overline{X}+\overline{X}-\mu)^2 = \sum_{i=1}^n(X_i-\overline{X})^2 + n(\overline{X}-\mu)^2,$$

since the sum $\sum_{i=1}^{n}(X_i - \overline{X})(\overline{X} - \mu) = (\overline{X} - \mu)\sum_{i=1}^{n}(X_i - \overline{X}) = 0$. In other words, $\sum_i(X_i - \mu)^2 = S_{XX} + n(\overline{X} - \mu)^2$.

Then use the general fact that if vector

$$\mathbf{X} - \mu\mathbf{1} = \begin{pmatrix} X_1 \\ \vdots \\ X_n \end{pmatrix} - \mu\begin{pmatrix} 1 \\ \vdots \\ 1 \end{pmatrix}$$

has IID entries $X_i - \mu \sim \mathrm{N}(0, \sigma^2)$, then for any real orthogonal $n \times n$ matrix A, vector

$$\begin{pmatrix} Y_1 \\ \vdots \\ Y_n \end{pmatrix} = A^{\mathrm{T}}(\mathbf{X} - \mu\mathbf{1})$$

has again IID components $Y_i \sim \mathrm{N}(0, \sigma^2)$ (see Problem 2.38).

We take any orthogonal A with the first column

$$\begin{pmatrix} 1/\sqrt{n} \\ \vdots \\ 1/\sqrt{n} \end{pmatrix};$$

to construct such a matrix you simply complete this column to an orthonormal basis in $\mathbb{R}^n$. For instance, the family $\mathbf{e}_2, \ldots, \mathbf{e}_n$ will do, where column $\mathbf{e}_k$ has its first $k-1$ entries $1/\sqrt{k(k-1)}$ followed by $-(k-1)/\sqrt{k(k-1)}$ and $n-k$ entries 0:

$$\mathbf{e}_k = \begin{pmatrix} \left.\begin{matrix} 1/\sqrt{k(k-1)} \\ \vdots \\ 1/\sqrt{k(k-1)} \end{matrix}\right\} \quad k-1 \\ -(k-1)/\sqrt{k(k-1)} \\ 0 \\ \vdots \\ 0 \end{pmatrix}, \quad k = 2, \ldots, n.$$

Then $Y_1 = \left(A^{\mathrm{T}}(\mathbf{X} - \mu\mathbf{1})\right)_1 = \sqrt{n}(\overline{X} - \mu)$, and $Y_2, \ldots, Y_n$ are independent of Y_1.

Because the orthogonal matrix preserves the length of a vector, we have that

$$\sum_{i=1}^{n} Y_i^2 = \sum_{i=1}^{n}(X_i - \mu)^2 = n(\overline{X} - \mu)^2 + \sum_{i=1}^{n}(X_i - \overline{X})^2 = Y_1^2 + S_{XX},$$

i.e. $S_{XX} = \sum_{i=2}^{n} Y_i^2$. Then $S_{XX}/\sigma^2 = \sum_{i=2}^{n} Y_i^2/\sigma^2 \sim \chi_{n-1}^2$, independently of Y_1.

Remark   Some authors call the statistic

$$s_{XX} = \sqrt{\frac{S_{XX}}{n-1}} \tag{3.33}$$

the *sample standard deviation*. The term the *standard error* is often used for $s_{XX}/\sqrt{n}$ which is an estimator of $\sigma/\sqrt{n}$. We will follow this tradition.

> Statisticians do all the standard deviations.
> Statisticians do all the standard errors.
> (From the series *'How they do it'*.)

3.6   Mean square errors. The Rao–Blackwell Theorem. The Cramér–Rao inequality

> Statistics show that of those who contract
> the habit of eating, very few survive.
> W. Irwin (1876–1959), American editor and writer

When we assess the quality of an estimator θ^* of a parameter θ, it is useful to consider the *mean square error* (MSE) defined as

$$\mathbb{E}_\theta(\theta^*(\mathbf{X}) - \theta)^2; \tag{3.34}$$

for an unbiased estimator, this gives the variance $\mathrm{Var}_\theta \theta^*(\mathbf{X})$. In general,

$$
\begin{aligned}
\mathbb{E}_\theta(\theta^*(\mathbf{X}) - \theta)^2 &= \mathbb{E}_\theta(\theta^*(\mathbf{X}) - \mathbb{E}_\theta\theta^*(\mathbf{X}) + \mathbb{E}_\theta\theta^*(\mathbf{X}) - \theta)^2 \\
&= \mathbb{E}_\theta(\theta^*(\mathbf{X}) - \mathbb{E}_\theta\theta^*(\mathbf{X}))^2 + (\mathbb{E}_\theta\theta^*(\mathbf{X}) - \theta)^2 \\
&\quad + 2(\mathbb{E}_\theta\theta^*(\mathbf{X}) - \theta)\mathbb{E}_\theta(\theta^*(\mathbf{X}) - \mathbb{E}_\theta\theta^*(\mathbf{X})) \\
&= \mathrm{Var}_\theta\theta^*(\mathbf{X}) + (\mathrm{Bias}_\theta\theta^*(\mathbf{X}))^2, \tag{3.35}
\end{aligned}
$$

where $\mathrm{Bias}_\theta\theta^* = \mathbb{E}_\theta\theta^* - \theta$.

In general, there is a simple way to decrease the MSE of a given estimator. It is to use the *Rao–Blackwell* (RB) *Theorem*:

> If T is a sufficient statistic and θ^* an estimator for θ then $\widehat{\theta^*} = \mathbb{E}(\theta^*|T)$ has
>
> $$\mathbb{E}(\widehat{\theta^*} - \theta)^2 \leq \mathbb{E}(\theta^* - \theta)^2, \; \theta \in \Theta. \tag{3.36}$$
>
> Moreover, if $\mathbb{E}(\theta^*)^2 < \infty$ for some $\theta \in \Theta$, then, for this θ, the inequality is strict unless θ^* is a function of T.

The proof is short: as $\mathbb{E}\widehat{\theta^*} = \mathbb{E}[\mathbb{E}(\theta^*|T)] = \mathbb{E}\theta^*$, both θ^* and $\widehat{\theta^*}$ have the same bias. By the conditional variance formula

$$\mathrm{Var}\,\theta^* = \mathbb{E}[\mathrm{Var}\,(\theta^*|T)] + \mathrm{Var}[\mathbb{E}(\theta^*|T)] = \mathbb{E}[\mathrm{Var}\,(\theta^*|T)] + \mathrm{Var}\,\widehat{\theta^*}.$$

Hence, $\mathrm{Var}\,\theta^* \geq \mathrm{Var}\,\widehat{\theta^*}$ and so $\mathbb{E}(\theta^* - \theta)^2 \geq \mathbb{E}(\widehat{\theta^*} - \theta)^2$. The equality is attained iff $\mathrm{Var}\,(\theta^*|T) = 0$.

Remark   (1) The quantity $\mathbb{E}(\theta^*|T)$ depends on a value of $\theta^*(\mathbf{x})$ but not on θ. Thus $\widehat{\theta}^*$ is correctly defined.

(2) If θ^* is unbiased, then so is $\widehat{\theta}^*$.

(3) If θ^* is itself a function of T, then $\widehat{\theta}^* = \theta^*$.

The RB theorem bears the names of two distinguished academics. D. Blackwell (1919–) is a US mathematician, a leading proponent of a game theoretical approach in statistics and other disciplines. Blackwell was one of the first African-American mathematicians to be employed by a leading university in the USA. C.R. Rao (1920–) is an Indian mathematician who studied in India and Britain (he took his Ph.D. at Cambridge University and was Fisher's only formal Ph.D. student in statistics), worked for a long time in India and currently lives and works in the USA.

Example 3.12   (i) Let $X_i \sim \mathrm{U}(0, \theta)$. Then $\theta^* = 2\overline{X} = 2\sum_i X_i/n$ is an unbiased estimator for θ, with

$$\mathrm{Var}(2\overline{X}) = \frac{4}{n}\mathrm{Var}X_1 = \frac{\theta^2}{3n}.$$

We know that the sufficient statistic T has the form $T(\mathbf{X}) = \max_i X_i$, with the PDF

$$f_T(x) = n\frac{x^{n-1}}{\theta^n}I(0 < x < \theta).$$

Hence,

$$\widehat{\theta}^* = \mathbb{E}\ (\theta^*|T) = \frac{2}{n}\sum_i \mathbb{E}\ (X_i|T) = 2\mathbb{E}(X_1|T)$$

$$= 2\left(\max X_i \times \frac{1}{n} + \frac{\max X_i}{2} \times \frac{n-1}{n}\right) = \frac{n+1}{n}T$$

and $\widehat{\theta}^*$ should have an MSE less than or equal to that of θ^*. Surprisingly, giving away a lot of information about the sample leads to an improved MSE! In fact, the variance $\mathrm{Var}\ \widehat{\theta}^* = (n+1)^2(\mathrm{Var}\ T)/n^2$, where

$$\mathrm{Var}\ T = \int_0^\theta n\frac{x^{n-1}}{\theta^n}x^2\mathrm{d}x - \left(\int_0^\theta n\frac{x^{n-1}}{\theta^n}x\mathrm{d}x\right)^2$$

$$= \theta^2\left[\frac{n}{n+2} - \left(\frac{n}{n+1}\right)^2\right] = \theta^2\frac{n}{(n+1)^2(n+2)}.$$

So $\mathrm{Var}\ \widehat{\theta}^* = \theta^2/[n(n+2)]$ which is $< \theta^2/3n$ for $n \geq 2$ and goes faster to 0 as $n \to \infty$.

(ii) Let $X_i \sim \mathrm{U}(\theta, \theta+1)$. Then $\widehat{\theta} = \sum_i X_i/n - \frac{1}{2}$ is an unbiased estimator for θ; here

$$\mathrm{Var}\ \overline{X} = \frac{1}{n}\mathrm{Var}\ X_1 = \frac{1}{12n}.$$

This form of the estimator emerges when we equate the value $\widehat{\theta} + 1/2$ with sample mean $\sum_{i=1}^{n} X_i/n$. Such a trick forms the basis of the so-called *method of moments* in statistics. The method of moments was popular in the past but presently has been superseded by the method of maximum likelihood.

We know that the sufficient statistic T has the form $(\min_i X_i, \max_i X_i)$. Hence,

$$\widehat{\theta}^* = \mathbb{E}(\widehat{\theta}|T) = \frac{1}{n}\sum_i \mathbb{E}(X_i|T) - \frac{1}{2}$$

$$= \mathbb{E}(X_1|T) - \frac{1}{2} = \frac{1}{2}(\min X_i + \max X_i - 1).$$

The estimator $\widehat{\theta}^*$ is unbiased:

$$\mathbb{E}\widehat{\theta}^* = \frac{1}{2}\left(\frac{n}{n+1} + \theta + \theta + \frac{1}{n+1} - 1\right) = \theta,$$

and it should have an MSE less than or equal to that of θ^*. Again, giving away excessive information about $\mathbf{X}$ leads to a lesser MSE. In fact, the variance $\operatorname{Var}\widehat{\theta}^* = \frac{1}{4}\operatorname{Var}(\min X_i + \max X_i)$ equals

$$\frac{1}{4}\mathbb{E}(\min X_i + \max X_i)^2 - \frac{1}{4}(\mathbb{E}\min X_i + \mathbb{E}\max X_i)^2$$

$$= \frac{1}{4}\int_\theta^{\theta+1}\int_x^{\theta+1} f_{\min X_i,\max X_i}(x,y)(x+y)^2 dy dx - \frac{1}{4}(2\theta+1)^2.$$

Cf. equations (3.30) and (3.31). Here,

$$f_{\min X_i,\max X_i}(x,y) = -\frac{d^2}{dxdy}(y-x)^n$$

$$= n(n-1)(y-x)^{n-2}, \quad \theta < x < y < \theta+1.$$

Writing

$$I = \int_\theta^{\theta+1}\int_x^{\theta+1}(y-x)^{n-2}(x+y)^2 dy dx,$$

we have

$$\frac{1}{4}n(n-1)I = \theta^2 + \theta + \frac{n^2+3n+4}{4(n+1)(n+2)}.$$

This yields for $n \geq 3$

$$\operatorname{Var}\widehat{\theta}^* = \frac{n^2+3n+4}{4(n+1)(n+2)} - \frac{1}{4} = \frac{1}{2(n+1)(n+2)} < \frac{1}{12n} = \operatorname{Var}\overline{X}.$$

Indeed, the above integral I equals

$$\int_{\theta}^{\theta+1}\int_{x}^{\theta+1} (y-x)^{n-2}[(y-x)^2 + 4x(y-x) + 4x^2]dydx$$

$$= \int_{\theta}^{\theta+1}\left[\frac{(\theta+1-x)^{n+1}}{n+1} + \frac{4x(\theta+1-x)^n}{n} + \frac{4x^2(\theta+1-x)^{n-1}}{n-1}\right]dx$$

$$= \frac{1}{(n-1)n(n+1)(n+2)}\Big[n(n-1)+4\theta(n-1)(n+2)$$

$$+ 4(n-1)+4\theta^2(n+1)(n+2)+8\theta(n+2)+8\Big].$$

Hence,

$$\frac{1}{4}n(n-1)I = \theta^2 + \theta + \frac{n^2+3n+4}{4(n+1)(n+2)},$$

as claimed. ∎

Example 3.13 Suppose that $X_1,\ldots,X_n$ are independent RVs uniformly distributed over $(\theta, 2\theta)$. Find a two-dimensional sufficient statistic $T(X)$ for θ. Show that an unbiased estimator of θ is $\hat{\theta} = 2X_1/3$.

Find an unbiased estimator of θ which is a function of $T(X)$ and whose mean square error is no more that of $\hat{\theta}$.

Here, the likelihood function is

$$f(X;\theta) = \prod_{i=1}^{n}\frac{1}{\theta}I(\theta < x_i < 2\theta) = \frac{1}{\theta^n}I(\min_i x_i > \theta,\ \max_i x_i < 2\theta),$$

and hence, by the factorisation criterion

$$T = (\min_i X_i,\ \max_i X_i)$$

is sufficient. Clearly, $\mathbb{E}X_1 = 3\theta/2$, so $\theta^* = 2X_1/3$ is an unbiased estimator. Define

$$\widehat{\theta^*} = \mathbb{E}\left(\frac{2}{3}X_1\Big|\min_i X_i = a,\ \max_i X_i = b\right)$$

$$= \frac{2a}{3n} + \frac{2b}{3n} + \frac{n-2}{n}\frac{2}{3}\frac{a+b}{2} = \frac{a+b}{3}.$$

In fact, X_1 equals a or b with probability $1/n$ each; otherwise (when $X_1 \neq a, b$ which holds with probability $(n-2)/n$) the conditional expectation of X_1 equals $(a+b)/2$ because of the symmetry.

Consequently, by the RB Theorem,

$$\widehat{\theta^*} = \frac{1}{3}\left(\min_i X_i + \max_i X_i\right)$$

is the required estimator. ∎

We would like of course to have an estimator with a minimal MSE (a *minimum MSE estimator*). An effective tool to find such estimators is given by the *Cramér–Rao* (CR) *inequality*, or CR *bound*.

> *Assume that a PDF/PMF $f(\cdot; \theta)$ depends smoothly on parameter θ and the following condition holds: $\forall \theta \in \Theta$*

$$\int \frac{\partial}{\partial \theta} f(x; \theta) dx = 0 \text{ or } \sum_{x \in \mathsf{V}} \frac{\partial}{\partial \theta} f(x; \theta) = 0. \tag{3.37}$$

> *Consider IID observations $X_1, \ldots, X_n$, with joint PDF/PMF $f(\mathbf{x}; \theta) = f(x_1; \theta) \ldots f(x_n; \theta)$. Take an unbiased estimator $\theta^*(\mathbf{X})$ of θ satisfying the condition: $\forall \theta \in \Theta$*

$$\int_{\mathbb{R}^n} \theta^*(\mathbf{x}) \frac{\partial}{\partial \theta} f(\mathbf{x}; \theta) d\mathbf{x} = 1, \text{ or } \sum_{x \in \mathsf{V}^n} \theta^*(\mathbf{x}) \frac{\partial}{\partial \theta} f(\mathbf{x}; \theta) = 1. \tag{3.38}$$

> *Then for any such estimator, the following bound holds:*

$$\operatorname{Var} T \geq \frac{1}{nA(\theta)}, \tag{3.39}$$

> *where*

$$A(\theta) = \int \frac{\left(\partial f(x; \theta)/\partial \theta\right)^2}{f(x; \theta)} dx \text{ or } \sum_{x \in \mathsf{V}} \frac{\left(\partial f(x; \theta)/\partial \theta\right)^2}{f(x; \theta)}. \tag{3.40}$$

The quantity $A(\theta)$ is often called the *Fisher information* and features in many areas of probability theory and statistics.

Remark In practice, condition (3.37) means that we can interchange the derivation $\partial/\partial \theta$ and the integration/summation in the (trivial) equality

$$\frac{\partial}{\partial \theta} \int f(x; \theta) dx = 0 \text{ or } \frac{\partial}{\partial \theta} \sum_{x \in \mathsf{V}} f(x; \theta) = 0.$$

The equality holds as

$$\int f(x; \theta) dx = 1 \text{ or } \sum_{x \in \mathsf{V}} f(x; \theta) = 1.$$

A sufficient condition for such an interchange is that

$$\int \left| \frac{\partial}{\partial \theta} f(x; \theta) \right| dx < \infty \text{ or } \sum_{x \in \mathsf{V}} \left| \frac{\partial}{\partial \theta} f(x; \theta) \right| < \infty,$$

which is often assumed by default. Similarly, equation (3.38) means that we can interchange the derivation $\partial/\partial \theta$ and the integration/summation in the equality

$$1 = \frac{\partial}{\partial \theta} \theta = \frac{\partial}{\partial \theta} \mathbb{E} \theta^*(\mathbf{X}),$$

as

$$\mathbb{E}\theta^*(\mathbf{X}) = \int_{\mathbb{R}^n} \theta^*(\mathbf{x})f(\mathbf{x}; \theta)d\mathbf{x} \ \ \text{or} \ \ \sum_{\mathbf{x}\in V^n} \theta^*(\mathbf{x})f(\mathbf{x}; \theta).$$

Observe that the derivative $\partial f(\mathbf{x}; \theta)/\partial\theta$ can be written as

$$f(\mathbf{x}; \theta) \sum_{i=1}^n \frac{\partial f(x_i; \theta)/\partial\theta}{f(x_i; \theta)} = f(\mathbf{x}; \theta) \sum_{i=1}^n \frac{\partial \ln\, f(x_i; \theta)}{\partial\theta}. \tag{3.41}$$

We will prove bound (3.39) in the continuous case only (the proof in the discrete case simply requires the integrals to be replaced by sums). Set

$$D(x, \theta) = \frac{\partial}{\partial\theta} \ln\, f(x; \theta).$$

By condition (3.37),

$$\int f(x; \theta)D(x, \theta)dx = 0,$$

and we have that $\forall\ \theta \in \Theta$

$$\int f(\mathbf{x}; \theta)\sum_{i=1}^n D(x_i, \theta)d\mathbf{x} = n\int \frac{\partial}{\partial\theta}f(x; \theta)dx = 0.$$

On the other hand, by virtue of (3.41) equation (3.38) has the form

$$\int_{\mathbb{R}^n} \theta^*(\mathbf{x})f(\mathbf{x}; \theta)\sum_{i=1}^n D(x_i, \theta)d\mathbf{x} = 1,\ \theta \in \Theta.$$

The two last equations imply that

$$\int_{\mathbb{R}^n} (\theta^*(\mathbf{x}) - \theta)\sum_{i=1}^n D(x_i, \theta)f(\mathbf{x}; \theta)d\mathbf{x} = 1,\ \theta \in \Theta.$$

In what follows we omit subscript $\mathbb{R}^n$ in n-dimensional integrals in $d\mathbf{x}$ (integrals in dx are of course one-dimensional). We interpret the last equality as $\langle g_1, g_2\rangle = 1$, where

$$\langle g_1, g_2\rangle = \int g_1(\mathbf{x})g_2(\mathbf{x})\eta(\mathbf{x})d\mathbf{x}$$

is a scalar product, with

$$g_1(\mathbf{x}) = \theta^*(\mathbf{x}) - \theta, \ \ g_2(\mathbf{x}) = \sum_{i=1}^n D(x_i, \theta) \ \ \text{and} \ \ \eta(\mathbf{x}) = f(\mathbf{x}; \theta) \geq 0$$

($\eta(\mathbf{x})$ is the weight function determining the scalar product).
    Now we use the CS inequality $\left(\langle g_1, g_2\rangle\right)^2 \leq \langle g_1, g_1\rangle\langle g_2, g_2\rangle$. We obtain thus

$$1 \leq \left[\int (\theta^*(\mathbf{x}) - \theta)^2 f(\mathbf{x}; \theta)d\mathbf{x}\right]\left\{\int \left[\sum_{i=1}^n D(x_i, \theta)\right]^2 f(\mathbf{x}; \theta)d\mathbf{x}\right\}.$$

Here, the first factor gives Var $\theta^*(\mathbf{X})$. The second factor will give precisely $nA(\theta)$. In fact:

$$\int \left(\sum_{i=1}^{n} D(x_i, \theta)\right)^2 f(\mathbf{x}; \theta)d\mathbf{x} = \sum_{i=1}^{n} \int D(x_i, \theta)^2 f(\mathbf{x}; \theta)d\mathbf{x}$$

$$+2 \sum_{1 \le i_1 < i_2 \le n} \int D(x_{i_1}, \theta)D(x_{i_2}, \theta)f(\mathbf{x}; \theta)d\mathbf{x}.$$

Each term in the first sum gives $A(\theta)$:

$$\int D(x_i, \theta)^2 f(\mathbf{x}; \theta)d\mathbf{x} = \int D(x, \theta)^2 f(x; \theta)dx,$$

while each term in the second sum gives zero, as $\int D(x, \theta)f(x; \theta)dx = 0$:

$$\int D(x_{i_1}, \theta)D(x_{i_1}, \theta)f(\mathbf{x}; \theta)d\mathbf{x} = \left[\int D(x, \theta)f(x; \theta)dx\right]^2 = 0.$$

This completes the proof.

To conclude this section, we give a short account of Rao's stay in Cambridge. Rao arrived in Cambridge in 1945 to do work in the University Museum of Archaeology and Anthropology on analysing objects (human skulls and bones) brought back by a British expedition from a thousand-year old site in North Africa. He had to measure them carefully and apply what is called the Mahalanobis distance to derive conclusions about the anthropological characteristics of an ancient population. Rao was then 25 years old and had 18 papers in statistics published or accepted for publication. His first paper, which had just been published, contained both the RB Theorem and CR inequality. Soon after his arrival in Cambridge he met Fisher and began also working in Fisher's Laboratory of Genetics on various characteristics of mice; he had to breed them, mate them in a determined way and record genetic parameters of the litter produced (kinky tails, ruffled hair and a disposition to keep shaking all the time). As is described in Rao's biography [Kr], his final duty in each experiment was to dispose of mice not needed in further work; according to the customs of the time, young mice were put in ether and mature ones had their heads hit against a table (a practice that would nowadays undoubtedly cause an objection from Animal Rights activists). Rao was too sensitive to do this particular job and had some friends who did it for him; otherwise he utterly enjoyed his work (and the rest of his time) in Cambridge. Another of his friends was Abdus Salam, the future Nobel Prize winner in physics and then a student at St John's College. Salam had doubts about his future in research and, seeing Rao's determination, expressed keen interests in statistics. However, it was Rao who persuaded him not to change his field . . .

The work in Fisher's Laboratory formed the basis of Rao's Ph.D. thesis which he passed successfully in 1948, by which time he had 40 published papers. After receiving his Ph.D. from such a prestigious university as Cambridge, it was supposed that back in India he would receive offers of matrimonial alliances from many rich families. But a month after returning home from Cambridge he became engaged to his future wife, who was then 23 and had her own academic degrees. The marriage was arranged by Rao's mother who was very progressive: she did not mind him marrying a highly educated woman, although at that time such brides were generally not wanted in families with eligible sons. Their marriage has been perfectly happy, which makes one wonder why the (completely unarranged) marriages of other famous statisticians in Europe and America (including ones repeatedly mentioned in this book) ended badly.

Rao's contributions in statistics are now widely recognised. It was not so in the beginning, particularly with the RB Theorem. Even in 1953, eight years after Rao's paper and five years after Blackwell's, some statisticians were referring to the procedure described in the theorem as 'Blackwellisation'. When Rao pointed out that he was the first to discover the result, a lecturer on this topic said that this term is easier on the tongue than 'Raoisation'. However, in a later paper the statistician in question proposed the term 'Rao–Blackwellisation' that is now used. On the issue of the CR inequality (the term proposed by Neyman), Rao remembers a call from an airline employee at the Teheran airport: 'Good news, Mr Cramer Rao, we found your bag', after a piece of his luggage was lost on a flight.

Rao likes using humour in serious situations. In India, as in many countries, birth control is an important issue, and providing women with reasons not to have too many children is one of the perennial tasks of local and central administration. In one of his articles on this topic, Rao points out that every fourth baby born in the world is Chinese and then makes the following statement to his Indian audience: 'Look before you leap to your next baby, if you already have three. The fourth will be a Chinese!' Hopefully, some readers of this book will find this instructive in dealing with sample means. . ..

3.7   Exponential families

> Sex, Lies and Exponential Families
> (From the series *'Movies that never made it to the Big Screen'*.)

It is interesting to investigate when the equality in CR bound (3.39) is attained. Here again, the CS inequality is crucial. We know that for equality we need functions g_1 and g_2 to be linearly dependent: $g_1(\mathbf{x}) = \lambda g_2(\mathbf{x})$. Then

$$\int g_1(\mathbf{x})g_2(\mathbf{x})\eta(\mathbf{x})d\mathbf{x} = \lambda \int g_2(\mathbf{x})^2 \eta(\mathbf{x})d\mathbf{x}.$$

In our case, $\int g_1(\mathbf{x})g_2(\mathbf{x})\eta(\mathbf{x})d\mathbf{x} = 1$ and so

$$\lambda = \left[\int g_2(\mathbf{x})^2 \eta(\mathbf{x})d\mathbf{x} \right]^{-1} = \frac{1}{nA(\theta)}.$$

Thus, we obtain the relation:

$$\theta^*(\mathbf{X}) - \theta = \frac{1}{nA(\theta)} \sum_{i=1}^{n} D(x_i, \theta),$$

or

$$\theta^*(\mathbf{X}) = \theta + \frac{1}{nA(\theta)} \sum_{i=1}^{n} D(x_i, \theta) = \frac{1}{n}\sum_{i=1}^{n}\left[\theta + \frac{D(x_i, \theta)}{A(\theta)}\right].$$

The LHS of the last equation does not depend on θ. Hence, each term $\theta + D(x_i, \theta)/A(\theta)$ should be independent of θ:

$$\theta + \frac{D(x, \theta)}{A(\theta)} = C(x).$$

In other words,

$$D(x, \theta) = A(\theta)(C(x) - \theta), \quad \theta \in \Theta, \tag{3.42}$$

and the estimator θ^* has the summatory form:

$$\theta^*(\mathbf{x}) = \frac{1}{n}\sum_{i=1}^{n} C(x_i). \tag{3.43}$$

Now solving equation (3.42):

$$\frac{\partial}{\partial\theta} \ln f(x; \theta) = A(\theta)\left[C(x) - \theta\right],$$

where $A(\theta) = B''(\theta)$, we obtain

$$\ln f(x; \theta) = B'(\theta)[C(x) - \theta] + B(\theta) + H(x).$$

Hence,

$$f(x; \theta) = \exp\left[B'(\theta)[C(x) - \theta] + B(\theta) + H(x)\right]. \tag{3.44}$$

Such families (of PDFs or PMFs) are called *exponential*.

Therefore, the following statement holds:

> *Equality in the CR inequality is attained iff family $\{f(x; \theta)\}$ is exponential, i.e. is given by equation (3.44), where $B''(\theta) > 0$. In this case the minimum MSE estimator of θ is given by (3.43) and its variance equals $1/(nB''(\theta))$. Thus the Fisher information equals $B''(\theta)$.*

Example 3.14   Let X_i be IID $N(\mu, \sigma^2)$ with a given σ^2 and unknown μ. Write $c = -[\ln(2\pi\sigma^2)]/2$ (which is a constant as σ^2 is fixed). Then

$$\ln f(x; \mu) = -\frac{(x-\mu)^2}{2\sigma^2} + c = \frac{\mu(x-\mu)}{\sigma^2} + \frac{\mu^2}{2\sigma^2} - \frac{x^2}{2\sigma^2} + c.$$

We see that the family of PDFs $f(; \mu)$ is exponential; in this case:

$$C(x) = x, \ B(\mu) = \frac{\mu^2}{2\sigma^2}, \ B'(\mu) = \frac{\mu}{\sigma^2},$$

$$A(\mu) = B''(\mu) = \frac{1}{\sigma^2}, \ H(x) = -\frac{x^2}{2\sigma^2} + c.$$

Note that here $A(\mu)$ does not depend on μ.

Hence, in the class of unbiased estimators $\theta^*(\mathbf{x})$ such that

$$\sum_{i=1}^{n} \int \theta^*(\mathbf{x}) \frac{(x_i - \mu)}{\sigma^2 (2\pi\sigma^2)^{n/2}} \exp\left[-\sum_j \frac{(x_j - \mu)^2}{2\sigma^2}\right] d\mathbf{x} = 1,$$

the minimum MSE estimator for μ is

$$\bar{x} = \frac{1}{n} \sum_i x_i,$$

the sample mean, with Var $\bar{X} = \sigma^2/n$. ∎

Example 3.15   If we now assume that X_i is IID $N(\mu, \sigma^2)$ with a given μ and unknown σ^2, then again the family of PDFs $f(\cdot\;; \sigma^2)$ is exponential:

$$\ln f(x; \sigma^2) = -\frac{1}{2\sigma^2}((x - \mu)^2 - \sigma^2) - \frac{1}{2} \ln (2\pi\sigma^2) - \frac{1}{2},$$

with

$$C(x) = (x - \mu)^2, \; B(\sigma^2) = -\frac{1}{2} \ln (\sigma^2), \; B'(\sigma^2) = -\frac{1}{2\sigma^2},$$

$$A(\sigma^2) = B''(\sigma^2) = \frac{1}{2}\left(\frac{1}{\sigma^2}\right)^2,$$

and $H(x) = -[\ln (2\pi) + 1]/2$. We conclude that, in the class of unbiased estimators $\theta^*(\mathbf{x})$ such that

$$\sum_{i=1}^{n} \int_{\mathbb{R}^n} \theta^*(\mathbf{x}) \frac{(x_i - \mu)^2 - \sigma^2}{2\sigma^4 (2\pi\sigma^2)^{n/2}} \exp\left[-\sum_j \frac{(x_j - \mu)^2}{2\sigma^2}\right] d\mathbf{x} = 1,$$

the minimum MSE estimator for σ^2 is Σ^2/n, where

$$\Sigma^2 = \sum_i (x_i - \mu)^2,$$

with Var $(\Sigma^2/n) = 2\sigma^4/n$. ∎

Example 3.16   The Poisson family, with $f_\lambda(k) = e^{-\lambda}\lambda^k/k!$, is also exponential:

$$\ln f_\lambda(k) = k \ln \lambda - \ln (k!) - \lambda = (k - \lambda) \ln \lambda + \lambda(\ln \lambda - 1) - \ln (k!),$$

with

$$C(k) = k, \; B(\lambda) = \lambda(\ln \lambda - 1), \; B'(\lambda) = \ln \lambda, \; A(\lambda) = B''(\lambda) = \frac{1}{\lambda},$$

and $H(k) = -\ln (k!)$. Thus the minimum MSE estimator is

$$\bar{\lambda} = \frac{1}{n} \sum_{i=1}^{n} k_i, \; \text{with Var } \bar{\lambda} = \frac{\lambda}{n}.$$

We leave it to the reader to determine in what class the minimum MSE is guaranteed. ∎

Example 3.17 The family where RVs $X_i \sim \mathrm{Exp}\,(\lambda)$ is exponential, but relative to $\theta = 1/\lambda$ (which is not surprising as $\mathbb{E}X_i = 1/\lambda$). In fact, here

$$f(x; \theta) = \left\{ \exp\left[-\frac{1}{\theta}(x - \theta) - \ln \theta - 1 \right] \right\} I(x > 0),$$

with

$$C(x) = x, \ B(\theta) = -\ln \theta, \ B'(\theta) = -\frac{1}{\theta}, \ A(\theta) = B''(\theta) = \frac{1}{\theta^2}, \ H(x) = -1.$$

Therefore, in the class of unbiased estimators $\theta^*(\mathbf{x})$ of $1/\lambda$, with

$$\sum_{i=1}^{n} \int_{\mathbb{R}_+^n} \theta^*(\mathbf{x})(\lambda^2 x_i - \lambda)\lambda^n \exp\left(-\lambda \sum_j x_j \right) d\mathbf{x} = 1,$$

the sample mean $\bar{x} = \sum_i x_i/n$ is the minimum MSE estimator, and it has $\mathrm{Var}\,\bar{X} = \theta^2/n = 1/(n\lambda^2)$. ∎

An important property is that for exponential families the minimum MSE estimator $\theta^*(\mathbf{x})$ coincides with the MLE $\widehat{\theta}(\mathbf{x})$, i.e.

$$\widehat{\theta}(\mathbf{x}) = \frac{1}{n} \sum_i C(x_i). \tag{3.45}$$

More precisely, we write the stationarity equation as

$$\frac{\partial}{\partial \theta} \ell(\theta; \mathbf{x}) = \sum_{i=1}^{n} D(x_i, \theta) f(\mathbf{x}; \theta) = 0,$$

which, under the condition that $f(\mathbf{x}; \theta) > 0$, becomes

$$\sum_{i=1}^{n} D(x_i, \theta) = 0. \tag{3.46}$$

In the case of an exponential family, with

$$f(x; \theta) = \exp\left(B'(\theta)[C(x) - \theta] + B(\theta) + H(x) \right),$$

we have

$$D(x, \theta) = \frac{\partial}{\partial \theta} \ln f(x; \theta) = B''(\theta)(C(x) - \theta).$$

We see that if $\theta^* = \theta^*(\mathbf{x}) = \sum_{i=1}^{n} C(x_i)/n$, then

$$D(x_i, \theta^*) = B''(\theta^*)[C(x_i) - \theta^*] = B''(\theta^*) \left[C(x_i) - \frac{1}{n} \sum_j C(x_j) \right],$$

and hence

$$\sum_{i=1}^{n} D(x_i, \theta^*) = B''(\theta^*) \sum_{i=1}^{n} \left[C(x_i) - \frac{1}{n} \sum_j C(x_j) \right] = 0.$$

Thus minimum MSE estimator θ^* solves the stationarity equation. Therefore, if an exponential family $\{f(x; \theta)\}$ is such that any solution to stationarity equation (3.29) gives a global likelihood maximum, then θ^* is the MLE.

The CR inequality is named after C.H. Cramér (1893–1985), a prominent Swedish analyst, number theorist, probabilist and statistician, and C.R. Rao. One story is that the final form of the inequality was proved by Rao, then a young (and inexperienced) lecturer at the Indian Statistical Institute, overnight in 1943 in response to a student enquiry about some unclear places in his presentation.

3.8   Confidence intervals

> Statisticians do it with 95% confidence.
>> (From the series *'How they do it'*.)

So far we developed ideas related to *point estimation*. Another useful idea is to consider *interval estimation*. Here, one works with *confidence intervals (CIs)* (in the case of a vector parameter $\theta \in \mathbb{R}^d$, confidence sets like squares, cubes, circles or ellipses, balls, etc.). Given $\gamma \in (0, 1)$, a $100\gamma\%$ CI for a scalar parameter $\theta \in \Theta \subseteq \mathbb{R}$ is any pair of functions $a(\mathbf{x})$ and $b(\mathbf{x})$ such that $\forall\ \theta \in \Theta$ the probability

$$\mathbb{P}_\theta(a(\mathbf{X}) \leq \theta \leq b(\mathbf{X})) = \gamma. \tag{3.47}$$

We want to stress that: (i) the randomness here is related to endpoints $a(\mathbf{x}) < b(\mathbf{x})$, not to θ, (ii) a and b should not depend on θ. A CI may be chosen in different ways; naturally, one is interested in 'shortest' intervals.

Confidence intervals regularly appear in the Tripos papers: MT-IB 1998-212E (this problem requires further knowledge of the material and will be discussed later), 1995-203G (i), 1993-203J, 1992-406D, and also SP-IB 1992-203H.

Example 3.18   The first standard example is a CI for the unknown mean of a normal distribution with a known variance. Assume that $X_i \sim \mathrm{N}(\mu, \sigma^2)$, where σ^2 is known and $\mu \in \mathbb{R}$ has to be estimated. We want to find a 99% CI for μ. We know that $\sqrt{n}(\overline{X} - \mu)/\sigma \sim \mathrm{N}(0, 1)$. Therefore, if we take $z_- < z_+$ such that $\Phi(z_+) - \Phi(z_-) = 0.99$, then the equation

$$\mathbb{P}\left(z_- < \frac{\sqrt{n}}{\sigma}(\overline{X} - \mu) < z_+\right) = 0.99$$

can be re-written as

$$\mathbb{P}\left(\overline{X} - \frac{z_+\sigma}{\sqrt{n}} < \mu < \overline{X} - \frac{z_-\sigma}{\sqrt{n}}\right) = 0.99,$$

i.e. gives the interval

$$\left(\overline{X} - \frac{z_+\sigma}{\sqrt{n}}, \overline{X} - \frac{z_-\sigma}{\sqrt{n}}\right),$$

centred at $\overline{X} - (z_- + z_+)\sigma/(2\sqrt{n})$ and of width $(z_+ - z_-)\sigma/\sqrt{n}$.

We still have a choice of z_- and z_+; to obtain the shortest interval we would like to choose $z_+ = -z_- := z$, as the $N(0,1)$ PDF is symmetric and has its peak at the origin. Then the interval becomes

$$\left(\overline{X} - \frac{z\sigma}{\sqrt{n}}, \overline{X} + \frac{z\sigma}{\sqrt{n}} \right),$$

and z will be the upper 0.005 point of the standard normal distribution, with $\Phi(a) = 1 - 0.005 = 0.995$. From the normal percentage tables: $z = 2.5758$. Hence, the answer:

$$\left(\overline{X} - \frac{2.5758\sigma}{\sqrt{n}}, \overline{X} + \frac{2.5758\sigma}{\sqrt{n}} \right). \quad \blacksquare$$

Example 3.19   The next example is to determine the CI for the unknown variance of a normal distribution with a known mean. Assume that $X_i \sim N(\mu, \sigma^2)$, where μ is known and $\sigma^2 > 0$ has to be estimated. We want to find a 98% CI for σ^2. Then $\sum_i (X_i - \mu)^2 / \sigma^2 \sim \chi_n^2$. Denote by $F_{\chi_n^2}$ the CDF $\mathbb{P}(X < x)$ of a RV $X \sim \chi_n^2$. Take $h^- < h^+$ such that $F_{\chi_n^2}(h^+) - F_{\chi_n^2}(h^-) = 0.98$. Then the condition

$$\mathbb{P}\left(h^- < \frac{1}{\sigma^2}\left(\sum_i (X_i - \mu)^2 \right) < h^+ \right) = 0.98$$

can be re-written as

$$\mathbb{P}\left(\frac{\sum_i (X_i - \mu)^2}{h^+} < \sigma^2 < \frac{\sum_i (X_i - \mu)^2}{h^-} \right) = 0.98.$$

This gives the interval

$$\left(\frac{1}{h^+} \sum_i (X_i - \mu)^2, \frac{1}{h^-} \sum_i (X_i - \mu)^2 \right).$$

Again we have a choice of h^- and h^+; a symmetric, or equal-tailed, option is to take $F_{\chi_n^2}(h^-) = 1 - F_{\chi_n^2}(h^+) = (1 - 0.98)/2 = 0.01$. From the χ^2 percentage tables, for $n = 38$, $h^- = 20.69$, $h^+ = 61.16$. See, for example, [LiS], pp. 40–41 (this is the standard reference to statistical tables used in a number of examples and problems below). $\blacksquare$

In Examples 3.18 and 3.19 we managed to find an RV $Z = Z(T(\mathbf{X}), \theta)$ which is a function of a sufficient statistic and the unknown parameter θ (μ in Example 3.18 and σ^2 in Example 3.19) and has a distribution not depending on this parameter. Namely,

$$Z = \sqrt{n}(\overline{X} - \mu)/\sigma \sim N(0, 1) \quad \text{in Example 3.18}$$

and

$$Z = \sum_i (X_i - \mu)^2 / \sigma^2 \sim \chi_n^2 \quad \text{in Example 3.19}.$$

We then produced values $y_\pm$ such that $\mathbb{P}(y_- < Z < y_+) = \gamma \ \gamma(z_\pm)$ in Example 3.18 and $h^\pm$ in Example 3.19 and solved the equations $Z(T, \theta) = y_\pm$ to find roots

$$a(\mathbf{X}) = a(T(\mathbf{X}, y_-)) \text{ and } b(\mathbf{X}) = b(T(\mathbf{X}, y_-)).$$

The last step is not always straightforward, in which case various approximations may be useful.

Example 3.20   In this (more challenging) example the above idea is, in a sense, pushed to its limits. Suppose $X_i \sim \mathrm{Po}\,(\lambda)$, and we want to find a $100\gamma\%$ CI for λ. Here we know that $n\overline{X} \sim \mathrm{Po}\,(n\lambda)$, which still depends on λ. The CDF $F = F_{\overline{X}}$ for $\overline{X}$ jumps at points k/n, $k = 0, 1, \ldots$, and has the form

$$F\,(x; \lambda) = I(x \geq 0) \sum_{r=0}^{\rceil nx \lceil} e^{-n\lambda} \frac{(n\lambda)^r}{r!}, \tag{3.48}$$

where, for $y > 0$,

$$\rceil y \lceil = \begin{cases} y - 1, & y \text{ is an integer,} \\ [y], & \text{the integer part of } y, \text{ if } y \text{ is not integer.} \end{cases}$$

It is differentiable and monotone decreasing in λ:

$$\frac{\partial}{\partial \lambda} F(k/n, \lambda) = -ne^{-n\lambda} \frac{(n\lambda)^{k-1}}{(k-1)!} < 0.$$

Thus if we look for bounds $a < \lambda < b$, it is equivalent to the functional inequalities $F(x, b) < F(x; \lambda) < F(x; a)$, $\forall x > 0$ of the form k/n, $k = 1, 2, \ldots$. We want a and b to be functions of $\mathbf{x}$, more precisely, of the sample mean $\bar{x}$. The symmetric, or equal-tailed, $100\gamma\%$-CI is with endpoints $a(\overline{X})$ and $b(\overline{X})$ such that

$$\mathbb{P}\big(\lambda \leq a(\overline{X})\big) = \mathbb{P}\big(\lambda \geq b(\overline{X})\big) = \frac{1 - \gamma}{2}.$$

Write

$$\mathbb{P}(\lambda > b(\overline{X})) = \mathbb{P}(F(\overline{X}; \lambda) < F(\overline{X}; b(\overline{X})))$$

and use the following fact (a modification of formula (2.57)):

> If, for an RV X with CDF F and a function $g : \mathbb{R} \to \mathbb{R}$, there exists a unique point $c \in \mathbb{R}$ such that $F(c) = g(c)$, $F(y) < g(y)$ for $y < c$ and $F(y) \geq g(y)$ for $y \geq c$, then

$$\mathbb{P}(F(X) < g(X)) = g(c). \tag{3.49}$$

Next, by equation (3.48):

$$\mathbb{P}\left(F(\overline{X}; \lambda) < F(\overline{X}; b(\overline{X}))\right) = F\left(\frac{k}{n}; b\left(\frac{k}{n}\right)\right),$$

provided that there exists k such that $F(k/n; \lambda) = F(k/n; b(k/n))$. In other words, if you choose $b = b(\bar{x})$ so that

$$\sum_{l=0}^{\lceil n\bar{x} \rceil} e^{-nb} \frac{(nb)^l}{l!} = \frac{1-\gamma}{2}, \tag{3.50}$$

it will guarantee that $\mathbb{P}(\lambda > b(\overline{X})) \le (1 - \gamma)/2$.

Similarly, choosing $a = a(\bar{x})$ so that

$$\sum_{l=n\bar{x}}^{\infty} e^{-na} \frac{(na)^l}{l!} = \frac{1-\gamma}{2} \tag{3.51}$$

will guarantee that $\mathbb{P}(\lambda < a(\overline{X})) \le (1 - \gamma)/2$. Hence, $(a(\overline{X}), b(\overline{X}))$ is the (equal-tailed) $100\gamma\%$ CI for λ.

The distributions of $a(\overline{X})$ and $b(\overline{X})$ can be identified in terms of quantiles of the χ^2 distribution. In fact, $\forall\ k = 0, 1, \dots$,

$$\frac{d}{ds} \sum_{k \le l < \infty} e^{-s} \frac{s^l}{l!} = \sum_{k \le l < \infty} \left[-e^{-s} \frac{s^l}{l!} + e^{-s} \frac{s^{l-1}}{(l-1)!} \right] = e^{-s} \frac{s^{k-1}}{(k-1)!} = 2 f_{Y_1}(2s),$$

where $Y_1 \sim \chi^2_{2k}$. That is, $\sum_{k \le l < \infty} e^{-s} s^l / l! = \mathbb{P}(Y_1 < 2s)$. We see that

$$2na = h^-_{m^-}((1-\gamma)/2),$$

the lower $(1 - \gamma)/2$-quantile of the χ^2-distribution with $m^- = 2n\overline{X}$ degrees of freedom.

Similarly:

$$\frac{d}{ds} \sum_{0 \le l \le k} e^{-s} \frac{s^l}{l!} = \sum_{0 \le l \le k} \left[-e^{-s} \frac{s^l}{l!} + e^{-s} \frac{s^{l-1}}{(l-1)!} \right] = -e^{-s} \frac{s^k}{k!} = -2 f_{Y_2}(2s)$$

for $Y_2 \sim \chi^2_{2k+2}$. That is,

$$\mathbb{P}(Y_2 > 2s) = \sum_{0 \le l \le k} e^{-s} \frac{s^l}{l!}.$$

It means that

$$2nb = h^+_{m^+}((1-\gamma)/2),$$

the upper $((1 + \gamma)/2)$-quantile of the χ^2 distribution with $m^+ = 2n\overline{X} + 2$ degrees of freedom.

These answers look cumbersome. However, for large n, we can think that

$$\overline{X} \sim N(\lambda, \lambda/n), \text{ i.e. } \sqrt{n/\lambda}(\overline{X} - \lambda) \sim N(0, 1).$$

Then if we take $\gamma = 0.99$ and, as before, $-a_- = a_+ = 2.5758$, we have that

$$\mathbb{P}\left(a_- < \frac{\sqrt{n}}{\sqrt{\lambda}}(\overline{X} - \lambda) < a_+ \right) \approx 0.99.$$

We can solve this equation for λ (or rather $\sqrt{\lambda}$). In fact, $(\overline{X} - \lambda)/\sqrt{\lambda}$ decreases with λ, and from the equation

$$\frac{\overline{X} - \lambda}{\sqrt{\lambda}} = \frac{a_{\pm}}{\sqrt{n}}$$

we find:

$$\sqrt{\lambda_{\mp}} = \sqrt{\frac{a_{\pm}^2}{4n} + \overline{X}} - \frac{a_{\pm}}{2\sqrt{n}}, \text{ i.e. } \lambda_{\mp} = \left(\sqrt{\frac{a_{\pm}^2}{4n} + \overline{X}} - \frac{a_{\pm}}{2\sqrt{n}}\right)^2$$

(the negative roots have to be discarded). Hence, we obtain that the probability

$$\mathbb{P}\left(\left(\sqrt{\frac{2.5758^2}{4n} + \overline{X}} - \frac{2.5758}{2\sqrt{n}}\right)^2 < \lambda < \left(\sqrt{\frac{2.5758^2}{4n} + \overline{X}} + \frac{2.5758}{2\sqrt{n}}\right)^2\right)$$

is ≈ 0.99. Hence,

$$\left(\left(\sqrt{\frac{2.5758^2}{4n} + \overline{X}} - \frac{2.5758}{2\sqrt{n}}\right)^2, \left(\sqrt{\frac{2.5758^2}{4n} + \overline{X}} + \frac{2.5758}{2\sqrt{n}}\right)^2\right)$$

gives an approximate answer whose accuracy increases with n.

Confidence intervals for the mean of a Poisson distribution attracted particular attention in many books, beginning with [P]. In this book, the term 'confidence belt' is used, to stress that the data serve a range of values of both n and $\overline{X}$. ∎

3.9   Bayesian estimation

> Bayesian Instinct
> Trading Priors
>> (From the series *'Movies that never made it to the Big Screen'.*)

A useful alternative to the above version of the estimation theory is where θ is treated as a random variable with values in Θ and some (given) *prior* PDF or PMF $\pi(\theta)$. After observing a sample $\mathbf{x}$, we can produce the *posterior* PDF or PMF $\pi(\theta|\mathbf{x})$. Owing to the Bayes Theorem, $\pi(\theta|\mathbf{x})$ is defined by

$$\pi(\theta|\mathbf{x}) \propto \pi(\theta) f(\mathbf{x}; \theta). \tag{3.52}$$

More precisely,

$$\pi(\theta|\mathbf{x}) = \frac{1}{\overline{f}(\mathbf{x})} \pi(\theta) f(\mathbf{x}; \theta), \tag{3.53}$$

where

$$\overline{f}(\mathbf{x}) = \int_{\Theta} \pi(\theta) f(\mathbf{x}; \theta) d\theta \text{ or } \sum_{\theta \in \Theta} \pi(\theta) f(\mathbf{x}; \theta).$$

Pictorially speaking, (posterior) $\propto$ (prior) $\times$ (likelihood), where the constant of proportionality is simply chosen to normalise the total mass to 1.

Remark   Note that the likelihood $f(\mathbf{x}; \theta)$ in equation (3.52) and (3.53) is considered as a conditional PDF/PMF of $\mathbf{X}$, given θ. This is the Bayesian interpretation of the likelihood, as opposed to the Fisherian interpretation where it is considered as a function of θ for fixed $\mathbf{x}$.

In Examples 3.21–3.24 we calculate posterior distributions in some models.

Example 3.21   Let $X_i \sim \text{Bin}(m, \theta)$ and the prior distribution for θ is Bet (a, b) for some known a, b:

$$\pi(\theta) \propto \theta^{a-1}(1 - \theta)^{b-1} I(0 < \theta < 1);$$

see Example 3.5. Then the posterior is

$$\pi(\theta|\mathbf{x}) \propto \theta^{\sum x_i + a - 1}(1 - \theta)^{nm - \sum x_i + b - 1} I(0 < \theta < 1),$$

which is Bet $(\sum_i x_i + a, nm - \sum_i x_i + b) = $ Bet $(n\bar{x} + a, n(m - \bar{x}) + b)$, with the proportionality constant $1/B(n\bar{x} + a, n(m - \bar{x}) + b)$. In other words, a Beta prior generates a Beta posterior. One says that the Beta family is *conjugate* for binomial samples.   ∎

The Unbelievable Conjugacy of Beta
(From the series *'Movies that never made it to the Big Screen'*.)

Example 3.22   The Beta family is also conjugate for negative binomial samples, where $X_i \sim \text{NegBin}(r, \theta)$, with known r. In fact, here the posterior

$$\pi(\theta|\mathbf{x}) \propto \theta^{nr + a - 1}(1 - \theta)^{n\bar{x} + b - 1},$$

i.e. is Bet $(nr + a, n\bar{x} + b)$.   ∎

Example 3.23   Another popular example of a conjugate family is Gamma, for Poisson or exponential samples. Indeed, if $X_i \sim \text{Po}(\lambda)$ and $\pi(\lambda) \propto \lambda^{\tau - 1} e^{-\lambda/\alpha}$, then the posterior

$$\pi(\lambda|\mathbf{x}) \propto \lambda^{n\bar{x} + \tau - 1} e^{-\lambda(\alpha n + 1)/\alpha}$$

has the PDF Gam $(\tau + n\bar{x}, (n\alpha + 1)/\alpha)$.
Similarly, if $X_i \sim \text{Exp}(\lambda)$ and $\pi(\lambda) \propto \lambda^{\tau - 1} e^{-\lambda\alpha}$, then the posterior

$$\pi(\lambda|\mathbf{x}) \propto \lambda^{n + \tau - 1} e^{-\lambda(n\bar{x} + \alpha)}$$

again has the PDF Gam $(\tau + n, n\bar{x} + \alpha)$.   ∎

Example 3.24 The normal distributions also emerge as a conjugate family. Namely, assume that $X_i \sim N(\mu, \sigma^2)$ where σ^2 is known, and $\pi(\mu) \propto \exp\left[-(\mu - a)^2/(2b^2)\right]$, for some known $a \in \mathbb{R}$ and $b > 0$. Then, by using the solutions to Problems 2.42 and 2.43, posterior $\pi(\mu|\mathbf{x})$ is

$$\propto \exp\left[-\frac{1}{2}\left(\frac{n}{\sigma^2} + \frac{1}{b^2}\right)\mu^2 + \left(\frac{a}{b^2} + \frac{n\bar{x}}{\sigma^2}\right)\mu\right] \propto \exp\left[-\frac{1}{2b_1^2}(\mu - a_1)^2\right],$$

where

$$b_1^2 = \left(\frac{1}{b^2} + \frac{n}{\sigma^2}\right)^{-1} \text{ and } a_1 = b_1^2\left(\frac{a}{b^2} + \frac{n\bar{x}}{\sigma^2}\right).$$

See MT-IB 1998-403E, 1997-212G. ∎

A further step is to introduce a *loss function* (LF) measuring the loss incurred in estimating a given parameter θ. This is a function $L(\theta, a)$, $\theta, a \in \Theta$, where θ is the true and a is the guessed value. For instance, a *quadratic* LF is $L(\theta, a) = (\theta - a)^2$, an *absolute error* LF is $L(\theta, a) = |\theta - a|$ etc.

We then consider the *posterior expected loss*

$$R(\mathbf{x}, a) = \int_\Theta \pi(\theta|\mathbf{x})L(\theta, a)\mathrm{d}\theta \text{ or } \sum_{\theta \in \Theta} \pi(\theta|\mathbf{x})L(\theta, a), \tag{3.54}$$

while guessing value a. We want to choose $\hat{a} = \hat{a}(\mathbf{x})$ minimising $R(\mathbf{x}, a)$:

$$\hat{a} = \arg\min_a R(\mathbf{x}, a). \tag{3.55}$$

The minimiser, $\hat{a}$, is called an *optimal Bayes estimator*, or optimal estimator for short. (Some authors say 'optimal point estimator'.) For the quadratic loss,

$$R(\mathbf{x}, a) = \int_\Theta \pi(\theta|\mathbf{x})(\theta - a)^2\mathrm{d}\theta.$$

By differentiation, $R(\mathbf{x}, a)$ is minimised at

$$\hat{a} = \int_\Theta \theta\pi(\theta|\mathbf{x})\mathrm{d}\theta,$$

i.e. at the *posterior mean* $\mathbb{E}(\theta|\mathbf{x})$. Furthermore, the minimal value of the posterior expected loss is

$$\min_a R(\mathbf{x}, a) = R(\mathbf{x}, \hat{a}) = \int [\theta - \mathbb{E}(\theta|\mathbf{x})]^2 \pi(\theta|\mathbf{x})\mathrm{d}\theta,$$

i.e. equals $\mathbb{E}[(\theta - \mathbb{E}(\theta|\mathbf{x}))^2|\mathbf{x}]$, the posterior variance. For the absolute error loss,

$$R(\mathbf{x}, a) = \int_\Theta \pi(\theta|\mathbf{x})|\theta - a|\mathrm{d}\theta = \int_{-\infty}^a \pi(\theta|\mathbf{x})(a - \theta)\mathrm{d}\theta + \int_a^\infty \pi(\theta|\mathbf{x})(\theta - a)\mathrm{d}\theta$$

which is minimised at the *posterior median*, i.e. the value $\widehat{a}$ for which

$$\int_{-\infty}^{a} \pi(\theta|\mathbf{x})d\theta = \int_{\widehat{a}}^{\infty} \pi(\theta|\mathbf{x})d\theta = 1/2.$$

A straightforward but important remark is that, in general, $\widehat{a}(\mathbf{x})$ also minimises the unconditional expected loss among *all* estimators $d : \mathbf{x} \to \Theta$. Here, it is instructive to slightly change the terminology: an estimator is considered as a *decision rule* (you observe $\mathbf{x}$ and decide that the value of the parameter is $d(\mathbf{x})$). Given a value θ and a decision rule d, the quantity

$$r(\theta, d) = \mathbb{E}_\theta L(\theta, d(\mathbf{X})) = \begin{cases} \sum_{\mathbf{x}} L(\theta, d(\mathbf{x}))f(\mathbf{x}; \theta), \\ \int L(\theta, d(\mathbf{x}))f(\mathbf{x}; \theta)d\mathbf{x}, \end{cases}$$

represents the *risk* under decision rule d when the parameter value is θ. We want to minimise the *Bayes risk*

$$r^B(d) = \int_\Theta r(\theta, d)\pi(\theta)d\theta \quad \text{or} \quad \sum_\Theta r(\theta, d)\pi(\theta). \tag{3.56}$$

The remark is that

$$\widehat{a} = \arg \min_d r^B(d). \tag{3.57}$$

For that reason, the optimal Bayes estimator is also called the optimal rule.

Formally, in equation (3.55) you minimise for every given $\mathbf{x}$ while in (3.57) you minimise the sum or the integral over all values of $\mathbf{x}$. It is important to see that both procedures lead to the same answer. The above remark asserts the minimiser in equation (3.55) yields the minimum of $r^B(d)$. But what about the inverse statement that every decision rule minimising $r^B(d)$ coincides with the optimal Bayes estimator? This is also true: by changing the order of summation/integration over $\mathbf{x}$ and θ, write

$$r(\theta, d) = \int \mathbb{E}_{\pi(\theta|(\mathbf{x}))} L(\theta, d(\mathbf{x}))\overline{f}(\mathbf{x})d\mathbf{x} \quad \text{or} \quad \sum_{\mathbf{x}} \mathbb{E}_{\pi(\theta|\mathbf{x})} L(\theta, d(\mathbf{x}))\overline{f}(\mathbf{x}). \tag{3.58}$$

Here $\overline{f}(\mathbf{x})$ is the marginal PDF/PMF of $\mathbf{X}$:

$$\overline{f}(\mathbf{x}) = \sum_\Theta \pi(\theta)f(\mathbf{x}; \theta) = \int_\Theta \pi(\theta)f(\mathbf{x}; \theta)d\theta,$$

and $\pi(\theta|\mathbf{x})$ is posterior PDF/PMF of θ for given $\mathbf{x}$.

Because $\overline{f}(\mathbf{x}) \geq 0$, the minimum in d of the sum/integral on the RHS of equation (3.58) can only be achieved when summands or values of the integrand $\mathbb{E}_{\pi(\theta|\mathbf{x})} L(\theta, d(\mathbf{x}))$ attain their minima in d. But this exactly means that the minimising decision rule equals $\widehat{a}$.

In Examples 3.25–3.27 we calculate Bayes' estimators in some models.

Example 3.25 A standard example is where $X_i \sim N(\mu, \sigma^2)$, σ^2 known and the prior for μ is $N(0, \tau^2)$, with known $\tau^2 > 0$. For the posterior, we have

$$\pi(\mu|\mathbf{x}) \propto \pi(\mu)f(\mathbf{x}; \mu) \propto \exp\left[-\frac{1}{2\sigma^2}\sum_i (x_i - \mu)^2\right] \times \exp\left[-\frac{\mu^2}{2\tau^2}\right]$$

$$\propto \exp\left[-\frac{1}{2}\left(\frac{n}{\sigma^2} + \frac{1}{\tau^2}\right)\left(\mu - \frac{\sum_i x_i/\sigma^2}{n/\sigma^2 + \tau^{-2}}\right)^2\right].$$

That is

$$\pi(\mu|\mathbf{x}) \sim N\left(\frac{n\bar{x}\tau^2}{n\tau^2 + \sigma^2}, \frac{\sigma^2\tau^2}{n\tau^2 + \sigma^2}\right).$$

The mean of a normal distribution equals its median. Thus under both quadratic and absolute error LFs, the optimal Bayes estimator for μ is

$$\frac{n\bar{x}\tau^2}{n\tau^2 + \sigma^2}.$$

If the prior is $N(\nu, \tau^2)$ with a general mean ν, then the Bayes estimater for μ under both the quadratic and absolute error loss is

$$\frac{\nu\sigma^2 + n\bar{x}\tau^2}{n\tau^2 + \sigma^2}. \quad \blacksquare$$

Example 3.26 Next, let $X_i \sim \text{Po}(\lambda)$, where the prior for λ is exponential, of known rate $\tau > 0$. The posterior

$$\pi(\lambda|\mathbf{x}) \propto e^{-n\lambda}\lambda^{n\bar{x}}e^{-\tau\lambda}$$

is $\text{Gam}(n\bar{x} + 1, n + \tau)$. Under the quadratic loss, the optimal Bayes estimator is

$$\frac{n\bar{x} + 1}{n + \tau}.$$

On the other hand, under the absolute error loss the optimal Bayes estimator equals the value $\hat{\lambda} > 0$ for which

$$\frac{(n + \tau)^{n\bar{x}+1}}{(n\bar{x})!}\int_0^{\hat{\lambda}} \lambda^{n\bar{x}}e^{-(n+\tau)\lambda}d\lambda = \frac{1}{2}. \quad \blacksquare$$

Example 3.27 It is often assumed that the prior is uniform on a given interval. For example, suppose $X_1, \ldots, X_n$ are IID RVs from a distribution uniform on $(\theta - 1, \theta + 1)$, and that the prior for θ is uniform on (a, b). Then the posterior is

$$\pi(\theta|\mathbf{x}) \propto I(a < \theta < b)I(\max x_i - 1 < \theta < \min x_i + 1)$$

$$= I(a \vee (\max x_i - 1) < \theta < b \wedge (\min x_i + 1)),$$

where $\alpha \vee \beta = \max [\alpha, \beta]$, $\alpha \wedge \beta = \min [\alpha, \beta]$. So, the posterior is uniform over this interval. Then the quadratic and absolute error LFs give the same Bayes estimator:

$$\widehat{\theta} = \frac{1}{2}[a \vee (\max \, x_i - 1) + b \wedge (\min \, x_i + 1)].$$

Another example is where θ is uniform on $(0,1)$ and $X_1, \ldots, X_n$ take two values, 0 and 1, with probabilities θ and $1 - \theta$. Here, the posterior is

$$\pi(\theta|\mathbf{x}) \propto \theta^{n\bar{x}}(1 - \theta)^{n - n\bar{x}},$$

i.e. the posterior is Bet $(n\bar{x} + 1, n - n\bar{x} + 1)$:

$$\pi(\theta|\mathbf{x}) \propto \pi(\theta) f(\mathbf{x}; \theta) = \frac{\theta^{n\bar{x}}(1 - \theta)^{n - n\bar{x}}}{\int_0^1 \mathrm{d}\widetilde{\theta} \, \widetilde{\theta}^{n\bar{x}}(1 - \widetilde{\theta})^{n - n\bar{x}}}, 0 < \theta < 1.$$

So, for the quadratic loss $\ell(\theta, d) = (\theta - d)^2$, the optimal Bayes estimator is given by

$$\widehat{d} = \frac{\int_0^1 \mathrm{d}\theta \theta^{t+1}(1 - \theta)^{n-t}}{\int_0^1 \mathrm{d}\theta \theta^t (1 - \theta)^{n-t}} = \frac{(t+1)!(n-t)!(n+1)!}{(n+2)!t!(n-t)!} = \frac{t+1}{n+2}.$$

Here $t := \sum_i x_i$, and we used the identity

$$\int_0^1 x^{m-1}(1 - x)^{n-1} \mathrm{d}x = (m - 1)!(n - 1)!/(m + n - 1)!$$

which is valid for all integers m and n. ∎

Calculations of Bayes' estimators figure prominently in the Tripos questions. See MT-IB 1999-112D (ii), 1998-403E, 1997-212G, 1996-203G (ii), 1995-103G (ii), 1993-403J (needs further knowledge of the course).

Next, we remark on another type of LF, a 0, 1-*loss* where $L(\theta, a) = 1 - \delta_{\theta,a}$ equals 0 when $\theta = a$ and 1 otherwise. Such a function is natural when the set Θ of possible values of θ is finite. For example, assume that Θ consists of two values, say 0 and 1. Let the prior probabilities be p_0 and p_1 and the corresponding PMF/PDF f_0 and f_1. The posterior probabilities $\pi(\cdot|\mathbf{x})$ are

$$\pi(0|\mathbf{x}) = \frac{p_0 f_0(\mathbf{x})}{p_0 f_0(\mathbf{x}) + p_1 f_1(\mathbf{x})}, \quad \pi(1|\mathbf{x}) = \frac{p_1 f_1(\mathbf{x})}{p_0 f_0(\mathbf{x}) + p_1 f_1(\mathbf{x})},$$

and the Bayes' estimator $\widehat{a}$, with values 0 and 1, should minimise the expected posterior loss $R(\mathbf{x}, a) = \pi(1 - a|\mathbf{x}), a = 0, 1$. That is

$$\widehat{a}(=\widehat{a}(\mathbf{x})) = \begin{cases} 0, \text{ if } \pi(0|\mathbf{x}) > \pi(1|\mathbf{x}), \\ 1, \text{ if } \pi(1|\mathbf{x}) > \pi(0|\mathbf{x}); \end{cases}$$

in the case $\pi(0|\mathbf{x}) = \pi(1|\mathbf{x})$ any choice will give the same expected posterior loss $1/2$. In other words,

$$\widehat{a} = \begin{cases} 1 \\ 0 \end{cases} \text{ when } \frac{\pi(1|\mathbf{x})}{\pi(0|\mathbf{x})} = \frac{p_1 f_1(\mathbf{x})}{p_0 f_0(\mathbf{x})} \begin{array}{c} > \\ < \end{array} 1, \text{ i.e. } \frac{f_1(\mathbf{x})}{f_0(\mathbf{x})} \begin{array}{c} > \\ < \end{array} k = \frac{p_0}{p_1}. \qquad (3.59)$$

We see that the 'likelihood ratio' $f_1(\mathbf{x})/f_0(\mathbf{x})$ is conspicuous here; we will encounter it many times in the next chapter.

To conclude the theme of Bayesian estimation, consider the following model. Assume RVs $X_1,\ldots,X_n$ have $X_i \sim \mathrm{Po}\,(\theta_i)$:

$$f(x;\theta_i) = \frac{\theta_i^x}{x!}e^{-\theta_i}, \quad x=0,1,\ldots.$$

Here, parameter θ_i is itself random, and has a CDF B (on $\mathbb{R}_+ = (0,\infty)$). We want to estimate θ_i. A classical application is where X_i is the number of accidents involving driver i in a given year.

First, if we know B, then the estimator $T_i(\mathbf{X})$ of θ_i minimising the mean square error $\mathbb{E}(T_i - \theta_i)^2$ does not depend on X_j with $j \neq i$:

$$T_i(\mathbf{X}) = T(X_i).$$

Here,

$$T(x) = \int \theta f(x;\theta)\mathrm{d}B(\theta)/g(x)$$

and

$$g(x) = \int f(x;\theta)\mathrm{d}B(\theta).$$

Substituting the Poisson PMF for $f(x;\theta)$ yields

$$T(x) = (x+1)\frac{g(x+1)}{g(x)}, \quad x=0,1,\ldots.$$

Hence,

$$T_i(\mathbf{X}) = (X_i+1)\frac{g(X_i+1)}{g(X_i)}. \tag{3.60}$$

But what if we do not know B and have to estimate it from sample $(\mathbf{X})$?

A natural guess is to replace B with its sample histogram

$$\widehat{B}(\theta) = \frac{1}{n}\#\{i: X_i < \theta\}, \quad \theta \geq 0,$$

jumping at integer point x by amount $\widehat{b}(x) = n_x/n$, where

$$n_x = \#\{i: X_i = x\}, \quad x=0,1,\ldots. \tag{3.61}$$

Then substitute $\widehat{b}(x)$ instead of $g(x)$. This yields the following estimator of θ_i:

$$\widehat{T}_i(\mathbf{X}) = (X_i+1)\frac{n_{X_i+1}}{n_{X_i}}. \tag{3.62}$$

This idea (according to some sources, it goes back to British mathematician A. Turing (1912–1954)) works surprisingly well. See [E2], [LS]. The surprise is that estimator $\widehat{T}_i$ uses observations X_j, $j \neq i$, that have nothing to do with X_i (apart from the fact that they have Poisson PMF and the same prior CDF). This observation was a starting point for the so-called empirical Bayes methodology developed by H. Robbins (1915–2001), another outstanding American statistician who began his career in pure mathematics. Like J.W. Tukey, Robbins did his Ph.D. in topology. In 1941 he and R. Courant published their book [CouR], which has remained a must for anyone interested in mathematics until the present day. Robbins is also credited with a number of aphorisms and jokes (one of which is 'Not a single good deed shall go unpunished.').

Robbins used formula (3.62) to produce a reliable estimator of the number S_0 of accidents incurred in the next year by the n_0 drivers who did not have accidents in the observed year. It is clear that 0 is an underestimate for S_0 (as it assumes that future is the same as past). On the other hand, $n_0 = \sum_{i \in \mathbb{Z}_+} i n_i$ gives an overestimate, since X_0 did not contribute to the sum $\sum_i i n_i$. A good estimator of S_0 is X_1, the number of drivers who recorded a single accident. In general, $(i+1) n_{i+1}$ accurately estimates S_i, the number of accidents incurred in the future year by the n_i drivers who recorded i accidents in the observed year.

Robbins introduced the term 'the number of unseen, or missing, species'. For example, one can count the number n_x of words used exactly x times in Shakespeare's known literary canon, $x = 1, 2, \ldots$; this gives Table 3.5 (see [ETh1]).

Table 3.5.

x					n_x					
	1	2	3	4	5	6	7	8	9	10
0+	14376	4343	2292	1463	1043	837	638	519	430	364
10+	305	259	242	223	187	181	179	130	127	128
20+	104	105	99	112	93	74	83	76	72	63

So, 14 376 words appeared just once, 4343 twice, etc. In addition, 2387 words appeared more than 30 times each. The total number of distinct words used in the canon equals 31 534 (in this counting, words 'tree' and 'trees' count separately). The missing species here are words that Shakespeare knew but did not use; let their number be n_0. Then the total number of words known to Shakespeare is, obviously, $n = 31534 + n_0$. The length of the canon (the number of words counted with their multiplicities) equals $N = 884\,647$.

Assume as before that X_i, the number of times word i appears in the canon, is $\sim \mathrm{Po}(\theta_i)$, where θ_i is random. Again, if CDF B of θ is known, the posterior expected value $\mathbb{E}(\theta_i | X_i = x)$ of θ_i gives the estimator minimising the mean square error and equals

$$(x+1) \frac{g(x+1)}{g(x)}, \tag{3.63}$$

where

$$g(x) = \int \frac{\theta^x}{x!} e^{-\theta} dB(\theta), \quad x = 0, 1, \ldots .$$

If B is unknown, we substitute the (still unknown) histogram

$$\widehat{B}(\theta) = \frac{1}{n} \#\{i : \theta_i < \theta\}, \quad \theta \geq 0.$$

The estimator of θ_i then becomes

$$\widehat{\mathbb{E}}(\theta_i | X_i = x) = (x+1) \frac{n_{x+1}}{n_x}. \tag{3.64}$$

For $x = 1$, it gives the following value for the expectation

$$\widehat{E}(\theta_i | X_i = 1) = 2 \times \frac{4343}{14\,374} = 0.604.$$

We immediately conclude that the single-time words are overrepresented, in the sense that if somewhere there exists a new Shakespeare canon equal in volume to the present one then the $14\,378$ words appearing once in the present canon will appear in the new one only $0.604 \times 14\,376 = 8683$ times.

Next, set

$$r_0 = \sum_{i=1}^n \theta_i \mathbf{1}(X_i = 0) \bigg/ \sum_{i=1}^n \theta_i .$$

The numerator is estimated by

$$\widehat{E}(\theta_i | X_i = 0) n_0 = \frac{n_1}{n_0} n_0 = n_1 = 14\,376$$

and the denominator by $N = 884\,647$. Then

$$\widehat{r}_0 = \frac{14\,376}{884\,647} = 0.016.$$

So, with a stretch of imagination one deduces that, should a new Shakespearean text appear, the probability that its first word will not be from the existing canon is 0.016; the same conclusion holds for the second word, etc. In fact, in 1985 the Bodleian Library in Oxford announced the discovery of a previously unknown poem that some experts attributed to Shakespeare. The above analysis was applied in this case [ETh2] and gave an interesting insight.

4 Hypothesis testing

4.1 Type I and type II error probabilities. Most powerful tests

Statisticians do it with only a 5% chance of being rejected.
(From the series '*How they do it*').

Testing statistical hypothesis, or *hypotheses testing*, is another way to make a judgement about the distribution (PDF or PMF) of an 'observed' random variable X or a sample

$$\mathbf{X} = \begin{pmatrix} X_1 \\ \vdots \\ X_n \end{pmatrix}.$$

Traditionally, one speaks here about *null* and *alternative hypotheses*. The simplest case is where we have to choose between two possibilities: the PDF/PMF f of X is f_0 or f_1. We say that $f = f_0$ will represent a *simple null hypothesis* and $f = f_1$ a *simple alternative*. This introduces a certain imparity between f_0 and f_1, which will also be manifested in further actions.

Suppose the observed value of X is x. A 'scientific' way to proceed is to partition the set of values of X (let it be $\mathbb{R}$) into two complementary domains, $\mathcal{C}$ and $\overline{\mathcal{C}} = \mathbb{R} \setminus \mathcal{C}$, and reject H_0 (i.e. accept H_1) when $x \in \mathcal{C}$ while accepting H_0 when $x \in \overline{\mathcal{C}}$. The test then is identified with domain $\mathcal{C}$ (called a *critical region*). In other words, we want to employ a two-valued decision function d (the indicator of set $\mathcal{C}$) such that when $d(x) = 1$ we reject and when $d(x) = 0$ we accept the null hypothesis.

Suppose we decide to worry mainly about rejecting H_0 (i.e. accepting H_1) when H_0 is correct: in our view this represents the principal danger. In this case we say that a *type I* error had been committed, and we want errors of this type to be comfortably rare. A less dangerous error would be of *type II*: to accept H_0 when it is false (i.e. H_1 takes place): we want it to occur reasonably infrequently, after the main goal that type I error is certainly rare has been guaranteed. Formally, we fix a threshold for *type I error probability* (TIEP), $\alpha \in (0, 1)$, and then try to minimise the *type II error probability* (TIIEP).

In this situation, one often says that H_0 is a *conservative* hypothesis, not to be rejected unless there is a clear evidence against it.

For example, if you are a doctor and see your patient's histology data, you say the hypothesis is that the patient has a tumour while the alternative is that he/she hasn't. Under

H_1 (no tumour), the data should group around 'regular' average values whereas under H_0 (tumour) they drift towards 'abnormal' ones. Given that the test is not 100% accurate, the data should be considered as random. You want to develop a scientific method of diagnosing the disease. And your main concern is not to miss a patient with a tumour since this might have serious consequences. On the other hand, if you commit a type II error (alarming the patient falsely), it might result in some relatively mild inconvenience (repeated tests, possibly a brief hospitalisation).

The question then is how to choose the critical region $\mathcal{C}$. Given $\mathcal{C}$, the TIEP equals

$$\mathbb{P}_0(\mathcal{C}) = \int f_0(x)I(x \in \mathcal{C})dx \text{ or } \sum_x f_0(x)I(x \in \mathcal{C}), \tag{4.1}$$

and is called the *size* of the critical region $\mathcal{C}$ (also the size of the test). Demanding that probability $\mathbb{P}_0(\mathcal{C})$ does not exceed a given α (called the *significance level*) does not determine C uniquely: we can have plenty of such regions. Intuitively, C must contain outcomes x with small values of $f_0(x)$, regardless of where they are located. But we want to be more precise. For instance, if PDF f_0 in H_0 is $N(\mu_0, \sigma_0^2)$, could C contain points near the mean-value μ_0 where PDF f_0 is relatively high? Or should it be a half-infinite interval $(\mu_0 + c, \infty)$ or $(-\infty, \mu_0 - c)$ or perhaps the union of the two? For example, we can choose c such that the integral

$$\frac{1}{\sqrt{2\pi}\sigma_0} \int_{\mu_0+c}^{\infty} e^{-(x-\mu_0)^2/2\sigma_0^2}dx \text{ or } \frac{1}{\sqrt{2\pi}\sigma_0} \int_{-\infty}^{\mu_0-c} e^{-(x-\mu_0)^2/2\sigma_0^2}dx$$

or their sum is $\leq \alpha$. In fact, it is the alternative PDF/PMF, f_1, that narrows the choice of C (and makes it essentially unique), because we wish the TIIEP

$$\mathbb{P}_1(\overline{\mathcal{C}}) = \int f_1(x)I(x \notin \mathcal{C})dx \text{ or } \sum_x f_1(x)I(x \notin \mathcal{C}) \tag{4.2}$$

to be minimised, for a given significance level. The complementary probability

$$\mathbb{P}_1(\mathcal{C}) = \int f_1(x)I(x \in \mathcal{C})dx \text{ or } \sum_x f_1(x)I(x \in \mathcal{C}) \tag{4.3}$$

is called the *power* of the test; it has to be maximised.

A test of the maximal power among tests of size $\leq \alpha$ is called a *most powerful* (MP) test for a given significance level α. Colloquially, one calls it an MP test of size α (because its size actually equals α as will be shown below).

4.2   **Likelihood ratio tests. The Neyman–Pearson Lemma and beyond**

> Statisticians are in need of good inference
> because at young age many of their hypotheses were rejected.
>> (From the series '*Why they are misunderstood*'.)

A natural (and elegant) idea is to look at the *likelihood ratio* (LR):

$$\Lambda_{H_1, H_0}(x) = \frac{f_1(x)}{f_0(x)}.$$
(4.4)

If, for a given x, $\Lambda_{H_1, H_0}(x)$ is large, we are inclined to think that H_0 is unlikely, i.e. to reject H_0; otherwise we do not reject H_0. Then comes the idea that we should look at regions of the form

$$\left\{ x : \Lambda_{H_1, H_0}(x) > k \right\}$$
(4.5)

and choose k to adapt to size α. The single value $x \in \mathbb{R}$ can be replaced here by a vector $\mathbf{x} = (x_1, \ldots, x_n) \in \mathbb{R}^n$, with $f_0(\mathbf{x}) = \prod f_0(x_i)$ and $f_1(\mathbf{x}) = \prod f_1(x_i)$. Critical region $\mathcal{C}$ then becomes a subset of $\mathbb{R}^n$.

This idea basically works well, as is shown in the famous Neyman–Pearson (NP) Lemma below. This statement is called after J. Neyman and E.S. Pearson (1895–1980), a prominent UK statistician. Pearson was the son of K. Pearson (1857–1936), who is considered the creator of modern statistical thinking. Both father and son greatly influenced the statistical literature of the period; their names will repeatedly appear in this part of the book.

It is interesting to compare the lives of the authors of the NP Lemma. Pearson spent all his active life at the University College London, where he followed his father. On the other hand, Neyman lived through a period of civil unrest, revolution and war in parts of the Russian Empire. See [Rei]. In 1920, as a Pole, he was jailed in the Ukraine and expected a harsh sentence (there was war with Poland and he was suspected of being a spy). He was saved after long negotiations between his wife and a Bolshevik official who in the past had been a medical student under the supervision of Neyman's father-in-law; eventually the Neymans were allowed to escape to Poland. In 1925 Neyman came to London and began working with E.S. Pearson. One of their joint results was the NP Lemma.

About his collaboration with Pearson, Neyman wrote: 'The initiative for cooperative studies was Pearson's. Also, at least during the early stages, he was the leader. Our cooperative work was conducted through voluminous correspondence and at sporadic get-togethers, some in England, others in France and some others in Poland. This cooperation continued over the decade 1928–38.'

The setting for the NP Lemma is as follows. Assume H_0: $f = f_0$ is to be tested against H_1: $f = f_1$, where f_1 and f_0 are two distinct PDFs/PMFs. The NP Lemma states:

> $\forall \, k > 0$, the test with the critical region $\mathcal{C} = \{\mathbf{x} : f_1(\mathbf{x}) > k f_0(\mathbf{x})\}$ has the highest power $\mathbb{P}_1(\mathcal{C})$ among all tests (i.e. critical regions $\mathcal{C}^*$) of size $\mathbb{P}_0(\mathcal{C}^*) \le \alpha = \mathbb{P}_0(\mathcal{C})$.

In other words, the test with $\mathcal{C} = \{f_1(\mathbf{x}) > k f_0(\mathbf{x})\}$ is MP among all tests of significance level $\alpha = \mathbb{P}_0(\mathcal{C})$. Here α appears as a function of k: $\alpha = \alpha(k)$, $k > 0$.

Proof. Writing $I_{\mathcal{C}}$ and $I_{\mathcal{C}^*}$ for the indicator functions of $\mathcal{C}$ and $\mathcal{C}^*$, we have

$$0 \le \left[I_{\mathcal{C}}(\mathbf{x}) - I_{\mathcal{C}^*}(\mathbf{x}) \right]\left[f_1(\mathbf{x}) - k f_0(\mathbf{x}) \right],$$

as the two brackets never have opposite signs (for $\mathbf{x} \in \mathcal{C}$: $f_1(\mathbf{x}) > kf_0(\mathbf{x})$ and $I_{\mathcal{C}}(\mathbf{x}) \geq I_{\mathcal{C}^*}(\mathbf{x})$ while for $\mathbf{x} \notin \mathcal{C}$: $f_1(\mathbf{x}) \leq kf_0(\mathbf{x})$ and $I_{\mathcal{C}}(\mathbf{x}) \leq I_{\mathcal{C}^*}(\mathbf{x})$). Then

$$0 \leq \int \left[I_{\mathcal{C}}(\mathbf{x}) - I_{\mathcal{C}^*}(\mathbf{x})\right]\left[f_1(\mathbf{x}) - kf_0(\mathbf{x})\right]d\mathbf{x},$$

or

$$0 \leq \sum_{\mathbf{x}} \left[I_{\mathcal{C}}(\mathbf{x}) - I_{\mathcal{C}^*}(\mathbf{x})\right]\left[f_1(\mathbf{x}) - kf_0(\mathbf{x})\right].$$

But the RHS here equals

$$\mathbb{P}_1(\mathcal{C}) - \mathbb{P}_1(\mathcal{C}^*) - k[\mathbb{P}_0(\mathcal{C}) - \mathbb{P}_0(\mathcal{C}^*)].$$

So

$$\mathbb{P}_1(\mathcal{C}) - \mathbb{P}_1(\mathcal{C}^*) \geq k[\mathbb{P}_0(\mathcal{C}) - \mathbb{P}_0(\mathcal{C}^*)].$$

Thus, if $\mathbb{P}_0(\mathcal{C}^*) \leq \mathbb{P}_0(\mathcal{C})$, then $\mathbb{P}_1(\mathcal{C}^*) \leq \mathbb{P}_1(\mathcal{C})$.   □

The test with $\mathcal{C} = \{\mathbf{x} : f_1\mathbf{x}) > kf_0(\mathbf{x})\}$ is often called the *likelihood ratio (LR) test* or the *NP test*.

The NP Lemma (either with or without the proof) and its consequences are among the most popular Tripos topics in Cambridge University IB Statistics. See MT-IB 1999-203D, 1999-212D, 1998-112E, 1996-403G (i), 1995-403G (i), 1993-103J (i, ii).

Remark   (1) The statement of the NP Lemma remains valid if the inequality $f_1(\mathbf{x}) > kf_0(\mathbf{x})$ is replaced by $f_1(\mathbf{x}) \geq kf_0(\mathbf{x})$.

(2) In practice, we have to solve an 'inverse' problem: for a given $\alpha \in (0, 1)$ we want to find an MP test of size $\leq \alpha$. That is we want to construct k as a function of α, not the other way around. In all the (carefully selected) examples in this section, this is not a problem as the function $k \mapsto \alpha(k)$ is one-to-one and admits a *bona fide* inversion $k = k(\alpha)$. Here, for a given α we can always find $k > 0$ such that $\mathbb{P}(\mathcal{C}) = \alpha$ for $\mathcal{C} = \{f_1(\mathbf{x}) > kf_0(\mathbf{x})\}$ (and finding such k is a part of the task).

However, in many examples (especially related to the discrete case), the NP test may not exist for every value of size α. To circumvent this difficulty, we have to consider more general *randomised* tests in which the decision function d may take not only values 0, 1, but also intermediate values from $(0, 1)$. Here, if the observed value is x, then we reject H_0 with probability $d(x)$. (As a matter of fact, the 'randomised' NP Lemma guarantees that there will always be an MP test in which d takes at most three values: 0, 1 and possibly one value between.)

To proceed formally, we want first to extend the concept of the TIEP and TIIEP. This is a straightforward generalisation of equations (4.1) and (4.2):

$$\mathbb{P}_0(d) = \int d(\mathbf{x})f_0(\mathbf{x})d\mathbf{x} \text{ or } \sum_{\mathbf{x}} d(\mathbf{x})f_0(\mathbf{x}) \qquad (4.6)$$

and

$$\mathbb{P}_1(\bar{d}) = \int [1 - d(\mathbf{x})] f_1(\mathbf{x}) d\mathbf{x} \text{ or } \sum_{\mathbf{x}} [1 - d(\mathbf{x})] f_1(\mathbf{x}). \qquad (4.7)$$

The power of the test with a decision function d is

$$\mathbb{P}_1(d) = \int d(\mathbf{x}) f_1(\mathbf{x}) d\mathbf{x} \text{ or } \sum_{\mathbf{x}} d(\mathbf{x}) f_1(\mathbf{x}). \qquad (4.8)$$

Then, as before, we fix $\alpha \in (0, 1)$ and consider all randomised tests d of size $\leq \alpha$, looking for the one amongst them which maximises $\mathbb{P}_1(d)$.

In the randomised version, the NP Lemma states:

For any pair of PDFs/PMFs f_0 and f_1 and $\alpha \in (0, 1)$, there exist unique $k > 0$ and $\rho \in [0, 1]$ such that the test of the form

$$d_{NP}(x) = \begin{cases} 1, & f_1(x) > k f_0(x), \\ \rho, & f_1(x) = k f_0(x), \\ 0, & f_1(x) < k f_0(x), \end{cases} \qquad (4.9)$$

has $\mathbb{P}_0(d_{NP}) = \alpha$. This test maximises the power among all randomised tests of size α:

$$\mathbb{P}_1(d_{NP}) = \max \left[\mathbb{P}_1(d^*) : \mathbb{P}_0(d^*) \leq \alpha \right]. \qquad (4.10)$$

That is d_{NP} is the MP randomised test of size $\leq \alpha$ for H_0: $f = f_0$ against H_1: $f = f_1$.

As before, the test d_{NP} described in formula (4.9) is called the (randomised) NP test. We want to stress once more that constant ρ may (and often does) coincide with 0 or 1, in which case d_{NP} becomes non-randomised.

The proof of the randomised NP Lemma is somewhat longer than in the non-randomised case, but still quite elegant. Assume first that we are in the continuous case and work with PDFs f_0 and f_1 such that $\forall\, k > 0$

$$\int f_0(\mathbf{x}) I\, (f_1(\mathbf{x}) = k f_0(\mathbf{x})) \, d\mathbf{x} = 0. \qquad (4.11)$$

In this case value ρ will be 0.

In fact, consider the function

$$G : k \mapsto \int f_0(\mathbf{x}) I\, (f_1(\mathbf{x}) > k f_0(\mathbf{x})) \, d\mathbf{x}.$$

Then $G(k)$ is monotone non-increasing in k, as the integration domain shrinks with k. Further, function G is right-continuous: $\lim_{r \searrow k+} G(r) = G(k)\ \forall\, k > 0$. In fact, when $r \searrow k+$,

$$I\, (f_1(\mathbf{x}) > r f_0(\mathbf{x})) \nearrow I\, (f_1(\mathbf{x}) > k f_0(\mathbf{x})),$$

as every point $\mathbf{x}$ with $f_1(\mathbf{x}) > kf_0(\mathbf{x})$ is eventually included in the (expanding) domains $\{f_1(\mathbf{x}) > rf_0(\mathbf{x})\}$. The convergence of the integrals is then a corollary of standard theorems of analysis. Moreover, in a similar fashion one proves that G has left-side limits. That is

$$G(k-) = \lim_{l \nearrow k-} G(l)$$

exists $\forall\, k > 0$ and equals

$$\int f_0(\mathbf{x}) I\,(f_1(\mathbf{x}) \geq kf_0(\mathbf{x}))\, d\mathbf{x}.$$

Under assumption (4.11), the difference

$$G(k-) - G(k) = \int f_0(\mathbf{x}) I\,(f_1(\mathbf{x}) = kf_0(\mathbf{x}))\, d\mathbf{x}$$

vanishes, and G is continuous. Finally, we observe that

$$G(0+) = \lim_{k \to 0+} G(k) = 1, \; G(\infty) = \lim_{k \to \infty} G(k) = 0. \tag{4.12}$$

Hence G crosses every level $\alpha \in (0, 1)$ at some point $k = k(\alpha)$ (possibly not unique). Then the (non-randomised) test with

$$d_{\mathrm{NP}}(\mathbf{x}) = \begin{cases} 1, f_1(\mathbf{x}) > kf_0(\mathbf{x}), \\ 0, f_1(\mathbf{x}) \leq kf_0(\mathbf{x}) \end{cases}$$

has size $\mathbb{P}_0(d) = G(k) = \alpha$ and fits formula (4.9) with $\rho = 0$.

This is the MP test of significance level α. In fact, let d^* be any other (possibly randomised) test of size $\mathbb{P}_0(d^*) \leq \alpha$. We again have that $\forall\, \mathbf{x}$

$$0 \leq [d_{\mathrm{NP}}(\mathbf{x}) - d^*(\mathbf{x})][f_1(\mathbf{x}) - kf_0(\mathbf{x})], \tag{4.13}$$

since if $d_{\mathrm{NP}}(\mathbf{x}) = 1$, then both brackets are ≥ 0 and if $d_{\mathrm{NP}}(\mathbf{x}) = 0$, they are both ≤ 0. Hence,

$$0 \leq \int [d_{\mathrm{NP}}(\mathbf{x}) - d^*(\mathbf{x})][f_1(\mathbf{x}) - k(\alpha)f_0(\mathbf{x})]d\mathbf{x},$$

but the RHS again equals $\mathbb{P}_1(d) - \mathbb{P}_1(d^*) - k[\mathbb{P}_0(d) - \mathbb{P}_0(d^*)]$. This implies that

$$\mathbb{P}_1(d) - \mathbb{P}_1(d^*) \geq k[\mathbb{P}_0(d) - \mathbb{P}_0(d^*)], \text{ i.e. } \mathbb{P}_1(d) \geq \mathbb{P}_1(d^*)$$

if $\mathbb{P}_0(d) \geq \mathbb{P}_0(d^*)$.

We are now prepared to include the general case, without assumption (4.11). Again consider the function

$$G: k \mapsto \int f_0(\mathbf{x}) I\,(f_1(\mathbf{x}) > kf_0(\mathbf{x}))\, d\mathbf{x} \text{ or } \sum_{\mathbf{x}} f_0(\mathbf{x}) I\,(f_1(\mathbf{x}) > kf_0(\mathbf{x}))\,. \tag{4.14}$$

It is still monotone non-decreasing in k, right-continuous and with left-side limits. (There exists a convenient French term 'càdlàg' (continue à droite, limits à gauche).) Also, equation (4.12) holds true. However, the difference $G(k-) - G(k)$ may be positive, as it equals the integral or the sum

$$\int f_0(\mathbf{x})I\left(f_1(\mathbf{x}) = kf_0(\mathbf{x})\right) d\mathbf{x} \quad \text{or} \quad \sum_{\mathbf{x}} f_0(\mathbf{x})I\left(f_1(\mathbf{x}) = kf_0(\mathbf{x})\right)$$

that do not necessarily vanish. It means that, given $\alpha \in (0, 1)$, we can only guarantee that $\exists\, k = k(\alpha) > 0$ such that

$$G(k-) \geq \alpha \geq G(k).$$

If $G(k-) = G(k)$ (i.e. G is continuous at k), the previous analysis is applicable. Otherwise, set

$$\rho = \frac{\alpha - G(k)}{G(k-) - G(k)}. \tag{4.15}$$

Then $\rho \in [0, 1]$, and we can define the test d_{NP} by formula (4.9). (If $\rho = 0$ or 1, d_{NP} is non-randomised.) See Figure 4.1.

The size

$$\mathbb{P}_0(d_{\mathrm{NP}}) = \int f_0(\mathbf{x}) d_{\mathrm{NP}}(\mathbf{x}) d\mathbf{x}$$

equals

$$\int f_0(\mathbf{x})I\left(f_1(\mathbf{x}) > kf_0(\mathbf{x})\right) d\mathbf{x} + \rho \int f_0(\mathbf{x})I\left(f_1(\mathbf{x}) = kf_0(\mathbf{x})\right) d\mathbf{x}$$

$$= G(k) + \frac{\alpha - G(k)}{G(k-) - G(k)} (G(k-) - G(k)) = \alpha.$$

It remains to check that d_{NP} is the MP test of size $\leq \alpha$. This is done as before, as inequality (4.13) still holds.

There is a useful corollary of the randomised NP Lemma:

> If $d(\mathbf{x})$ is an MP test of size $\alpha = \mathbb{P}_0(d)$. Then its power $\beta = \mathbb{P}_1(d)$ cannot be less than α.

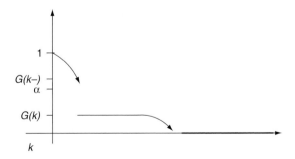

Figure 4.1

Proof. In fact, consider a (randomised) test with $d^*(\mathbf{x}) \equiv \alpha$. It has $\mathbb{P}_0(d^*) = \alpha = \mathbb{P}_1(d^*)$. As d is MP, $\beta \geq \mathbb{P}_1(d^*) = \alpha$. $\square$

NP tests work for some problems with *composite* (i.e. not simple) null hypotheses and alternatives. A typical example of composite hypotheses is where $\Theta = \mathbb{R}$, $H_0: \theta \leq \theta_0$ and $H_1: \theta > \theta_0$ for some given θ_0. Quite often one of H_0 and H_1 is simple (e.g. $H_0: \theta = \theta_0$) and the other composite ($\theta > \theta_0$ or $\theta < \theta_0$ or $\theta \neq \theta_0$).

A general pair of composite hypotheses is $H_0: \theta \in \Theta_0$ against $H_1: \theta \in \Theta_1$, where Θ_0 and Θ_1 are disjoint parts of the parameter set Θ. To construct a test, we again design a critical region $\mathcal{C} \subset \mathbb{R}^n$ such that H_0 is rejected when $\mathbf{x} \in \mathcal{C}$. As in the case of simple hypotheses, the probability

$$\mathbb{P}_\theta(\mathcal{C}) = \mathbb{P}_\theta(\text{reject } H_0), \quad \theta \in \Theta_0, \tag{4.16}$$

is treated as TIEP. Similarly, the probability

$$\mathbb{P}_\theta(\overline{\mathcal{C}}) = \mathbb{P}_\theta(\text{accept } H_0), \quad \theta \in \Theta_1, \tag{4.17}$$

is treated as the TIIEP, while

$$\mathbb{P}_\theta(\mathcal{C}) = \mathbb{P}_\theta(\text{reject } H_0), \quad \theta \in \Theta_1, \tag{4.18}$$

is treated as the power of the test with critical region $\mathcal{C}$. Here, they all are functions of θ, either on Θ_0 or on Θ_1. (In fact, equations (4.16) and (4.18) specify the same function $\theta \mapsto \mathbb{P}_\theta(\mathcal{C})$ considered on a pair of complementary sets Θ_0 and Θ_1.)

To state the problem, we adopt the same idea as before. Namely, we fix a significance level $\alpha \in (0, 1)$ and look for a test (i.e. a critical region) such that: (i)

$$\mathbb{P}_\theta(\mathcal{C}) \leq \alpha, \quad \theta \in \Theta_0, \tag{4.19}$$

(ii) $\forall$ test with critical region $\mathcal{C}^*$ satisfying condition (4.19),

$$\mathbb{P}_\theta(\mathcal{C}) \geq \mathbb{P}_\theta(\mathcal{C}^*), \quad \theta \in \Theta_1. \tag{4.20}$$

Such a test is called *uniformly most powerful* (UMP) of level α (for testing $H_0: \theta \in \Theta_0$ against $H_1: \theta \in \Theta_1$).

An important related concept is a family of PDFs/PMFs $f(\,\cdot\,; \theta)$, $\theta \in \Theta \subseteq \mathbb{R}$, with a *monotone likelihood ratio* (MLR). Here, we require that for any pair $\theta_1, \theta_2 \in \Theta$ with $\theta_1 < \theta_2$,

$$\Lambda_{\theta_2, \theta_1}(\mathbf{x}) = \frac{f(\mathbf{x}; \theta_2)}{f(\mathbf{x}; \theta_1)} = g_{\theta_2, \theta_1}(T(\mathbf{x})). \tag{4.21}$$

where T is a real-valued statistic (i.e. a scalar function depending on $\mathbf{x}$ only) and $g_{\theta_2, \theta_1}(y)$ is a monotone non-decreasing function ($g_{\theta_2, \theta_1}(y) \leq g_{\theta_2, \theta_1}(y')$ when $y \leq y'$). In this definition, we can also require that $g_{\theta_2, \theta_1}(y)$ is a monotone non-increasing function of y; what is important is that the direction of monotonicity is the same for any $\theta_1 < \theta_2$.

Example 4.1   A *hypergeometric distribution* Hyp(N, D, n). You have a stack of N items and select n of them at random, $n \leq N$, for a check, without replacement. If the stack contains $D \leq N$ defective items, the number of defects in the selected sample has the PMF

$$f_D(x) = \frac{\binom{D}{x}\binom{N-D}{n-x}}{\binom{N}{n}}, \quad x = \max[0, n+D-N], \ldots, \min[D, n]. \tag{4.22}$$

Here, $\theta = D$. The ratio

$$\frac{f_{D+1}(x)}{f_D(x)} = \frac{D+1}{N-D} \frac{N-D-n+x}{D+1-x}$$

is monotone increasing in x; hence we have an MLR family, with $T(x) = x$.

The hypergeometric distribution has a number of interesting properties and is used in several areas of theoretical and applied probability and statistics. In this book it appears only in this example. However, we give below a useful equation for the PGF $\mathbb{E}s^X$ of an RV $X \sim$ Hyp(N, D, n):

$$\mathbb{E}s^X = \sum_{k=0}^{\min[n, D]} \frac{(-D)_k (-n)_k}{(-N)_k} \frac{(1-s)^k}{k!}$$

In this formula, $0 \leq \max[n, D] \leq N$, and $(a)_k$ is the so-called Pochhammer symbol: $(a)_k = a(a+1)\cdots(a+k-1)$. The series on the RHS can be written as ${}_2F_1(-D, -n; -N; 1-s)$, or ${}_2F_1\left({-D, \ -n \atop -N; \ 1-s}\right)$, where ${}_2F_1$ is the Gauss hypergeometric function. In [GriS2], a different (and elegant) recipe is given for calculating the PGF of Hyp(N, D, n).

Amazingly, in many books the range of the hypergeometric distribution (i.e. the set of values of x for which $f_D(x) > 0$) is presented in a rather confusing (or perhaps, amusing) fashion. In particular, the left hand end $(n+D-N)_+ = \max[0, n+D-N]$ is often not even mentioned (or worse, replaced by 0). In other cases, the left and right hand ends finally appear after an argument suggesting that these values have been learned only in the course of writing (which is obviously the wrong impression).   ∎

Example 4.2   A *binomial distribution* Bin (n, θ). Here, you put the checked item back in the stack. Then

$$f(x; \theta) = \binom{n}{x} \theta^x (1-\theta)^{n-x}, \quad x = 0, 1, \ldots, n,$$

where $\theta = D/N \in [0, 1]$. This is an MLR family, again with $T(x) = x$.   ∎

Example 4.3   In general, any family of PDFs/PMFs of the form

$$f(x; \theta) = C(\theta)H(x)\exp[Q(\theta)R(x)],$$

where Q is a strictly increasing or strictly decreasing function of θ, has an MLR. Here,

$$\frac{f(\mathbf{x}; \theta_2)}{f(\mathbf{x}; \theta_1)} = \frac{C(\theta_2)^n}{C(\theta_1)^n} \exp\left[(Q(\theta_2) - Q(\theta_1)) \sum_i R(x_i)\right],$$

which is monotone in $T(\mathbf{x}) = \sum_i R(x_i)$. In particular, the family of normal PDFs with a fixed variance has an MLR (with respect to $\theta = \mu \in \mathbb{R}$) as well as the family of normal PDFs with a fixed mean (with respect to $\theta = \sigma^2 > 0$). Another example is the family of exponential PDFs $f(x; \theta) = \theta e^{-\theta x} I(x \geq 0)$, $\theta > 0$. ∎

Example 4.4 Let X be an RV and consider the null hypothesis $H_0: X \sim N(0, 1)$ against $H_1: X \sim f(x) = \frac{1}{4} e^{-|x|/2}$ (double exponential). We are interested in the MP test of size α, $\alpha < 0.3$.

Here we have

$$f(x|H_1)/f(x|H_0) = C \exp\left[\frac{1}{2}(x^2 - |x|)\right].$$

So $f(x|H_1)/f(x|H_0) > K$ iff $x^2 - |x| > K'$ iff $|x| > t$ or $|x| < 1 - t$, for some $t > 1/2$. We want $\alpha = \mathbb{P}_{H_0}(|X| > t) + \mathbb{P}_{H_0}(|X| < 1 - t)$ to be < 0.3. If $t \leq 1$,

$$\mathbb{P}_{H_0}(|X| > t) \geq \mathbb{P}_{H_0}(|X| > 1) = 0.3174 > \alpha.$$

So, we must have $t > 1$ to get

$$\alpha = P_{H_0}(|X| > t) = \Phi(-t) + 1 - \Phi(t).$$

So, $t = \Phi^{-1}(1 - \alpha/2)$, and the test rejects H_0 if $|X| > t$. The power is

$$\mathbb{P}_{H_1}(|X| > t) = 1 - \frac{1}{4}\int_{-t}^{t} e^{-|x|/2}dx = e^{-t/2}. \quad ∎$$

The following theorem extends the NP Lemma to the case of families with an MLR, for the *one-side* null hypothesis $H_0: \theta \leq \theta_0$ against the one-side alternative $H_1: \theta > \theta_0$:

> Let $f(\mathbf{x}; \theta)$, $\theta \in \mathbb{R}$, be an MLR family of PDFs/PMFs. Fix $\theta_0 \in \mathbb{R}$ and $k > 0$ and choose any $\theta_1 > \theta_0$. Then the test of H_0 against H_1, with the critical region
>
> $$\mathcal{C} = \{\mathbf{x}: f(\mathbf{x}; \theta_1) > k f(\mathbf{x}; \theta_0)\}, \tag{4.23}$$
>
> is UMP of significance level $\alpha = \mathbb{P}_{\theta_0}(\mathcal{C})$.

This statement looks rather surprising because the rôle of value θ_1 is not clear. In fact, as we will see, θ_1 is needed only to specify the critical region (and consequently, size α). More precisely, owing to the MLR, $\mathcal{C}$ will be written in the form

$$\{f(\mathbf{x}; \theta') > k(\theta') f(\mathbf{x}; \theta_0)\}$$

$\forall \theta' > \theta_0$, for a suitably chosen $k(\theta')$. This fact will be crucial for the proof.

Proof. Without loss of generality, assume that all functions g_{θ_2, θ_1} in the definition of the MLR are monotone non-decreasing. First, owing to the NP Lemma, the test (4.23) is NP of size $\leq \alpha = \mathbb{P}_{\theta_0}(\mathcal{C})$ for the simple null hypothesis $f = f(\,\cdot\,; \theta_0)$ against the simple alternative $f = f(\,\cdot\,; \theta_1)$. That is

$$\mathbb{P}_{\theta_1}(\mathcal{C}) \geq \mathbb{P}_{\theta_1}(\mathcal{C}^*) \; \forall \; \mathcal{C}^* \text{ with } \mathbb{P}_{\theta_0}(\mathcal{C}^*) \leq \alpha. \tag{4.24}$$

By using the MLR property, test (4.23) is equivalent to

$$\mathcal{C} = \{\mathbf{x}: \; T(\mathbf{x}) > c\}$$

for some value c. But then, again by the MLR property, $\forall \; \theta' > \theta_0$, test (4.23) is equivalent to

$$\mathcal{C} = \{f(\mathbf{x}; \theta') > k(\theta')f(\mathbf{x}; \theta_0)\}, \tag{4.25}$$

for some value $k(\theta')$. Now, again owing to the NP Lemma, test (4.23) (and hence test (4.25)) is MP of size $\leq \mathbb{P}_{\theta_0}(\mathcal{C}) = \alpha$ for the simple null hypothesis $f = f(\,\cdot\,; \theta_0)$ against the simple alternative $f = f(\,\cdot\,; \theta')$. That is

$$\mathbb{P}_{\theta'}(\mathcal{C}) \geq \mathbb{P}_{\theta'}(\mathcal{C}^*) \; \forall \; \mathcal{C}^* \text{ with } \mathbb{P}_{\theta_0}(\mathcal{C}^*) \leq \alpha.$$

In other words, we established that test (4.23) is UMP of significance level α, for the simple null hypothesis $f = f(\,\cdot\,; \theta_0)$ against the one-side alternative that $f = f(\,\cdot\,; \theta')$ for some $\theta' > \theta_0$. Formally,

$$\mathbb{P}_{\theta'}(\mathcal{C}) \geq \mathbb{P}_{\theta'}(\mathcal{C}^*) \text{ whenever } \theta' > \theta_0 \text{ and } \mathbb{P}_{\theta_0}(\mathcal{C}^*) \leq \alpha.$$

But then the same inequality will hold under the additional restriction (on $\mathcal{C}^*$) that $\mathbb{P}_{\theta}(\mathcal{C}^*) \leq \alpha \; \forall \; \theta \leq \theta_0$. The last fact means precisely that test (4.25) (and hence test (4.23)) gives a UMP test of significance level α for $H_0: \; \theta \leq \theta_0$ versus $H_1: \; \theta > \theta_0$. $\square$

Analysing the proof, one can see that the same assertion holds when H_0 is $\theta < \theta_0$ and $H_1: \; \theta \geq \theta_0$. As in the case of simple hypotheses, the inverse problem (to find constants $k(\theta')$, $\theta' > \theta_0$, for a given $\alpha \in (0, 1)$) requires randomisation of the test. The corresponding assertion guarantees the existence of a randomised UMP test with at most three values of the decision function.

UMP test for MLR families occasionally appears in Tripos papers: see MT-IB 1999-212D.

4.3   Goodness of fit. Testing normal distributions, 1: homogeneous samples

Fit, Man, Test, Woman.
(From the series 'Movies that never made it to the Big Screen'.)

The NP Lemma and its modifications are rather exceptional examples in which the problem of hypothesis testing can be efficiently solved. Another collection of (practically important) examples is where RVs are normally distributed. Here, the hypotheses testing can be successfully done (albeit in a somewhat incomplete formulation). This is based on the Fisher Theorem that if $X_1, \ldots, X_n$ is a sample of IID RVs $X_i \sim N(\mu, \sigma^2)$, then

$$\frac{\sqrt{n}}{\sigma}(\overline{X} - \mu) \sim N(0, 1) \text{ and } \frac{1}{\sigma^2} S_{XX} \sim \chi^2_{n-1}, \text{ independently.}$$

Here $\overline{X} = \sum_i X_i/n$ and $S_{XX} = \sum_i (X_i - \overline{X})^2$. See Section 3.5.

A sample

$$\mathbf{X} = \begin{pmatrix} X_1 \\ \vdots \\ X_n \end{pmatrix}, \text{ with } X_i \sim N(\mu, \sigma^2),$$

is called homogeneous normal; a non-homogeneous normal sample is where $X_i \sim N(\mu_i, \sigma_i^2)$, i.e. the parameters of the distribution of X_i varies with i. One also says that this is a single sample (normal) case.

Testing a given mean, unknown variance   Consider a homogeneous normal sample $\mathbf{X}$ and the null hypothesis H_0: $\mu = \mu_0$ against H_1: $\mu \neq \mu_0$. Here μ_0 is a given value. Our test will be based on *Student's* t-statistic

$$T(\mathbf{X}) = \frac{\sqrt{n}(\overline{X} - \mu_0)/\sigma}{\sqrt{S_{XX}/(n-1)\sigma^2}} = \frac{\sqrt{n}(\overline{X} - \mu_0)}{s_{XX}}, \tag{4.26}$$

where s_{XX} is the sample standard deviation (see equation (1.33)). According to the definition of the t distribution in Example 3.3 (see equation (3.6)), $T(\mathbf{X}) \sim t_{n-1}$ under H_0. A remarkable fact here is that calculating $T(\mathbf{x})$ does not require knowledge of σ^2. Therefore, the test will work regardless of whether or not we know σ^2.

Hence, a natural conclusion is that if under H_0 the absolute value $|T(\mathbf{x})|$ of t-statistic $T(\mathbf{x})$ is large, then H_0 is to be rejected. More precisely, given $\gamma \in (0, 1)$, we will denote by $t_{n-1}(\gamma)$ the upper γ point (quantile) of the t_{n-1} distribution, defined as the value a for which

$$\int_a^\infty f_{t_{n-1}}(x) dx = \gamma$$

(the lower γ point is of course $-t_{n-1}(\gamma)$). Then we reject H_0 when, with $\mu = \mu_0$, we have $|T(\mathbf{x})| > t_{n-1}(\alpha/2)$.

This routine is called the *Student*, or t-*test* (the t distribution is also often called the Student distribution). It was proposed in 1908 by W.S. Gossett (1876–1937), a UK statistician who worked in Ireland and England and wrote under the pen name 'Student'. Gossett's job was with the Guinness Brewery, where he was in charge of experimental brewing (he was educated as a chemist). In this capacity he spent an academic year in K. Pearson's Biometric Laboratory in London, where he learned Statistics. Gossett was known as a meek and shy man; there was a joke that he was the only person who managed to be friendly with both K. Pearson and Fisher at the same time. (Fisher was not only famous for his research results but also renowned as a very irascible personality. When he was confronted (or even mildly asked) about inconsistencies or obscurities in his writings and sayings he often got angry and left the audience. Once Tukey (who was himself famous for his 'brashness') came to his office and began a discussion about points made by Fisher. After five minutes the conversation became heated, and Fisher said: 'All right. You can leave my office now.' Tukey said 'No, I won't do that as I respect you too much.' 'In that case', Fisher replied, 'I'll leave.')

Returning to the t-test: one may ask what to do when $|t| < t_{n-1}(\alpha/2)$. The answer is rather diplomatic: you then don't reject H_0 at significance level α (as it is still treated as a conservative hypothesis).

The t-statistic can also be used to construct a confidence interval (CI) for μ (again with or without knowing σ^2). In fact, in the equation

$$\mathbb{P}\left(-t_{n-1}(\alpha/2) < \frac{\sqrt{n}(\overline{X} - \mu)}{s_{XX}} < t_{n-1}(\alpha/2)\right) = 1 - \alpha$$

the inequalities can be imposed on μ:

$$\mathbb{P}\left(\overline{X} - \frac{1}{\sqrt{n}}s_{XX}t_{n-1}(\alpha/2) < \mu < \overline{X} + \frac{1}{\sqrt{n}}s_{XX}t_{n-1}(\alpha/2)\right) = 1 - \alpha.$$

Here $\mathbb{P}$ stands for $\mathbb{P}_{\mu,\sigma^2}$, the distribution of the IID sample with $X_i \sim N(0, \sigma^2)$. This means that a $100(1 - \alpha)\%$ equal-tailed CI for μ is

$$\left(\overline{X} - \frac{1}{\sqrt{n}}s_{XX}t_{n-1}(\alpha/2), \overline{X} + \frac{1}{\sqrt{n}}s_{XX}t_{n-1}(\alpha/2)\right). \tag{4.27}$$

The t-statistic and the t-test appeared frequently in the Tripos questions: see MT-IB 1998-212E, 1995-103G (i), 1994-203F (ii,d), 1993-203J (ii), 1992-406D.

> Some statisticians don't drink because they are t-test totallers.
> (From the series *'Why they are misunderstood'*.)

Example 4.5   The durability is to be tested, of two materials a and b used for soles of ladies' shoes. A paired design is proposed, where each of 10 volunteers had one sole made of a and one of b. The wear (in suitable units) is shown in Table 4.1.

Table 4.1.

					Volunteer					
	1	2	3	4	5	6	7	8	9	10
a	14.7	9.7	11.3	14.9	11.9	7.2	9.6	11.7	9.7	14.0
b	14.4	9.2	10.8	14.6	12.1	6.1	9.7	11.4	9.1	13.2
Difference	0.3	0.5	0.5	0.3	−0.2	1.1	−0.1	0.3	0.6	0.8

Assuming that differences $X_1, \ldots, X_{10}$ are IID $N(\mu, \sigma^2)$, with μ and σ^2 unknown, one tests $H_0: \mu = 0$ against $H_1: \mu \neq 0$. Here $\bar{x} = 0.41$, $S_{xx} = \sum_i x_i^2 - n\bar{x}^2 = 1.349$ and the t-statistic

$$t = \sqrt{10} \times 0.41 / \sqrt{1.349/9} = 3.35.$$

In a size 0.05 test, H_0 is rejected as $t_9(0.025) = 2.262 < 3.35$, and one concludes that there is a difference between the mean wear of a and b. This is a *paired samples* t-test. A 95% confidence interval for the mean difference is

$$\left(0.41 - \frac{\sqrt{1.349/9} \times 2.262}{\sqrt{10}}, 0.41 + \frac{\sqrt{1.349/9} \times 2.262}{\sqrt{10}} \right) = (0.133, 0.687). \quad \blacksquare$$

Historically, the invention of the t-test was an important point in the development of the subject of statistics. As a matter of fact, testing a similar hypothesis about the variance of the normal distribution is a simpler task, as we can use statistic S_{XX} only.

Testing a given variance, unknown mean We again take a homogeneous normal sample $\mathbf{X}$ and consider the null hypothesis $H_0: \sigma^2 = \sigma_0^2$ against $H_1: \sigma^2 \neq \sigma_0^2$, where σ_0^2 is a given value. As was said above, the test is based on the statistic

$$\frac{1}{\sigma_0^2} S_{XX} = \frac{1}{\sigma_0^2} \sum_i (X_i - \overline{X})^2, \tag{4.28}$$

which is $\sim \chi_{n-1}^2$ under H_0. Hence, given $\alpha \in (0, 1)$, we reject H_0 in an equal-tailed two-sided test of level α when the value S_{xx}/σ_0^2 is either less than $h_{n-1}^-(\alpha/2)$ (which would favour $\sigma^2 < \sigma_0^2$) or greater than $h_{n-1}^+(\alpha/2)$ (in which case σ^2 is probably $> \sigma_0^2$). Here and in what follows $h_m^+(\gamma)$ stands for the upper γ point (quantile) of χ_m^2, i.e. the value of a such that

$$\int_a^\infty f_{\chi_m^2}(x) dx = \gamma.$$

Similarly, $h_m^-(\gamma)$ is the lower γ point (quantile), i.e. the value of a for which

$$\int_0^a f_{\chi_m^2}(x) dx = \gamma;$$

as was noted before, $h_m^-(\gamma) = h_m^+(1 - \gamma)$, $0 < \gamma < 1$. By $f_{\chi_m^2}(x)$ we denote here and below the χ^2 PDF with m degrees of freedom.

This test is called the normal χ^2 test. It works without any reference to μ which may be known or unknown.

The normal χ^2 test allows us to construct a confidence interval for σ^2, regardless of whether we know μ or not. Namely, we re-write the equation

$$\mathbb{P}\left(h_{n-1}^-(\alpha/2) < \frac{1}{\sigma^2} S_{XX} < h_{n-1}^+(\alpha/2) \right) = 1 - \alpha$$

as

$$\mathbb{P}\left(\frac{S_{XX}}{h_{n-1}^+(\alpha/2)} < \sigma^2 < \frac{S_{XX}}{h_{n-1}^-(\alpha/2)} \right) = 1 - \alpha$$

(here $\mathbb{P}$ is $\mathbb{P}_{\mu,\sigma^2}$, the sample distribution with $X_i \sim N(\mu, \sigma^2)$). Then, clearly, a $100(1-\alpha)\%$ equal-tailed CI for σ^2 is

$$\left(\frac{S_{XX}}{h_{n-1}^+(\alpha/2)}, \frac{S_{XX}}{h_{n-1}^-(\alpha/2)} \right). \tag{4.29}$$

Example 4.6   At a new call centre, the manager wishes to ensure that callers do not wait too long for their calls to be answered. A sample of 30 calls at the busiest time of the day gives a mean waiting time of 8 seconds (judged acceptable). At the same time, the sample value of $S_{xx}/(n-1)$ is 16, which is considerably higher than 9, the value from records of other call centres. The manager tests H_0: $\sigma^2 = 9$ against H_1: $\sigma^2 > 9$, assuming that call waiting times are IID $N(\mu, \sigma^2)$ (an idealised model).

We use a χ^2 test. The value $S_{xx}/\sigma^2 = 29 \times 16/9 = 51.55$. For $\alpha = 0.05$ and $n = 30$, the upper α point $h_{n-1}^+(\alpha)$ is 42.56. Hence, at the 5% level we reject H_0 and conclude that the variance at the new call centre is > 9.

The confidence interval for σ^2 is

$$\left(\frac{16 \times 29}{45.72}, \frac{16 \times 29}{16.05} \right) = (10.15, 28.91),$$

as $h_{29}^+(0.025) = 45.72$ and $h_{29}^+(0.975) = 16.05$.   ∎

Both the t test and normal χ^2 tests are examples of so-called *goodness of fit tests*. Here, we have a null hypothesis H_0 corresponding to a 'thin' subset Θ_0 in the parameter set Θ. In Examples 4.5 and 4.6 it was a half-line $\{\mu = \mu_0, \sigma^2 > 0 \text{ arbitrary}\}$ or a line $\{\mu \in \mathbb{R}, \sigma^2 = \sigma_0^2\}$ embedded into the half-plane $\Theta = \{\mu \in \mathbb{R}, \sigma^2 > 0\}$. The alternative was specified by the complement $\Theta_1 = \Theta \setminus \Theta_0$. We have to find a test statistic (in the examples, T or S_{XX}), with a 'standard' distribution under H_0. Then we reject H_0 at a given significance level α when the value of the statistic does not lie in the high-probability domain specified for this α. See Figure 4.2. Such a domain was the interval

$$(-t_{n-1}(\alpha/2), t_{n-1}(\alpha/2))$$

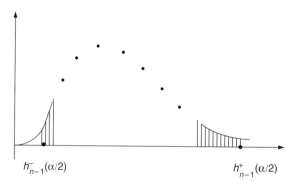

$h_{n-1}^-(\alpha/2)$ $h_{n-1}^+(\alpha/2)$

Figure 4.2

for the t-test and

$$\left(h_{n-1}^-(\alpha/2),\, h_{n-1}^+(\alpha/2)\right)$$

for the χ^2 test. In this case one says that the data are significant at level α. Otherwise, the data at this level are declared insignificant, and H_0 is not rejected.

Remark   In this section we followed the tradition of disjoint null and alternative hypotheses ($\Theta_0 \cap \Theta_1 = \emptyset$). Beginning with the next section, the null hypothesis H_0 in a goodness of fit test will be specified by a 'thin' set of parameter values Θ_0 while the alternative H_1 will correspond to a 'full' set $\Theta \supset \Theta_0$. This will not affect examples considered below as the answers will be the same. Also a number of problems in Chapter 5 associate the alternative with the complement $\Theta \setminus \Theta_0$ rather than with the whole set Θ; again the answers are unaffected.

4.4   The Pearson χ^2 test. The Pearson Theorem

>   Statisticians do it with significance.
>   (From the series '*How they do it*'.)

Historically, the idea of the goodness of fit approach goes back to K. Pearson and dates to 1900. The idea was further developed in the 1920s within the framework of the so-called *Pearson chi-square*, or *Pearson*, or χ^2, *test* based on the *Pearson chi-square*, or *Pearson*, or χ^2, *statistic*. A feature of the Pearson test is that it is 'universal', in the sense that it can be used to check the hypothesis that $\mathbf{X}$ is a random sample from any given PMF/PDF. The hypothesis is rejected when the value of the Pearson statistic does not fall into the interval of highly-probable values. The universality is manifested in the fact that the formula for the Pearson statistic and its distribution do not depend on the form of the tested PMF/PDF. However, the price paid is that the test is only asymptotically accurate, as n, the size of the sample, tends to ∞.

Suppose we test the hypothesis that an IID random sample $\mathbf{X} = (X_1, \ldots, X_n)$ comes from a PDF/PMF f^0. We partition $\mathbb{R}$, the space of values for RVs X_i, into k disjoint sets (say intervals) $D_1, \ldots, D_k$ and calculate the probabilities

$$p_l^0 = \int_{D_l} f^0(x)\mathrm{d}x \quad \text{or} \quad \sum_{D_l} f^0(x). \tag{4.30}$$

The null hypothesis H_0 is that $\forall\, \ell$, the probability $\mathbb{P}(X_i \in D_l)$ is equal to p_ℓ^0, the value given in equation (4.30). The alternative is that they are unrestricted (apart from $p_l \geq 0$ and $p_1 + \cdots + p_k = 1$). The Pearson χ^2 statistic is

$$P(\mathbf{x}) = \sum_{l=1}^{k} \frac{(n_l - e_l)^2}{e_l}, \quad \mathbf{x} = \begin{pmatrix} x_1 \\ \vdots \\ x_n \end{pmatrix}, \tag{4.31}$$

where $n_l \, (= n_l(\mathbf{x}))$ is the number of values x_i falling in D_l, with $n_1 + \cdots + n_k = n$, and $e_l = n p_l^0$ is the expected number under H_0. Letter P here is used to stress Pearson's pioneering contribution. Then, given $\alpha \in (0, 1)$, we reject the null hypothesis at significance level α when the value p of P exceeds $h_{k-1}^+(\alpha)$, the upper α quantile of the χ_{k-1}^2 distribution.

This routine is based on the *Pearson Theorem*:

> *Suppose that $X_1, X_2, \ldots$ is a sequence of IID RVs. Let $D_1, \ldots, D_k$ be a partition of $\mathbb{R}$ into pair-wise disjoint sets and set $q_l = \mathbb{P}(X_i \in D_l), l = 1, \ldots, k$, with $q_1 + \cdots + q_k = 1$. Next, $\forall \, l = 1, \ldots, k$ and $n \geq 1$, define*
> *$N_{l,n}$ = the number of RVs X_i among $X_1, \ldots, X_n$ such that $X_i \in D_l$, with $N_{1,n} + \cdots + N_{k,n} = n$, and*

$$P_n = \sum_{l=1}^{k} \frac{(N_{l,n} - n q_l)^2}{n q_l}. \tag{4.32}$$

> *Then $\forall \, \alpha > 0$:*

$$\lim_{n \to \infty} \mathbb{P}(P_n > \alpha) = \int_{\alpha}^{\infty} f_{\chi_{k-1}^2}(x) \mathrm{d}x. \tag{4.33}$$

Proof. We use the fact that relation (4.33) is equivalent to the convergence of the characteristic functions $\psi_{P_n}(t) = \mathbb{E} e^{\mathrm{i} t P_n}$:

$$\lim_{n \to \infty} \psi_{P_n}(t) = \int_0^{\infty} f_{\chi_{k-1}^2}(x) e^{\mathrm{i} t x} \mathrm{d}x, \quad t \in \mathbb{R}. \tag{4.34}$$

Set

$$Y_{l,n} = \frac{N_{l,n} - n q_l}{\sqrt{n q_l}}, \quad \text{so that } P_n = \sum_{l=1}^{k} Y_{l,n}^2 \tag{4.35}$$

and

$$\sum_{l=1}^{k} Y_{l,n} \sqrt{q_l} = \frac{1}{\sqrt{n}} \sum_l (N_{l,n} - n q_l) = \frac{1}{\sqrt{n}} \left(\sum_l N_{l,n} - n \sum_l q_l \right) = 0. \tag{4.36}$$

Our aim is to determine the limiting distribution of the random vector

$$\mathbf{Y}_n = \begin{pmatrix} Y_{1,n} \\ \vdots \\ Y_{k,n} \end{pmatrix}.$$

Take the unit vector

$$\kappa = \begin{pmatrix} \sqrt{q_1} \\ \vdots \\ \sqrt{q_k} \end{pmatrix}$$

and consider a real orthogonal $k \times k$ matrix A with kth column κ. Such a matrix always exists: you simply complete vector κ to an orthonormal basis in $\mathbb{R}^k$ and use this basis to form the columns of A. Consider the random vector

$$\mathbf{Z}_n = \begin{pmatrix} Z_{1,n} \\ \vdots \\ Z_{k,n} \end{pmatrix}$$

defined by

$$\mathbf{Z}_n = A^{\mathrm{T}} \mathbf{Y}_n.$$

Its last entry vanishes:

$$Z_{k,n} = \sum_{l=1}^{k} Y_{l,n} A_{l,k} = \sum_{l=1}^{k} Y_{l,n} \sqrt{q_l} = 0.$$

At the same time, owing to the orthogonality of A:

$$P_n = \sum_{l=1}^{k} Z_{l,n}^2 = \sum_{l=1}^{k-1} Z_{l,n}^2. \tag{4.37}$$

Equation (4.36) gives insight into the structure of RV P_n. We see that to prove limiting relation (4.33) it suffices to check that RVs $Z_{1,n}, \ldots, Z_{k-1,n}$ become asymptotically independent $N(0, 1)$. Again using the CHFs, it is enough to prove that the joint CHF converges to the product:

$$\lim_{n \to \infty} \mathbb{E} e^{it_1 Z_{1,n} + \cdots + it_{k-1} Z_{k-1,n}} = \prod_{l=1}^{k-1} e^{-t_l^2/2}. \tag{4.38}$$

To prove relation (4.38), we return to the RVs $N_{l,n}$. Write

$$\mathbb{P}\left(N_{1,n} = n_1, \ldots, N_{k,n} = n_k\right) = \frac{n!}{n_1! \ldots n_k!} q_1^{n_1} \cdots q_k^{n_k},$$

$\forall$ non-negative integers $n_1, \ldots, n_k$ with $n_1 + \cdots + n_k = n$. Then the joint CHF

$$\mathbb{E} e^{i \sum_l t_l N_{l,n}} = \sum_{n_1, \ldots, n_k: \sum_l n_l = n} \frac{n!}{n_1! \cdots n_k!} q_1^{n_1} \cdots q_k^{n_k} e^{i \sum_l t_l n_l} = \left(\sum_{l=1}^{k} q_l e^{it_l}\right)^n.$$

Passing to $Y_{1,n}, \ldots, Y_{k,n}$ gives

$$\mathbb{E} e^{i \sum_l t_l Y_{l,n}} = e^{-i\sqrt{n} \sum_l t_l \sqrt{q_l}} \left(\sum_{l=1}^{k} q_l e^{it_l/\sqrt{nq_l}}\right)^n.$$

and

$$\ln \mathbb{E} e^{i \sum_l t_l Y_{l,n}} = n \ln \left(\sum_{l=1}^k q_l e^{it_l / \sqrt{nq_l}} \right) - i \sqrt{n} \sum_l t_l \sqrt{q_l}$$

$$= n \ln \left[1 + \frac{i}{\sqrt{n}} \sum_{l=1}^k t_l \sqrt{q_l} - \frac{1}{2n} \sum_{l=1}^k t_l^2 + O\left(n^{-3/2}\right) \right] - i \sqrt{n} \sum_l t_l \sqrt{q_l}$$

$$= -\frac{1}{2} \sum_{l=1}^k t_l^2 + \frac{1}{2} \left(\sum_{l=1}^k t_l \sqrt{q_l} \right)^2 + O\left(n^{-1/2}\right),$$

by the Taylor expansion.

As $n \to \infty$, this converges to

$$-\frac{1}{2} ||\mathbf{t}||^2 + \frac{1}{2} \left(A^\mathsf{T} \mathbf{t}\right)_k^2 = -\frac{1}{2} ||A^\mathsf{T} \mathbf{t}||^2 + \frac{1}{2} \left(A^\mathsf{T} \mathbf{t}\right)_k^2$$

$$= -\frac{1}{2} \sum_{l=1}^{k-1} \left(A^\mathsf{T} \mathbf{t}\right)_l^2, \quad \mathbf{t} = \begin{pmatrix} t_1 \\ \vdots \\ t_k \end{pmatrix} \in \mathbb{R}^n.$$

Here A is the above $k \times k$ orthogonal matrix, with $A^\mathsf{T} = A^{-1}$, and $|| \ ||^2$ stands for the square of the norm (or length) of a vector from $\mathbb{R}^k$ (so that $||\mathbf{t}||^2 = \sum_l t_l^2$ and $||A^\mathsf{T} \mathbf{t}||^2 = \sum_l \left(A^\mathsf{T} \mathbf{t}\right)_l^2$). Consequently, with $\langle \mathbf{t}, \mathbf{Y}_n \rangle = \sum_{l=1}^k t_l Y_{l,n}$, we have

$$\lim_{n \to \infty} \mathbb{E} e^{i \langle \mathbf{t}, \mathbf{Y}_n \rangle} = \prod_{l=1}^{k-1} e^{-(A^\mathsf{T} \mathbf{t})_l^2 / 2}. \tag{4.39}$$

Then, for RVs $Z_{l,n}$, in a similar notation

$$\mathbb{E} e^{i \sum_l t_l Z_{l,n}} = \mathbb{E} e^{i \langle \mathbf{t}, \mathbf{Z}_n \rangle} = \mathbb{E} e^{i \langle \mathbf{t}, A^\mathsf{T} \mathbf{Y}_n \rangle} = \mathbb{E} e^{i \langle A \mathbf{t}, \mathbf{Y}_n \rangle}$$

which, by (4.38), should converge as $n \to \infty$ to

$$\prod_{l=1}^{k-1} e^{-(A^\mathsf{T} A \mathbf{t})_l^2 / 2} = \prod_{l=1}^{k-1} e^{-t_l^2 / 2}.$$

This completes the proof.

We stress once again that the Pearson χ^2 test becomes accurate only in the limit $n \to \infty$. However, as the approximation is quite fast, one uses this test for moderate values of n, quoting the Pearson Theorem as a ground.

It is interesting to note that K. Pearson was also a considerable scholar in philosophy. His 1891 book 'The Grammar of Science' gave a lucid exposition of scientific philosophy of the Vienna school of the late nineteenth century and was critically reviewed in the works of Lenin (which both of the present authors had to learn during their university years). However, Lenin made a clear distinction between Pearson and Mach, the main representative of this philosophical direction (and a prominent physicist of the period) and clearly considered Pearson superior to Mach.

Example 4.7   In his famous experiments Mendel crossed 556 smooth yellow male peas with wrinkled green female peas. From this progeny, we define:

$N_1 =$ the number of smooth yellow peas,
$N_2 =$ the number of smooth green peas,
$N_3 =$ the number of wrinkled yellow peas,
$N_4 =$ the number of wrinkled green peas,

and we consider the null hypothesis for the proportions

$$H_0: (p_1, p_2, p_3, p_4) = \left(\frac{9}{16}, \frac{3}{16}, \frac{3}{16}, \frac{1}{16} \right),$$

as predicted by Mendel's theory. The alternative is that the p_i are unrestricted (with $p_i \geq 0$) and $\sum_{i=1}^{4} p_i = 1$.

It was observed that $(n_1, n_2, n_3, n_4) = (301, 121, 102, 32)$, with

$$e = (312.75, 104.25, 104.25, 34.75), \quad P = 3.39888,$$

and the upper 0.25 point $= h_3^+(0.25) = 4.10834$. We therefore have no reason to reject Mendel's predictions, even at the 25% level.

Note that the above null hypothesis that $f = f^0$ specifies probabilities p_l^0 completely. In many cases, we have to follow a less precise null hypothesis that the p_l^0 belong to a given family $\{p_l^0(\theta), \ \theta \in \Theta\}$. In this situation we apply the previous routine after estimating the value of the parameter from the same sample. That is the χ^2 statistic is calculated as

$$P(\mathbf{x}) = \sum_{l=1}^{k} \frac{(n_l - \widehat{e}_l)^2}{\widehat{e}_l}, \tag{4.40}$$

where $\widehat{e}_l = n p_l^0(\widehat{\theta})$ and $\widehat{\theta} \ (= \widehat{\theta}(\mathbf{x}))$ is an estimator of θ.

Usually, $\widehat{\theta}$ is taken to be the MLE. Then the value $\pi = \sum_{i=1}^{k}(n_l - \widehat{e})^2/\widehat{e}$ of statistic P is compared not with $h_{k-1}^+(\alpha)$ but with $h_{k-1-|\Theta|}^+(\alpha)$, where $|\Theta|$ is the dimension of set Θ. That is at significance level $\alpha \in (0, 1)$ we reject H_0: p_l^0 belong to family $\{p_l^0(\theta), \ \theta \in \Theta\}$ when $\pi > h_{k-1-|\Theta|}^+(\alpha)$. (Typically, the value $h_{k_1}^+(\alpha)$ is higher than $h_{k_2}^+(\alpha)$ when $k_1 > k_2$.) This is based on a modified Pearson Theorem similar to the one above.   ∎

The Pearson χ^2 statistic and Pearson test for goodness of fit are the subject of MT-IB 1997-412G and 1994-103F(i). However, their infrequent appearance in Tripos questions should not be taken as a sign that this topic is considered as non-important. Moreover, it gives rise to the material discussed in the remaining sections which is very popular among Tripos examples.

4.5   Generalised likelihood ratio tests. The Wilks Theorem

Silence of the Lambdas
(From the series '*Movies that never made it to the Big Screen*'.)

The idea of using the MLEs for calculating a test statistic is pushed further forward when one discusses the so-called generalised likelihood ratio test. Here, one considers a null hypothesis H_0 that $\theta \in \Theta_0$, where $\Theta_0 \subset \Theta$, and the general alternative H_1: $\theta \in \Theta$. A particular example is the case of H_0: $f = f^{(0)}$, where probabilities $p_l^0 = \mathbb{P}(D_l)$ have been completely specified; in this case Θ is the set of all PMFs $(p_1, \ldots, p_k)$ and Θ_0 reduced to a single point $(p_1^{(0)}, \ldots, p_k^{(0)})$. The case where p_l^0 depend on parameter θ is a more general example. Now we adopt a similar course of action: consider the maxima

$$\max \left[f(\mathbf{x}; \theta) : \theta \in \Theta_0 \right]$$

and

$$\max \left[f(\mathbf{x}; \theta) : \theta \in \Theta \right]$$

and take their ratio (which is ≥ 1):

$$\Lambda_{H_1:H_0}(\mathbf{x}) = \frac{\max \left[f(\mathbf{x}; \theta) : \theta \in \Theta \right]}{\max \left[f(\mathbf{x}; \theta) : \theta \in \Theta_0 \right]}. \tag{4.41}$$

$\Lambda_{H_1:H_0}$ is called the *generalised likelihood ratio* (GLR) for H_0 and H_1; sometimes the denominator is called the likelihood of H_0 and the numerator the likelihood of H_1.

In some cases, the GLR $\Lambda_{H_1:H_0}$ has a recognised distribution. In general, one takes

$$R = 2 \ln \Lambda_{H_1:H_0}(\mathbf{X}). \tag{4.42}$$

Then if, for a given $\alpha \in (0, 1)$, the value of R in formula (4.42) exceeds the upper point $h_p(\alpha)$, H_0 is rejected at level α. Here p, the number of degrees of freedom of the χ^2 distribution, equals $|\Theta| - |\Theta_0|$, the difference of the dimensions of sets Θ and Θ_0.

This routine is called the *generalised likelihood ratio test* (GLRT) and is based on the *Wilks Theorem*, which is a generalisation of the Pearson Theorem on asymptotical properties of the χ^2 statistic. Informally, this theorem is as follows.

> *Suppose that $\mathbf{X}$ is a random sample with IID components X_i and with PDF/PMF $f(\,\cdot\,; \theta)$ depending on a parameter $\theta \in \Theta$, where Θ is an open domain in a Euclidean space of dimension $|\Theta|$. Suppose that the MLE $\widehat{\theta}(\mathbf{X})$ is an asymptotically normal RV as $n \to \infty$. Fix a subset $\Theta_0 \subset \Theta$ such that Θ_0 is an open domain in a Euclidean space of a lesser dimension $|\Theta_0|$. State the null hypothesis H_0 that $\theta \in \Theta_0$. Then (as in the Pearson Theorem), under H_0, the RV R in formula (4.42) has asymptotically the χ_p^2 distribution with $p = |\Theta| - |\Theta_0|$ degrees of freedom. More precisely, $\forall\ \theta \in \Theta_0$ and $h > 0$:*
>
> $$\lim_{n \to \infty} \mathbb{P}_\theta(R > h) = \int_h^\infty f_{\chi_p^2}(x) \mathrm{d}x.$$

There is also a version of the Wilks Theorem for independent, but not IID RVs $X_1, X_2, \ldots$. The theorem was named after S.S. Wilks (1906–1964), an American statistician who for a while worked with Fisher. We will not prove the Wilks Theorem but illustrate its rôle in several examples.

Example 4.8   A simple example is where $X_i \sim \mathrm{N}(\mu, \sigma^2)$, with σ^2 known or unknown. Suppose first that σ^2 is known. As in the previous section, we fix a value μ_0 and test H_0: $\mu = \mu_0$ against H_1: $\mu \in \mathbb{R}$ unrestricted. Then, under H_1, $\bar{x}$ is the MLE of μ, and

$$\max\left[f(\mathbf{x}; \mu) : \mu \in \mathbb{R}\right] = \frac{1}{(\sqrt{2\pi}\sigma)^n} \exp\left(-\frac{S_{xx}}{2\sigma^2}\right),$$

where

$$S_{xx} = \sum_i (x_i - \bar{x})^2,$$

whereas under H_0,

$$f(\mathbf{x}; \mu_0) = \frac{1}{(\sqrt{2\pi}\sigma)^n} \exp\left(-\frac{\Sigma^2_{\mu_0}}{2\sigma^2}\right),$$

where

$$\Sigma^2_{\mu_0} = \sum_i (x_i - \mu_0)^2.$$

Hence,

$$\Lambda_{H_1;H_0} = \exp\left(\frac{1}{2\sigma^2}\left(\Sigma^2_{\mu_0} - S_{xx}\right)\right),$$

and

$$2\ln \Lambda_{H_1;H_0} = \frac{1}{\sigma^2}\left(\Sigma^2_{\mu_0} - S_{xx}\right) = \frac{n}{\sigma^2}(\bar{x} - \mu)^2 \sim \chi^2_1.$$

We see that in this example, $2\ln \Lambda_{H_1;H_0}(\mathbf{X})$ has precisely a χ^2_1 distribution (i.e. the Wilks Theorem is exact). According to the GLRT, we reject H_0 at level α when $2\ln \Lambda_{H_1;H_0}(\mathbf{x})$ exceeds $h_1^+(\alpha)$, the upper α quantile of the χ^2_1 distribution. This is equivalent to rejecting H_0 when $|\bar{x} - \mu|\sqrt{n}/\sigma$ exceeds $z^+(\alpha/2)$, the upper $\alpha/2$ quantile of the $\mathrm{N}(0, 1)$ distribution.

If σ^2 is unknown, then we have to use the MLE $\hat{\sigma}^2$ equal to $\Sigma^2_{\mu_0}/n$ under H_0 and S_{XX}/n under H_1. In this situation, σ^2 is considered as a *nuisance parameter*: it does not enter the null hypothesis and yet is to be taken into account. Then

$$\mathrm{under} H_1 : \max\left[f(\mathbf{x}; \mu, \sigma^2) : \mu \in \mathbb{R}, \sigma^2 > 0\right] = \frac{1}{(2\pi S_{xx}/n)^{n/2}} e^{-n/2},$$

and

$$\mathrm{under}\ H_0 : \max\left[f(\mathbf{x}; \mu_0, \sigma^2) : \sigma^2 > 0\right] = \frac{1}{(2\pi \Sigma^2_{\mu_0}/n)^{n/2}} e^{-n/2}.$$

Hence,

$$\Lambda_{H_1;H_0} = \left(\frac{\Sigma^2_{\mu_0}}{S_{xx}}\right)^{n/2} = \left(1 + \frac{n(\bar{x} - \mu_0)^2}{S_{xx}}\right)^{n/2},$$

and we reject H_0 when $n(\bar{x} - \mu_0)^2/S_{xx}$ is large. But this is again precisely the t-test, and we do not need the Wilks Theorem.

We see that the t-test can be considered as an (important) example of a GLRT. ∎

Example 4.9   Now let μ be known and test H_0: $\sigma^2 = \sigma_0^2$ against $\sigma^2 > 0$ unrestricted. The MLE of σ^2 is Σ^2/n, where $\Sigma^2 = \Sigma_i(x_i - \mu)^2$. Hence,

$$
\text{under } H_1 : \max\left[f(\mathbf{x}; \mu, \sigma^2) : \sigma^2 > 0\right] = \left(\frac{\sqrt{n}}{\sqrt{2\pi\Sigma^2}}\right)^n e^{-n/2},
$$

and

$$
\text{under } H_0 : f(\mathbf{x}; \mu, \sigma_0^2) = \frac{1}{(2\pi\sigma_0^2)^{n/2}} \exp\left[-\frac{1}{2\sigma_0^2}\Sigma^2\right],
$$

with

$$
\Lambda_{H_1 ; H_0} = \left(\frac{n\sigma_0^2}{\Sigma^2}\right)^{n/2} \exp\left[-\frac{n}{2} + \frac{\Sigma^2}{2\sigma_0^2}\right],
$$

and

$$
2\ln \Lambda_{H_1 ; H_0} = n\ln \frac{\sigma_0^2 n}{\Sigma^2} - n + \frac{\Sigma^2}{\sigma_0^2} = -n\ln \frac{\Sigma^2}{n\sigma_0^2} + \frac{\Sigma^2 - n\sigma_0^2}{\sigma_0^2}
$$

$$
= -n\ln\left(1 + \frac{\Sigma^2 - n\sigma_0^2}{n\sigma_0^2}\right) + \frac{\Sigma^2 - n\sigma_0^2}{\sigma_0^2}.
$$

We know that under H_0, $\Sigma^2/\sigma^2 \sim \chi_n^2$. By the LLN, the ratio Σ^2/n converges under H_0 to $\mathbb{E}(X_1 - \mu)^2 = \operatorname{Var} X_1 = \sigma_0^2$. Hence, for large n, the ratio $(\Sigma^2 - n\sigma_0^2)/n\sigma_0^2$ is close to 0. Then we can use the Taylor expansion of the logarithm. Thus,

$$
2\ln \Lambda_{H_1 ; H_0} \approx -n\left[\frac{\Sigma^2 - n\sigma_0^2}{n\sigma_0^2} - \frac{1}{2}\left(\frac{\Sigma^2 - n\sigma_0^2}{n\sigma_0^2}\right)^2\right] + \frac{\Sigma^2 - n\sigma_0^2}{\sigma_0^2}
$$

$$
= \frac{n}{2}\left(\frac{\Sigma^2 - n\sigma_0^2}{n\sigma_0^2}\right)^2 = \left(\frac{\Sigma^2 - n\sigma_0^2}{\sqrt{2n\sigma_0^4}}\right)^2.
$$

The next fact is that, by the CLT, as $n \to \infty$,

$$
\frac{\Sigma^2 - n\sigma_0^2}{\sqrt{2n\sigma_0^4}} \sim \mathrm{N}(0, 1).
$$

This is because RV $\Sigma^2(\mathbf{X})$ is the sum of IID RVs $(X_i - \mu)^2$, with

$$
\mathbb{E}(X_i - \mu)^2 = \sigma^2 \text{ and } \operatorname{Var}(X_i - \mu)^2 = 2\sigma_0^4.
$$

Hence, under H_0, as $n \to \infty$, the square

$$
\left(\frac{\Sigma^2 - n\sigma_0^2}{\sqrt{2n\sigma_0^4}}\right)^2 \sim \chi_1^2, \text{ i.e. } 2\ln \Lambda_{H_1 ; H_0} \sim \chi_1^2,
$$

in agreement with the Wilks Theorem.

We see that for n large, the GLRT test is to reject H_0 at level α when

$$\frac{(\Sigma^2 - n\sigma_0^2)^2}{2n\sigma_0^4} > h_1^+(\alpha).$$

Of course, in this situation there is the better way to proceed: we might use the fact that

$$\frac{\Sigma^2}{\sigma_0^2} \sim \chi_n^2.$$

So we reject H_0 when $\Sigma^2/\sigma_0^2 > h_n^+(\alpha/2)$ or $\Sigma^2/\sigma_0^2 < h_n^-(\alpha/2)$. Here we use the same statistic as in the GLRT, but with a different critical region.

Alternatively, we can take

$$\frac{S_{XX}}{\sigma_0^2} \sim \chi_{n-1}^2,$$

where $S_{XX} = \sum_{i=1}^n (X_i - \overline{X})^2$. The normal χ^2 test:

$$\text{reject } H_0 \text{ when } \frac{1}{\sigma_0^2} S_{xx} > h_{n-1}^+(\alpha/2) \text{ or } < h_{n-1}^-(\alpha/2)$$

is again more precise than the GLRT in this example.

In a similar way we treat the case where H_0: $\sigma^2 = \sigma_0^2$, H_1: $\sigma^2 \neq \sigma_0^2$ and μ is unknown (i.e. is a nuisance parameter). Here $\widehat{\mu} = \overline{x}$, $\widehat{\sigma}^2 = S_{XX}/n$, and

$$\text{under } H_1: \max\left[f(\mathbf{x}; \mu, \sigma^2): \mu \in \mathbb{R}, \sigma^2 > 0\right] = \frac{1}{(2\pi S_{xx}^2/n)^{n/2}} e^{-n/2},$$

and

$$\text{under } H_0: \max\left[f(\mathbf{x}; \mu, \sigma_0^2): \mu \in \mathbb{R}\right] = \frac{1}{(2\pi\sigma_0^2)^{n/2}} \exp\left[-\frac{1}{2\sigma_0^2} S_{xx}\right],$$

with the GLR statistic

$$\Lambda_{H_1;H_0} = (ne)^{-n/2} \left(\frac{\sigma_0^2}{S_{xx}}\right)^{n/2} \exp\left(\frac{1}{2}\frac{S_{xx}}{\sigma_0^2}\right).$$

We see that $\Lambda_{H_1;H_0}$ is large when S_{xx}/σ_0^2 is large or close to 0. Hence, in the GLRT paradigm, H_0 is rejected when S_{XX}/σ_0^2 is either small or large. But $S_{XX}/\sigma_0^2 \sim \chi_{n-1}^2$. We see that in this example the standard χ^2 test and the GLRT use the same table (but operate with different critical regions). ∎

Another aspect of the GLRT is its connection with Pearson's χ^2 test. Consider the null hypothesis H_0: $p_i = p_i(\theta)$, $1 \leq i \leq k$, for some parameter $\theta \in \Theta_0$, where set Θ_0 has dimension (the number of independent co-ordinates) $|\Theta_0| = k_0 < k - 1$. The alternative H_1 is that probabilities p_i are unrestricted (with the proviso that $p_i \geq 0$ and $\sum_i p_i = 1$).

Example 4.10   Suppose that RVs $X_1, \ldots, X_m \sim N(\mu_X, \sigma_X^2)$ and $Y_1, \ldots, Y_n \sim N(\mu_Y, \sigma_Y^2)$, where μ_X, σ_X^2, μ_Y and σ_Y^2 are all unknown. Let H_0 be that $\sigma_X^2 = \sigma_Y^2$ and H_1 that $\sigma_X^2 > \sigma_Y^2$. Derive the form of the likelihood ratio test and specify the distribution of the relevant statistic.

Here

$$L_{xy}(H_0) = \max \left[f_X(x; \mu_X, \sigma^2) f_Y(y; \mu_Y, \sigma^2) : \mu_X \in \mathbb{R}, \ \mu_Y \in \mathbb{R}, \ \sigma > 0 \right]$$

$$= \max \left[\frac{1}{(2\pi\sigma^2)^{(n+m)/2}} \exp\left(-\frac{S_{xx} + S_{yy}}{2\sigma^2} \right) : \sigma > 0 \right].$$

Note that for $g(x) = x^a e^{-bx}$ with $a, b > 0$:

$$\max_{x>0} g(x) = g\left(\frac{a}{b} \right) = \left(\frac{a}{b} \right)^a e^{-a}.$$

Hence,

$$L_{xy}(H_0) = \frac{1}{(2\pi\widehat{\sigma}_0^2)^{(m+n)/2}} e^{-(m+n)/2}, \qquad \widehat{\sigma}_0^2 = \frac{S_{xx} + S_{yy}}{m+n}.$$

Similarly, under H_1

$$L_{xy}(H_1) = \max \left[f_X(x; \mu_X, \sigma_X^2) f_Y(y; \mu_Y, \sigma_Y^2) : \mu_X, \mu_Y \in \mathbb{R}, \ \sigma_X, \sigma_Y > 0 \right]$$

$$= \frac{1}{(2\pi\widehat{\sigma}_X^2)^{m/2} (2\pi\widehat{\sigma}_Y^2)^{n/2}} e^{-(m+n)/2},$$

with $\widehat{\sigma}_X^2 = S_{xx}/m$, $\widehat{\sigma}_Y^2 = S_{yy}/n$ (provided that $\widehat{\sigma}_X^2 > \widehat{\sigma}_Y^2$). As a result,

$$\text{if } \frac{S_{xx}}{m} > \frac{S_{yy}}{n}, \text{ then } \Lambda = \left(\frac{S_{xx} + S_{yy}}{m+n} \right)^{(m+n)/2} \left(\frac{S_{xx}}{m} \right)^{-m/2} \left(\frac{S_{yy}}{n} \right)^{-n/2},$$

and

$$\text{if } \frac{S_{xx}}{m} \leq \frac{S_{yy}}{n}, \text{ then } \Lambda = 1.$$

Further,

$$\text{if } \frac{S_{xx}}{S_{yy}} > \frac{m}{n} \text{ then } 2 \ln \Lambda = c + f\left(\frac{S_{xx}}{S_{yy}} \right).$$

Here

$$f(u) = (m+n) \ln (1+z) - m \ln u,$$

and c is a constant. Next,

$$f'(u) = \frac{m+n}{1+u} - \frac{m}{u} = \frac{nu - m}{u(1+u)},$$

i.e. f increases when $u > m/n$ increases. As a result, we reject H_0 if S_{xx}/S_{yy} is large. Under H_0,

$$\sigma_X^2 = \sigma_Y^2 = \sigma^2, \text{ and } \frac{S_{xx}}{\sigma^2} \sim \chi_{m-1}^2, \ \frac{S_{yy}}{\sigma^2} \sim \chi_{n-1}^2, \text{ independently.}$$

Therefore,

$$\frac{S_{xx}/(m-1)}{S_{yy}/(n-1)} \sim F_{m-1,n-1}. \quad \blacksquare$$

Example 4.11   There are $k+1$ probabilities $p_0, \ldots, p_k$. A null hypothesis H_0 is that they are of the form

$$p_i(\theta) = \binom{k}{i} \theta^i (1-\theta)^{k-i}, \ 0 \le i \le k,$$

where $\theta \in (0, 1)$. (Here $|\Theta_0| = 1$.) The alternative H_1 is that these probabilities form a k-dimensional variety.

The MLE under H_0 maximises $\sum_{i=0}^{k} n_i \log p_i(\theta)$, where $n_0, \ldots, n_k$ are occurrence numbers of values $0, \ldots, k$. Let $\widehat{\theta}$ be the maximiser. An easy calculation shows that under H_0, $\widehat{\theta} = \sum_{i=1}^{k} i n_i / (kn)$. Under H_1 the MLE is $\widehat{p}_i = n_i / n$, where $n = n_0 + \cdots + n_k$. Then the logarithm of the GLR yields

$$2 \ln \Lambda = 2 \ln \frac{\prod_{i=0}^{k} \widehat{p}_i^{\,n_i}}{\prod_{i=0}^{k} \left(p_i(\widehat{\theta})\right)^{n_i}} = 2 \sum_i n_i \ln \frac{n_i}{np_i(\widehat{\theta})}.$$

The number of degrees of freedom is $k + 1 - 1 - |\Theta_0| = k - 1$. We reject H_0 when $2 \ln \Lambda > h_{k-1}^+(\alpha)$.

Write $np_i(\widehat{\theta}) = e_i$, $\delta_i = n_i - e_i$, $0 \le i \le k$, with $\sum_i \delta_i = 0$. A straightforward calculation is:

$$2 \ln \Lambda = 2 \sum_i n_i \ln \frac{n_i}{e_i} = 2 \sum_i (e_i + \delta_i) \ln \left(1 + \frac{\delta_i}{e_i}\right)$$

$$\approx 2 \sum_i \left(\delta_i + \frac{\delta_i^2}{e_i} - \frac{\delta_i^2}{2e_i}\right) = \sum_i \frac{\delta_i^2}{e_i} = \sum_i \frac{(n_i - e_i)^2}{e_i}.$$

This is precisely Pearson's χ^2 statistic.   ∎

It has to be said that, generally speaking, the GLRT remains a universal and powerful tool for testing goodness of fit. We discuss two examples where it is used for testing homogeneity in non-homogeneous samples.

Example 4.12   Let $X_1, \ldots, X_n$ be independent RVs, $X_i \sim \text{Po}(\lambda_i)$ with unknown mean $\lambda_i, i = 1, \ldots, n$. Find the form of the generalised likelihood ratio statistic for testing $H_0: \lambda_1 = \ldots = \lambda_n$, and show that it may be approximated by $Z = \sum_{i=1}^{n} (X_i - \overline{X})^2 / \overline{X}$, where $\overline{X} = n^{-1} \sum_{i=1}^{n} X_i$. If, for $n = 7$, you find that the value of this statistic was 26.9, would you accept H_0?

For $X_i \sim \text{Po}(\lambda_i)$,

$$f_{\mathbf{X}}(\mathbf{x}; \lambda) = \prod_{i=1}^{n} \frac{\lambda_i^{x_i}}{x_i!} e^{-\lambda_i}.$$

Under H_0, the MLE $\widehat{\lambda}$ for λ is $\overline{x} = \sum_i x_i / n$, and under H_1, $\widehat{\lambda}_i = x_i$. Then the GLR

$$\Lambda_{H_1; H_0}(\mathbf{x}) = \frac{f(\mathbf{x}; \widehat{\lambda}_1, \ldots, \widehat{\lambda}_n)}{f(\mathbf{x}; \widehat{\lambda})} = \frac{\prod_i x_i^{x_i}}{\prod_i \overline{x}^{x_i}},$$

and

$$2 \ln \Lambda_{H_1:H_0}(\mathbf{x}) = 2 \sum_i x_i \ln \frac{x_i}{\bar{x}}$$

$$= 2 \sum_i [\bar{x} + (x_i - \bar{x})] \ln \left(1 + \frac{x_i - \bar{x}}{\bar{x}}\right) \approx \sum_i \frac{(x_i - \bar{x})^2}{\bar{x}},$$

as $\ln(1+x) \approx x - x^2/2$.

Finally, we know that for $Z \sim \chi_6^2$, $\mathbb{E}Z = 6$ and $\operatorname{Var} Z = 12$. The value 26.9 appears too large. So we would reject H_0. ■

Example 4.13 Suppose you are given a collection of np independent random variables organised in n samples, each of length p:

$$X^{(1)} = (X_{11}, \dots, X_{1p})$$
$$X^{(2)} = (X_{21}, \dots, X_{2p})$$
$$\vdots \qquad \vdots$$
$$X^{(n)} = (X_{n1}, \dots, X_{np}).$$

The RV X_{ij} has a Poisson distribution with an unknown parameter λ_j, $1 \le j \le p$. You are required to test the hypothesis that $\lambda_1 = \dots = \lambda_p$ against the alternative that values $\lambda_j > 0$ are unrestricted. Derive the form of the likelihood ratio test statistic. Show that it may be approximated by

$$\frac{n}{\bar{X}} \sum_{j=1}^p (\bar{X}_j - \bar{X})^2$$

with

$$\bar{X}_j = \frac{1}{n} \sum_{i=1}^n X_{ij}, \quad \bar{X} = \frac{1}{np} \sum_{i=1}^n \sum_{j=1}^p X_{ij}.$$

Explain how you would test the hypothesis for large n.

In this example, the likelihood

$$\prod_{i=1}^n \prod_{j=1}^p \frac{e^{-\lambda_j} \lambda_j^{x_{ij}}}{x_{ij}!}$$

is maximised under H_0 at $\hat{\lambda} = \bar{x}$, and has the maximal value proportional to $e^{-np\bar{x}} \bar{x}^{np\bar{x}}$. Under H_1, $\hat{\lambda}_j = \bar{x}_j$ and the maximal value is proportional to the product $e^{-np\bar{x}} \prod_{j=1}^p \bar{x}_j^{n\bar{x}_j}$. Hence,

$$\Lambda_{H_1:H_0} = \frac{1}{\bar{x}^{np\bar{x}}} \prod_{j=1}^p \bar{x}_j^{n\bar{x}_j}.$$

We reject H_0 when $2\ln \Lambda_{H_1;H_0}$, is large. Now, $\ln \Lambda_{H_1,H_0}$ has the form

$$\left(\sum_{j=1}^{p} n\bar{x}_j \ln \bar{x}_j - np\bar{x}\ln \bar{x}\right) = n\left[\sum_j \left(\bar{x}_j \ln \bar{x}_j - \bar{x}_j \ln \bar{x}\right)\right]$$

$$= n\left[\sum_j \left(\bar{x}+\bar{x}_j - \bar{x}\right)\ln\left(\frac{\bar{x}+\bar{x}_j-\bar{x}}{\bar{x}}\right)\right]$$

$$= n\left[\sum_j \left(\bar{x}+\bar{x}_j - \bar{x}\right)\ln\left(1+\frac{\bar{x}_j-\bar{x}}{\bar{x}}\right)\right].$$

Under H_0, $n\bar{X}_j \sim$ Po $(n\lambda)$ and $np\bar{X} \sim$ Po $(np\lambda)$. By the LLN, both $\bar{X}_j$ and $\bar{X}$ converge to the constant λ, hence $(\bar{X}_j - \bar{X})/\bar{X}$ converges to 0. This means that for n large enough, we can take a first-order expansion as a good approximation for the logarithm. Hence, for n large, $2\ln \Lambda_{H_1,H_0}$ is

$$\approx 2n\sum_{j=1}^{p}(\bar{x}+\bar{x}_j-\bar{x})\left[\frac{\bar{x}_j-\bar{x}}{\bar{x}}-\frac{1}{2}\left(\frac{\bar{x}_j-\bar{x}}{\bar{x}}\right)^2\right] \approx n\sum_{i=1}^{p}\frac{(\bar{x}_j-\bar{x})^2}{\bar{x}}.$$

Observe that the ratios $Y_{l,n} = (\bar{X}_l - \bar{X})/\sqrt{\bar{X}}$ resemble the ratios $Y_{l,n}$ from definition (4.35). In fact, it is possible to check that as $n\to\infty$,

$$\sqrt{n}\frac{\bar{X}_l - \bar{X}}{\sqrt{\bar{X}}} \sim \text{N}(0,1).$$

(This is the main part of the proof of Wilks Theorem.) We also have, as in equation (4.36), that the $Y_{l,n}$ satisfy a linear relation

$$\sum_{l=1}^{p}Y_{l,n}\sqrt{\bar{X}} = \sum_{l=1}^{p}(\bar{X}_l - \bar{X}) = 0.$$

Then as in the proof of the Pearson Theorem, as $n\to\infty$,

$$2\ln \Lambda_{H_1,H_0} \sim \chi^2_{p-1}. \quad \blacksquare$$

Concluding this section, we provide an example of a typical Cambridge-style Mathematical Tripos question.

Example 4.14   What is meant by a generalised likelihood ratio test? Explain in detail how to perform such a test.

The GLRT is designed to test H_0: $\theta \in \Theta_0$ against the general alternative H_1: $\theta \in \Theta$, where $\Theta_0 \subset \Theta$. Here we use the GLR

$$\Lambda_{H_1;H_0}(\mathbf{x}) = \frac{\max\left[f(\mathbf{x};\theta): \theta \in \Theta\right]}{\max\left[f(\mathbf{x};\theta): \theta \in \Theta_0\right]}, \text{ where } \mathbf{x} = \begin{pmatrix} x_1 \\ \vdots \\ x_n \end{pmatrix}.$$

The GLRT rejects H_0 for large values of $\Lambda_{H_1;H_0}$.

If random sample $\mathbf{X}$ has IID components $X_1, \ldots, X_n$ and n is large then, under H_0, $2\ln\Lambda_{H_1;H_0}(\mathbf{X}) \sim \chi_p^2$, where the number of degrees of freedom $p = |\Theta| - |\Theta_0|$ (the Wilks Theorem). Therefore, given $\alpha \in (0, 1)$, we reject H_0 in a test of size α if

$$2\ln\Lambda_{H_1;H_0}(\mathbf{x}) > h_p^+(\alpha). \quad \blacksquare$$

It is interesting to note that the GLRT was proposed by Neyman and E. Pearson in 1928. The NP Lemma, however, was proved only in 1933.

The GLRT test is a subject of the following Tripos questions: MT-IB 1995-403G (ii), 1992-206D.

4.6   Contingency tables

> Statisticians do it by tables when it counts.
> (From the series 'How they do it'.)

A popular example of the GLRT is where you try to test independence of different 'categories' or 'attributes' assigned to several types of individuals or items.

Example 4.15   An IQ test was proposed to split people approximately into three equal groups: excellent (A), good (B), moderate (C). The following table gives the numbers of people who obtained grades A, B, C in three selected regions in Endland, Gales and Grogland. The data are given in Table 4.2.

Table 4.2.

Region	Grade			Total n_{i+}
	A	B	C	
Endland	3009	2832	3008	8849
Gales	3047	3051	2997	9095
Grogland	2974	3038	3018	9030
Total n_{+j}	9030	8921	9023	26974

It is required to derive a GLRT of independence of this classification. Here, $e_{ij} = n_{i+}n_{+j}/n_{++}$ are

$$
\begin{array}{ccc}
2962.35 & 2926.59 & 2960.06 \\
3044.70 & 3007.95 & 3042.34. \\
3022.94 & 2986.45 & 3020.60
\end{array}
$$

The value of the Pearson statistic (defined in equation (4.45) below) is

$$4.56828 + 1.29357 + 1.68437 = 7.54622,$$

with the number of degrees of freedom $(3-1) \times (3-1) = 4$. At the 0.05 level $h_4^+(0.05) = 9.488$. Hence, there is no reason to reject the hypothesis that all groups are homogeneous at this level (let alone the 0.01 level). ■

In general, you have a *contingency table* with r rows and c columns. The independence means that the probability p_{ij} that say a type i item falls into category, or receives attribute, j is of the form $\alpha_i \beta_j$ where

$$\alpha_i, \beta_j \geq 0 \text{ and } \sum_{i=1}^{r} \alpha_i = \sum_{j=1}^{c} \beta_j = 1.$$

This will be our null hypothesis H_0. The alternative H_1 corresponds to a general constraint $p_{ij} \geq 0$ and $\sum_{i=1}^{r} \sum_{j=1}^{c} p_{ij} = 1$. Here we set

$$p_{i+} = \sum_j p_{ij}, \ p_{+j} = \sum_i p_{ij} \text{ and } p_{++} = \sum_i p_{i+} = \sum_j p_{+j} = 1$$

(the notation comes from early literature).

The model behind Example 4.15 is that you have n items or individuals (26 974 in the example), and n_{ij} of them fall in cell (i, j) of the table, so that $\sum_{i,j} n_{ij} = n$. (Comparing with Example 4.1, we now have a model of selection with replacement which generalises Example 4.2.) Set

$$n_{i+} = \sum_j n_{ij}, \ n_{+j} = \sum_i n_{ij} \text{ and } n_{++} = \sum_i n_{i+} = \sum_j n_{+j} = n.$$

The RVs N_{ij} have jointly a *multinomial distribution* with parameters $(n; (p_{ij}))$:

$$\mathbb{P}(N_{ij} = n_{ij} \ \forall \ i, j) = n! \prod \frac{1}{n_{ij}!} p_{ij}^{n_{ij}} \qquad (4.43)$$

(the p_{ij} are often called 'cell probabilities'). It is not difficult to recognise the background for the GLRT, where you have $|\Theta_1| = rc - 1$ (as the sum $p_{++} = 1$) and $|\Theta_0| = r - 1 + c - 1$, with

$$|\Theta_1| - |\Theta_0| = (r-1)(c-1).$$

Under H_1, the MLE for p_{ij} is $\widehat{p}_{ij} = n_{ij}/n$. In fact, $\forall \ n_{ij}$ with $n_{++} = n$, the PMF (i.e. the likelihood) is

$$f_{(N_{ij})}\left((n_{ij}); (p_{ij})\right) = n! \prod_{i,j} \frac{p_{ij}^{n_{ij}}}{n_{ij}!},$$

and the LL is

$$\ell\left((n_{ij}); (p_{ij})\right) = \sum_{i,j} n_{ij} \ln p_{ij} + \ln(n!) - \sum_{i,j} \ln(n_{ij}!).$$

We want to maximise ℓ in p_{ij} under the constraints

$$p_{ij} \geq 0, \sum_{i,j} p_{ij} = 1.$$

Omitting the term

$$\ln(n!) - \sum_{i,j} \ln(n_{ij}!),$$

write the Lagrangian

$$\mathcal{L} = \sum_{i,j} n_{ij} \ln p_{ij} - \lambda \left(\sum_{i,j} p_{ij} - 1 \right).$$

Its maximum is attained when $\forall\, i, j$:

$$\frac{\partial}{\partial p_{ij}} \mathcal{L} = \frac{n_{ij}}{p_{ij}} - \lambda = 0,$$

i.e. when $p_{ij} = n_{ij}/\lambda$. Adjusting the constraint $p_{++} = 1$ yields $\lambda = n$.

Under H_0, the MLE for α_i is $\widehat{\alpha}_i = n_{i+}/n$; for β_j the MLE $\widehat{\beta}_j = n_{+j}/n$. In fact, here the likelihood is

$$f_{(N_{ij})}\left((n_{ij}); (\alpha_i), (\beta_j)\right) = n! \prod_{i,j} \frac{(\alpha_i \beta_j)^{n_{ij}}}{n_{ij}!} = n! \prod_i \alpha_i^{n_{i+}} \prod_j \beta_j^{n_{+j}} \bigg/ \prod_{i,j} n_{ij}!,$$

and the LL is

$$\ell\left((n_{ij}); (\alpha_i), (\beta_j)\right) = \sum_i n_{i+} \ln \alpha_i + \sum_j n_{+j} \ln \beta_j + \ln(n!) - \sum_{i,j} \ln(n_{ij}!).$$

This is to be maximised in α_i and β_j under the constraints

$$\alpha_i, \beta_j \geq 0, \sum_i \alpha_i = \sum_j \beta_j = 1.$$

The Lagrangian now is

$$\mathcal{L} = \sum_i n_{i+} \ln \alpha_i + \sum_j n_{+j} \ln \beta_j - \lambda \left(\sum_i \alpha_i - 1 \right) - \mu \left(\sum_j \beta_j - 1 \right)$$

(with the term $\left[\ln(n!) - \sum \ln(n_{ij}!) \right]$ again omitted). The stationarity condition

$$\frac{\partial}{\partial \alpha_i} \mathcal{L} = \frac{\partial}{\partial \beta_j} \mathcal{L} = 0, \ 1 \leq i \leq r, \ 1 \leq j \leq c,$$

yields that $\widehat{\alpha}_i = n_{i+}/\lambda$ and $\widehat{\beta}_j = n_{+j}/\mu$. Adjusting the constraints gives that $\lambda = \mu = n$.
    Then the GLR

$$\Lambda_{H_1:H_0} = \prod_{i,j} \left(\frac{n_{ij}}{n} \right)^{n_{ij}} \bigg/ \prod_i \left(\frac{n_{i+}}{n} \right)^{n_{i+}} \prod_j \left(\frac{n_{+j}}{n} \right)^{n_{+j}},$$

and the statistic $2\ln\Lambda_{H_1:H_0}$ equals

$$2\sum_{i,j}n_{ij}\ln\frac{n_{ij}}{n}-2\sum_{i}n_{i+}\ln\frac{n_{i+}}{n}-2\sum_{j}n_{+j}\ln\frac{n_{+j}}{n}=2\sum_{i,j}n_{ij}\ln\frac{n_{ij}}{e_{ij}},\tag{4.44}$$

where $e_{ij}=\left(n_{i+}n_{+j}\right)/n$. Writing $n_{ij}=e_{ij}+\delta_{ij}$ and expanding the logarithm, the RHS is

$$2\sum_{i,j}(e_{ij}+\delta_{ij})\ln\left(1+\frac{\delta_{ij}}{e_{ij}}\right)=2\sum_{i,j}(e_{ij}+\delta_{ij})\left(\frac{\delta_{ij}}{e_{ij}}-\frac{1}{2}\frac{\delta_{ij}^2}{e_{ij}^2}+\cdots\right),$$

which is

$$\approx\sum_{i,j}\frac{(\delta_{ij})^2}{e_{ij}}=\sum_{i=1}^{r}\sum_{j=1}^{c}\frac{(n_{ij}-e_{ij})^2}{e_{ij}},$$

by the same approximation as before.

Therefore, we can repeat the general GLRT routine: form the statistic

$$2\ln\Lambda=\sum_{i=1}^{r}\sum_{j=1}^{c}\frac{(n_{ij}-e_{ij})^2}{e_{ij}},\tag{4.45}$$

and reject H_0 at level α when the value of the statistic exceeds $h^+_{(r-1)(c-1)}(\alpha)$, the upper α point of $\chi^2_{(r-1)(c-1)}$.

Contingency tables also give rise to another model where you fix not only n_{++}, the total number of observations, but also some margins, for instance, all (or a part) of row sums n_{i+}. Then the random counts N_{ij} in the ith row with fixed n_{i+} become distributed multinomially with parameters $(n_{i+};p_{i1},\ldots,p_{ic})$, independently of the rest. The null hypothesis here is that $p_{ij}=p_j$ does not depend on the row label i, $\forall\,j=1,\ldots,c$. The alternative is that p_{ij} are unrestricted but $p_{i+}=1$, $i=1,\ldots,r$. This situation is referred to as testing homogeneity.

Example 4.16 In an experiment 150 patients were allocated to three groups of 45, 45 and 60 patients each. Two groups were given a new drug at different dosage levels and the third group received a placebo. The responses are in Table 4.3.

Table 4.3.

	Improved	No difference	Worse
Placebo	16	20	9
Half dose	17	18	10
Full dose	26	20	14

We state H_0 as

the probabilities p_{Improved}, $p_{\text{No difference}}$ and p_{Worse} are the same for all three patient groups, and H_1 as

these probabilities may vary from one group to another.

Then under H_1 the likelihood is

$$f_{(N_{ij})}\big((n_{ij})|(p_{ij})\big) = \prod_{i=1}^{r} \frac{n_{i+}!}{n_{i1}!\ldots n_{ic}!} p_{i1}^{n_{i1}}\cdots p_{ic}^{n_{ic}},$$

and the LL is

$$\ell\big((n_{ij}); (p_{ij})\big) = \sum_{i,j} n_{ij}\ln p_{ij} + (\text{terms not depending on } p_{ij}).$$

Here, the MLE $\widehat{p}_{ij}$ of p_{ij} is n_{ij}/n_{i+}. Similarly, under H_0 the LL is:

$$\ell\big((n_{ij}); (p_{ij})\big) = \sum_{j} n_{+j}\ln p_j + (\text{terms not depending on } p_j),$$

and the MLE $\widehat{p}_j$ of p_j equals n_{+j}/n_{++}. Then, as before, the GLR is

$$2\ln\Lambda_{H_1;H_0} = 2\sum_{i,j} n_{ij}\ln \frac{\widehat{p}_{ij}}{\widehat{p}_j} = 2\sum_{i,j} n_{ij}\ln \frac{n_{ij}}{e_{ij}} \approx \sum_{i,j} \frac{(n_{ij}-e_{ij})^2}{e_{ij}},$$

with $e_{ij} = (n_{i+}n_{+j})/n_{++}$. The number of degrees of freedom is equal to $(r-1)(c-1)$, as $|\Theta_1| = r(c-1)$ $(c-1$ independent variables p_{ij} in each of r rows) and $|\Theta_0| = c-1$ (the variables are constant along the columns).

In the example under consideration: $r=3$, $c=3$, $(r-1)(c-1)=4$, $h_4^+(0.05)=9.488$. The array of data is shown in Table 4.4, which yields the array of expected values shown in Table 4.5.

Table 4.4.

n_{ij}	Improved	No difference	Worse	
Placebo	16	20	9	45
Half dose	17	18	10	45
Full dose	26	20	14	60
	59	58	33	150

Table 4.5.

e_{ij}	Improved	No difference	Worse
Placebo	17.7	17.4	9.9
Half dose	17.7	17.4	9.9
Full dose	23.6	23.2	13.2

Thus $\Lambda_{H_1;H_0}$ is calculated as

$$\frac{(2.4)^2}{23.6} + \frac{(1.7)^2}{17.7} + \frac{(0.7)^2}{17.7} + \frac{(2.6)^2}{17.4} + \frac{(0.6)^2}{17.4}$$
$$+ \frac{(3.2)^2}{23.2} + \frac{(0.9)^2}{9.9} + \frac{(0.1)^2}{9.9} + \frac{(0.8)^2}{13.2} = 1.41692.$$

This value is < 9.488, hence insignificant at the 5% level.   ∎

We finish this section with a short discussion of an often observed *Simpson paradox*. It demonstrates that contingency tables are delicate structures and can be damaged when one pools data. However, there is nothing mystic about it. Consider the following example. A new complex treatment has been made available, to cure a potentially lethal illness. Not surprisingly, doctors decided to use it primarily in more serious cases. Consequently, the new data cover many more of these cases as opposed to the old data spread evenly across a wider range of patients. This may lead to a deceptive picture (see, e.g., Table 4.6).

Table 4.6.

	Didn't recover	Recovered	Recovery %
Previous treatment	7500	5870	43.9
New treatment	12520	1450	10.38

You may think that the new treatment is four times worse than the old one. However, the point is that the new treatment is applied considerably more often than the old one in hospitals where serious cases are usually dealt with. On the other hand, in clinics where cases are typically less serious the new treatment is rare. The corresponding data are shown in Tables 4.7 and 4.8.

It is now evident that the new treatment is better in both categories.

Table 4.7. *Hospitals*

	Didn't recover	Recovered	Recovery %
Previous treatment	1100	70	5.98
New treatment	12500	1200	8.76

Table 4.8. *Clinics*

	Didn't recover	Recovered	Recovery %
Previous treatment	6400	5800	47.54
New treatment	20	250	92.60

In general, the method of contingency tables is not free of logical difficulties. Suppose you have the data in Table 4.9 coming from 100 two-child families.

Table 4.9.

		1st Child		
		Boy	Girl	
2nd Child	Boy	30	20	50
	Girl	20	30	50
		50	50	100

Consider two null hypotheses

H_0^1: The probability of a boy is $1/2$ and the genders of the children are independent.
H_0^2: The genders of the children are independent.

Each of these hypotheses is tested against the alternative: the probabilities are unrestricted (which means that p_{BB}, p_{BG}, p_{GB}, p_{GG} only obey $p_{..} \geq 0$ and $p_{BB} + p_{BG} + p_{GB} + p_{GG} = 1$; this includes possible dependence). The value of the Pearson statistic equals

$$\frac{(30-25)^2}{25} + \frac{(20-25)^2}{25} + \frac{(20-25)^2}{25} + \frac{(30-25)^2}{25} = 4.$$

The number of degrees of freedom, for H_0^1 equals 3, with the 5%-quantile 7.815. For H_0^2, the number of degrees of freedom is 1; the same percentile equals 3.841.

We see that at significance level 5%, there is no evidence to reject H_0^1 but strong evidence to reject H_0^2, although logically H_0^1 implies H_0^2. We thank A. Hawkes for this example (see [Haw]).

The contingency table GLRT appeared in MT-IB 1998-412E.

4.7   Testing normal distributions, 2: non-homogeneous samples

> Variance is what any two statisticians are at.
> (From the series *'Why they are misunderstood'*.)

A typical situation here is when we have several (in the simplest case, two) samples coming from normal distributions with parameters that may vary from one sample to another. The task then is to test that the value of a given parameter is the same for all groups.

Two-sample normal distribution tests   Here we have two independent samples,

$$\mathbf{X} = \begin{pmatrix} X_1 \\ \vdots \\ X_m \end{pmatrix} \quad \text{and} \quad \mathbf{Y} = \begin{pmatrix} Y_1 \\ \vdots \\ Y_n \end{pmatrix}$$

where the X_i are IID $N(\mu_1, \sigma_1^2)$ and the Y_j are IID $N(\mu_2, \sigma_2^2)$.

(a) *Testing equality of means, common known variance* In this model one assumes that $\sigma_1^2 = \sigma_2^2 = \sigma^2$ is known and test H_0: $\mu_1 = \mu_2$ against H_1: μ_1, μ_2 unrestricted. One uses the GLRT (which works well). Under H_1, the likelihood $f_{\mathbf{X}}(\mathbf{x}; \mu_1, \sigma^2) f_{\mathbf{Y}}(\mathbf{y}; \mu_2, \sigma^2)$ is

$$\frac{1}{\left(\sqrt{2\pi}\sigma\right)^{m+n}} \exp\left[-\frac{1}{2\sigma^2}\sum_i (x_i - \mu_1)^2 - \frac{1}{2\sigma^2}\sum_j (y_j - \mu_2)^2\right]$$

$$= \frac{1}{\left(\sqrt{2\pi}\sigma\right)^{m+n}} \exp\left[-\frac{1}{2\sigma^2}\left(S_{xx} + m(\bar{x} - \mu_1)^2 + S_{yy} + n(\bar{y} - \mu_2)^2\right)\right]$$

and is maximised at

$$\widehat{\mu}_1 = \bar{x} \text{ and } \widehat{\mu}_2 = \bar{y} \text{ where } \bar{x} = \frac{1}{m}\sum_i x_i, \ \bar{y} = \frac{1}{n}\sum_j y_j.$$

Under H_0, the likelihood $f_{\mathbf{X}}(\mathbf{x}; \mu, \sigma^2) f_{\mathbf{Y}}(\mathbf{y}; \mu, \sigma^2)$ equals

$$\frac{1}{\left(\sqrt{2\pi}\sigma\right)^{m+n}} \exp\left\{-\frac{1}{2\sigma^2}\left[S_{xx} + m(\bar{x} - \mu)^2 + S_{yy} + n(\bar{y} - \mu)^2\right]\right\}$$

and is maximised at

$$\widehat{\mu} = \frac{m\bar{x} + n\bar{y}}{m + n}.$$

Then the GLR

$$\Lambda_{H_1;H_0} = \exp\left[\frac{1}{2\sigma^2}\frac{mn(\bar{x} - \bar{y})^2}{m + n}\right],$$

and we reject H_0 when $|\bar{x} - \bar{y}|$ is large.
Under H_0,

$$\bar{X} - \bar{Y} \sim N\left(0, \sigma^2\left(\frac{1}{m} + \frac{1}{n}\right)\right).$$

Hence, under H_0 the statistic

$$Z = \frac{\bar{X} - \bar{Y}}{\sigma\sqrt{1/m + 1/n}} \sim N(0, 1),$$

and we reject H_0 at level α when the value of $|Z|$ is $> z_+(\alpha/2) = \Phi^{-1}(1 - \alpha/2)$. The test is called the normal z-test. Note that Wilks Theorem is exact here.

In what follows $z_+(\gamma)$ denotes the upper γ point (quantile) of the normal $N(0,1)$ distribution, i.e. the value for which

$$\frac{1}{\sqrt{2\pi}} \int_{z_+(\gamma)}^{\infty} e^{-x^2/2} dx = \gamma.$$

As was noted, it is given by $\Phi^{-1}(1 - \gamma)$, $0 < \gamma < 1$.

(b) *Testing equality of means, common unknown variance* Here the assumption is that $\sigma_1^2 = \sigma_2^2 = \sigma^2$ is unknown: it gives a single nuisance parameter. Again, H_0: $\mu_1 = \mu_2$ and H_1: μ_1, μ_2 unrestricted. As we will see, the result will be a t-test.

Under H_1 we have to maximise

$$\frac{1}{\left(\sqrt{2\pi}\sigma\right)^{m+n}} \exp\left\{-\frac{1}{2\sigma^2}\left[S_{xx} + m(\bar{x} - \mu_1)^2 + S_{yy} + n(\bar{y} - \mu_2)^2\right]\right\},$$

in μ_1, μ_2 and σ^2. As before, $\hat{\mu}_1 = \bar{x}$, $\hat{\mu}_2 = \bar{y}$. The MLE for σ^2 is

$$\frac{S_{xx} + S_{yy}}{m+n} \quad \text{where} \quad S_{xx} = \sum_i (x_i - \bar{x})^2, \ S_{yy} = \sum_j (y_j - \bar{y})^2.$$

(The corresponding calculation is similar to the one producing $\hat{\sigma}^2 = S_{xx}/n$ in the single-sample case.) Hence, under H_1,

$$\max\left[f(\mathbf{x}; \mu_1, \sigma^2)f(\mathbf{y}; \mu_2, \sigma^2): \mu_1, \mu_2 \in \mathbb{R}, \ \sigma^2 > 0\right]$$

$$= \left(2\pi \frac{S_{xx} + S_{yy}}{m+n}\right)^{-(m+n)/2} e^{-(m+n)/2}.$$

Under H_0, the MLE for μ is, as before,

$$\hat{\mu} = \frac{m\bar{x} + n\bar{y}}{m+n},$$

and the MLE for σ^2 is

$$\hat{\sigma}^2 = \frac{1}{m+n}\left[\sum_i (x_i - \hat{\mu})^2 + \sum_j (y_j - \hat{\mu})^2\right]$$

$$= \frac{1}{m+n}\left[S_{xx} + S_{yy} + \frac{mn}{m+n}(\bar{x} - \bar{y})^2\right].$$

Hence, under H_0,

$$\max\left[f(\mathbf{x}; \mu, \sigma^2)f(\mathbf{y}; \mu, \sigma^2): \mu \in \mathbb{R}, \ \sigma^2 > 0\right]$$

$$= \left\{\frac{2\pi}{m+n}\left[S_{xx} + S_{yy} + \frac{mn}{m+n}(\bar{x} - \bar{y})^2\right]\right\}^{-(m+n)/2} e^{-(m+n)/2}.$$

This yields the following expression for the GLR:

$$\Lambda_{H_1;H_0} = \left(\frac{(m+n)(S_{xx} + S_{yy}) + mn(\bar{x} - \bar{y})^2}{(m+n)(S_{xx} + S_{yy})}\right)^{(m+n)/2}$$

$$= \left(1 + \frac{mn(\bar{x} - \bar{y})^2}{(m+n)(S_{xx} + S_{yy})}\right)^{(m+n)/2}.$$

Hence, in the GLRT, we reject H_0 when

$$\frac{|\bar{x} - \bar{y}|}{\sqrt{(S_{xx} + S_{yy})(1/m + 1/n)}}$$

is large. It is convenient to multiply the last expression by $\sqrt{n+m-2}$. This produces the following statistic:

$$T = \frac{\overline{X} - \overline{Y}}{\sqrt{\frac{S_{XX}+S_{YY}}{m+n-2}\left(\frac{1}{m} + \frac{1}{n}\right)}} \sim t_{n+m-2}.$$

In fact, under H_0,

$$\frac{\overline{X} - \overline{Y}}{\sigma\sqrt{1/m + 1/n}} \sim N(0, 1), \quad \frac{1}{\sigma^2}S_{XX} \sim \chi^2_{m-1} \text{ and } \frac{1}{\sigma^2}S_{YY} \sim \chi^2_{n-1},$$

independently. Thus

$$\frac{S_{XX} + S_{YY}}{\sigma^2} \sim \chi^2_{m+n-2}.$$

The Wilks Theorem is of course valid in the limit $m, n \to \infty$, but is not needed here.

The t-test for equality of means in two normal samples was set in MT-IB 1995-103G (i).

Example 4.17 Seeds of a particular variety of plant were randomly assigned to either a nutritionally rich environment (the treatment) or standard conditions (the control). After a predetermined period, all plants were harvested, dried and weighed, with weights in grams shown in Table 4.10.

Table 4.10.

Control	4.17	5.58	5.18	6.11	4.50	4.61	5.17	4.53	5.33	5.14
Treatment	4.81	4.17	4.41	3.59	5.87	3.83	6.03	4.89	4.32	4.69

Here control observations $X_1, \ldots, X_m$ are IID $N(\mu_X, \sigma^2)$, and treatment observations $Y_1, \ldots, Y_n$ are IID $N(\mu_Y, \sigma^2)$. One tests

$$H_0: \mu_X = \mu_Y \text{ against } H_1: \mu_X, \mu_Y \text{ unrestricted.}$$

We have

$$m = n = 10, \bar{x} = 5.032, S_{xx} = 3.060, \bar{y} = 4.661, \ S_{yy} = 5.669.$$

Then

$$\widehat{\sigma}^2 = \frac{S_{xx} + S_{yy}}{m + n - 2} = 0.485$$

and

$$|t| = \frac{|\bar{x} - \bar{y}|}{\sqrt{\widehat{\sigma}^2(1/m + 1/n)}} = 1.19.$$

With $t_{18}(0.025) = 2.101$, we do not reject H_0 at level 95%, concluding that there is no difference between the mean weights due to environmental conditions.

Now suppose that the value of the variance σ^2 is known to be 0.480 and is not estimated from the sample. Calculating

$$z = |\bar{x} - \bar{y}| \left/ \sigma \sqrt{\frac{1}{m} + \frac{1}{n}} \right.$$

yields the value $z = 1.974$. Since $\Phi(z) = 0.97558$, which is just over 0.975, we formally have to reject H_0 (although, admittedly, the data are not convincing). ∎

(c) *Testing equality of variances, known means* First, consider the case where μ_1 and μ_2 are known (and not necessarily equal). The null hypothesis is H_0: $\sigma_1^2 = \sigma_2^2$ and the alternative H_1: σ_1^2, σ_2^2 unrestricted. Under H_0, the likelihood $f(\mathbf{x}; \mu_1, \sigma^2) f(\mathbf{y}; \mu_2, \sigma^2)$ is maximised at $\hat{\sigma}^2 = (\Sigma_{xx}^2 + \Sigma_{yy}^2)/(m + n)$ and attains the value

$$\left[\frac{2\pi (\Sigma_{xx}^2 + \Sigma_{yy}^2)}{m + n} \right]^{-(m+n)/2} e^{-(m+n)/2}.$$

Here $\Sigma_{xx}^2 = \sum_i (x_i - \mu_1)^2$ and $\Sigma_{yy}^2 = \sum_j (y_j - \mu_2)^2$.

Under H_1, the likelihood $f(\mathbf{x}; \mu_1, \sigma^2) f(\mathbf{y}; \mu_2, \sigma^2)$ is maximised at $\hat{\sigma}_1^2 = \Sigma_{xx}^2/m$, $\hat{\sigma}_2^2 = \Sigma_{yy}^2/n$ and attains the value

$$\left(\frac{2\pi\Sigma_{xx}^2}{m} \right)^{-m/2} \left(\frac{2\pi\Sigma_{yy}^2}{n} \right)^{-n/2} e^{-(m+n)/2}.$$

Then the GLR

$$\Lambda_{H_1 : H_0} = \frac{m^{m/2} n^{n/2}}{(m+n)^{(m+n)/2}} \left(\frac{\Sigma_{xx}^2 + \Sigma_{yy}^2}{\Sigma_{xx}^2} \right)^{m/2} \left(\frac{\Sigma_{xx}^2 + \Sigma_{yy}^2}{\Sigma_{yy}^2} \right)^{n/2}$$

$$\propto \left(1 + \frac{\Sigma_{yy}^2}{\Sigma_{xx}^2} \right)^{m/2} \left(1 + \frac{\Sigma_{xx}^2}{\Sigma_{yy}^2} \right)^{n/2}.$$

The last expression is large when $\Sigma_{yy}^2/\Sigma_{xx}^2$ is large or small. But under H_0,

$$\frac{1}{\sigma^2} \Sigma_{XX}^2 \sim \chi_m^2 \text{ and } \frac{1}{\sigma^2} \Sigma_{YY}^2 \sim \chi_n^2, \text{ independently.}$$

Thus, under H_0, the ratio $\Sigma_{YY}^2/\Sigma_{XX}^2 \sim F_{n,m}$. Hence, we reject H_0 at level α when the value of $\Sigma_{yy}^2/\Sigma_{xx}^2$ is either greater than $\varphi_{n,m}^+(\alpha/2)$ or less than $\varphi_{n,m}^-(\alpha/2)$. Here, and below, $\varphi_{n,m}^+(\gamma)$ denotes the upper and $\varphi_{n,m}^-(\gamma)$ the lower γ point (quantile) of the Fisher distribution $F_{n,m}$, i.e. the value a for which

$$\int_a^\infty f_{F_{n,m}}(x) dx = \gamma \quad \text{or} \quad \int_0^a f_{F_{n,m}}(x) dx = \gamma,$$

with $\varphi_{n,m}^-(\gamma) = \varphi_{n,m}^+(1 - \gamma), 0 < \gamma < 1$.

Again, the strict application of the GLRT leads to a slightly different critical region (unequally tailed test).

(d) *Testing equality of variances, unknown means* Now assume that μ_1 and μ_2 are unknown nuisance parameters and test H_0: $\sigma_1^2 = \sigma_2^2$ against H_1: σ_1^2, σ_2^2 unrestricted. Under H_0, the likelihood $f(\mathbf{x}; \mu_1, \sigma^2)f(\mathbf{y}; \mu_2, \sigma^2)$ is maximised at

$$\widehat{\mu}_1 = \overline{x}, \ \widehat{\mu}_2 = \overline{y}, \ \widehat{\sigma}^2 = \frac{1}{m+n}(S_{xx} + S_{yy}),$$

where it attains the value

$$\left(2\pi \frac{S_{xx} + S_{yy}}{m+n}\right)^{-(m+n)/2}.$$

Under H_1, the likelihood is $f(\mathbf{x}; \mu_1, \sigma_1^2)f(\mathbf{y}; \mu_2, \sigma_2^2)$. Its maximum value is attained at

$$\widehat{\mu}_1 = \overline{x}, \ \widehat{\mu}_2 = \overline{y}, \ \widehat{\sigma}_1^2 = \frac{1}{m}S_{xx}, \ \widehat{\sigma}_2^2 = \frac{1}{n}S_{yy},$$

and equals

$$\left(\frac{1}{2\pi S_{xx}/m}\right)^{m/2}\left(\frac{1}{2\pi S_{yy}/n}\right)^{n/2} e^{-(m+n)/2}.$$

Then

$$\Lambda_{H_1;H_0} = \left(\frac{S_{xx} + S_{yy}}{S_{xx}}\right)^{m/2}\left(\frac{S_{xx} + S_{yy}}{S_{yy}}\right)^{n/2}\frac{m^{m/2}n^{n/2}}{(m+n)^{(m+n)/2}},$$

and we reject H_0 when

$$\left(1 + \frac{S_{xx}}{S_{yy}}\right)^{n/2}\left(1 + \frac{S_{yy}}{S_{xx}}\right)^{m/2}$$

is large.

But, as follows from Example 3.4 (see equation (3.7)),

$$\frac{S_{XX}/(m-1)}{S_{YY}/(n-1)} \sim \mathrm{F}_{m-1,n-1}.$$

So, at level α we reject H_0 in the 'upper tail' test when

$$\frac{(n-1)S_{xx}}{(m-1)S_{yy}} > \varphi^+_{m-1,n-1}(\alpha)$$

and in the 'lower tail' test when

$$\frac{(m-1)S_{yy}}{(n-1)S_{xx}} > \varphi^+_{n-1,m-1}(\alpha).$$

These tests are determined by the upper α points of the corresponding F-distributions. We also can use a two-tailed test where H_0 is rejected if, say,

$$\frac{(n-1)S_{xx}}{(m-1)S_{yy}} > \varphi^+_{m-1,n-1}(\alpha/2) \quad \text{or} \quad \frac{(m-1)S_{yy}}{(n-1)S_{xx}} > \varphi^+_{n-1,m-1}(\alpha/2).$$

The particular choice of the critical region can be motivated by the form of the graph of the PDF $f_{F_{m,n}}$ for given values of m and n. See Figure 3.3.

The F-test for equality of variances was proposed by Fisher. The statistic $(n-1)S_{xx}/\big((m-1)S_{yy}\big)$ is called the F-*statistic*.

Remark   We can repeat the routine for the situation where alternative H_1, instead of being $\sigma_1^2 \ne \sigma_2^2$, is $\sigma_1^2 \ge \sigma_2^2$ (in which case we speak of *comparison of variances*). Then the GLR $\Lambda_{H_1:H_0}$ is taken to be

$$\begin{cases} \left(\frac{S_{xx}+S_{yy}}{S_{xx}}\right)^{m/2}\left(\frac{S_{xx}+S_{yy}}{S_{yy}}\right)^{n/2}\dfrac{m^{m/2}n^{n/2}}{(m+n)^{(m+n)/2}}, & \text{if } \dfrac{1}{m}S_{xx} > \dfrac{1}{n}S_{yy}, \\ 1, \text{ if } \dfrac{1}{m}S_{xx} < \dfrac{1}{n}S_{yy}, \end{cases}$$

and at level α we reject H_0 when

$$\frac{(n-1)S_{xx}}{(m-1)S_{yy}} > \varphi^+_{m-1,n-1}(\alpha).$$

A similar modification can be made in other cases considered above.

The F-test is useful when we have $\mathbf{X}$ from $\mathrm{Exp}\,(\lambda)$ and $\mathbf{Y}$ from $\mathrm{Exp}\,(\mu)$ and test H_0: $\lambda = \mu$. A relevant Tripos question is MT-IB 1997-112G.

In the context of a GLRT, the F-test also appears when we test H_0: $\mu_1 = \cdots = \mu_n$, where $X_i \sim \mathrm{N}(\mu_i, \sigma^2)$ with the same variance σ^2. See MT-IB 1999-403D, 1995-403G (ii). See also 1994-203F (ii,d).

Example 4.18   To determine the concentration of nickel in solution, one can use an alcohol method or an aqueous method. One wants to test whether the variability of the alcohol method is greater than that of the aqueous method. The observed nickel concentrations (in tenths of a per cent) are shown in Table 4.11.

Table 4.11.

Alcohol method	4.28	4.32	4.25	4.29	4.31	4.35	4.32	4.33
	4.28	4.27	4.38	4.28				
Aqueous method	4.27	4.32	4.29	4.30	4.31	4.30	4.30	4.32
	4.28	4.32						

The model is that the values $X_1, \ldots, X_{12}$ obtained by the alcohol method and $Y_1, \ldots, Y_{10}$ obtained by the aqueous method are independent, and $X_i \sim \mathrm{N}(\mu_X, \sigma_X^2)$, $Y_j \sim \mathrm{N}\,(\mu_Y, \sigma_Y^2)$. The null hypothesis is H_0: $\sigma_X^2 = \sigma_Y^2$ against the alternative H_1: $\sigma_X^2 \ge \sigma_Y^2$.

Here

$$m = 12, \ \bar{x} = 4.311, \ S_{xx} = 0.01189,$$

and

$$n = 10, \ \bar{y} = 4.301, \ S_{yy} = 0.00269.$$

This gives $(S_{xx}/11) \big/ (S_{yy}/9) = 3.617$. From the χ^2 percentage tables, $\varphi^+_{11,9}(0.05) = 3.10$ and $\varphi^+_{11,9}(0.01) = 5.18$. Thus, we reject H_0 at the 5% level but accept at the 1% level. This provides some evidence that $\sigma^2_X > \sigma^2_Y$ but further investigation is needed to reach a higher degree of certainty. For instance, as $\varphi^+_{11,9}(0.025) \approx 3.92$, H_0 should be accepted at the 2.5% level.   ∎

Non-homogeneous normal samples

Here **X** has $X_i \sim N(\mu_i, \sigma^2_i)$, with parameters varying from one RV to another.
   (a) *Testing equality of means, known variances* Assuming that $\sigma^2_1, \ldots, \sigma^2_n$ are known (not necessarily equal), we want to test

$$H_0: \ \mu_1 = \cdots = \mu_n \text{ against } H_1: \ \mu_i \in \mathbb{R} \text{ are unrestricted.}$$

Again use the GLRT. Under H_1, the likelihood is

$$f(\mathbf{x}; \mu_1, \ldots, \mu_n; \sigma^2_1, \ldots, \sigma^2_n) = \left(\frac{1}{2\pi}\right)^{n/2} \frac{1}{\prod_i \sigma_i} \exp\left[-\frac{1}{2}\sum_i (x_i - \mu_i)^2 / \sigma^2_i\right],$$

maximised at $\widehat{\mu}_i = x_i$, with the maximal value $(2\pi)^{-n/2} \big/ (\prod_i \sigma_i)$.
   Under H_0, the likelihood

$$f(\mathbf{x}; \mu, \sigma^2_1, \ldots, \sigma^2_n) = \left(\frac{1}{2\pi}\right)^{n/2} \frac{1}{\prod_i \sigma_i} \exp\left[-\frac{1}{2}\sum_i (x_i - \mu)^2 / \sigma^2_i\right]$$

attains the maximal value

$$\left(\frac{1}{2\pi}\right)^{n/2} \prod_i \frac{1}{\sigma_i} \exp\left[-\frac{1}{2}\sum_i (x_i - \widehat{\mu})^2 / \sigma^2_i\right]$$

at the weighted mean

$$\widehat{\mu} = \widehat{\mu}(\mathbf{x}) = \sum_i x_i / \sigma^2_i \bigg/ \sum_i 1/\sigma^2_i.$$

Then the GLR

$$\Lambda_{H_1; H_0} = \exp\left[\frac{1}{2}\sum_i (x_i - \widehat{\mu})^2 / \sigma^2_i\right].$$

We reject H_0 when the sum $\sum_i (x_i - \widehat{\mu})^2/\sigma_i^2$ is large. More precisely,

$$2 \ln \Lambda_{H_1;H_0} = \sum_i \frac{(X_i - \widehat{\mu}(\mathbf{X}))^2}{\sigma_i^2} \sim \chi_{n-1}^2$$

(another case where the Wilks Theorem is exact). So, H_0 is rejected at level α when $\sum_i (x_i - \widehat{\mu}(\mathbf{x}))^2/\sigma_i^2$ exceeds $h_{n-1}^+(\alpha)$, the upper α point of χ_{n-1}^2.

(b) *Testing equality of means, unknown variances* Consider the same null hypothesis H_0: $\mu_1 = \cdots = \mu_n$ and alternative H_1: μ_i unrestricted, in the case of unknown variances (equal or not). Then, under H_1, we have to maximise the likelihood in σ_i^2 as well. That is in addition to $\widehat{\mu}_i = x_i$, we have to set $\widehat{\sigma}_i = 0$, which is not feasible. This is an example where the GLR routine is not applicable.

However, let us assume that normal sample $\mathbf{X}$ has been divided into groups so that at least one of them contains more than a single RV, and it is known that within a given group the means are the same. In addition, let the variance be the same for all RVs X_i. Then one uses a routine called *ANOVA* (analysis of variance) to test the null hypothesis that all means are the same. In a sense, this is a generalisation of the test in subsection (a) above.

So, assume that we have k groups of n_i observations in group i, with $n = n_1 + \cdots + n_k$. Set

$$X_{ij} = \mu_i + \epsilon_{ij}, \quad j = 1, \ldots, n_i, \quad i = 1, \ldots, k.$$

We assume that $\mu_1, \ldots, \mu_k$ are fixed unknown constants and $\epsilon_{ij} \sim N(0, \sigma^2)$, independently. The variance σ^2 is unknown. The null hypothesis is

$$H_0: \mu_1 = \cdots = \mu_k = \mu,$$

and the alternative H_1 is that the μ_i are unrestricted.

Under H_1, the likelihood

$$\frac{1}{(2\pi\sigma^2)^{n/2}} \exp\left[-\frac{1}{2\sigma^2} \sum_{i=1}^{k} \sum_{j=1}^{n_i} (x_{ij} - \mu_i)^2 \right]$$

attains its maximum at

$$\widehat{\mu}_i = \bar{x}_{i+}, \quad \widehat{\sigma}^2 = \frac{1}{n} \sum_i \sum_j \left(x_{ij} - \bar{x}_{i+}\right)^2 \text{ where } \bar{x}_{i+} = \frac{1}{n_i} \sum_j x_{ij}.$$

The maximum value equals

$$\left[2\pi \frac{1}{n} \sum_i \sum_j (x_{ij} - \bar{x}_{i+})^2 \right]^{-n/2} e^{-n/2}.$$

It is convenient to denote

$$s_1 = \sum_i \sum_j (x_{ij} - \bar{x}_{i+})^2;$$

this sum is called the *within group sum of squares*. If we assume that at least one among numbers $n_1, \ldots, n_k$, say n_i, is > 1, then the corresponding term $\sum_j (X_{ij} - \overline{X}_{i+})^2$ is > 0 with probability 1. Then the random value $S_1 = \sum_{i,j} (X_{ij} - \overline{X}_{i+})^2$ of the within group sum of squares is also > 0 with probability 1.

Under H_0, the likelihood

$$\frac{1}{(2\pi\sigma^2)^{n/2}} \exp\left[-\frac{1}{2\sigma^2} \sum_{i=1}^{k} \sum_{j=1}^{n_i} (x_{ij} - \mu)^2 \right]$$

is maximised at

$$\hat{\mu} = \bar{x}_{++}, \quad \hat{\sigma}^2 = \frac{1}{n} \sum_i \sum_j (x_{ij} - \bar{x}_{++})^2$$

and attains the value

$$\left[2\pi \frac{1}{n} \sum_i \sum_j (x_{ij} - x_{++})^2 \right]^{-n/2} e^{-n/2},$$

where

$$\bar{x}_{++} = \frac{1}{n} \sum_{i,j} x_{ij}.$$

We write

$$s_0 = \sum_{i,j} (x_{ij} - \bar{x}_{++})^2;$$

this is the *total sum of squares*.

The GLR is

$$\Lambda_{H_1;H_0} = \left(\frac{s_0}{s_1} \right)^{n/2}.$$

Write s_0 in the form

$$\sum_{i=1}^{k} \sum_{j=1}^{n_i} \left(x_{ij} - \bar{x}_{i+} + \bar{x}_{i+} - \bar{x}_{++} \right)^2 = \sum_{i,j} (x_{ij} - \bar{x}_{i+})^2 + \sum_i n_i (\bar{x}_{i+} - \bar{x}_{++})^2$$

(the cross-term sum vanishes as $\sum_j (x_{ij} - \bar{x}_{i+}) = 0 \ \forall \ i = 1, \ldots, k$). Then write

$$s_2 = \sum_i n_i (\bar{x}_{i+} - \bar{x}_{++})^2;$$

this sum is called the *between groups sum of squares*.

Hence, $s_0 = s_1 + s_2$, and

$$\Lambda_{H_1;H_0} = \left(1 + \frac{s_2}{s_1} \right)^{n/2}.$$

So, we reject H_0 when s_2/s_1 is large, or equivalently,

$$\frac{s_2/(k-1)}{s_1/(n-k)}$$

is large.

Now the analysis of the distributions of X_{ij}, $\overline{X}_{i+}$ and $\overline{X}_{++}$ is broken up into three steps. First, $\forall\ i$, under both H_0 and H_1:

$$\frac{1}{\sigma^2}\sum_{j=1}^{n_i}\left(X_{ij}-\overline{X}_{i+}\right)^2 \sim \chi^2_{n_i-1},$$

according to the Fisher Theorem (see Section 3.5). Then, summing independent χ^2 distributed RVs:

$$\frac{S_1}{\sigma^2} = \frac{1}{\sigma^2}\sum_i\sum_j\left(X_{ij}-\overline{X}_{i+}\right)^2 \sim \chi^2_{n-k}.$$

Next, $\forall\ i$, again under both H_0 and H_1,

$$\sum_{j=1}^{n_i}\left(X_{ij}-\overline{X}_{i+}\right)^2 \text{ and } \overline{X}_{i+}$$

are independent; see again the Fisher Theorem. Then

$$S_1 = \sum_{i,j}\left(X_{ij}-\overline{X}_{i+}\right)^2 \text{ and } S_2 = \sum_i\left(\overline{X}_{i+}-\overline{X}_{++}\right)^2$$

are independent.

Finally, under H_0, the X_{ij} are IID $N(\mu,\sigma^2)$. Then

$$\overline{X}_{++} \sim N\left(\mu,\frac{\sigma^2}{n}\right) \text{ and } \frac{n}{\sigma^2}(\overline{X}_{++}-\mu)^2 \sim \chi^2_1.$$

Also, the $\overline{X}_{i+}$ are independent $N(\mu,\sigma^2/n_i)$ and $n_i\left(\overline{X}_{i+}-\mu\right)^2/\sigma^2 \sim \chi^2_1$. Hence,

$$\sum_i\frac{n_i}{\sigma^2}\left(\overline{X}_{i+}-\mu\right)^2 \sim \chi^2_k.$$

Moreover, $\overline{X}_{++}$ and $\overline{X}_{i+}-\overline{X}_{++}$ are independent, as they are jointly normal and with $\mathrm{Cov}\left(\overline{X}_{++},\overline{X}_{i+}-\overline{X}_{++}\right)=0$. Writing

$$\sum_i\frac{n_i}{\sigma^2}\left(\overline{X}_{i+}-\mu\right)^2 = \frac{1}{\sigma^2}\sum_i n_i\left(\overline{X}_{i+}-\overline{X}_{++}\right)^2 + \frac{n}{\sigma^2}\left(\overline{X}_{++}-\mu\right)^2,$$

we conclude that on the RHS,

$$\frac{1}{\sigma^2}\sum_i n_i\left(\overline{X}_{i+}-\overline{X}_{++}\right)^2 = \frac{S_2}{\sigma^2} \sim \chi^2_{k-1} \text{ and } \frac{n}{\sigma^2}\left(\overline{X}_{++}-\mu\right)^2 \sim \chi^2_1,$$

independently. On the other hand, write S_2 in the form

$$S_2 = \sum_i n_i \left[(\overline{X}_{i+} - \mu_i) + (\mu_i - \bar{\mu}) + (\bar{\mu} - \overline{X}_{++}) \right]^2,$$

where $\bar{\mu} = \sum_i \mu_i n_i / n$. Then a straightforward calculation shows that

$$\mathbb{E}S_2 = (k-1)\sigma^2 + \sum_i n_i(\mu_i - \bar{\mu})^2.$$

We conclude that S_2 under H_1 tends to be inflated.

All in all, we see that the statistic

$$Q = \frac{S_2/(k-1)}{S_1/(n-k)}$$

is $\sim F_{k-1,n-k}$ under H_0 and tends to be larger under H_1. Therefore, we reject H_0 at level α when the value of Q is bigger than $\varphi^+_{k-1,n-k}(\alpha)$, the upper α point of $F_{k-1,n-k}$. This is summarised in Table 4.12.

Table 4.12.

	Degrees of freedom	Sum of squares	Mean square
Between groups	$k-1$	S_2	$s_2/(k-1)$
Within groups	$n-k$	S_1	$s_1/(n-k)$
Total	$n-1$	S_0	

Example 4.19 In a psychology study of school mathematics teaching methods, 45 pupils were divided at random into 5 groups of 9. Groups A and B were taught in separate classes by the normal method, and groups C, D and E were taught together. Each day, every pupil from group C was publicly praised, every pupil from group D publicly reproved, and the members of group E ignored. At the end of the experiment, all the pupils took a test, and their results, in percentage of full marks, are shown in Table 4.13.

Table 4.13.

A (control)	34	28	48	40	48	46	32	30	48
B (control)	42	46	26	38	26	38	40	42	32
C (praised)	56	60	58	48	54	60	56	56	46
D (reproved)	38	56	52	52	38	48	48	46	44
E (ignored)	42	28	26	38	30	30	20	36	40

Psychologists are interested in whether there are significant differences between the groups. Values X_{ij} are assumed independent, with $X_{ij} \sim N(\mu_i, \sigma^2)$, $i = 1, \ldots, 5$, $j = 1, \ldots, 9$. The null hypothesis is H_0: $\mu_i \equiv \mu$, the alternative H_1: μ_i unrestricted.

Table 4.14.

	Total	$\bar{x}$	$\sum_j(x_{ij}-\bar{x}_{i+})^2$
A	354	39.3	568.0
B	330	36.7	408.0
C	494	54.9	192.9
D	422	46.9	304.9
E	290	32.2	419.6

Table 4.15.

	DF	SS	MS	F-ratio
Between groups	4	2891.48	722.869	722.869/47.33=15.273
Within groups	40	1893.3	47.33	
Total	44	4784.78		

Form the results we draw up Tables 4.14 and 4.15. The observed value of statistic Q is 15.3. As $\varphi^+_{4,40}(0.001)=5.7$, the value falls deep into the rejection region for $\alpha=0.001$. Hence, H_0 is rejected, and the conclusion is that the way the children are psychologically treated has a strong impact on their mathematics performance. ∎

The ANOVA test appeared in the question MT-IB 1995-403G.
    Statisticians are fond of curvilinear shapes and often own as a pet a large
    South American snake called ANOCOVA.
        (From the series 'Why they are misunderstood'.)

    Nowadays in any Grand Slam you will be tested by an -Ova.
        (A chat at a Wimbledon women's final.)

(c) *Testing equality of variances, known mean* Now assume that $\mu_1=\cdots=\mu_n=\mu$ is known. Let us try to test $H_0: \sigma_1^2=\cdots=\sigma_n^2$ against $H_1: \sigma_i^2>0$ are unrestricted. Under H_1, the likelihood is

$$f(\mathbf{x};\mu,\sigma_1^2,\ldots,\sigma_n^2)=\left(\frac{1}{2\pi}\right)^{n/2}\frac{1}{\prod_i\sigma_i}\exp\left[-\frac{1}{2}\sum_i(x_i-\mu)^2/\sigma_i^2\right].$$

It attains the maximal value

$$\frac{1}{(2\pi)^{n/2}[\prod_i(x_i-\mu)^2]^{1/2}}e^{-n/2}$$

at $\widehat{\sigma}_i^2 = (x_i - \mu)^2$. Under H_0, the likelihood is maximised at $\widehat{\sigma}^2 = \sum_i (x_i - \mu)^2 \big/ n$ and attains the value

$$\frac{1}{(2\pi)^{n/2} \left[\sum_i (x_i - \mu)^2 / n \right]^{n/2}} e^{-n/2}.$$

This gives the GLR

$$\Lambda_{H_1 : H_0} = \left\{ \frac{\prod_i (x_i - \mu)^2}{\left[\sum_i (x_i - \mu)^2 / n \right]^n} \right\}^{-1/2}.$$

The difficulty here is in finding the distribution of the GLR under H_0. Consequently, there is no effective test for the null hypothesis. However, the problem is very important in a number of applications, in particular, in modern financial mathematics. In financial mathematics, the variance acquired a lot of importance (and is considered as an essentially negative factor).

> The statistician's attitude to variation
> is like that of an evangelist to sin;
> he sees it everywhere to a greater or lesser extent.
> (From the series *'Why they are misunderstood'*.)

4.8 Linear regression. The least squares estimators

> Ordinary Least Squares People
> (From the series *'Movies that never made it to the Big Screen'*.)

The next topic to discuss is *linear regression*. The model here is that a random variable of interest, Y, is known to be of the form $Y = g(x) + \epsilon$, where: (i) ϵ is an RV of zero mean (for example, $\epsilon \sim N(0, \sigma^2)$ with a known or unknown variance), (ii) $g(x) = g(x, \theta)$ is a function of a given type (e.g. a linear form $\gamma + \beta x$ or an exponential $k e^{\alpha x}$ (made linear after taking logs), where some or all components of parameter θ are unknown), and (iii) x is a given value (or sometimes a random variable). We will mainly deal with a multidimensional version in which Y and ϵ are replaced with random vectors

$$\mathbf{Y} = \begin{pmatrix} Y_1 \\ \vdots \\ Y_n \end{pmatrix} \quad \text{and} \quad \boldsymbol{\epsilon} = \begin{pmatrix} \epsilon_1 \\ \vdots \\ \epsilon_n \end{pmatrix},$$

and function $g(x)$ with a vector function $\mathbf{g}(\mathbf{x})$ of a vector argument $\mathbf{x}$. The aim is to specify $g(x)$ or $\mathbf{g}(\mathbf{x})$, i.e. to infer the values of the unknown parameters.

An interesting example of how to use linear regression is related to the Hubble law of linear expansion in astronomy. In 1929, E.P. Hubble (1889–1953), an American astronomer, published an important paper reporting that the Universe is expanding (i.e. galaxies are moving away from Earth, and the more distant the galaxy the greater speed with which it is receding). The constant of proportionality which arises in this linear dependence was named the Hubble constant, and its calculation became one of the central challenges in astronomy: it would allow us to assess the age of the Universe.

To solve this problem, one uses linear regression, as data available are scarce; related measurements on galaxies are long and determined. Since 1929, there have been several rounds of meticulous calculations, and the Hubble constant has been subsequently re-estimated. Every new round of calculation so far has produced a greater age for the Universe; it will be interesting to see whether this trend continues.

We mainly focus on a *simple linear regression* in which each component Y_i of $\mathbf{Y}$ is determined by

$$Y_i = \gamma + \beta x_i + \epsilon_i, \quad i = 1, \ldots, n, \tag{4.46}$$

and $\epsilon_1, \ldots, \epsilon_n$ are IID, with $\mathbb{E}\epsilon_i = 0$, Var $\epsilon_i = \sigma^2$. Here γ and β are unknown, while $\mathbf{x} = (x_1, \ldots, x_n)$ is a given vector. RVs ϵ_i are considered to represent a 'noise' and are often called errors (of observation or measurement). Then, of course, $Y_1, \ldots, Y_n$ are independent, with $\mathbb{E}Y_i = \gamma + \beta x_i$ and Var $\epsilon = \sigma^2$. A convenient re-parametrisation of equation (4.46) is

$$Y_i = \alpha + \beta(x_i - \bar{x}) + \epsilon_i, \quad i = 1, \ldots, n, \tag{4.47}$$

with

$$\bar{x} = \frac{1}{n} \sum_i x_i, \ \alpha = \gamma + \beta\bar{x} \text{ and } \sum_i (x_i - \bar{x}) = 0.$$

In this model, we receive data $(x_1, y_1), \ldots, (x_n, y_n)$, where the values x_i are known and the values y_i are realisations of RVs Y_i. Determining values of γ and β (or equivalently, α and β) means drawing a *regression line*

$$y = \gamma + \beta x \text{ (or } y = \alpha + \beta(x - \bar{x})),$$

that is the best linear approximation for the data. In the case $\gamma = 0$ one deals with regression through the origin.

A natural idea is to consider the pair of the *least squares estimators* (LSEs) $\hat{\alpha}$ and $\hat{\beta}$ of α and β, i.e. the values minimising the sum

$$\sum [y_i - \alpha - \beta(x_i - \bar{x})]^2.$$

This sum measures the deviation of data $(x_1, y_1), \ldots, (x_n, y_n)$ from the attempted line $y = (\alpha - \beta\bar{x}) + \beta x$. By solving the stationarity equations

$$\frac{\partial}{\partial \alpha} \sum [y_i - \alpha - \beta(x_i - \bar{x})]^2 = 0, \ \frac{\partial}{\partial \beta} \sum [y_i - \alpha - \beta(x_i - \bar{x})]^2 = 0,$$

we find that the LSEs are given by

$$\hat{\alpha} = \bar{y}, \ \hat{\beta} = \frac{S_{xy}}{S_{xx}}. \tag{4.48}$$

Here

$$\bar{y}=\frac{1}{n}\sum_i y_i, \quad S_{xx}=\sum_i (x_i-\bar{x})^2,$$

$$S_{xy}=\sum_i (x_i-\bar{x})y_i=\sum_i (x_i-\bar{x})(y_i-\bar{y}),$$

(4.49)

with $\widehat{\gamma}=\widehat{\alpha}-\widehat{\beta}\bar{x}=\bar{y}-S_{xy}\bar{x}/S_{xx}$. We obviously have to assume that $S_{xx}\neq 0$, i.e. not all x_i are the same. Note that

$$\bar{y}=\widehat{\gamma}+\widehat{\beta}\bar{x},$$

i.e. $(\bar{x},\bar{y})$ lies on the regression line. Furthermore, $\widehat{\alpha}$ and $\widehat{\beta}$ are linear functions of $y_1,\ldots,y_n$.

The last remark implies a number of straightforward properties of the LSEs, which are listed in the following statements

(i) $\mathbb{E}\widehat{\alpha}=\alpha,\ \mathbb{E}\widehat{\beta}=\beta.$
(ii) $\mathrm{Var}\ \widehat{\alpha}=\sigma^2/n,\ \mathrm{Var}\ \widehat{\beta}=\sigma^2/S_{xx}.$
(iii) $\mathrm{Cov}\ (\widehat{\alpha},\widehat{\beta})=0.$
(iv) *The LSE $(\widehat{\alpha},\widehat{\beta})$ gives the best linear unbiased estimator (BLUE) of (α,β), i.e. the least variance unbiased estimator among those linear in $Y_1,\ldots,Y_n$.*

In fact, (i)

$$\mathbb{E}\widehat{\alpha}=\frac{1}{n}\sum [\alpha+\beta(x_i-\bar{x})]=\alpha$$

and

$$\mathbb{E}\widehat{\beta}=\frac{1}{S_{xx}}\sum_i (x_i-\bar{x})\mathbb{E}Y_i=\frac{1}{S_{xx}}\sum_i (x_i-\bar{x})[\alpha+\beta(x_i-\bar{x})]=\beta.$$

Next, (ii)

$$\mathrm{Var}\ \widehat{\alpha}=\frac{1}{n^2}\sum \mathrm{Var}\ Y_i=\frac{\sigma^2}{n}$$

and

$$\mathrm{Var}\ \widehat{\beta}=\frac{1}{S_{xx}^2}\sum_i (x_i-\bar{x})^2\mathrm{Var}\ Y_i=\frac{\sigma^2}{S_{xx}}.$$

Further, (iii)

$$\mathrm{Cov}\ (\widehat{\alpha},\widehat{\beta})=\frac{1}{nS_{xx}}\sum \mathrm{Cov}\ (Y_i,(x_i-\bar{x})Y_i)=\frac{1}{nS_{xx}}\sigma^2\sum_i (x_i-\bar{x})=0.$$

Finally, (iv) let $\widetilde{A}=\sum_i c_i Y_i$ and $\widetilde{B}=\sum_i d_i Y_i$ be linear unbiased estimators of α and β, respectively, with

$$\sum_i c_i[\alpha+\beta(x_i-\bar{x})]\equiv\alpha,\quad \text{and}\quad \sum_i d_i[\alpha+\beta(x_i-\bar{x})]\equiv\beta.$$

Then we must have

$$\sum_i c_i = 1, \quad \sum_i c_i(x_i - \bar{x}) = 0, \quad \sum_i d_i = 0 \quad \text{and} \quad \sum_i d_i(x_i - \bar{x}) = 1.$$

Minimising $\text{Var}\,\widetilde{A} = \sigma^2 \sum_i c_i^2$ and $\text{Var}\,\widetilde{B} = \sigma^2 \sum_i d_i^2$ under the above constraints is reduced to minimising the Lagrangians

$$\mathcal{L}_1(c_1, \ldots, c_n; \alpha) = \sigma^2 \sum_i c_i^2 - \lambda \left(\sum_i c_i - 1 \right) - \mu \sum_i c_i(x_i - \bar{x})$$

and

$$\mathcal{L}_2(d_1, \ldots, d_n; \beta) = \sigma^2 \sum_i d_i^2 - \lambda \sum_i d_i - \mu \left(\sum_i d_i(x_i - \bar{x}) - 1 \right).$$

The minimisers for both $\mathcal{L}_1$ and $\mathcal{L}_2$ are

$$c_i = d_i = \frac{1}{2\sigma^2}[\lambda + \mu(x_i - \bar{x})].$$

Adjusting the corresponding constraints then yields

$$c_i = \frac{1}{n} \quad \text{and} \quad d_i = \frac{x_i - \bar{x}}{S_{xx}},$$

i.e. $\widetilde{A} = \widehat{\alpha}$ and $\widetilde{B} = \widehat{\beta}$.

The relevant Tripos question is MT-IB 1996-103G.

4.9   Linear regression for normal distributions

> Statisticians must stay away from children's toys
> because they regress easily.
>> (From the series *'Why they are misunderstood'*.)

Note that so far we have not used any assumption about the form of the common distribution of errors $\epsilon_1, \ldots, \epsilon_n$. If, for instance, $\epsilon_i \sim N(0, \sigma^2)$ then the LSE pair $(\widehat{\alpha}, \widehat{\beta})$ coincides with the MLE. In fact, in this case

$$Y_i \sim N(\alpha + \beta(x_i - \bar{x}), \sigma^2),$$

and minimising the sum

$$\sum_i [y_i - \alpha - \beta(x_i - \bar{x})]^2$$

is the same as maximising the log-likelihood

$$\ell(\mathbf{x}, \mathbf{y}; \mu, \sigma^2) = -\frac{n}{2} \ln(2\pi\sigma^2) - \frac{1}{2\sigma^2} \sum_i [y_i - \alpha - \beta(x_i - \bar{x})]^2.$$

This fact explains analogies with earlier statements (like Fisher Theorem), which emerge when we analyse the normal linear regression in more detail.

Indeed, assume that $\epsilon_i \sim N(0, \sigma^2)$, independently. Consider the minimal value of sum $\sum_i [Y_i - \alpha - \beta(x_i - \bar{x})]^2$:

$$R = \sum \left[Y_i - \widehat{\alpha} - \widehat{\beta}(x_i - \bar{x}) \right]^2 . \tag{4.50}$$

R is called the *residual sum of squares* (RSS). The following theorem holds:

(i) $\widehat{\alpha} \sim N\left(\alpha, \sigma^2/n\right)$.
(ii) $\widehat{\beta} \sim N\left(\beta, \sigma^2/S_{xx}\right)$.
(iii) $R/\sigma^2 \sim \chi^2_{n-2}$.
(iv) $\widehat{\alpha}$, $\widehat{\beta}$ and R are independent.
(v) $R/(n-2)$ is an unbiased estimator of σ^2.

To prove (i) and (ii) observe that $\widehat{\alpha}$ and $\widehat{\beta}$ are linear combinations of independent normal RVs Y_i. To prove (iii) and (iv), we follow the same strategy as in the proof of the Fisher and Pearson Theorems.

Let A be an $n \times n$ real orthogonal matrix whose first and second columns are

$$\begin{pmatrix} 1/\sqrt{n} \\ \vdots \\ 1/\sqrt{n} \end{pmatrix} \quad \text{and} \quad \begin{pmatrix} (x_1 - \bar{x})/\sqrt{S_{xx}} \\ \vdots \\ (x_n - \bar{x})/\sqrt{S_{xx}} \end{pmatrix}$$

and the remaining columns are arbitrary (chosen to maintain orthogonality). Such a matrix always exists: the first two columns are orthonormal (owing to equation $\sum_i (x_i - \bar{x}) = 0$), and we can always complete them with $n - 2$ vectors to form an orthonormal basis in $\mathbb{R}^n$. Then consider the random vector $\mathbf{Z} = A^T \mathbf{Y}$, with first two entries

$$Z_1 = \sqrt{n}\, \bar{Y} = \sqrt{n}\, \widehat{\alpha} \sim N(\sqrt{n}\alpha, \sigma^2)$$

and

$$Z_2 = \frac{1}{\sqrt{S_{xx}}} \sum_i (x_i - \bar{x}) Y_i = \sqrt{S_{xx}}\, \widehat{\beta} \sim N(\sqrt{S_{xx}}\beta, \sigma^2).$$

In fact, owing to orthogonality of A, we have that all entries $Z_1, \ldots, Z_n$ are independent normal RVs with the same variance. Moreover,

$$\sum_{i=1}^{n} Z_i^2 = \sum_{i=1}^{n} Y_i^2 .$$

At the same time,

$$\sum_{i=3}^{n} Z_i^2 = \sum_{i=1}^{n} Z_i^2 - Z_1^2 - Z_2^2 = \sum_{i=1}^{n} Y_i^2 - n\widehat{\alpha}^2 - S_{xx}\widehat{\beta}^2$$

$$= \sum_{i=1}^{n} (Y_i - \bar{Y})^2 + \widehat{\beta}^2 S_{xx} - 2\widehat{\beta} S_{xY}$$

$$= \sum_{i=1}^{n} \left(Y_i - \widehat{\alpha} - \widehat{\beta}(x_i - \bar{x}) \right)^2 = R.$$

Hence, $\widehat{\alpha}$, $\widehat{\beta}$ and R are independent.

If matrix $A = (A_{ij})$, then (i) $\forall\, j \geq 2$, $\sum_i A_{ij} = 0$, as columns $2, \ldots, n$ are orthogonal to column 1, (ii) $\forall\, j \geq 3$, $\sum_i A_{ij}(x_i - \bar{x}) = 0$, as columns $3, \ldots, n$ are orthogonal to column 2. Then $\forall\, j \geq 3$

$$\mathbb{E}Z_j = \sum_i \mathbb{E}Y_i A_{ij} = \sum_i (\alpha + \beta(x_i - \bar{x})) A_{ij}$$

$$= \alpha \sum_i A_{ij} + \beta \sum_i (x_i - \bar{x}) A_{ij} = 0$$

and

$$\operatorname{Var} Z_j = \sigma^2 \sum_i a_{ij}^2 = \sigma^2.$$

Hence,

$$Z_3, \ldots, Z_n \sim \mathrm{N}(0, \sigma^2) \text{ independently.}$$

But then $R/\sigma^2 \sim \chi_{n-2}^2$. Now (v) follows immediately, and $\mathbb{E}R = (n-2)\sigma^2$.
It is useful to remember that

$$R = \sum_{i=1}^n Y_i^2 - n\widehat{\alpha}^2 - S_{xx}\widehat{\beta}^2. \tag{4.51}$$

Now we can test the null hypothesis $H_0\colon \beta = \beta_0$ against $H_1\colon \beta \in \mathbb{R}$ unrestricted. In fact, we have that under H_0,

$$\widehat{\beta} \sim \mathrm{N}\left(\beta_0, \frac{\sigma^2}{S_{xx}}\right) \text{ and } \frac{1}{\sigma^2} R \sim \chi_{n-2}^2, \text{ independently.}$$

Then

$$T = \frac{\widehat{\beta} - \beta_0}{\sqrt{R}\Big/\sqrt{(n-2)S_{xx}}} \sim t_{n-2}. \tag{4.52}$$

Hence, an α size test rejects H_0 when the value t of $|T|$ exceeds $t_{n-2}(\alpha/2)$, the upper $\alpha/2$ point of the t_{n-2}-distribution. A frequently occurring case is $\beta_0 = 0$ when we test whether we need the term $\beta(x_i - \bar{x})$ at all.

Similarly, we can use the statistic

$$T = \frac{\widehat{\alpha} - \alpha_0}{\sqrt{R}/\sqrt{(n-2)n}} \sim t_{n-2}. \tag{4.53}$$

to test $H_0\colon \alpha = \alpha_0$ against $H_1\colon \alpha \in \mathbb{R}$ unrestricted.

The above tests lead to the following $100(1 - \gamma)\%$ confidence intervals for α and β:

$$\left(\widehat{\alpha} - \frac{\sqrt{R}}{\sqrt{(n-2)n}} t_{n-2}(\gamma/2),\ \widehat{\alpha} + \frac{\sqrt{R}}{\sqrt{(n-2)n}} t_{n-2}(\gamma/2)\right), \tag{4.54}$$

$$\left(\widehat{\beta} - \frac{\sqrt{R}}{\sqrt{(n-2)S_{xx}}} t_{n-2}(\gamma/2),\ \widehat{\beta} + \frac{\sqrt{R}}{\sqrt{(n-2)S_{xx}}} t_{n-2}(\gamma/2)\right). \tag{4.55}$$

A similar construction works for the sum $\alpha + \beta$. Here we have

$$\widehat{\alpha} + \widehat{\beta} \sim \mathrm{N}\left(\alpha + \beta, \sigma^2 \left(\frac{1}{n} + \frac{1}{S_{xx}}\right)\right), \quad \text{independently of } R.$$

Then

$$T = \frac{\widehat{\alpha} + \widehat{\beta} - (\alpha + \beta)}{\left[\left(\frac{1}{n} + \frac{1}{S_{xx}}\right) R/(n-2)\right]^{1/2}} \sim t_{n-2}.$$

Hence,

$$\left\{\widehat{\alpha} + \widehat{\beta} - t_{n-2}(\gamma/2)\left[\left(\frac{1}{n} + \frac{1}{S_{xx}}\right) R/(n-2)\right]^{1/2}, \right.$$

$$\left. \widehat{\alpha} + \widehat{\beta} + t_{n-2}(\gamma/2)\left[\left(\frac{1}{n} + \frac{1}{S_{xx}}\right) R/(n-2)\right]^{1/2}\right\} \quad (4.56)$$

gives the $100(1 - \gamma)\%$ CI for $\alpha + \beta$.

Next, we introduce a value x and construct the so-called *prediction interval* for a random variable $Y \sim \mathrm{N}(\alpha + \beta(x - \bar{x}), \sigma^2)$, independent of $Y_1, \ldots, Y_n$. Here, $\bar{x} = \sum_{i=1}^{n} x_i/n$, $x_1, \ldots, x_n$ are the points where observations $Y_1, \ldots, Y_n$ have been made, and x is considered as a new point (where no observation has so far been made). It is natural to consider

$$\widehat{Y} = \widehat{\alpha} + \widehat{\beta}(x - \bar{x})$$

as a value predicted for Y by our model, with

$$\mathbb{E}\widehat{Y} = \alpha + \beta(x - \bar{x}) = \mathbb{E}Y, \quad (4.57)$$

and

$$\mathrm{Var}\left(\widehat{Y} - Y\right) = \mathrm{Var}\,\widehat{Y} + \mathrm{Var}\,Y$$

$$= \mathrm{Var}\,\widehat{\alpha} + (x - \bar{x})^2 \mathrm{Var}\,\widehat{\beta} + \sigma^2 \quad (4.58)$$

$$= \sigma^2 \left[1 + \frac{1}{n} + \frac{(x - \bar{x})^2}{S_{xx}}\right].$$

Hence,

$$\widehat{Y} - Y \sim \mathrm{N}\left(0, \sigma^2 \left[1 + \frac{1}{n} + \frac{(x - \bar{x})^2}{S_{xx}}\right]\right),$$

and we find a $100(1 - \gamma)\%$ prediction interval for Y is centred at $\widehat{Y}$ and has length

$$2\widehat{\sigma}\sqrt{1 + \frac{1}{n} + \frac{(x - \bar{x})^2}{S_{xx}}} t_{n-2}\left(\frac{\gamma}{2}\right).$$

Finally, we construct the CI for the linear combination $\alpha + l\beta$. Here, by the similar argument, we obtain that

$$\left(\widehat{\alpha} + l\widehat{\beta} - \frac{\sqrt{R}}{\sqrt{n-2}}\sqrt{\frac{1}{n} + \frac{l^2}{S_{xx}}}t_{n-2}(\gamma/2), \widehat{\alpha} + l\widehat{\beta} + \frac{\sqrt{R}}{\sqrt{n-2}}\sqrt{\frac{1}{n} + \frac{l^2}{S_{xx}}}t_{n-2}(\gamma/2)\right) \quad (4.59)$$

is a $100(1-\gamma)\%$ CI for $\alpha + l\beta$.

In particular, when we select $l = x - \overline{x}$, we obtain a CI for the mean value $\mathbb{E}Y$ for a given observation point x. That is

$$\alpha + \beta(x - \overline{x}), \quad \overline{x} = \frac{1}{n}\sum_i x_i.$$

A natural question is: how well does the estimated regression line fit the data? A bad fit could occur because of a large value of σ^2, but σ^2 is unknown. To make a judgement, one performs several observations $Y_{i1}, \ldots, Y_{im_i}$ at each value x_i:

$$Y_{ij} = \alpha + \beta(x_i - \overline{x}) + \epsilon_{ij}, \quad j = 1, \ldots, m_i, \ i = 1, \ldots, n. \quad (4.60)$$

Then the null hypothesis H_0 that the mean value $\mathbb{E}Y$ is linear in x is tested as follows. The average

$$\overline{Y}_{i+} = \frac{1}{m_i}\sum_{j=1}^{m_i} Y_{ij}$$

equals

$$\alpha + \beta(x_i - \overline{x}) + \overline{\epsilon}_{i+}, \quad \text{where} \quad \overline{\epsilon}_{i+} \sim N\left(0, \frac{\sigma^2}{m_i}\right).$$

Under H_0, we measure the 'deviation from linearity' by the RSS $\sum_i m_i(\overline{Y}_{i+} - \widehat{\alpha} - \widehat{\beta}(x_i - \overline{x}))^2$ which obeys

$$\frac{1}{\sigma^2}\sum_i m_i(\overline{Y}_{i+} - \widehat{\alpha} - \widehat{\beta}(x_i - \overline{x}))^2 \sim \chi^2_{n-2}. \quad (4.61)$$

Next, regardless of whether or not H_0 is true, $\forall \ i = 1, \ldots, n$, the sum

$$\frac{1}{\sigma^2}\sum_j (Y_{ij} - \overline{Y}_{i+})^2 \sim \chi^2_{m_i-1},$$

independently of $\overline{Y}_{1+}, \ldots, \overline{Y}_{n+}$. Hence

$$\frac{1}{\sigma^2}\sum_{i,j} (Y_{ij} - \overline{Y}_{i+})^2 \sim \chi^2_{m_1+\cdots+m_n-n}. \quad (4.62)$$

independently of $\overline{Y}_{1+}, \ldots, \overline{Y}_{n+}$. Under H_0, the statistic

$$
\frac{\sum_i m_i \left[\overline{Y}_{i+} - \widehat{\alpha} - \widehat{\beta}(x_i - \overline{x}) \right]^2 / (n-2)}{\sum_{i,j} (Y_{ij} - \overline{Y}_{i+})^2 / (\sum_i m_i - n)} \sim F_{n-2, m_1 + \cdots + m_n - n}.
\tag{4.63}
$$

Then, given $\alpha \in (0, 1)$, H_0 is rejected when the value of statistic (4.63) is $> \varphi^+_{n-2, m_1 + \cdots + m_n - n}(\alpha)$.

The linear regression for a normal distribution appears in the following Tripos examples: MT-IB 1999-412D, 1998-203E, 1993-403J, 1992-406D. See also SP-IB 1992-203H(i).

5 Cambridge University Mathematical Tripos examination questions in IB Statistics (1992–1999)

> Statisticians will probably do it.
>> (From the series 'How they do it'.)

> The manipulation of statistical formulas
> is no substitute for knowing what one is doing.
>> H.M. Blalock (1926–), American social scientist

The problems and solutions below are listed in inverse chronological order (but the order within a given year is preserved).

Problem 5.1 (MT-IB 1999-103D short) Let $X_1, \ldots, X_6$ be a sample from a uniform distribution on $[0, \theta]$, where $\theta \in [1, 2]$ is an unknown parameter. Find an unbiased estimator for θ of variance less than 1/10.

Solution Set

$$M = \max [X_1, \ldots, X_6].$$

Then M is a sufficient statistic for θ, and we can use it to get an unbiased estimator for θ. We have

$$F_M(y; \theta) = \mathbb{P}_\theta(M < y) = \mathbb{P}_\theta(X_1 < y, \ldots, X_6 < y)$$

$$= \prod_{i=1}^{6} \mathbb{P}_\theta(X_i < y) = (F_X(y; \theta))^6, \quad 0 \leq y \leq \theta.$$

Then the PDF of M is

$$f_M(y; \theta) = \frac{d}{dy} F_M(y; \theta) = \frac{6y^5}{\theta^6}, 0 \leq y \leq \theta,$$

and the mean value equals

$$\mathbb{E}M = \int_0^\theta dy y \frac{6y^5}{\theta^6} = \frac{6y^7}{7\theta^6} \Big|_0^\theta = \frac{6\theta}{7}.$$

So, an unbiased estimator for θ is $7M/6$. Next,

$$\mathrm{Var}\,\frac{7M}{6} = \mathbb{E}\left(\frac{7M}{6}\right)^2 - \left(\mathbb{E}\frac{7M}{6}\right)^2 = \int\limits_0^\theta \frac{7^2 \times 6y^7}{6^2\theta^6}\,\mathrm{d}y - \theta^2 = \frac{7^2\theta^2}{8\times 6} - \theta^2 = \frac{\theta^2}{48}.$$

For $\theta \in [1,2]$,

$$\frac{\theta^2}{48} \in \left[\frac{1}{48}, \frac{1}{12}\right],$$

i.e. $\theta^2/48 < 1/10$, as required. Hence the answer:

$$\text{the required estimator} = \frac{7}{6} \times \max[X_1, \ldots, X_6]. \quad \square$$

Problem 5.2 (MT-IB 1999-112D long) Let $X_1, \ldots, X_n$ be a sample from a uniform distribution on $[0, \theta]$, where $\theta \in (0, \infty)$ is an unknown parameter.

  (i) Find a one-dimensional sufficient statistic M for θ and construct a 95% confidence interval for θ based on T.

  (ii) Suppose now that θ is an RV having prior density

$$\pi(\theta) \propto I(\theta \geq a)\theta^{-k},$$

where $a > 0$ and $k > 2$. Compute the posterior density for θ and find the optimal Bayes estimator $\widehat{\theta}$ under the quadratic loss function $(\theta - \widehat{\theta})^2$.

Solution (i) Set

$$M = \max\,[X_1, \ldots, X_n].$$

Then M is a sufficient statistic for θ. In fact, the likelihood

$$f(\mathbf{x}; \theta) = \begin{cases} \dfrac{1}{\theta^n}, & \theta > x_{\max}, \\ 0, & \text{otherwise.} \end{cases}$$

So, if we have θ and $x_{\max}$, we can calculate $f(\,\cdot\,; \theta)$ which means that M is sufficient. (Formally, it follows from the factorisation criterion.)

  Now, $\mathbb{P}_\theta(\theta \geq M) = 1$. So, we can construct the confidence interval with the lower bound M and an upper bound $b(M)$ such that $\mathbb{P}(\theta \leq b(M)) = 0.95$. Write

$$\mathbb{P}(\theta \leq b(M)) = 1 - \mathbb{P}(b(M) < \theta) = 1 - \mathbb{P}(M < b^{-1}(\theta)) = 1 - 0.05;$$

as the PDF is

$$f_M(t; \theta) = n\frac{t^{n-1}}{\theta^n} I(0 < x < \theta),$$

we obtain

$$\frac{(b^{-1}(\theta))^n}{\theta^n} = 0.05, \quad \text{whence } b^{-1}(\theta) = \theta(0.05)^{1/n}.$$

Then $M = \theta(0.05)^{1/n}$ gives $\theta = M/(0.05)^{1/n}$. Hence, the 95% CI for θ is $(T, T/(0.05)^{1/n})$.

(ii) The Bayes formula

$$\pi(\theta|\mathbf{x}) = \frac{f(\mathbf{x};\theta)\pi(\theta)}{\int f(\mathbf{x}|\phi)\pi(\phi)\mathrm{d}\phi}$$

yields for the posterior PDF:

$$\pi(\theta|\mathbf{x}) \propto f(\mathbf{x};\theta)\pi(\theta) = \frac{1}{\theta^{n+k}}I(\theta \geq c),$$

where

$$c = c(\mathbf{x}) = \max\left[a, x_{\max}\right].$$

Thus,

$$\pi(\theta|\mathbf{x}) = (n+k-1)c^{n+k-1}\theta^{-n-k}I(\theta \geq c).$$

Next, with the quadratic LF we have to minimise

$$\int \pi(\theta|\mathbf{x})(\theta - \widehat{\theta})^2\mathrm{d}\theta.$$

This gives the posterior mean

$$\widehat{\theta}^* = \arg\min_{\widehat{\theta}} \int \pi(\theta|\mathbf{x})(\theta - \widehat{\theta})^2\mathrm{d}\theta = \int \theta\pi(\theta|\mathbf{x})\mathrm{d}\theta.$$

Now

$$\int_c^\infty \theta^{-n-k+1}\mathrm{d}\theta = \frac{1}{-n-k+2}\theta^{-n-k+2}\Big|_c^\infty = \frac{1}{n+k-2}c^{2-n-k},$$

and after normalising, we obtain

$$\widehat{\theta}^* = \frac{k+n-1}{k+n-2}c = \frac{k+n-1}{k+n-2}\max[a, x_1, \ldots, x_n]. \quad \square$$

Problem 5.3   (MT-IB 1999-203D short) Write a short account of the standard procedure used by statisticians for hypothesis testing. Your account should explain, in particular, why the null hypothesis is considered differently from the alternative and also say what is meant by a likelihood ratio test.

Solution   Suppose we have data

$$\mathbf{x} = \begin{pmatrix} x_1 \\ \vdots \\ x_n \end{pmatrix}$$

from a PDF/PMF f. We make two mutually excluding hypotheses about f, H_0 (a null hypothesis) and H_1 (an alternative).

These hypotheses have different statuses. H_0 is treated as a conservative hypothesis, not to be rejected unless there is a clear evidence against it, for example:

(i) H_0: $f = f_0$ against H_1: $f = f_1$; both f_0, f_1 specified. (This case is covered by the NP Lemma.)

(ii) H_0: $f = f_0$ (a specified PDF/PMF) against H_1: f unrestricted. (This includes the Pearson Theorem leading to χ^2 tests.)

(iii) $f = f(\cdot; \theta)$ is determined by the value of a parameter; H_0: $\theta \in \Theta_0$ against H_1: $\theta \in \Theta_1$, where $\Theta_0 \cap \Theta_1 = \emptyset$ (e.g. families with a monotone likelihood ratio).

(iv) H_0: $\theta \in \Theta_0$ against H_1: $\theta \in \Theta$, where $\Theta_0 \subset \Theta$, and Θ has more degrees of freedom than Θ_0.

A test is specified by a critical region $\mathcal{C}$ such that if $\mathbf{x} \in \mathcal{C}$, then H_0 is rejected, while if $\mathbf{x} \notin \mathcal{C}$, H_0 is not rejected (which highlights the conservative nature of H_0). A type I error occurs when H_0 is rejected while being true. Further, the type I error probability is defined as $\mathbb{P}(\mathcal{C})$ under H_0; we say that a test has size a (or $\leq a$), if $\max_{H_0} \mathbb{P}(\mathcal{C}) \leq a$. We choose a at our discretion (e.g. 0.1, 0.01, etc.) establishing an accepted chance of rejecting H_0 wrongly. Then we look for a test of a given size a which minimises the type II error probability $1 - \mathbb{P}(\mathcal{C})$, i.e. maximises the power $\mathbb{P}(\mathcal{C})$ under H_1.

To define an appropriate critical region, one considers the likelihood ratio

$$\frac{\max f(\mathbf{x}|H_1)}{\max f(\mathbf{x}|H_0)},$$

where the suprema are taken over PDFs/PMFs representing H_0 and H_1, respectively. The critical region is then defined as the set of data samples $\mathbf{x}$, where the likelihood ratio is large, depending on the given size a. $\square$

Problem 5.4 (MT-IB 1999-212D long) State and prove the NP Lemma. Explain what is meant by a uniformly most powerful test.

Let $X_1, \ldots, X_n$ be a sample from the normal distribution of mean θ and variance 1, where $\theta \in \mathbb{R}$ is an unknown parameter. Find a UMP test of size $1/100$ for

$$H_0: \theta \leq 0, \qquad H_1: \theta > 0,$$

expressing your answer in terms of an appropriate distribution function. Justify carefully that your test is uniformly most powerful of size $1/100$.

Solution The NP Lemma is applicable when both the null hypothesis and the alternative are simple, i.e. H_0: $f = f_0$, H_1: $f = f_1$, where f_1 and f_0 are two PDFs/PMFs defined on the same region. The NP Lemma states: $\forall\ k > 0$, *the test with critical region* $\mathcal{C} = \{\mathbf{x}: f_1(\mathbf{x}) > kf_0(\mathbf{x})\}$ *has the highest power* $\mathbb{P}_1(\mathcal{C})$ *among all tests (i.e. critical regions) of size* $\mathbb{P}(\mathcal{C})$.

For the proof of the NP Lemma: see Section 4.2.

The UMP test, of size a for H_0: $\theta \in \Theta_0$ against H_1: $\theta \in \Theta_1$, has the critical region $\mathcal{C}$ such that: (i) $\max[\mathbb{P}_\theta(\mathcal{C}) : \theta \in \Theta_0] \leq a$ and (ii) $\forall$ $\mathcal{C}^*$ with $\max[\mathbb{P}_\theta(\mathcal{C}^*) : \theta \in \Theta_0] \leq a$: $\mathbb{P}_\theta(\mathcal{C}) \geq \mathbb{P}_\theta(\mathcal{C}^*)$ $\forall$ $\theta \in \Theta_1$.

In the example where $X_i \sim N(\theta, 1)$, H_0: $\theta \leq 0$ and H_1: $\theta > 0$, we fix some $\theta_1 > 0$ and consider the simple null hypothesis that $\theta = 0$ against the simple alternative that $\theta = \theta_1$. The log-likelihood

$$\ln \frac{f(\mathbf{x}; \theta_1)}{f(\mathbf{x}; 0)} = \theta_1 \sum_i x_i - \frac{n}{2}\theta_1^2$$

is large when $\sum_i x_i$ is large. Choose $k_1 > 0$ so that

$$\frac{1}{100} = \mathbb{P}_0\left(\sum_i X_i > k_1\right) = \mathbb{P}_0\left(\frac{\sum_i X_i}{\sqrt{n}} > \frac{k_1}{\sqrt{n}}\right) = 1 - \Phi\left(\frac{k_1}{\sqrt{n}}\right),$$

i.e. $k_1/\sqrt{n} = z_+(0.01) = \Phi^{-1}(0.99)$. Then

$$\mathbb{P}_\theta\left(\sum_i X_i > k_1\right) < \frac{1}{100}$$

$\forall$ $\theta < 0$. Thus, the test with the critical region

$$\mathcal{C} = \left\{\mathbf{x} : \sum_i X_i > k_1\right\}$$

has size 0.01 for $H_0 : \theta \leq 0$.

Now, $\forall$ $\theta' > 0$, $\mathcal{C}$ can be written as

$$\mathcal{C} = \left\{\mathbf{x} : \frac{f(\mathbf{x}; \theta')}{f(\mathbf{x}; 0)} > k'\right\}$$

with some $k' = k'(\theta') > 0$. By the NP Lemma, $\mathbb{P}_{\theta'}(\mathcal{C}^*) \leq \mathbb{P}_{\theta'}(\mathcal{C})$, $\forall$ $\mathcal{C}^*$ with $\mathbb{P}_0(\mathcal{C}^*) \leq 0.01$. Similarly $\forall$ $\theta' > 0$

$$\mathbb{P}_{\theta'}(\mathcal{C}^*) \leq \mathbb{P}_{\theta'}(\mathcal{C}) \ \forall \ \mathcal{C}^* \text{ such that } \mathbb{P}_\theta(\mathcal{C}^*) \leq \frac{1}{100} \ \forall \ \theta \leq 0.$$

So, $\mathcal{C} = \{\mathbf{x} : \sum_i X_i > k_1\}$ is size 0.01 UMP for H_0 against H_1. $\quad\square$

Problem 5.5 (MT-IB 1999-403D short) Students of mathematics in a large university are given a percentage mark in their annual examination. In a sample of nine students the following marks were found:

28   32   34   39   41   42   42   46   56

Students of history also receive a percentage mark. A sample of five students reveals the following marks:

53   58   60   61   68

Do these data support the hypothesis that the marks for mathematics are more variable than the marks for history? Quantify your conclusion. Comment on your modelling assumptions.

distribution	N(0, 1)	$F_{9,5}$	$F_{8,4}$	χ^2_{14}	χ^2_{13}	χ^2_{12}
95% percentile	1.65	4.78	6.04	23.7	22.4	21.0

Solution Take independent RVs

$$X_i(= X_i^M) \sim N(\mu_1, \sigma_1^2), \quad i = 1, \ldots, 9,$$

and

$$Y_j(= Y_j^H) \sim N(\mu_2, \sigma_2^2), \quad j = 1, \ldots, 5.$$

If $\sigma_1^2 = \sigma_2^2$, then

$$F = \frac{1}{8} \sum_{i=1}^{9} (X_i - \overline{X})^2 \Big/ \frac{1}{4} \sum_{i=1}^{5} (Y_i - \overline{Y})^2 \sim F_{8,4},$$

where

$$\overline{X} = \frac{1}{9} \sum_i X_i, \quad \overline{Y} = \frac{1}{5} \sum_j Y_j.$$

We have $\overline{X} = 40$ and the values shown in Table 5.1, with $\sum_i (X_i - \overline{X})^2 = 546$.
Similarly, $\overline{Y} = 60$, and we have the values shown in Table 5.2, with $\sum_j (Y_j - \overline{Y})^2 = 118$.

Table 5.1.

X_i	28	32	34	39	41	42	42	46	56
$X_i - \overline{X}$	-12	-8	-6	-1	1	2	2	6	16
$(X_i - \overline{X})^2$	144	64	36	1	1	4	4	36	256

Table 5.2.

Y_j	53	58	60	61	68
$Y_j - \overline{Y}$	-7	-2	0	1	8
$(Y_j - \overline{Y})^2$	49	4	0	1	64

Then

$$F = \frac{1}{8} 546 \Big/ \frac{1}{4} 118 = \frac{273}{118} \approx 2.31.$$

But $\varphi^+_{8,4}(0.05) = 6.04$. So, we have no evidence to reject $H_0 : \sigma_1^2 = \sigma_2^2$ at the 95% level, i.e. we do not accept that $\sigma_1^2 > \sigma_2^2$. □

Problem 5.6   (MT-IB 1999-412D long) Consider the linear regression model

$$Y_i = \alpha + \beta x_i + \epsilon_i, \qquad \epsilon_i \sim N(0, \sigma^2), \quad i = 1, \ldots, n,$$

where $x_1, \ldots, x_n$ are known, with $\sum_{i=1}^n x_i = 0$, and where $\alpha, \beta \in \mathbb{R}$ and $\sigma^2 \in (0, \infty)$ are unknown. Find the maximum likelihood estimators $\hat{\alpha}, \hat{\beta}, \hat{\sigma}^2$ and write down their distributions.

Consider the following data:

x_i	-3	-2	-1	0	1	2	3
Y_i	-5	0	3	4	3	0	-5

Fit the linear regression model and comment on its appropriateness.

Solution   RV Y_i has the PDF

$$f_{Y_i}(y; \alpha, \beta, \sigma^2) = (2\pi\sigma^2)^{-1/2} e^{-(y - \alpha - \beta x_i)^2 / 2\sigma^2},$$

with

$$\ln f_Y(\mathbf{y}; \alpha, \beta, \sigma^2) = -\frac{n}{2} \ln \sigma^2 - \sum_i (y_i - \alpha - \beta x_i)^2 / (2\sigma^2).$$

To find the MLEs $\hat{\alpha}$ and $\hat{\beta}$, consider the stationary points:

$$0 = \frac{\partial}{\partial \alpha} \sum_i (y_i - \alpha - \beta x_i)^2 = -2n(\bar{y} - \alpha), \text{ whence } \hat{\alpha} = \bar{Y},$$

$$0 = \frac{\partial}{\partial \beta} \sum_i (y_i - \alpha - \beta x_i)^2 = -2 \sum_i x_i (y_i - \alpha - \beta x_i), \text{ whence } \hat{\beta} = \frac{S_{xY}}{S_{xx}},$$

where $\bar{Y} = \sum_i Y_i / n$, $S_{xY} = \sum_i x_i Y_i$ and $S_{xx} = \sum_i x_i^2$. The fact that they give the global maximum follows from the uniqueness of the stationary point and the fact that $f_Y(\mathbf{y}; \alpha, \beta, \sigma^2) \to 0$ as any of α, β and $\sigma^2 \to \infty$ or $\sigma^2 \to 0$.
Set $R = \sum_{i=1}^n (Y_i - \hat{\alpha} - \hat{\beta} x_i)^2$, then at $\hat{\sigma}^2$:

$$0 = \frac{\partial}{\partial \sigma^2} \left(-\frac{n}{2} \ln \sigma^2 - \frac{R}{2\sigma^2} \right) = -\frac{n}{\sigma^2} + \frac{R}{\sigma^4}, \text{ whence } \hat{\sigma}^2 = \frac{R}{n}.$$

The distributions are

$$\hat{\alpha} \sim N\left(\alpha, \frac{\sigma^2}{n} \right), \quad \hat{\beta} \sim N\left(\beta, \sigma^2 \Big/ \sum_i x_i^2 \right), \quad \text{and } R/\sigma^2 \sim \chi^2_{n-2}.$$

In the example, $\bar{Y} = 0$ and $S_{xY} = 0$, hence $\hat{\alpha} = \hat{\beta} = 0$. Further, $R = 84$, i.e. $\hat{\sigma}^2 = 14$.
This model is not particularly good as the data for (x_i, Y_i) show a parabolic shape, not linear. See Figure 5.1. □

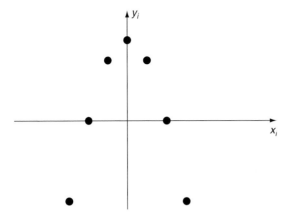

Figure 5.1

Problem 5.7 (MT-IB 1998-103E short) The independent observations X_1, X_2 are distributed as Poisson RVs, with means μ_1, μ_2 respectively, where

$$\ln \mu_1 = \alpha,$$
$$\ln \mu_2 = \alpha + \beta,$$

with α and β unknown parameters. Write down $\ell(\alpha, \beta)$, the log-likelihood function, and hence find the following:

(i) $\dfrac{\partial^2 \ell}{\partial \alpha^2}$, $\dfrac{\partial^2 \ell}{\partial \alpha \partial \beta}$, $\dfrac{\partial^2 \ell}{\partial \beta^2}$;

(ii) $\widehat{\beta}$, the maximum likelihood estimator of β.

Solution We have, $\forall$ integer $x_1, x_2 \geq 0$,

$$\ell(x_1, x_2; \alpha, \beta) = \ln \left(e^{-\mu_1} \frac{\mu_1^{x_1}}{x_1!} e^{-\mu_2} \frac{\mu_2^{x_2}}{x_2!} \right)$$

$$= -e^{\alpha} + x_1 \alpha - e^{\alpha+\beta} + x_2(\alpha+\beta) - \ln(x_1! x_2!).$$

So, (i)

$$\frac{\partial^2}{\partial \alpha^2} \ell = -e^{\alpha}(1 + e^{\beta}), \quad \frac{\partial^2}{\partial \alpha \partial \beta} \ell = -e^{\alpha+\beta}, \quad \frac{\partial^2}{\partial \beta^2} \ell = -e^{\alpha+\beta}.$$

(ii) Consider the stationary point

$$\frac{\partial}{\partial \alpha} \ell = 0 \Rightarrow x_1 + x_2 = e^{\widehat{\alpha}} + e^{\widehat{\alpha}+\widehat{\beta}}, \quad \frac{\partial}{\partial \beta} \ell = 0 \Rightarrow x_2 = e^{\widehat{\alpha}+\widehat{\beta}}.$$

Hence,

$$\widehat{\alpha} = \ln x_1, \quad \widehat{\beta} = \ln \frac{x_2}{x_1}.$$

The stationary point gives the global maximum, as it is unique and $\ell \to -\infty$ as $|\alpha|, |\beta| \to \infty$. □

Problem 5.8 (MT-IB 1998-112E long) The lifetime of certain electronic components may be assumed to follow the exponential PDF

$$f(t; \theta) = \frac{1}{\theta} \exp\left(-\frac{t}{\theta}\right), \quad \text{for } t \geq 0,$$

where t is the sample value of T.

Let $t_1, \ldots, t_n$ be a random sample from this PDF. Quoting carefully the NP Lemma, find the form of the most powerful test of size 0.05 of

$$H_0: \quad \theta = \theta_0 \quad \text{against} \quad H_1: \quad \theta = \theta_1$$

where θ_0 and θ_1 are given, and $\theta_0 < \theta_1$. Defining the function

$$G_n(u) = \int_0^u e^{-t} \frac{t^{n-1}}{(n-1)!} dt,$$

show that this test has power

$$1 - G_n\left[\frac{\theta_0}{\theta_1} G_n^{-1}(1 - \alpha)\right],$$

where $\alpha = 0.05$.

If for $n = 100$ you observed $\sum_i t_i/n = 3.1$, would you accept the hypothesis $H_0: \theta_0 = 2$? Give reasons for your answer, using the large-sample distribution of $(T_1 + \cdots + T_n)/n$.

Solution   The likelihood function for a sample vector $\mathbf{t} \in \mathbb{R}^n$ is

$$f(\mathbf{t}; \theta) = \frac{1}{\theta^n} \exp\left(-\frac{1}{\theta} \sum_i t_i\right) I(\min t_i > 0), \quad \mathbf{t} = \begin{pmatrix} t_1 \\ \vdots \\ t_n \end{pmatrix}.$$

By the NP Lemma, the MP test of size α will be with the critical region

$$\mathcal{C} = \left\{ \mathbf{t} : \frac{f_{\theta_1}(\mathbf{t})}{f_{\theta_0}(\mathbf{t})} > k \right\},$$

such that

$$\int_{\mathcal{C}} f_{\theta_0}(\mathbf{t}) d\mathbf{t} = 0.05.$$

As

$$\frac{f_{\theta_1}(\mathbf{t})}{f_{\theta_0}(\mathbf{t})} = \left(\frac{\theta_0}{\theta_1}\right)^n \exp\left[\left(\frac{1}{\theta_0} - \frac{1}{\theta_1}\right) \sum_i t_i\right] \text{ and } \frac{1}{\theta_0} > \frac{1}{\theta_1},$$

$\mathcal{C}$ has the form

$$\left\{ \mathbf{t} : \sum_i t_i > c \right\}$$

for some $c > 0$. Under H_0,

$$X = \frac{1}{\theta_0} \sum_{i=1}^{n} T_i \sim \text{Gam}\,(n, 1), \quad \text{with } \mathbb{P}_{\theta_0}(X < u) = G_n(u).$$

Hence, to obtain the MP test of size 0.05, we choose c so that

$$1 - G_n\left(\frac{c}{\theta_0}\right) = 0.05, \quad \text{i.e.} \quad \frac{c}{\theta_0} = G_n^{-1}(0.95).$$

Then the power of the test is

$$\int_c f_{\theta_1}(\mathbf{t})\mathrm{d}\mathbf{t} = 1 - G_n\left(\frac{1}{\theta_1}c\right),$$

which equals

$$1 - G_n\left[\frac{\theta_0}{\theta_1}G_n^{-1}(0.95)\right],$$

as required.

As $\mathbb{E}T_i = \theta$ and $\text{Var } T_i = n\theta^2$, for n large:

$$\frac{\sum_i T_i - n\theta}{\theta\sqrt{n}} \sim \text{N}(0, 1)$$

by the CLT. Under $H_0 \colon \theta_0 = 2$,

$$\frac{\sum_i T_i - n\theta_0}{\theta_0\sqrt{n}} = 5.5.$$

On the other hand, $z_+(0.05) = 1.645$. As $5.5 > 1.645$, we reject H_0. □

Problem 5.9 (MT-IB 1998-203E short) Consider the model

$$y_i = \beta_0 + \beta_1 x_i + \beta_2 x_i^2 + \epsilon_i, \qquad \text{for } 1 \le i \le n,$$

where $x_1, \ldots, x_n$ are given values, with $\sum_i x_i = 0$, and where $\epsilon_1, \ldots, \epsilon_n$ are independent normal errors, each with zero mean and unknown variance σ^2.

(i) Obtain equations for $(\widehat{\beta}_0, \widehat{\beta}_1, \widehat{\beta}_2)$, the MLE of $(\beta_0, \beta_1, \beta_2)$. Do not attempt to solve these equations.

(ii) Obtain an expression for $\widehat{\beta}_1^*$, the MLE of β_1 in the reduced model

$$H_0 \colon y_i = \beta_0 + \beta_1 x_i + \epsilon_i, \qquad 1 \le i \le n,$$

with $\sum_i x_i = 0$, and $\epsilon_1, \ldots, \epsilon_n$ distributed as above.

Solution   (i) As $Y_i \sim N(\beta_0 + \beta_1 x_i + \beta_2 x_i^2, \sigma^2)$, independently, the likelihood

$$f(\mathbf{y}|\beta_0, \beta_1, \beta_2, \sigma^2) = \prod_{i=1}^{n} \frac{1}{(2\pi\sigma^2)^{1/2}} \exp\left[-\frac{1}{2\sigma^2}(y_i - \beta_0 - \beta_1 x_i - \beta_2 x_i^2)^2\right]$$

$$= \left(\frac{1}{2\pi\sigma^2}\right)^{n/2} \exp\left[-\frac{1}{2\sigma^2}\sum_{i=1}^{n}(y_i - \beta_0 - \beta_1 x_i - \beta_2 x_i^2)^2\right].$$

We then obtain the equations for the stationary points of the LL:

$$\frac{\partial}{\partial\beta_0}\ell = \frac{1}{\sigma^2}\sum_{i=1}^{n}(y_i - \beta_0 - \beta_1 x_i - \beta_2 x_i^2) = 0,$$

$$\frac{\partial}{\partial\beta_1}\ell = \frac{1}{\sigma^2}\sum_{i=1}^{n}x_i(y_i - \beta_0 - \beta_1 x_i - \beta_2 x_i^2) = 0,$$

$$\frac{\partial}{\partial\beta_2}\ell = \frac{1}{\sigma^2}\sum_{i=1}^{n}x_i^2(y_i - \beta_0 - \beta_1 x_i - \beta_2 x_i^2) = 0.$$

In principle, σ^2 should also figure here as a parameter. The last system of equations still contains σ^2, but luckily the equation $\partial\ell/\partial\sigma^2 = 0$ could be dropped.

(ii) The same is true for the reduced model. The answer

$$\widehat{\beta}_1 = \frac{S_{xY}}{S_{xx}}$$

with $S_{xY} = \sum_i x_i Y_i$, $S_{xx} = \sum_i x_i^2$ is correct. Again, σ^2 will not appear in the expression for $\widehat{\beta}_1$. □

Problem 5.10   (MT-IB 1998-212E long) Let $(x_1, \ldots, x_n)$ be a random sample from the normal PDF with mean μ and variance σ^2.

(i)   Write down the log-likelihood function $\ell(\mu, \sigma^2)$.

(ii)   Find a pair of sufficient statistics, for the unknown parameters (μ, σ^2), carefully quoting the relevant theorem.

(iii) Find $(\widehat{\mu}, \widehat{\sigma}^2)$, the MLEs of (μ, σ^2). Quoting carefully any standard distributional results required, show how to construct a 95% confidence interval for μ.

Solution   (i) The LL is

$$\ell(\mathbf{x}; \mu, \sigma^2) = -\frac{n}{2}\ln(2\pi\sigma^2) - \frac{1}{2\sigma^2}\sum_{i=1}^{n}(x_i - \mu)^2.$$

(ii) By the factorisation criterion, $T(\mathbf{x})$ is sufficient for (μ, σ^2) iff $\ell(\mathbf{x}; \mu, \sigma^2) = \ln g(T(\mathbf{x}), \mu, \sigma^2) + \ln h(\mathbf{x})$ for some functions g and h. Now

$$\ell(\mathbf{x}; \mu, \sigma^2) = -\frac{n}{2}\ln(2\pi\sigma^2) - \frac{1}{2\sigma^2}\sum_i[(x_i - \overline{x}) + (\overline{x} - \mu)]^2.$$

The remaining calculations affect the sum $\sum_i$ only:

$$-\frac{1}{2\sigma^2}\sum_{i=1}^{n}[(x_i - \bar{x})^2 + 2(x_i - \bar{x})(\bar{x} - \mu) + (\bar{x} - \mu)^2]$$

$$= -\frac{1}{2\sigma^2}\left[\sum_{i=1}^{n}(x_i - \bar{x})^2 + n(\bar{x} - \mu)^2\right],$$

as $\sum_{i=1}^{n}(x_i - \bar{x}) = 0$.

Thus, with $T(\mathbf{X}) = (\bar{X}, \sum_i(X_i - \bar{X})^2)$, we satisfy the factorisation criterion (with $h \equiv 1$). So, $T(\mathbf{X}) = (\bar{X}, \sum_i(X_i - \bar{X})^2)$ is sufficient for (μ, σ^2).

(iii) The MLEs for (μ, σ^2) are found from

$$\frac{\partial}{\partial \mu}\ell = \frac{\partial}{\partial \sigma^2}\ell = 0$$

and are given by

$$\hat{\mu} = \bar{x}, \quad \hat{\sigma}^2 = \frac{S_{xx}}{n}, \quad \text{where } S_{xx} = \sum_i(x_i - \bar{x})^2.$$

We know that

$$\bar{X} \sim N\left(\mu, \frac{\sigma^2}{n}\right) \quad \text{and} \quad \frac{1}{\sigma^2}S_{XX} \sim \chi^2_{n-1}.$$

Then

$$\frac{\bar{X} - \mu}{\sigma/\sqrt{n}}\bigg/\frac{1}{\sigma}\sqrt{\frac{S_{XX}}{n-1}} \sim t_{n-1}.$$

So, if $t_{n-1}(0.025)$ is the upper point of t_{n-1}, then

$$\left(\bar{x} - \frac{1}{\sqrt{n}}s_{xx}t_{n-1}(0.025), \ \bar{x} + \frac{1}{\sqrt{n}}s_{xx}t_{n-1}(0.025)\right)$$

is the 95% CI for μ. Here $s_{xx} = \sqrt{S_{xx}/(n-1)}$. $\square$

Problem 5.11 (MT-IB 1998-403E short) Suppose that, given the real parameter θ, the observation X is normally distributed with mean θ and variance v, where v is known. If the prior density for θ is

$$\pi(\theta) \propto \exp\left[-(\theta - \mu_0)^2/2v_0\right],$$

where μ_0 and v_0 are given, show that the posterior density for θ is $\pi(\theta|x)$, where

$$\pi(\theta|x) \propto \exp\left[-(\theta - \mu_1)^2/2v_1\right],$$

and μ_1 and v_1 are given by

$$\mu_1 = \frac{\mu_0/v_0 + x/v}{1/v_0 + 1/v}, \quad \frac{1}{v_1} = \frac{1}{v_0} + \frac{1}{v}.$$

Sketch typical curves $\pi(\theta)$ and $\pi(\theta|x)$, with μ_0 and x marked on the θ-axis.

Solution   We have

$$f(x; \theta) \propto \exp\left[-\frac{1}{2v}(x-\theta)^2\right], \quad -\infty < x < \infty.$$

Then

$$\pi(\theta|x) \propto f(x; \theta)\pi(\theta) \propto \exp\left[-\frac{1}{2v}(x-\theta)^2 - \frac{1}{2v_0}(\theta-\mu_0)^2\right].$$

Write

$$\frac{(x-\theta)^2}{v} + \frac{(\theta-\mu_0)^2}{v_0} = \left(\frac{1}{v}+\frac{1}{v_0}\right)\theta^2 - 2\theta\left(\frac{x}{v}+\frac{\mu_0}{v_0}\right) + \frac{\mu_0^2}{v_0} + \frac{x^2}{v}$$

$$= \left(\frac{1}{v}+\frac{1}{v_0}\right)\left(\theta - \frac{x/v+\mu_0/v_0}{1/v_0+1/v}\right)^2$$

$$- \left(\frac{x/v+\mu_0/v_0}{1/v_0+1/v}\right)^2 + \frac{\mu_0^2}{v_0} + \frac{x^2}{v}$$

$$= \frac{1}{v_1}(\theta - \mu_1)^2 + \text{terms not containing } \theta,$$

where μ_1, v_1 are as required.
   Thus

$$\pi(\theta|x) \propto \exp\left[-\frac{1}{2v_1}(\theta-\mu_1)^2 - \frac{k}{2}\right] \propto \exp\left[-\frac{1}{2v_1}(\theta-\mu_1)^2\right].$$

   Both PDFs $\pi(\theta)$ and $\pi(\cdot|x)$ are normal; as Figure 5.2 shows, the variance of π is larger than that of $\pi(\cdot|x)$.   □

Problem 5.12   (MT-IB 1998-412E long) Let (n_{ij}) be the observed frequencies for an $r \times c$ contingency table, let $n = \sum_{i=1}^{r}\sum_{j=1}^{c} n_{ij}$ and let

$$\mathbb{E}(n_{ij}) = np_{ij}, \quad 1 \leq i \leq r, \quad 1 \leq j \leq c,$$

thus $\sum_i \sum_j p_{ij} = 1$.

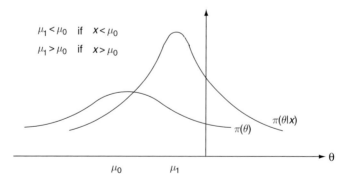

$\mu_1 < \mu_0$ if $x < \mu_0$
$\mu_1 > \mu_0$ if $x > \mu_0$

Figure 5.2

Under the usual assumption that (n_{ij}) is a multinomial sample, show that the likelihood ratio statistic for testing

$$H_0: p_{ij} = \alpha_i \beta_j,$$

for all (i, j) and for some vectors $\alpha = (\alpha_1, \ldots, \alpha_r)$ and $\beta = (\beta_1, \ldots, \beta_c)$, is

$$D = 2 \sum_{i=1}^{r} \sum_{j=1}^{c} n_{ij} \ln(n_{ij}/e_{ij}),$$

where you should define (e_{ij}). Show further that for $|n_{ij} - e_{ij}|$ small, the statistic D may be approximated by

$$Z^2 = \sum_{i=1}^{r} \sum_{j=1}^{c} (n_{ij} - e_{ij})^2 / e_{ij}.$$

In 1843 William Guy collected the data shown in Table 5.3 on 1659 outpatients at a particular hospital showing their physical exertion at work and whether they had pulmonary consumption (TB) or some other disease. For these data, Z^2 was found to be 9.84. What do you conclude?

Table 5.3.

Level of exertion at work	Disease	Type
	Pulmonary consumption	Other disease
Little	125	385
Varied	41	136
More	142	630
Great	33	167

(Note that this question can be answered without calculators or statistical tables.)

Solution With n_{ij} standing for counts in the contingency table, set

$$n = \sum_{i=1}^{r} \sum_{j=1}^{c} n_{ij}, \quad \mathbb{E}(n_{ij}) = n p_{ij}, \quad 1 \le i \le r, \ 1 \le j \le c.$$

We have the constraint: $\sum_{i,j} p_{ij} = 1$. Further,

$$H_0: p_{ij} = \alpha_i \beta_j; \quad \sum_i \alpha_i = \sum_j \beta_j = 1; \quad H_1: p_{ij} \text{ arbitrary}, \sum_{i,j} p_{ij} = 1.$$

Under H_1, in the multinomial distribution, the probability of observing a sample (n_{ij}) equals $\prod_{i,j} p_{ij}^{n_{ij}} / n_{ij}!$. The LL, as a function of arguments p_{ij}, is

$$\ell(p_{ij}) = \sum_{i,j} n_{ij} \ln p_{ij} + A,$$

where $A = -\sum_{ij} \ln(n_{ij}!)$ does not depend on p_{ij}. Hence, under the constraint $\sum_{i,j} p_{ij} = 1$, ℓ is maximised at

$$\hat{p}_{ij} = \frac{n_{ij}}{n}.$$

Under H_0, the LL equals

$$\ell(\alpha_i, \beta_j) = \sum_i n_{i+} \ln \alpha_i + \sum_j n_{+j} \ln \beta_j + B,$$

where B does not depend on α_i, β_j, and

$$n_{i+} = \sum_j n_{ij}, \quad n_{+j} = \sum_i n_{ij}.$$

Under the constraint $\sum \alpha_i = \sum \beta_j = 1$, ℓ is maximised at

$$\hat{\alpha}_i = \frac{n_{i+}}{n}, \quad \hat{\beta}_j = \frac{n_{+j}}{n}.$$

Thus,

$$\tilde{p}_{ij} = \frac{e_{ij}}{n}, \text{ where } e_{ij} = \frac{n_{i+} n_{+j}}{n}.$$

Then the LR statistic

$$2 \ln \frac{\max\left[\ell(p_{ij}) : \sum_{i,j} p_{ij} = 1\right]}{\max\left[\ell(\alpha_i, \beta_j) : \sum \alpha_i = \sum \beta_j = 1\right]} = 2 \sum_{i,j} n_{ij} \ln(n_{ij}/e_{ij})$$

coincides with D, as required.

With $n_{ij} - e_{ij} = \delta_{ij}$, then

$$D = 2 \sum_{i,j} (e_{ij} + \delta_{ij}) \ln\left(1 + \frac{\delta_{ij}}{e_{ij}}\right)$$

and, omitting the subscripts,

$$\approx 2 \sum (e + \delta) \left(\frac{\delta}{e} - \frac{1}{2}\frac{\delta^2}{e^2} + \cdots\right) = 2 \sum \delta + \sum \frac{\delta^2}{e} + \cdots$$

As $\sum \delta = 0$,

$$D \approx \sum \frac{\delta_{ij}^2}{e_{ij}} = Z^2.$$

The data given yield $Z^2 = 9.84$. With $r = 4$, $c = 2$, we use χ_3^2. The value 9.84 is too high for χ_3^2, so we reject H_0. The conclusion is that incidence of TB rather than other diseases is reduced when the level of exertion increases. □

Problem 5.13 (MT-IB 1997-103G short) In a large group of young couples, the standard deviation of the husbands' ages is four years, and that of the wives' ages is three years. Let D denote the age difference within a couple.

Under what circumstances might you expect to find that the standard deviation of age differences in the group is about 5 years?

Instead you find it to be 2 years. One possibility is that the discrepancy is the result of random variability. Give another explanation.

Solution Writing $D = H - W$, we see that if the ages H and W are independent, then

$$\text{Var } D = \text{Var } H + \text{Var } W = 4^2 + 3^2 = 5^2,$$

and the standard deviation is 5. Otherwise we would expect Var $D \neq 5$, so value 5 is taken under independence.

An alternative explanation for the value 2 is that H and W are correlated. If $H = \alpha W + \varepsilon$ with W and ε independent, then with Var $H = 16$ and Var $W = 9$

$$\text{Var } H = 16 = \alpha^2 \text{Var } W + \text{Var } \varepsilon,$$

$$\text{Var } (H - W) = 4 = (\alpha - 1)^2 \text{Var } W + \text{Var } \varepsilon.$$

Hence,

$$12 = (\alpha^2 - (\alpha - 1)^2)\text{Var } W,$$

and $\alpha = 7/6$. $\square$

Problem 5.14 (MT-IB 1997-112G long) Suppose that $X_1, \ldots, X_n$ and $Y_1, \ldots, Y_m$ form two independent samples, the first from an exponential distribution with the parameter λ, and the second from an exponential distribution with parameter μ.

(i) Construct the likelihood ratio test of $H_0 : \lambda = \mu$ against $H_1 : \lambda \neq \mu$.

(ii) Show that the test in part (i) can be based on the statistic

$$T = \frac{\sum_{i=1}^n X_i}{\sum_{i=1}^n X_i + \sum_{j=1}^m Y_j}.$$

(iii) Describe how the percentiles of the distribution of T under H_0 may be determined from the percentiles of an F-distribution.

Solution (i) For $X_i \sim \text{Exp}(\lambda)$, $Y_j \sim \text{Exp}(\mu)$,

$$f(\mathbf{x}; \lambda) = \prod_{i=1}^n (\lambda e^{-\lambda x_i}) \, I \, (\min x_i \geq 0); \text{ maximised at } \widehat{\lambda}^{-1} = \frac{1}{n} \sum_{i=1}^n x_i,$$

with

$$\max \left[f(\mathbf{x}; \lambda) : \lambda > 0 \right] = \widehat{\lambda}^n e^{-n},$$

and

$$f(\mathbf{y}; \mu) = \prod_{j=1}^{m} (\mu e^{-\mu y_j})\, I\,(\min y_j \ge 0); \text{ maximised at } \widehat{\mu}^{-1} = \frac{1}{m}\sum_{j=1}^{m} y_j,$$

with

$$\max\left[f(\mathbf{y}; \mu): \mu > 0\right] = \widehat{\mu}^m e^{-m}.$$

Under H_0, the likelihood is

$$f(\mathbf{x}; \theta) f(\mathbf{y}; \theta); \text{ maximised at } \widehat{\theta}^{-1} = \frac{1}{n+m}\left(\sum_{i=1}^{n} x_i + \sum_{j=1}^{m} y_j\right),$$

with

$$\max\left[f(\mathbf{x}; \theta) f(\mathbf{y}; \theta): \theta > 0\right] = \widehat{\theta}^{n+m} e^{-(n+m)}.$$

Then the test is: reject H_0 if the ratio

$$\frac{\widehat{\lambda}^n \widehat{\mu}^m}{\widehat{\theta}^{n+m}} = \left(\frac{\sum_i x_i + \sum_j y_j}{n+m}\right)^{n+m} \left(\frac{n}{\sum_i x_i}\right)^n \left(\frac{m}{\sum_j y_j}\right)^m$$

is large.

(ii) The logarithm has the form

$$(n+m)\ln\left(\sum_i x_i + \sum_j y_j\right) - n\ln\sum_i x_i - m\ln\sum_j y_j$$

plus terms not depending on $\mathbf{x}, \mathbf{y}$. The essential part is

$$-n\ln\frac{\sum_i x_i}{\sum_i x_i + \sum_j y_j} - m\ln\frac{\sum_j y_j}{\sum_i x_i + \sum_j y_j} = -n\ln T - m\ln(1-T).$$

Thus the test is indeed based on T. Furthermore, H_0 is rejected when T is close to 0 or 1.

(iii) Under H_0, $\lambda = \mu = \theta$, and

$$2\theta\sum_i X_i \sim \chi^2_{2n}, \quad 2\theta\sum_j Y_j \sim \chi^2_{2m}.$$

Hence,

$$T^{-1} = 1 + R\frac{m}{n}, \quad R = \frac{\sum_j Y_j/(2m)}{\sum_i X_i/(2n)} \sim F_{2m,2n}.$$

Thus the (equal-tailed) critical region is the union

$$\mathcal{C} = \left(0, \left(1+\frac{m}{n}\varphi^{+}_{2m,2n}(\alpha/2)\right)^{-1}\right) \cup \left(\left(1+\frac{m}{n}\varphi^{-}_{2m,2n}(\alpha/2)\right)^{-1}, 1\right),$$

and it is determined by percentiles of $F_{2m,2n}$. $\square$

Problem 5.15 (MT-IB 1997-203G short) Explain what is meant by a sufficient statistic.

Consider the independent RVs X_1, X_2, ..., X_n, where $X_i \sim N(\alpha + \beta c_i, \theta)$ for given constants c_i, $i = 1, 2, \ldots, n$, and unknown parameters α, β and θ. Find three sample quantities that together constitute a sufficient statistic.

Solution A statistic $T = T(\mathbf{x})$ is sufficient for a parameter θ if $f(\mathbf{x}; \theta) = g(T, \theta) h(\mathbf{x})$. For a data vector $\mathbf{x}$ with entries $x_1, \ldots, x_n$, we have

$$f(\mathbf{x}; \theta) = \prod_{i=1}^{n} f_{i,\theta}(x_i) = \prod_{i=1}^{n} \frac{1}{\sqrt{2\pi\theta}} \exp\left[-\frac{1}{2\theta}(x_i - \alpha - \beta c_i)^2\right]$$

$$= \frac{1}{(2\pi\theta)^{n/2}} \exp\left\{-\frac{1}{2\theta}\left[\sum_i x_i^2 - 2\sum_i x_i(\alpha + \beta c_i) + \sum_i (\alpha + \beta c_i)^2\right]\right\}$$

$$= \frac{1}{(2\pi\theta)^{n/2}} \exp\left[-\frac{1}{2\theta}\sum_i x_i^2 + \frac{\alpha}{\theta}\sum_i x_i + \frac{\beta}{\theta}\sum_i x_i c_i - \frac{1}{2\theta}\sum_i (\alpha + \beta c_i)^2\right].$$

Thus, the triple

$$T(\mathbf{x}) = \left(\sum_i x_i^2, \sum_i x_i, \sum_i c_i x_i\right)$$

is a sufficient statistic. □

Problem 5.16 (MT-IB, 1997-212G long) Let X_1, X_2, ..., X_n be a random sample from the $N(\theta, \sigma^2)$-distribution, and suppose that the prior distribution for θ is the $N(\mu, \tau^2)$-distribution, where σ^2, μ and τ^2 are known. Determine the posterior distribution for θ, given X_1, X_2, ..., X_n, and the optimal estimator of θ under (i) quadratic loss, and (ii) absolute error loss.

Solution For the first half, see Problems 2.42 and 2.43. For the second half: as was shown in Problem 2.3.8, the posterior distribution is

$$\pi(\theta|\mathbf{x}) = \frac{\pi(\theta)f(\mathbf{x}; \theta)}{\int \pi(\theta')f(\mathbf{x}; \theta')d\theta'} = \frac{1}{\sqrt{2\pi}\tau_n} \exp\left[-\frac{(\theta - \mu_n)^2}{2\tau_n^2}\right],$$

where

$$\frac{1}{\tau_n^2} = \frac{1}{\tau^2} + \frac{n}{\sigma^2}, \quad \mu_n = \frac{\mu/\tau^2 + n\bar{x}/\sigma^2}{1/\tau^2 + n/\sigma^2}.$$

As the normal distribution is symmetric about its mean, the best estimators in both cases are $\mathbb{E}(\theta|\mathbf{x}) = \mu_n$. □

Problem 5.17 (MT-IB 1997-403G short) X_1, X_2, ..., X_n form a random sample from a uniform distribution on the interval $(-\theta, 2\theta)$, where the value of the positive parameter θ is unknown. Determine the maximum likelihood estimator of the parameter θ.

Solution   The likelihood $f(\mathbf{x}; \theta)$ is

$$\frac{1}{(3\theta)^n} I(-\theta < x_1, \ldots, x_n < 2\theta) = \frac{1}{(3\theta)^n} I(-\theta < \min x_i) I(\max x_i < 2\theta).$$

Hence, the MLE is of the form

$$\widehat{\theta} = \max \left[-\min x_i, \ \frac{1}{2} \max x_i \right]. \quad \square$$

Problem 5.18   (MT-IB 1997-412G long) The χ^2-statistic is often used as a measure of the discrepancy between observed frequencies and expected frequencies under a null hypothesis. Describe the χ^2-statistic, and the χ^2 test for goodness of fit.

The number of directory enquiry calls arriving each day at a centre is counted over a period of K weeks. It may be assumed that the number of such calls on any day has a Poisson distribution, that the numbers of calls on different days are independent, and that the expected number of calls depends only on the day of the week. Let n_i, $i = 1, 2, \ldots, 7$, denote, respectively, the total number of calls received on a Monday, Tuesday, ..., Sunday.

Derive an approximate test of the hypothesis that calls are received at the same rate on all days of the week except Sundays.

Find also a test of a second hypothesis, that the expected numbers of calls received are equal for the three days from Tuesday to Thursday, and that the expected numbers of calls received are equal on Monday and Friday.

Solution   Suppose we have possible counts n_i of occurrence of states $i = 1, \ldots, n$, of expected frequencies e_i. The χ^2-statistic is given by

$$P = \sum_{i=1}^{n} \frac{(n_i - e_i)^2}{e_i}.$$

The χ^2 test for goodness of fit is for $H_0 : p_i = p_i(\theta)$, $\theta \in \Theta$, against $H_1 : p_i$ unrestricted, where p_i is the probability of occurrence of state i.

In the example, we assume that the fraction of calls arriving during all days except Sundays is fixed (and calculated from the data). Such an assumption is natural when the data array is massive. However, the fractions of calls within a given day from Monday to Saturday fluctuate, and we proceed as follows. Let $e^* = \frac{1}{6}(n_1 + \cdots + n_6)$, $e_1^* = \frac{1}{3}(n_2 + n_3 + n_4)$, $e_2^* = \frac{1}{2}(n_1 + n_5)$.

Under H_0: *on Monday–Saturday calls are received at the same rate*, the statistic is

$$\sum_{i=1}^{6} \frac{(n_i - e^*)^2}{e^*}$$

and has an approximately χ_5^2 distribution. (Here the number of degrees of freedom is five, since one parameter is fitted.)

In the second version, H_0 is that *on Tuesday, Wednesday and Thursday calls are received at one rate and on Monday and Friday at another rate.* Here, the statistic is

$$\sum_{i=2}^{4}\frac{(n_i-e_1^*)^2}{e_1^*}+\sum_{j=1,5}\frac{(n_j-e_2^*)^2}{e_2^*}$$

and has an approximately χ_3^2 distribution. $\square$

Problem 5.19 (MT-IB 1996-103G long) (i) Aerial observations x_1, x_2, x_3, x_4 are made of the interior angles $\theta_1, \theta_2, \theta_3, \theta_4$, of a quadrilateral on the ground. If these observations are subject to small independent errors with zero means and common variance σ^2, determine the least squares estimator of $\theta_1, \theta_2, \theta_3, \theta_4$.

(ii) Obtain an unbiased estimator of σ^2 in the situation described in part (i).

Suppose now that the quadrilateral is known to be a parallelogram with $\theta_1 = \theta_3$ and $\theta_2 = \theta_4$. What now are the least squares estimates of its angles? Obtain an unbiased estimator of σ^2 in this case.

Solution (i) The LSEs should minimise $\sum_{i=1}^{4}(\theta_i - x_i)^2$, subject to $\sum_{i=1}^{4}\theta_i = 2\pi$. The Lagrangian

$$L = \sum_i (\theta_i - x_i)^2 - \lambda\left(\sum_i \theta_i - 2\pi\right)$$

has

$$\frac{\partial}{\partial\theta_i}L = 0 \text{ when } 2(\theta_i - x_i) - \lambda = 0, \text{ i.e. } \widehat{\theta}_i = x_i + \frac{\lambda}{2}.$$

Adjusting $\sum\widehat{\theta}_i = 2\pi$ yields $\lambda = \frac{1}{2}(2\pi - \sum_i x_i)$, and

$$\widehat{\theta}_i = x_i + \frac{1}{4}\left(2\pi - \sum_i x_i\right).$$

(ii) From the least squares theory, X_i, $i = 1, 2, 3, 4$ has mean θ_i, variance σ^2, and the X_i are independent. Write

$$\mathbb{E}(X_i - \widehat{\theta}_i)^2 = \mathbb{E}\left[\frac{1}{4}\left(2\pi - \sum_i X_i\right)\right]^2 = \frac{1}{16}\mathbb{E}\left\{\left[\sum_i X_i - \mathbb{E}\left(\sum_i X_i\right)\right]^2\right\}$$

$$= \frac{1}{16}\mathrm{Var}\left(\sum_i X_i\right) = \frac{1}{16}\times 4\sigma^2 = \frac{\sigma^2}{4}.$$

Thus,

$$\mathbb{E}\left(\sum_i (X_i - \widehat{\theta}_i)^2\right) = \sigma^2,$$

and $\sum_i (x_i - \widehat{\theta}_i)^2$ is an unbiased estimator of σ^2.

If $\theta_1 = \theta_3$ and $\theta_2 = \theta_4$, the constraint becomes $2(\theta_1 + \theta_2) = 2\pi$, i.e. $\theta_1 + \theta_2 = \pi$. Then the Lagrangian

$$L = (\theta_1 - x_1)^2 + (\theta_2 - x_2)^2 + (\theta_1 - x_3)^2 + (\theta_2 - x_4)^2 - 2\lambda\,(\theta_1 + \theta_2 - \pi)$$

has

$$\frac{\partial}{\partial \theta_1} L = 2(\theta_1 - x_1) + 2(\theta_1 - x_3) - 2\lambda,$$

$$\frac{\partial}{\partial \theta_2} L = 2(\theta_2 - x_2) + 2(\theta_2 - x_4) - 2\lambda,$$

and is minimised at

$$\widehat{\theta}_1 = \frac{1}{2}(x_1 + x_3 + \lambda), \ \ \widehat{\theta}_2 = \frac{1}{2}(x_2 + x_4 + \lambda).$$

The constraint $\widehat{\theta}_1 + \widehat{\theta}_2 = \pi$ then gives $\lambda = (\pi - \sum_i x_i/2)$, and

$$\widehat{\theta}_1 = \frac{1}{2}(x_1 + x_3) + \frac{1}{4}\left(2\pi - \sum_i x_i\right) = \frac{1}{4}(x_1 + x_3 - x_2 - x_4) + \frac{\pi}{2},$$

and similarly

$$\widehat{\theta}_2 = \frac{1}{4}(x_2 + x_4 - x_1 - x_3) + \frac{\pi}{2}.$$

Now

$$\mathbb{E}\left(X_1 - \widehat{\theta}_1\right)^2 = \mathbb{E}\left(\frac{3X_1}{4} - \frac{X_3}{4} + \frac{X_2}{4} + \frac{X_4}{4} - \frac{\pi}{2}\right)^2$$

$$= \mathrm{Var}\left(\frac{3X_1}{4} - \frac{X_3}{4} + \frac{X_2}{4} + \frac{X_4}{4}\right)$$

$$= \left[\left(\frac{3}{4}\right)^2 + \left(\frac{1}{4}\right)^2 + \left(\frac{1}{4}\right)^2 + \left(\frac{1}{4}\right)^2\right]\sigma^2 = \frac{3}{4}\sigma^2.$$

The same holds for $i = 2, 3, 4$. Hence, $\sum(x_i - \widehat{\theta}_i)^2/3$ is an unbiased estimator of σ^2. ☐

Problem 5.20 (MT-IB 1996-203G long) (i) $X_1, \ X_2, \ \dots, \ X_n$ form a random sample from a distribution whose PDF is

$$f(x; \theta) = \begin{cases} 2x/\theta^2, & 0 \le x \le \theta \\ 0, & \text{otherwise,} \end{cases}$$

where the value of the positive parameter θ is unknown. Determine the MLE of the median of the distribution.

(ii) There is widespread agreement amongst the managers of the Reliable Motor Company that the number x of faulty cars produced in a month has a binomial distribution

$$\mathbb{P}(x = s) = \binom{n}{s} p^s (1 - p)^{n-s} \ \ (s = 0, 1, \dots, n; \ 0 \le p \le 1).$$

There is, however, some dispute about the parameter p. The general manager has a prior distribution for p which is uniform (i.e. with the PDF $f_p(x) = I(0 \le x \le 1)$), while the more pessimistic production manager has a prior distribution with density $f_p(x) = 2xI(0 \le x \le 1)$. Both PDFs are concentrated on $(0, 1)$.

In a particular month, s faulty cars are produced. Show that if the general manager's loss function is $(\widehat{p} - p)^2$, where $\widehat{p}$ is her estimate and p is the true value, then her best estimate of p is

$$\widehat{p} = \frac{s+1}{n+2}.$$

The production manager has responsibilities different from those of the general manager, and a different loss function given by $(1 - p)(\widehat{p} - p)^2$. Find his best estimator of p and show that it is greater than that of the general manager unless $s \ge n/2$.

You may assume that, for non-negative integers α, β,

$$\int_0^1 p^\alpha (1 - p)^\beta \, dp = \frac{\alpha!\beta!}{(\alpha + \beta + 1)!}.$$

Solution   (i) If m is the median, the equation

$$\int_0^m \frac{2x}{\theta^2} dx = \left. \frac{x^2}{\theta^2} \right|_0^m = \frac{1}{2}$$

gives $m = \theta/\sqrt{2}$. Then

$$f(x; m\sqrt{2}) = \frac{2x}{(\sqrt{2}m)^2} = \frac{x}{m^2}, \quad 0 \le x \le m\sqrt{2},$$

is maximised in m by $x/\sqrt{2}$, and $f(\mathbf{x}; m\sqrt{2}) = \prod f(x_i; m\sqrt{2})$ by

$$\widehat{m} = \frac{1}{\sqrt{2}} \max x_i.$$

(ii) As $\mathbb{P}_p(X = s) \propto p^s(1 - p)^{n-s}$, $s = 0, 1, \ldots n$, the posterior for the general manager (GM) is

$$\pi^{\mathrm{GM}}(p|s) \propto p^s(1 - p)^{n-s}I(0 < p < 1),$$

and for the production manager (PM)

$$\pi^{\mathrm{PM}}(p|s) \propto pp^s(1 - p)^{n-s}I(0 < p < 1).$$

Then the expected loss for the GM is minimised at the posterior mean:

$$\widehat{p}^{\mathrm{GM}} = \int_0^1 pp^s(1 - p)^{n-s}dp \bigg/ \int_0^1 p^s(1 - p)^{n-s}dp$$

$$= \frac{(s+1)!(n-s)!}{(n-s+s+2)!} \frac{(n-s+s+1)!}{s!(n-s)!} = \frac{s+1}{n+2}.$$

For the PM, the expected loss

$$\int_0^1 (1-p)(p-a)^2 \pi^{\mathrm{PM}}(p|s)\mathrm{d}p$$

is minimised at

$$a = \int_0^1 p(1-p)\pi^{\mathrm{PM}}(p|s)\mathrm{d}p \Big/ \int_0^1 (1-p)\pi^{\mathrm{PM}}(p|s)\mathrm{d}p,$$

which yields

$$\hat{p}^{\mathrm{PM}} = \int_0^1 p(1-p)pp^s(1-p)^{n-s}\mathrm{d}p \Big/ \int_0^1 p(1-p)pp^s(1-p)^{n-s}\mathrm{d}p$$

$$= \frac{(s+2)!(n-s+1)!}{(n-s+s+4)!}\, \frac{(n-s+s+3)!}{(s+1)!(n-s+1)!} = \frac{s+2}{n+4}.$$

We see that $(s+2)/(n+4) > (s+1)/(n+2)$ iff $s < n/2$.   □

Problem 5.21   (MT-IB 1996-403G long) (i) What is a simple hypothesis? Define the terms size and power for a test of one simple hypothesis against another.

State and prove the NP Lemma.

(ii) There is a single observation of an RV X which has a PDF $f(x)$. Construct the best test of size 0.05 for the null hypothesis

$$H_0: f(x) = \frac{1}{2} \ (-1 \le x \le 1)$$

against the alternative hypothesis

$$H_1: f(x) = \frac{3}{4}(1-x^2) \ (-1 \le x \le 1).$$

Calculate the power of your test.

Solution   (i) A simple hypothesis for a parameter θ is $H_0: \theta = \theta_0$. The size equals the probability of rejecting H_0 when it is true. The power equals the probability of rejecting H_0 when it is not true; for the simple alternative $H_1: \theta = \theta_1$ it is a number (in general, a function of the parameter varying within H_1).

The statement of the NP Lemma for $H_0: f = f_0$ against $H_1: f = f_1$ is as follows.

> *Among all tests of size $\le \alpha$, the test with maximum power is given by* $\mathcal{C} = \{x: f_1(x) > kf_0(x)\}$, *for k such that* $\mathbb{P}(x \in \mathcal{C}|H_0) = \int_{\mathcal{C}} f_0(x)\mathrm{d}x = \alpha.$

In other word, $\forall \ k > 0$ the test:

$$\text{reject } H_0 \text{ when } f_1(x) > kf_0(x)$$

has the maximum power among the tests of size $\le \alpha := \mathbb{P}_0(f_1(X) > kf_0(X))$.

The proof of the NP Lemma is given in Section 4.2.

(ii) The LR $f_1(x)/f_0(x) = 3(1 - x^2)/2$; we reject H_0 when it is $\geq k$, i.e. $|x| \leq (1 - 2k/3)^{1/2}$. We want

$$0.05 = \mathbb{P}\left(|x| \leq \left(1 - \frac{2k}{3}\right)^{1/2} \Big| H_0\right) = \left(1 - \frac{2k}{3}\right)^{1/2}.$$

That is the condition $|x| \leq (1 - 2k/3)^{1/2}$ is the same as $|x| \leq 0.05$. By the NP Lemma, the test

reject H_0 when $|x| \leq 0.05$

is the most powerful of size 0.05.
The power is then

$$\mathbb{P}(|x| \leq 0.05 | H_1) = \frac{3}{4}\int_{-0.05}^{0.05}(1 - x^2)dx = \frac{3}{2}\left(x - \frac{x^3}{3}\right)\Big|_0^{0.05} \approx 0.075. \quad \square$$

Problem 5.22 (MT-IB 1995-103G long) (i) Let $X_1, \ldots, X_m$ and $Y_1, \ldots, Y_n$ be independent random samples, respectively, from the $N(\mu_1, \sigma^2)$-and the $N(\mu_2, \sigma^2)$-distributions. Here the parameters μ_1, μ_2 and σ^2 are all unknown. Explain carefully how you would test the hypothesis H_0: $\mu_1 = \mu_2$ against H_1: $\mu_1 \neq \mu_2$.
(ii) Let $X_1, \ldots, X_n$ be a random sample from the distribution with the PDF

$$f(x; \theta) = e^{-(x-\theta)}, \text{ for } \theta < x < \infty,$$

where θ has a prior distribution the standard normal $N(0, 1)$. Determine the posterior distribution of θ.

Suppose that θ is to be estimated when the loss function is the absolute error loss, $L(a, \theta) = |a - \theta|$. Determine the optimal Bayes estimator and express it in terms of the function $c_n(x)$ defined by

$$2\Phi(c_n(x) - n) = \Phi(x - n), \text{ for } -\infty < x < \infty,$$

where

$$\Phi(x) = \frac{1}{\sqrt{2\pi}}\int_{-\infty}^{x} e^{-y^2/2}dy$$

is the standard normal distribution function.

Solution With

$$\overline{X} = \frac{1}{m}\sum_i X_i \sim N(\mu_1, \sigma^2/m) \text{ and } \overline{Y} = \frac{1}{n}\sum_i Y_j \sim N(\mu_2, \sigma^2/n),$$

we have that under H_0

$$\frac{1}{\sigma}(\overline{X} - \overline{Y})\left(\frac{1}{m} + \frac{1}{n}\right)^{-1/2} \sim N(0, 1).$$

Set

$$S_{XX} = \sum_{i=1}^{m}(X_i - \overline{X})^2 \sim \sigma^2 \chi^2_{m-1}, \quad S_{YY} = \sum_{j=1}^{n}(Y_j - \overline{Y})^2 \sim \sigma^2 \chi^2_{n-1}.$$

Then

$$\frac{1}{\sigma^2}(S_{XX} + S_{YY}) \sim \chi^2_{m+n-2},$$

and

$$t = (\overline{X} - \overline{Y})\left(\frac{1}{m} + \frac{1}{n}\right)^{-1/2} \bigg/ [(S_{XX} + S_{YY})/(m+n-2)]^{1/2} \sim t_{m+n-2}.$$

The MP test of size α rejects H_0 when $|t|$ exceeds $t_{m+n-2}(\alpha/2)$, the upper $\alpha/2$ point of the t_{m+n-2}-distribution.

(ii) By the Bayes' formula,

$$\pi(\theta|\mathbf{x}) \propto \pi(\theta)f(\mathbf{x};\theta) \propto e^{-\theta^2/2+n\theta-\sum_i x_i} I(\theta < \min x_i),$$

with the constant of proportionality

$$\left(\int e^{-\theta^2/2+n\theta-\sum_i x_i} I(\theta < \min x_i)d\theta\right)^{-1} = \frac{1}{\sqrt{2\pi}} \frac{\exp\left(-n^2/2 + \sum_i x_i\right)}{\Phi(\min x_i - n)}.$$

Under absolute error LF $L(a,\theta) = |a - \theta|$, the optimal Bayes estimator is the posterior median. That is we want s, where

$$\int_{-\infty}^{s} d\theta e^{-(\theta-n)^2/2} = \frac{1}{2}\int_{-\infty}^{\min x_i} e^{-(\theta-n)^2/2},$$

or, equivalently: $2\Phi(s - n) = \Phi(\min x_i - n)$, and $s = c_n(\min x_i)$, as required. $\square$

Problem 5.23 (MT-IB 1995-203G long) (i) Let $X_1, \ldots, X_n$ be a random sample from the distribution with the PDF

$$f(x;\theta) = \frac{2x}{\theta^2}, \quad \text{for } 0 \le x \le \theta.$$

Determine the MLE M of θ and show that $\left(M, M/(1-\gamma)^{\frac{1}{2n}}\right)$ is a $100\gamma\%$ CI for θ, where $0 < \gamma < 1$.

(ii) Let $X_1, \ldots, X_n$ be an independent random sample from the uniform distribution on $[0, \theta_1]$ and let $Y_1, \ldots, Y_n$ be an independent random sample from the uniform distribution on $[0, \theta_2]$. Derive the form of the likelihood ratio test of the hypothesis H_0: $\theta_1 = \theta_2$ against H_1: $\theta_1 \ne \theta_2$ and express this test in terms of the statistic

$$T = \frac{\max(M_X, M_Y)}{\min(M_X, M_Y)},$$

where $M_X = \max_{1 \le i \le n} X_i$ and $M_Y = \max_{1 \le i \le n} Y_i$.

By observing that under the hypothesis H_0 the distribution of T is independent of $\theta = \theta_1 = \theta_2$, or otherwise, determine exactly the critical region for the test of size α.

Solution   (i) The likelihood

$$f(\mathbf{x}; \theta) = \frac{2^n}{\theta^{2n}} \left(\prod_{i=1}^{n} x_i \right) I(\max x_i \le \theta)$$

is written as $g(M(\mathbf{x}, \theta))h(\mathbf{x})$, where $M(\mathbf{x}) = \max x_i$. Hence, $M = M(\mathbf{X}) = \max X_i$ is a sufficient statistic. It is also the MLE, with

$$\mathbb{P}(M \le u) = (\mathbb{P}(X_1 \le u))^n = \left(\frac{u}{\theta} \right)^{2n}, \quad 0 \le u \le \theta.$$

Then

$$\mathbb{P}\left(M \le \theta \le \frac{M}{(1 - \gamma)^{1/(2n)}} \right) = \mathbb{P}\left(\theta(1 - \gamma)^{1/(2n)} \le M \le \theta \right)$$

$$= 1 - \mathbb{P}\left(M < \theta(1 - \gamma)^{1/(2n)} \right)$$

$$= 1 - \frac{\left(\theta(1 - \gamma)^{1/(2n)} \right)^{2n}}{\theta^{2n}}$$

$$= 1 - (1 - \gamma) = \gamma.$$

Hence, $\left(M, M \big/ (1 - \gamma)^{1/(2n)} \right)$ is a $100\gamma\%$ CI for θ.

(ii) Under H_0,

$$f_{\mathbf{X},\mathbf{Y}} = \left(\frac{1}{\theta} \right)^{2n} I(0 \le x_1, \ldots, x_n, y_1, \ldots, y_n \le \theta),$$

is maximised at the MLE $\widehat{\theta} = \max(M_X, M_Y)$. Under H_1,

$$f_{\mathbf{X},\mathbf{Y}} = \left(\frac{1}{\theta_1} \right)^n \left(\frac{1}{\theta_2} \right)^n I(0 \le x_1, \ldots, x_n \le \theta_1) I(0 \le x_1, \ldots, x_n \le \theta_2),$$

is maximised at the MLE $(\widehat{\theta}_1, \widehat{\theta}_2) = (M_X, M_Y)$.

Then the ratio $\Lambda_{H_1;H_0}(\mathbf{x}, \mathbf{y})$ is

$$\frac{\left(\frac{1}{M_X} \right)^n \left(\frac{1}{M_Y} \right)^n}{\left[\frac{1}{\max(M_X, M_Y)} \right]^{2n}} = \frac{[\max(M_X, M_Y)]^{2n}}{(M_X M_Y)^n} = \left[\frac{\max(M_X, M_Y)}{\min(M_X, M_Y)} \right]^n = T^n.$$

So, we reject H_0 if $T(\mathbf{x}, \mathbf{y}) \ge k$ for some $k \ge 1$.

Now, under H_0

$$\mathbb{P}(M_X \le x) = \left(\frac{x}{\theta} \right)^n, \quad \text{i.e. } f_{M_X}(x) = n \frac{1}{\theta} \left(\frac{x}{\theta} \right)^{n-1},$$

and similarly for M_y. Then, for $0 < \alpha < 1$ and $k \geq 1$, we want α to be equal to

$$\mathbb{P}(T \geq k|H_0) = \frac{1}{\theta^2} \int_0^\theta \int_0^\theta n^2 \left(\frac{x}{\theta}\right)^{n-1} \left(\frac{y}{\theta}\right)^{n-1} I\left(\frac{\max[x,y]}{\min[x,y]} \geq k\right) dy dx$$

$$= 2n^2 \int_0^1 \int_0^{x/k} x^{n-1} y^{n-1} dy dx = 2n \int_0^1 x^{n-1} \frac{x^n}{k^n} dx = \frac{1}{k^n}.$$

So, $k = \alpha^{-1/n}$, and the critical region for the size α test is

$$\mathcal{C} = \left\{\mathbf{x}, \mathbf{y}: T > \alpha^{-1/n}\right\}. \quad \square$$

Problem 5.24   (MT-IB 1995-403G long) (i) State and prove the NP Lemma.

(ii) Let $X_1, \ldots, X_n$ be a random sample from the $N(\mu, \sigma^2)$-distribution. Prove that the random variables $\overline{X}$ (the sample mean) and $\sum_{i=1}^n (X_i - \overline{X})^2$ $((n-1) \times$ the sample variance) are independent and determine their distributions.

Suppose that

$$
\begin{matrix}
X_{11}, & \cdots & X_{1n}, \\
X_{21}, & \cdots & X_{2n}, \\
\vdots & & \vdots \\
X_{m1}, & \cdots & X_{mn},
\end{matrix}
$$

are independent RVs and that X_{ij} has the $N(\mu_i, \sigma^2)$-distribution for $1 \leq j \leq n$, where $\mu_1, \ldots, \mu_m, \sigma^2$ are unknown constants. With reference to your previous result, explain carefully how you would test the hypothesis $H_0: \mu_1 = \cdots = \mu_m$.

Solution   (Part (ii) only.) We claim:

(a) $\overline{X} = \sum_i X_i/n \sim N(\mu, \sigma^2/n)$;

(b) $\overline{X}$ and $S_{XX} = \sum_i (X_i - \overline{X})^2$ are independent;

(c) $S_{XX} = \sum_i (X_i - \overline{X})^2/\sigma^2 \sim \chi_{n-1}^2$.

To prove (a) note that linear combinations of normal RVs are normal, so, owing to independence, $\overline{X} \sim N(\mu, \sigma^2/n)$. Also, $(\overline{X} - \mu)^2/\sigma^2 \sim \chi_1^2$.

To prove (b) and (c) observe that

$$\sum_i (X_i - \mu)^2 = \sum_i [(X_i - \overline{X}) + (\overline{X} - \mu)]^2$$

$$= \sum_i [(X_i - \overline{X})^2 + 2(X_i - \overline{X})(\overline{X} - \mu) + (\overline{X} - \mu)^2]$$

$$= S_{XX} + n(\overline{X} - \mu)^2.$$

Given an orthogonal $n \times n$ matrix A, set

$$\mathbf{X} = \begin{pmatrix} X_1 \\ \vdots \\ X_n \end{pmatrix}, \quad \mathbf{Y} = \begin{pmatrix} Y_1 \\ \vdots \\ Y_n \end{pmatrix} \quad \text{where } \mathbf{Y} = A^{\mathsf{T}}(\mathbf{X} - \mu \mathbf{1}).$$

We want to choose A so that the first entry $Y_1 = \sqrt{n}(\overline{X} - \mu)$. That is A must be of the form

$$A = \begin{pmatrix} 1/\sqrt{n} & \cdots & \cdots & \cdots \\ 1/\sqrt{n} & \cdots & \cdots & \cdots \\ \vdots & \vdots & \vdots & \vdots \\ 1/\sqrt{n} & \cdots & \cdots & \cdots \end{pmatrix},$$

where the remaining columns are to be chosen so as to make the matrix orthogonal.

Then $Y_1 = \sqrt{n}(\overline{X} - \mu) \sim N(0, \sigma^2)$, and Y_1 is independent of $Y_2, \ldots, Y_n$. Since $\sum_i Y_i^2 = \sum_i (X_i - \mu)^2$, we have

$$\sum_{i=2}^{n} Y_i^2 = \sum_{i=1}^{n} (X_i - \mu)^2 - n(\overline{X} - \mu)^2 = S_{XX}.$$

Hence $S_{XX} = \sum_{i=2}^{n} Y_i^2$, where $Y_2, \ldots, Y_n$ are IID $N(0, \sigma^2)$ RVs. Therefore $S_{XX}/\sigma^2 \sim \chi_{n-1}^2$.

Now consider the RVs X_{ij} as specified. To test $H_0: \mu_1 = \cdots = \mu_m = \mu$ against H_1: $\mu_1, \ldots, \mu_m$ unrestricted, we use analysis of variance (one-way ANOVA). Write $N = mn$ and $X_{ij} = \mu_i + \epsilon_{ij}, j = 1, \ldots, n, i = 1, \ldots, m$, where ϵ_{ij} are IID $N(0, \sigma^2)$.

Apply the GLRT: the LR $\Lambda_{H_1;H_0}((x_{ij}))$ equals

$$\frac{\max_{\mu_1,\ldots,\mu_m,\sigma^2} (2\pi\sigma^2)^{-N/2} \exp\left[-\sum_{ij}(x_{ij} - \mu_i)^2/(2\sigma^2)\right]}{\max_{\mu,\sigma^2} (2\pi\sigma^2)^{-N/2} \exp\left[-\sum_{ij}(x_{ij} - \mu)^2/(2\sigma^2)\right]} = \left(\frac{S_0}{S_1}\right)^{N/2}.$$

Here

$$S_0 = \sum_{ij}(x_{ij} - \overline{x}_{++})^2, \quad S_1 = \sum_{ij}(x_{ij} - \overline{x}_{i+})^2,$$

and

$$\overline{x}_{i+} = \sum_{j=1}^{n} \frac{x_{ij}}{n}, \text{ the mean within group } i \text{ (and the MLE of } \mu_i \text{ under } H_1\text{)},$$

$$\overline{x}_{++} = \sum_{ij} \frac{x_{ij}}{N} = \sum_i \frac{n\overline{x}_{i+}}{N}, \text{ the overall mean (and the MLE of } \mu \text{ under } H_0\text{)}.$$

The test is of the form

$$\text{reject } H_0 \text{ when } \frac{S_0}{S_1} \text{ is large.}$$

Next,

$$S_0 = \sum_{i=1}^{m} \sum_{j=1}^{n} (x_{ij} - \overline{x}_{i+} + \overline{x}_{i+} - \overline{x}_{++})^2$$

$$= \sum_{i=1}^{m} \sum_{j=1}^{n} \left[(x_{ij} - \overline{x}_{i+})^2 + 2(x_{ij} - \overline{x}_{i+})(\overline{x}_{i+} - \overline{x}_{++}) + (\overline{x}_{i+} - \overline{x}_{++})^2 \right]$$

$$= \sum_{ij} (x_{ij} - \overline{x}_{i+})^2 + n \sum_{i} (\overline{x}_{i+} - \overline{x}_{++})^2 = S_1 + S_2,$$

where

$$S_2 = n \sum_{i} (\overline{x}_{i+} - \overline{x}_{++})^2.$$

Thus, S_0/S_1 is large when S_2/S_1 is large. S_1 is called the within groups (or within samples) and S_2 the between groups (between samples) sum of squares.

Next, whether or not H_0 is true,

$$\sum_{i} (X_{ij} - \overline{X}_{i+})^2 \sim \sigma^2 \chi_{n-1}^2 \ \forall \ i,$$

since $\mathbb{E} X_{ij}$ depends only on i. Hence,

$$S_1 \sim \sigma^2 \chi_{N-m}^2$$

as samples for different i are independent. Also, $\forall i$,

$$\sum_{j} (X_{ij} - \overline{X}_{i+})^2 \text{ is independent of } \overline{X}_{i+}.$$

Therefore, S_1 is independent of S_2. If H_0 is true,

$$S_2 \sim \sigma^2 \chi_{m-1}^2.$$

Further, if H_0 is not true, S_2 has

$$\mathbb{E} S_2 = (m-1)\sigma^2 + n \sum_{i=1}^{m} (\mu_i - \overline{\mu})^2,$$

where $\overline{\mu} = \sum_i \mu_i / m$.

Moreover, under H_1, the value of S_2 tends to be inflated. So, if H_0 is true, then

$$Q = \frac{S_2/(m-1)}{S_1/(N-m)} \sim F_{m-1, N-m},$$

while under H_1 the value of Q is inflated. Thus, in a size α test we reject H_0 when the value of statistic Q is larger than $\phi_{m-1, N-m}^+(\alpha)$, the upper α quantile of $F_{m-1, N-m}$. □

The Private Life of C.A.S. Anova

(From the series 'Movies that never made it to the Big Screen'.)

Problem 5.25 (MT-IB 1994-103F long) (i) At a particular time three high street restaurants are observed to have 43, 41 and 30 customers, respectively. Detailing carefully the underlying assumptions that you are making, explain how you would test the hypothesis that all three restaurants are equally popular against the alternative that they are not.

(ii) Explain the following terms in the context of hypothesis testing:

(a) simple hypothesis;
(b) composite hypothesis;
(c) type I and type II error probabilities;
(d) size of a test;
(e) power of a test.

Let $X_1, \ldots, X_n$ be a sample from the $N(\mu, 1)$-distribution. Construct the most powerful size α test of the hypothesis $H_0 : \mu = \mu_0$ against $H_1 : \mu = \mu_1$, where $\mu_1 > \mu_0$.

Find the test that minimises the larger of the two error probabilities. Justify your answer carefully.

Solution (i) In total, 114 customers have been counted. Assuming that each customer chooses one of the three restaurants with probabilities p_1, p_2, p_3, independently, we should work with a multinomial distribution. The null hypothesis $H_0 : p_1 = p_2 = p_3 = 1/3$, with the expected numbers 38. The value of the Pearson χ^2-statistic:

$$P = \sum \frac{(\text{observed} - \text{expected})^2}{\text{expected}} = \frac{25 + 9 + 64}{38} = \frac{98}{38} = 2.579.$$

Given $\alpha \in (0, 1)$, we compare P with $h_2^+(\alpha)$, the upper α quantile of χ_2^2. The size α test will reject H_0 when $P > h_2^+(\alpha)$.

(ii) (a) A simple hypothesis H_0 is that the PDF/PMF $f = f_0$, a completely specified probability distribution that enables explicit numerical probabilities to be calculated.

(b) A composite hypothesis is $f \in$ a set of PDFs/PMFs.

(c) The TIEP is $\mathbb{P}(\text{reject } H_0 | H_0 \text{ true})$ and the TIIEP is $\mathbb{P}(\text{accept } H_0 | H_0 \text{ not true})$. For a simple H_0, the TIEP is a number, and for a composite H_0 it is a function (of an argument running over the parameter set corresponding to H_0). The TIIEP has a similar nature.

(d) The size of a test is equal to the maximum of the TIEP taken over the parameter set corresponding to H_0. If H_0 is simple, it is simply the TIEP.

(e) Similarly, the power is 1 minus the TIIEP. It should be considered as a function on the set of parameters corresponding to H_1.

To construct the MP test, we use the NP Lemma. The LR

$$\frac{f(\mathbf{x}; \mu_1)}{f(\mathbf{x}; \mu_0)} = \exp\left(-\frac{1}{2}\sum_i \left[(x_i - \mu_1)^2 - (x_i - \mu_0)^2\right]\right)$$

$$= \exp\left[n(\mu_1 - \mu_0)\bar{x} + \frac{n}{2}(\mu_0^2 - \mu_1^2)\right]$$

is large when $\bar{x}$ is large. Thus the critical region for the MP test of size α is

$$\mathcal{C} = \{\mathbf{x} : \bar{x} > c\}$$

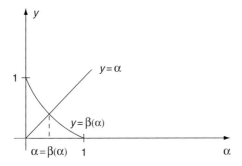

Figure 5.3

where

$$\alpha = \mathbb{P}_{\mu_0}\left(\overline{X} > c\right) = 1 - \Phi\left(\sqrt{n}(c - \mu_0)\right),$$

i.e.

$$c = \mu_0 + \frac{1}{\sqrt{n}}z_+(\alpha), \quad \text{where} \quad \frac{1}{\sqrt{2\pi}}\int_{z_+(\alpha)}^{\infty}e^{-x^2/2}dx = \alpha.$$

The test minimising the larger of the error probabilities must again be NP; otherwise there would be a better one. For size α, the TIIEP as a function of α is

$$\beta(\alpha) = \mathbb{P}_{\mu_1}\left(\overline{X} < c\right) = \Phi\left(z_+(\alpha) + \sqrt{n}(\mu_0 - \mu_1)\right),$$

where $z_+(\alpha)$ is the upper α quantile of $N(0, 1)$. Clearly, $\max[\alpha, \beta(\alpha)]$ is minimal when $\alpha = \beta(\alpha)$. See Figure 5.3.

We know that when α increases, $z_+(\alpha)$ decreases. We choose α with

$$\alpha = \Phi\left(z_+(\alpha) + \sqrt{n}(\mu_0 - \mu_1)\right), \text{i.e. } z_+(\alpha) = -\frac{\sqrt{n}}{2}(\mu_0 - \mu_1).$$

This yields

$$c = \mu_0 - \frac{1}{2}(\mu_0 - \mu_1) = \frac{\mu_0 + \mu_1}{2}. \quad \square$$

Problem 5.26 (MT-IB 1994-203F long) (i) Let $X_1, \ldots, X_m$ be a random sample from the $N(\mu_1, \sigma_1^2)$-distribution and let $Y_1, \ldots, Y_n$ be an independent sample from the $N(\mu_2, \sigma_2^2)$-distribution. Here the parameters μ_1, μ_2, σ_1^2 and σ_2^2 are unknown. Explain carefully how you would test the hypothesis H_0: $\sigma_1^2 = \sigma_2^2$ against H_1: $\sigma_1^2 \neq \sigma_2^2$.

(ii) Let $Y_1, \ldots, Y_n$ be independent RVs, where Y_i has the $N(\beta x_i, \sigma^2)$-distribution, $i = 1, \ldots, n$. Here $x_1, \ldots, x_n$ are known but β and σ^2 are unknown.

(a) Determine the maximum-likelihood estimators $(\widehat{\beta}, \widehat{\sigma}^2)$ of (β, σ^2).
(b) Find the distribution of $\widehat{\beta}$.

(c) By showing that $Y_i - \widehat{\beta} x_i$ and $\widehat{\beta}$ are independent, or otherwise, determine the joint distribution of $\widehat{\beta}$ and $\widehat{\sigma}^2$.

(d) Explain carefully how you would test the hypothesis H_0: $\beta = \beta_0$ against H_1: $\beta \neq \beta_0$.

Solution (i) Set

$$S_{xx} = \sum_{i=1}^{m}(x_i - \bar{x})^2, \text{ with } \frac{1}{\sigma_1^2} S_{XX} \sim \chi^2_{m-1},$$

and

$$S_{yy} = \sum_{j=1}^{n}(y_j - \bar{y})^2, \text{ with } \frac{1}{\sigma_2^2} S_{YY} \sim \chi^2_{n-1}.$$

Moreover, S_{XX} and S_{YY} are independent.

Then under H_0,

$$\frac{1}{m-1}S_{XX} \bigg/ \frac{1}{n-1}S_{YY} \sim F_{m-1,n-1},$$

and in a size α test we reject H_0 when $[S_{xx}/(m-1)]/[S_{yy}/(n-1)]$ is either $< \varphi^-_{m-1,n-1}(\alpha/2)$ or $> \varphi^+_{m-1,n-1}(\alpha/2)$, where $\varphi^{\pm}_{m-1,n-1}(\alpha/2)$ is the upper/lower quantile of $F_{m-1,n-1}$.

(ii) The likelihood

$$f_{\beta,\sigma^2}(\mathbf{Y}) = \left(\frac{1}{\sqrt{2\pi}\sigma}\right)^n \exp\left[-\frac{1}{2\sigma^2}\sum_i (Y_i - \beta x_i)^2\right].$$

(a) The MLE $(\widehat{\beta}, \widehat{\sigma}^2)$ of (β, σ^2) is found from

$$\frac{\partial}{\partial\beta} \ln f_{\beta,\sigma^2}(\mathbf{Y}) = \frac{1}{\sigma^2}\sum_i x_i(Y_i - \beta x_i) = 0,$$

$$\frac{\partial}{\partial\sigma^2} \ln f_{\beta,\sigma^2}(\mathbf{Y}) = -\frac{n}{2\sigma^2} + \frac{1}{2\sigma^4}\sum_i (Y_i - \beta x_i)^2 = 0,$$

and is

$$\widehat{\beta} = \frac{\sum_i x_i Y_i}{\sum_i x_i^2}, \quad \widehat{\sigma}^2 = \frac{1}{n}\sum_i (Y_i - \widehat{\beta} x_i)^2.$$

This gives the global minimum, as we minimise a convex quadratic function $-\ln f_{\beta,\sigma^2}(\mathbf{Y})$.

(b) As a linear combination of independent normals, $\widehat{\beta} \sim N(\beta, \sigma^2/\sum_i x_i^2)$.

(c) As $\widehat{\beta}$ and $Y_i - \widehat{\beta} x_i$ are jointly normal, they are independent iff their covariance vanishes. As $\mathbb{E}(Y_i - \widehat{\beta} x_i) = 0$,

$$\text{Cov}(Y_i - \widehat{\beta} x_i, \widehat{\beta}) = \mathbb{E}(Y_i - \widehat{\beta} x_i)\widehat{\beta} = \mathbb{E}(Y_i\widehat{\beta}) - x_i\mathbb{E}(\widehat{\beta}^2),$$

which is zero since

$$\mathbb{E}(Y_i\widehat{\beta}) = \sum_{j\neq i} x_j \frac{\mathbb{E}Y_i\mathbb{E}Y_j}{\sum_k x_k^2} + x_i \frac{\mathbb{E}Y_i^2}{\sum_k x_k^2}$$

$$= \sum_{j\neq i} \frac{\beta^2 x_i x_j^2}{\sum_k x_k^2} + x_i \frac{\sigma^2 + \beta^2 x_i^2}{\sum_k x_k^2}$$

$$= x_i\left(\beta^2 + \frac{\sigma^2}{\sum_k x_k^2}\right) = x_i\left(\operatorname{Var}\widehat{\beta} + (\mathbb{E}\widehat{\beta})^2\right) = x_i\mathbb{E}(\widehat{\beta}^2).$$

In a similar fashion we can check that $Y_1 - \widehat{\beta}x_1, \ldots, Y_n - \widehat{\beta}x_n$, $\widehat{\beta}$ are independent and normal. Then $\widehat{\sigma}^2$ and $\widehat{\beta}$ are independent.

Next, the sum $\sum_i(Y_i - \beta x_i)^2$ equals

$$\sum_i\left[Y_i - \widehat{\beta}x_i + (\widehat{\beta} - \beta)x_i\right]^2 = \sum_i(Y_i - \widehat{\beta}x_i)^2 + \left(\sum_i x_i^2\right)(\widehat{\beta} - \beta)^2.$$

As

$$\frac{1}{\sigma^2}\sum_i(Y_i - \beta x_i)^2 \sim \chi_n^2 \text{ and } \frac{1}{\sigma^2}\left(\sum_i x_i^2\right)(\widehat{\beta} - \beta)^2 \sim \chi_1^2,$$

we conclude that $\widehat{\sigma}^2 n/\sigma^2 \sim \chi_{n-1}^2$.

(d) Under H_0

$$T = \frac{(\widehat{\beta} - \beta_0)\sqrt{\sum_i x_i^2}}{\sqrt{\frac{1}{n-1}\sum_i(Y_i - \widehat{\beta}x_i)^2}} \sim t_{n-1}.$$

Hence, the test is

reject H_0 if $|T| > t_{n-1}(\alpha/2)$, the upper $\alpha/2$ point of t_{n-1}. □

Problem 5.27   (MT-IB 1994-403F long) (i) Let X be a random variable with the PDF

$$f(x; \theta) = e^{-(x-\theta)}, \quad \theta < x < \infty,$$

where θ has a prior distribution the exponential distribution with mean 1. Determine the posterior distribution of θ.

Find the optimal Bayes estimator of θ based on X under quadratic loss.

(ii) Let $X_1, \ldots, X_n$ be a sample from the PDF

$$f(x; \lambda, \mu) = \begin{cases} \dfrac{1}{\lambda + \mu}e^{-x/\lambda}, & x \geq 0, \\[2mm] \dfrac{1}{\lambda + \mu}e^{x/\mu}, & x < 0, \end{cases}$$

where $\lambda > 0$ and $\mu > 0$ are unknown parameters. Find (simple) sufficient statistics for (λ, μ) and determine the MLEs $(\widehat{\lambda}, \widehat{\mu})$ of (λ, μ).

Now suppose that $n = 1$. Is $\widehat{\lambda}$ an unbiased estimator of λ? Justify your answer.

Solution   (i) For the posterior PDF, write

$$\pi(\theta|\mathbf{x}) \propto e^{-\theta}e^{-(x-\theta)}I(x > \theta > 0) \propto I(0 < \theta < x).$$

So, the posterior is $U(0, x)$.

Under quadratic loss, the optimal estimator is the posterior mean, i.e. $x/2$.

(ii)

$$f(\mathbf{x}; \lambda, \mu) = \frac{1}{(\lambda + \mu)^n} \exp\left[-\sum_i x_i I(x_i \geq 0)/\lambda + \sum_i x_i I(x_i < 0)/\mu \right].$$

Hence

$$T(\mathbf{x}) = (S^+, S^-), \quad S^+ = \sum_i x_i I(x_i \geq 0), \quad S^- = \sum_j x_j I(x_j < 0)$$

is a sufficient statistic.

To find the MLE $(\widehat{\lambda}, \widehat{\mu})$, differentiate $\ell(\mathbf{x}; \lambda, \mu) = \ln f(\mathbf{x}; \lambda, \mu)$:

$$\frac{\partial}{\partial \lambda}\ell(\mathbf{x}; \lambda, \mu) = -\frac{n}{\lambda + \mu} + \frac{S^+}{\lambda^2} = 0,$$

$$\frac{\partial}{\partial \mu}\ell(\mathbf{x}; \lambda, \mu) = -\frac{n}{\lambda + \mu} - \frac{S^-}{\mu^2} = 0,$$

whence

$$\widehat{\lambda} = \frac{1}{n}\left(S^+ + \sqrt{-S^- S^+}\right), \quad \widehat{\mu} = \frac{1}{n}\left(-S^- + \sqrt{-S^- S^+}\right),$$

which is the only solution. These values maximise ℓ, i.e. give the MLE for (λ, μ) as $\ell \to -\infty$ when either λ or μ tend to 0 or ∞.

This argument works when both $S^+, -S^- > 0$. If one of them is 0, the corresponding parameter is estimated by 0. So, the above formula is valid $\forall$ samples $\mathbf{x} \in \mathbb{R}^n \setminus \{\mathbf{0}\}$.

For $n = 1$, $\widehat{\lambda} = xI(x \geq 0)$. As

$$\mathbb{E}\widehat{\lambda} = \frac{1}{\lambda + \mu}\int_0^\infty xe^{-x/\lambda}\mathrm{d}x$$

obviously depends on μ, it cannot give λ. So, $\widehat{\lambda}$ is biased.

An exception is $\mu = 0$. Then $\mathbb{E}\widehat{\lambda} = \lambda$, and it is unbiased. In general, $\mathbb{E}S^+ = n\lambda^2/(\lambda + \mu)$. So,

$$\mathbb{E}\widehat{\lambda} = \frac{\lambda^2}{\lambda + \mu} + \frac{1}{n}\mathbb{E}\sqrt{-S^- S^+}.$$

The second summand is ≥ 0 for $n > 1$ unless $\mu = 0$ and $S^- = 0$. Thus, in the exceptional case $\mu = 0$, $\mathbb{E}\widehat{\lambda} = \lambda \; \forall \; n \geq 1$.   □

Problem 5.28 (MT-IB 1993-103J long) (i) A sample $x_1, \ldots, x_n$ is taken from a normal distribution with an unknown mean μ and a known variance σ^2. Show how to construct the most powerful test of a given size $\alpha \in (0, 1)$ for a null hypothesis $H_0: \mu = \mu_0$ against an alternative $H_1: \mu = \mu_1$ ($\mu_0 \neq \mu_1$).

What is the value of α for which the power of this test is $1/2$?

(ii) State and prove the NP Lemma. For the case of simple null and alternative hypotheses, what sort of test would you propose for minimising the sum of probabilities of type I and type II errors? Justify your answer.

Solution   (i) Since both hypotheses are simple, the MP test of size $\leq \alpha$ is the LR test with the critical region

$$\mathcal{C} = \left\{ \mathbf{x} : \frac{f(\mathbf{x}|H_1)}{f(\mathbf{x}|H_0)} > k \right\}$$

where k is such that the TIEP $\mathbb{P}(\mathcal{C}|H_0) = \alpha$. We have

$$f(\mathbf{x}|H_0) = \prod_{i=1}^{n} \frac{1}{\sqrt{2\pi}\sigma} e^{-(x_i - \mu_0)^2/(2\sigma^2)}, \quad f(\mathbf{x}|H_1) = \prod_{i=1}^{n} \frac{1}{\sqrt{2\pi}\sigma} e^{-(x_i - \mu_1)^2/(2\sigma^2)},$$

and

$$\Lambda_{H_1;H_0}(\mathbf{x}) = \ln \frac{f(\mathbf{x}|H_1)}{f(\mathbf{x}|H_0)} = -\frac{1}{2\sigma^2} \sum_i [(x_i - \mu_1)^2 - (x_i - \mu_0)^2]$$

$$= \frac{n}{2\sigma^2} [2\bar{x}(\mu_1 - \mu_0) - (\mu_1^2 - \mu_0^2)].$$

Case 1: $\mu_1 > \mu_0$. Then we reject H_0 when $\bar{x} > k$, where $\alpha = \mathbb{P}(\bar{X} > k|H_0)$. Under H_0, $\bar{X} \sim N(\mu_0, \sigma^2/n)$, since $X_i \sim N(\mu_0, \sigma^2)$, independently. Then $\sqrt{n}(\bar{X} - \mu_0)/\sigma \sim N(0, 1)$. Thus we reject H_0 when $\bar{x} > \mu_0 + \sigma z_+(\alpha)/\sqrt{n}$ where $z_+(\alpha) = \Phi^{-1}(1 - \alpha)$ is the upper α point of $N(0, 1)$.

The power of the test is

$$\frac{1}{\sqrt{2\pi}\sigma} \int_{\sigma\sqrt{n}z_+(\alpha)+n\mu_0}^{\infty} \mathrm{d}y \, e^{-(y-n\mu_1)^2/(2\sigma^2 n)} = \frac{1}{\sqrt{2\pi}} \int_{z_+(\alpha)+\sqrt{n}(\mu_0-\mu_1)/\sigma}^{\infty} e^{-y^2/2} \mathrm{d}y.$$

It equals $1/2$ when $z_+(\alpha) = (\mu_1 - \mu_0)\sqrt{n}/\sigma$, i.e.

$$\alpha = \frac{1}{\sqrt{2\pi}} \int_{\sqrt{n}(\mu_1-\mu_0)/\sigma}^{\infty} e^{-y^2/2} \mathrm{d}y.$$

Case 2: $\mu_1 < \mu_0$. Then we reject H_0 when $\bar{x} < \mu_0 - \sigma z_+(\alpha)/\sqrt{n}$, by a similar argument. The power equals $1/2$ when

$$\alpha = \frac{1}{\sqrt{2\pi}} \int_{-\infty}^{\sqrt{n}(\mu_1-\mu_0)/\sigma} e^{-y^2/2} \mathrm{d}y.$$

Hence, the power is $1/2$ when

$$\alpha = \frac{1}{\sqrt{2\pi}} \int_{\sqrt{n}|\mu_0-\mu_1|/\sigma}^{\infty} e^{-y^2/2} \mathrm{d}y.$$

(ii) (Last part only) Assume the continuous case: $f(\cdot|H_0)$ and $f_1(\cdot|H_1)$ are PDFs. Take the test with the critical region $\mathcal{C} = \{x : f(x|H_1) > f(x|H_0)\}$, with error probabilities $\mathbb{P}(\mathcal{C}|H_0)$ and $1 - \mathbb{P}(\mathcal{C}|H_1)$. Then for any test with a critical region $\mathcal{C}^*$, the sum of error probabilities is

$$\mathbb{P}(\mathcal{C}^*|H_0) + 1 - \mathbb{P}(\mathcal{C}^*|H_1) = 1 + \int_{\mathcal{C}^*} [f(x|H_0) - f(x|H_1)] \, dx$$

$$= 1 + \int_{\mathcal{C}^* \cap \mathcal{C}} [f(x|H_0) - f(x|H_1)] \, dx + \int_{\mathcal{C}^* \setminus \mathcal{C}} [f(x|H_0) - f(x|H_1)] \, dx$$

$$\geq 1 + \int_{\mathcal{C}^* \cap \mathcal{C}} [f(x|H_0) - f(x|H_1)] \, dx$$

as the integral over $\mathcal{C}^* \setminus \mathcal{C}$ is ≥ 0.

Next,

$$1 + \int_{\mathcal{C}^* \cap \mathcal{C}} [f(x|H_0) - f(x|H_1)] \, dx$$

$$= 1 + \int_{\mathcal{C}} [f(x|H_0) - f(x|H_1)] \, dx - \int_{\mathcal{C} \setminus \mathcal{C}^*} [f(x|H_0) - f(x|H_1)] \, dx$$

$$\geq 1 + \int_{\mathcal{C}} [f(x|H_0) - f(x|H_1)] \, dx = \mathbb{P}(\mathcal{C}|H_0) + 1 - \mathbb{P}(\mathcal{C}|H_1),$$

as the integral over $\mathcal{C} \setminus \mathcal{C}^*$ is < 0. $\square$

It has to be said that the development of Statistics after the Neyman–Pearson Lemma was marred by long-lasting controversies to which a great deal of personal animosity was added. The most prominent (and serious in its consequences) was perhaps a Fisher–Neyman dispute that opened in public in 1935 and continued even after Fisher's death in 1962 (one of Neyman's articles was entitled 'The Silver jubilee of my dispute with Fisher').

Two authoritative books on the history of Statistics, [FiB] and [Rei], give different versions of how it all started and developed. According to [FiB], p. 263, it was Neyman who in 1934–1935 'sniped at Fisher in his lectures and blew on the unquenched sparks of misunderstanding between the (Fisher's and K. Pearson's) departments (at University College London) with apparent, if undeliberate, genius for making mischief.' On the other hand, [Rei] clearly lays the blame on Fisher, supporting it with quotations attributed to a number of people, such as J.R. Oppenheimer (the future head of the Los Alamos atomic bomb project), who supposedly said of Fisher in 1936: 'I took one look at him and decided I did not want to meet him' (see [Rei], p. 144). In this situation, it is plausible that F.N. David (1909–1993), a prominent British statistician, who knew all parties involved well, was right when she said: 'They (Fisher, Neyman, K. Pearson, E. Pearson) were all jealous of one another, afraid that somebody would get ahead.' And on the particular issue of the NP Lemma: 'Gosset didn't have a jealous bone in his body. He asked the question.

Egon Pearson to a certain extent phrased the question which Gosset had asked in statistical parlance. Neyman solved the problem mathematically' ([Rei], p. 133). According to [FiB], p. 451, the NP Lemma was 'originally built in part on Fisher's work,' but soon 'diverged from it. It came to be very generally accepted and widely taught, especially in the United States. It was not welcomed by Fisher ... '.

David was one of the great-nieces of Florence Nightingale. It is interesting to note that David was the first woman Professor of Statistics in the UK, while Florence Nightingale was the first woman Fellow of the Royal Statistical Society. At the end of her career David moved to California but for decades maintained her cottage and garden in south-east England.

It has to be said that the lengthy arguments about the NP Lemma and the theory (and practice) that stemmed from it, which have been produced in the course of several decades, reduced in essence to the following basic question. You observe a sample, $(x_1, \ldots, x_n)$. What can you (reasonably) say about a (supposedly) random mechanism that is behind it? According to a persistent opinion, the Fisherian approach will be conspicuous in the future development of statistical sciences (see, e.g. [E1]). But even recognised leaders of modern statistics do not claim to have a clear view on this issue (see a discussion following the main presentation in [E1]). However, it should not distract us too much from our humble goal.

Problem 5.29 (MT-IB 1993-203J long) (i) Explain what is meant by constructing a confidence interval for an unknown parameter θ from a given sample $x_1, \ldots, x_n$. Let a family of PDFs $f(x; \theta)$, $-\infty < \theta < \infty$, be given by

$$f(x; \theta) = \begin{cases} e^{-(x-\theta)}, & x \geq \theta, \\ 0, & x < \theta. \end{cases}$$

Suppose that $n = 4$ and $x_1 = -1.0$, $x_2 = 1.5$, $x_3 = 0.5$, $x_4 = 1.0$. Construct a 95% confidence interval for θ.

(ii) Let $f(x; \mu, \sigma^2)$ be a family of normal PDFs with an unknown mean μ and an unknown variance $\sigma^2 > 0$. Explain how to construct a 95% confidence interval for μ from a sample $x_1, \ldots, x_n$. Justify the claims about the distributions you use in your construction.

> My Left Tail
>
>   (From the series *'Movies that never made it to the Big Screen'*.)

Solution (i) To construct a $100(1 - \gamma)\%$CI, we need to find two estimators $a = a(\mathbf{x})$, $b = b(\mathbf{x})$ such that $\mathbb{P}(a(\mathbf{X}) \leq \theta \leq b(\mathbf{X})) \geq 1 - \gamma$.

In the example,

$$f(\mathbf{x}; \theta) = e^{-\sum_i x_i + n\theta} I(x_i \geq \theta \, \forall \, i).$$

So $\min X_i$ is a sufficient statistic and $\min X_i - \theta \sim \text{Exp}(n)$. Then we can take $a = b - \delta$ and $b = \min x_i$, where

$$\int_\delta^\infty n e^{-nx} dx = e^{-n\delta} = \gamma.$$

With $\gamma = 0.05$, $n = 4$ and $\min x_i = -1$, we obtain the CI

$$\left[-1 - \frac{\ln 20}{4}, -1 \right].$$

In a different solution, one considers

$$\sum_{i=1}^{n} X_i - n\theta \sim \text{Gam}\,(n, 1), \quad \text{or} \quad 2\sum_{i=1}^{n}(X_i - \theta) \sim \chi^2_{2n}.$$

Hence we can take

$$a = \left(\sum_i X_i - \frac{1}{2}h^+_{2n}(\gamma/2) \right) \Big/ n, \quad b = \left(\sum_i X_i - \frac{1}{2}h^-_{2n}(\gamma/2) \right) \Big/ n,$$

where $h^\pm_{2n}(\gamma/2)$ are the upper/lower $\gamma/2$ quantiles of χ^2_{2n}. With $\gamma = 0.05$, $n = 4$ and $\sum_i x_i = 2$, we obtain

$$\left[\frac{1}{2} - \frac{h^+_8(0.025)}{8}, \frac{1}{2} - \frac{h^-_8(0.025)}{8} \right].$$

The precise value for the first interval is $[-1.749, -1]$, and that for the second

$$\left[\frac{1}{2} - \frac{17.530}{8}, \frac{1}{2} - \frac{2.180}{8} \right] = [-1.6912, 0.2275].$$

The endpoint 0.2275 is of course of no use since we know that $\theta \le x_1 = -1.0$. Replacing 0.2275 by -1 we obtain a shorter interval.

(ii) Define

$$\widehat{\mu} = \frac{1}{n}\sum_{i=1}^{n} X_i, \quad \widehat{\sigma}^2 = \frac{1}{n-1}\sum_{i=1}^{n}(X_i - \overline{X})^2.$$

Then (a) $\sqrt{n}(\widehat{\mu} - \mu)/\sigma \sim N(0, 1)$ and (b) $(n-1)\widehat{\sigma}^2/\sigma^2 \sim \chi^2_{n-1}$, independently, implying that

$$\sqrt{n}\frac{\widehat{\mu} - \mu}{\widehat{\sigma}} \sim t_{n-1}.$$

Hence, an equal-tailed CI is

$$\left[\widehat{\mu} - \frac{\widehat{\sigma}}{\sqrt{n}}t_{n-1}(0.025), \widehat{\mu} + \frac{\widehat{\sigma}}{\sqrt{n}}t_{n-1}(0.025) \right],$$

where $t_{n-1}(\alpha/2)$ is the $\alpha/2$ point of t_{n-1}, $\alpha = 0.05$. The justification of claims (a), (b) has been given above. $\square$

Problem 5.30 (MT-IB 1993-403J long) (i) State and prove the factorisation criterion for sufficient statistics, in the case of discrete random variables.

(ii) A linear function $y = Ax + B$ with unknown coefficients A and B is repeatedly measured at distinct points $x_1, \ldots, x_k$: first n_1 times at x_1, then n_2 times at x_2 and so on; and finally n_k times at x_k. The result of the ith measurement series is a sample $y_{i1}, \ldots, y_{in_i}$, $i = 1, \ldots, k$. The errors of all measurements are independent normal variables, with mean zero and variance 1. You are asked to estimate A and B from the whole sample y_{ij}, $1 \leq j \leq n_i$, $1 \leq i \leq k$. Prove that the maximum likelihood and the least squares estimators of (A, B) coincide and find these.

Denote by $\widehat{A}$ the maximum likelihood estimator of A and by $\widehat{B}$ the maximum likelihood estimator of B. Find the distribution of $\left(\widehat{A}, \widehat{B}\right)$.

Solution   (Part (ii) only) Define

$$\langle n \rangle = \sum_i n_i, \quad \bar{x} = \frac{\sum_i x_i n_i}{\langle n \rangle}$$

and

$$u_i = x_i - \bar{x}, \quad \text{with } \langle un \rangle = \sum_i u_i n_i = 0, \ \langle u^2 n \rangle = \sum_i u_i^2 n_i > 0.$$

Let $\alpha = A$, $\beta = B + A\bar{x}$, then $y_{ij} = \alpha u_i + \beta + \epsilon_{ij}$ and $Y_{ij} \sim N(\alpha u_i + \beta, 1)$, i.e.

$$f_{Y_{ij}}(y_{ij}) = \frac{1}{\sqrt{2\pi}} \exp\left[-\frac{1}{2}(y_{ij} - \alpha u_i - \beta)^2\right].$$

To find the MLE, we need to maximise $\prod_{i,j} \exp\left[-\frac{1}{2}(y_{ij} - \alpha u_i - \beta)^2\right]$. This is equivalent to minimising the quadratic function $\sum_{i,j}(y_{ij} - \alpha u_i - \beta)^2$. The last problem gives precisely the least squares estimators. Therefore the MLEs and the LSEs coincide.

To find the LSEs, we solve

$$\frac{\partial}{\partial \beta} \sum_{ij}(Y_{ij} - \beta - \alpha u_i)^2 = 0 \Leftrightarrow \widehat{\beta} = \frac{1}{\langle n \rangle} \sum_{i,j} Y_{ij},$$

$$\frac{\partial}{\partial \alpha} \sum_{ij}(Y_{ij} - \beta - \alpha u_i)^2 = 0 \Leftrightarrow \widehat{\alpha} = \frac{1}{\langle u^2 n \rangle} \sum_{i,j} Y_{ij} u_i,$$

with

$$\widehat{\beta} \sim N\left(\beta, \frac{1}{\langle n \rangle}\right), \quad \widehat{\alpha} \sim N\left(\alpha, \frac{1}{\langle u^2 n \rangle}\right).$$

That is

$$\widehat{A} = \frac{1}{\langle u^2 n \rangle} \sum_{i,j} Y_{ij} u_i \sim N\left(A, \frac{1}{\langle u^2 n \rangle}\right),$$

$$\widehat{B} = \frac{1}{\langle n \rangle} \sum_{i,j} Y_{ij} - \frac{\bar{x}}{\langle u^2 n \rangle} \sum_{i,j} Y_{ij} u_i \sim N\left(B, \frac{1}{\langle n \rangle} + \frac{\bar{x}^2}{\langle u^2 n \rangle}\right).$$

Observe that $\left(\widehat{A}, \widehat{B}\right)$ are jointly normal, with covariance

$$\text{Cov}\left(\widehat{A}, \widehat{B}\right) = \frac{\sum_i n_i x_i}{\left(\sum_i n_i x_i\right)^2 - \left(\sum_i n_i\right)\left(\sum_i n_i x_i^2\right)}. \quad \square$$

Problem 5.31 (SP-IB 1992 103H long) (i) Let $x_1, \ldots, x_n$ be a random sample from the PDF $f(x; \theta)$. What is meant by saying that $t(x_1, \ldots, x_n)$ is sufficient for θ?

Let

$$f(x; \theta) = \begin{cases} e^{-(x-\theta)}, & x > \theta, \\ 0, & x \le \theta, \end{cases}$$

and suppose $n = 3$. Let $y_1 < y_2 < y_3$ be ordered values of x_1, x_2, x_3. Show that y_1 is sufficient for θ.

(ii) Show that the distribution of $Y_1 - \theta$ is exponential of parameter 3. Your client suggests the following possibilities as estimators of θ:

$$\bar{\theta}_1 = x_3 - 1,$$
$$\bar{\theta}_2 = y_1,$$
$$\bar{\theta}_3 = \tfrac{1}{3}(x_1 + x_2 + x_3) - 1.$$

How would you advise him?

Hint: General theorems used should be clearly stated, but need not be proved.

Solution (i) $T(\mathbf{x})$ is sufficient for θ iff the conditional distribution of $\mathbf{X}$ given $T(\mathbf{X})$ does not involve θ. That is, $\mathbb{P}_\theta(\mathbf{X} \in B | T = t)$ is independent of θ $\forall$ domain B in the sample space. The factorisation criterion states that T is sufficient iff the sample PDF $f(\mathbf{x}; \theta) = g(T(\mathbf{x}), \theta) h(\mathbf{x})$ for some functions g and h.

For $f(x; \theta)$ as specified, with $n = 3$, $\mathbf{x} = (x_1, x_2, x_3)$.

$$f(\mathbf{x}; \theta) = \prod_{i=1}^{3} e^{-(x_i - \theta)} I(x_i > \theta) = e^{3\theta} e^{-\sum_i x_i} I(\min x_i > \theta), \qquad \mathbf{x} = \begin{pmatrix} x_1 \\ x_2 \\ x_3 \end{pmatrix}.$$

So, $T = \min X_i := Y_1$ is a sufficient statistic: here $g(y, \theta) = e^{3\theta} I(y > \theta)$, $h(\mathbf{x}) = e^{-\sum_i x_i}$.

(ii) $\forall\ y > 0$: $\mathbb{P}_\theta(Y_1 - \theta > y) = \mathbb{P}_\theta(X_1, X_2, X_3 > y + \theta) = \prod_{i=1}^{3} \mathbb{P}_\theta(X_i > y + \theta) = e^{-3y}$.

Hence, $Y_1 \sim \text{Exp}(3)$.

Now, $\mathbb{E}(X_3 - 1) = \mathbb{E}(X_3 - \theta) + \theta - 1 = \theta$, Var $X_3 = 1$.

Next, $Y_1 = \min X_i$ is the MLE maximising $f(x; \theta)$ in θ; it is biased as

$$\mathbb{E} Y_1 = \mathbb{E}(Y_1 - \theta) + \theta = \frac{1}{3} + \theta.$$

The variance Var $\bar{\theta}_2$ equals $1/9$.

Finally,

$$\mathbb{E}\left(\frac{1}{3}(X_1 + X_2 + X_3) - 1\right) = \theta, \qquad \text{Var}\left(\frac{1}{3}(X_1 + X_2 + X_3) - 1\right) = \frac{1}{3}.$$

We see that Y_1 has the least variance of the three. We could advise the client to use $\bar{\theta}_2$ bearing in mind the bias. However, the better choice is $\bar{\theta}_2 - 1/3$. □

Remark   The RB Theorem suggests to use the estimator $\widehat{\theta} = \mathbb{E}(\bar{\theta}_1 | \bar{\theta}_2 = t) = \mathbb{E}(\bar{\theta}_3 | \bar{\theta}_2 = t)$. A straightforward computation yields that $\widehat{\theta} = t - 1/3$. Hence, $\mathbb{E}\widehat{\theta} = \theta$ and Var $\widehat{\theta} = 1/9$. That is, the RB procedure does not create the bias and reduces the variance to the minimum.

Problem 5.32   (SP-IB 1992, 203H long) (i) Derive the form of the MLEs of α, β and σ^2 in the linear model

$$Y_i = \alpha + \beta x_i + \epsilon_i,$$

$1 \le i \le n$, where $\epsilon \sim N(0, \sigma^2)$ and $\sum_{i=1}^{n} x_i = 0$.
   (ii) What is the joint distribution of the maximum likelihood estimators $\widehat{\alpha}, \widehat{\beta}$ and $\widehat{\sigma}^2$? Construct 95% confidence intervals for

(a) σ^2,
(b) $\alpha + \beta$.

Solution   (i) We have that $Y_i \sim N(\alpha + \beta x_i, \sigma^2)$. Then

$$f(\mathbf{y}; \alpha, \beta, \sigma^2) = \frac{1}{(\sqrt{2\pi\sigma^2})^n} \exp\left[-\frac{1}{2\sigma^2} \sum_i (y_i - \alpha - \beta x_i)^2 \right],$$

and

$$\ell(\mathbf{y}; \alpha, \beta, \sigma^2) = \ln f(\mathbf{y}; \alpha, \beta, \sigma^2) = -\frac{n}{2} \ln (2\pi\sigma^2) - \frac{1}{2\sigma^2} \sum_i (y_i - \alpha - \beta x_i)^2.$$

The minimum is attained at

$$\frac{\partial}{\partial \alpha} \ell(\mathbf{y}; \alpha, \beta, \sigma^2) = \frac{\partial}{\partial \beta} \ell(\mathbf{y}; \alpha, \beta, \sigma^2) = \frac{\partial}{\partial \sigma^2} \ell(\mathbf{y}; \alpha, \beta, \sigma^2) = 0,$$

i.e. at

$$\widehat{\alpha} = \bar{y}, \quad \widehat{\beta} = \frac{\mathbf{x}^\mathsf{T}\mathbf{Y}}{(\mathbf{x}^\mathsf{T}\mathbf{x})}, \quad \widehat{\sigma}^2 = \frac{1}{n} \sum_i \widehat{\epsilon}_i^2, \quad \text{where } \widehat{\epsilon}_i = Y_i - \widehat{\alpha} - \widehat{\beta} x_i.$$

(ii) Then we have

$$\widehat{\alpha} \sim N\left(\alpha, \frac{\sigma^2}{n}\right), \quad \widehat{\beta} \sim N\left(\beta, \frac{\sigma^2}{\mathbf{x}^\mathsf{T}\mathbf{x}}\right), \quad \left(\frac{n}{\sigma^2}\right) \widehat{\sigma}^2 \sim \chi^2_{n-2}.$$

Also,

$$\text{Cov}\,(\widehat{\alpha}, \widehat{\beta}) = \text{Cov}\left(\frac{1}{n} \sum_i Y_i, \frac{1}{\mathbf{x}^\mathsf{T}\mathbf{x}} \sum_i x_i Y_i \right)$$

$$= \frac{1}{n\mathbf{x}^\mathsf{T}\mathbf{x}} \sum_i x_i \text{Cov}\,(Y_i, Y_i) = \frac{\sigma^2}{n\mathbf{x}^\mathsf{T}\mathbf{x}} \sum_i x_i = 0,$$

$$\mathrm{Cov}\,(\widehat{\alpha}, \widehat{\epsilon}_i) = \mathrm{Cov}\,(\widehat{\alpha}, Y_i - \widehat{\alpha} - \widehat{\beta}x_i)$$

$$= \frac{1}{n} \sum_{j=1}^{n} \left(\delta_{ij} - \frac{1}{n} \right) \mathrm{Cov}\,(Y_j, Y_j) = 0,$$

and

$$\mathrm{Cov}\,(\widehat{\beta}, \widehat{\epsilon}_i) = \mathrm{Cov}\,(\widehat{\beta}, Y_i - \widehat{\alpha} - \widehat{\beta}x_i)$$

$$= \frac{1}{\mathbf{x}^{\mathsf{T}}\mathbf{x}} \sum_{j=1}^{n} x_j \left(\delta_{ij} - \frac{1}{\mathbf{x}^{\mathsf{T}}\mathbf{x}} x_j x_i \right) \mathrm{Cov}\,(Y_j, Y_j)$$

$$= \frac{\sigma^2}{\mathbf{x}^{\mathsf{T}}\mathbf{x}} (x_i - x_i) = 0.$$

So, $\widehat{\alpha}$, $\widehat{\beta}$ and $\widehat{\epsilon}_1, \ldots, \widehat{\epsilon}_n$ are independent. Hence, $\widehat{\alpha}$, $\widehat{\beta}$ and $\widehat{\sigma}^2$ are also independent. Therefore, (a) the 95% CI for σ^2 is

$$\left(\frac{n\widehat{\sigma}^2}{h^+}, \frac{n\widehat{\sigma}^2}{h^-} \right)$$

where h^- is the lower and h^+ the upper 0.025 point of χ^2_{n-2}.
  Finally, (b)

$$\widehat{\alpha} + \widehat{\beta} \sim \mathrm{N}\left(\alpha + \beta, \sigma^2 \left(\frac{1}{n} + \frac{1}{\mathbf{x}^{\mathsf{T}}\mathbf{x}} \right) \right)$$

and is independent of $\widehat{\sigma}^2$. Then

$$\frac{(\widehat{\alpha} + \widehat{\beta}) - (\alpha + \beta)}{\sqrt{\frac{1}{n} + \frac{1}{\mathbf{x}^{\mathsf{T}}\mathbf{x}}}} \Bigg/ \widehat{\sigma}\sqrt{\frac{n}{n-2}} \sim \mathrm{t}_{n-2}.$$

Hence, the 95% CI for $\alpha + \beta$ is

$$\left(\widehat{\alpha} + \widehat{\beta} - t\widehat{\sigma}\sqrt{\frac{1}{n} + \frac{1}{\mathbf{x}^{\mathsf{T}}\mathbf{x}}}\sqrt{\frac{n}{n-2}}, \ \widehat{\alpha} + \widehat{\beta} + t\widehat{\sigma}\sqrt{\frac{1}{n} + \frac{1}{\mathbf{x}^{\mathsf{T}}\mathbf{x}}}\sqrt{\frac{n}{n-2}} \right),$$

where t is the upper 0.025 quantile of t_{n-2}. $\square$

Problem 5.33 (SP-IB 403H 1992 long) (i) Describe briefly a procedure for obtaining a Bayesian point estimator from a statistical experiment. Include in your description definitions of the terms:

(a) prior;
(b) posterior.

(ii) Let $X_1, \ldots, X_n$ be independent identically distributed random variables, each having distribution Gam (k, λ). Suppose k is known, and, *a priori*, λ is exponential of parameter μ. Suppose a penalty of $(a - \lambda)^2$ is incurred on estimating λ by a. Calculate the posterior for λ and find an optimal Bayes estimator for λ.

Solution   (i) We are given a sample PDF/PMF $f(\mathbf{x}; \theta)$ and a prior PDF/PMF $\pi(\theta)$ for the parameter. We then consider the posterior distribution

$$\pi(\theta|\mathbf{x}) \propto \pi(\theta) f(\mathbf{x}; \theta),$$

normalised to have the total mass 1. The Bayes' estimator is defined as the minimiser of the expected loss $\mathbb{E}_{\pi(\theta|\mathbf{x})} L(a, \theta)$ for a given function $L(a, \theta)$ specifying the loss incurred when θ is estimated by a.

(ii) For the quadratic loss, where $L(a, \theta) = (a - \theta)^2$, the Bayes' estimator is given by the posterior mean. In the example given, the posterior is Gamma:

$$\pi(\lambda|\mathbf{x}) \propto e^{-\mu\lambda} \lambda^{kn} \prod_i e^{-\lambda x_i} \sim \text{Gam}\left(kn + 1, \mu + \sum_i x_i\right).$$

Hence, the Bayes' estimator is given by its mean $(kn + 1)\big/\left(\mu + \sum_i x_i\right)$. $\square$

Problem 5.34   (MT-IB 1992-106D long) Let $X_1, X_2, \ldots, X_n$ be an independent sample from a normal distribution with unknown mean μ and variance σ^2. Show that the pair $(\overline{X}, \overline{S}^2)$, where

$$\overline{X} = \frac{1}{n} \sum_{i=1}^{n} X_i, \quad \overline{S}^2 = \frac{1}{n} \sum_{i=1}^{n} (X_i - \overline{X})^2,$$

is a sufficient statistic for (μ, σ^2).

Given $\lambda > 0$, consider $\lambda \overline{S}^2$ as an estimator of σ^2. For what values of λ is $\lambda \overline{S}^2$

(i)   maximum likelihood,
(ii)   unbiased?

Which value of λ minimises the mean square error

$$\mathbb{E}(\lambda \overline{S}^2 - \sigma^2)^2?$$

Solution   The likelihood

$$f(\mathbf{x}; \mu, \sigma^2) = \left(\frac{1}{\sqrt{2\pi\sigma^2}}\right)^n \exp\left[-\frac{1}{2\sigma^2} \sum_{i=1}^{n} (x_i - \mu)^2\right]$$

$$= \left(\frac{1}{\sqrt{2\pi\sigma^2}}\right)^n \exp\left[-\frac{1}{2\sigma^2} \sum_{i=1}^{n} (x_i - \overline{x} + \overline{x} - \mu)^2\right]$$

$$= \left(\frac{1}{\sqrt{2\pi\sigma^2}}\right)^n \exp\left[-\frac{1}{2\sigma^2} \sum_{i=1}^{n} (x_i - \overline{x})^2 - \frac{n}{2\sigma^2} (\overline{x} - \mu)^2\right].$$

Hence, by the factorisation criterion, $(\overline{X}, \overline{S}^2)$ is a sufficient statistic.

Maximising in μ and σ^2 is equivalent to solving

$$\frac{\partial}{\partial\mu}\ln f(\mathbf{x};\mu,\sigma^2)=\frac{\partial}{\partial\sigma^2}\ln f(\mathbf{x};\mu,\sigma^2)=0,$$

which yields $\widehat{\mu}=\bar{x},\widehat{\sigma}^2=\sum_i(x_i-\bar{x})^2/n$. Hence, (i) $\lambda\overline{S}^2$ is MLE for $\lambda=1$.
For (ii)

$$\mathbb{E}n\overline{S}^2=\sum_{i=1}^n\mathbb{E}(X_i-\overline{X})^2=\sum_{i=1}^n\mathbb{E}(X_i-\mu)^2-n\mathbb{E}(\overline{X}-\mu)^2$$

$$=n\mathrm{Var}\,X-\frac{n}{n^2}n\mathrm{Var}\,X=(n-1)\sigma^2.$$

Hence, $\lambda\overline{S}^2$ is unbiased for $\lambda=n/(n-1)$.
Finally, set $\phi(\lambda)=\mathbb{E}(\lambda\overline{S}^2-\sigma^2)^2$. The differentiating yields

$$\phi'(\lambda)=2\mathbb{E}(\lambda\overline{S}^2-\sigma^2)\overline{S}^2.$$

Next, $n\overline{S}^2\sim(Y_1^2+\cdots+Y_{n-1}^2)$, where $Y_i\sim N(0,\sigma^2)$, independently. Hence,

$$\mathbb{E}\left(\overline{S}^2\right)^2=n^{-2}\mathbb{E}(Y_1^2+\cdots+Y_{n-1}^2)^2$$

$$=n^{-2}\left[(n-1)\mathbb{E}Y_1^4+(n-1)(n-2)\left(\mathbb{E}Y_1^2\right)^2\right]$$

$$=n^{-2}\left[3(n-1)\sigma^4+(n-1)(n-2)\sigma^4\right]=n^{-2}(n^2-1)\sigma^4.$$

As $\mathbb{E}\overline{S}^2=(n-1)\sigma^2/n$, equation $\phi'(\lambda)=0$ gives

$$\lambda=\frac{\sigma^2\mathbb{E}\overline{S}^2}{\mathbb{E}\left(\overline{S}^2\right)^2}=\frac{n}{n+1},$$

which is clearly the minimiser. $\quad\square$

The Mystery of Mean Mu and Squared Stigma

(From the series *'Movies that never made it to the Big Screen'*.)

Problem 5.35 (MT-IB 1992-206D long) Suppose you are given a collection of np independent random variables organised in n samples, each of length p:

$$X^{(1)}=(X_{11},\ldots,X_{1p})$$
$$X^{(2)}=(X_{21},\ldots,X_{2p})$$
$$\vdots\qquad\qquad\vdots$$
$$X^{(n)}=(X_{n1},\ldots,X_{np}).$$

The RV X_{ij} has a Poisson distribution with an unknown parameter λ_j, $1\le j\le p$. You are required to test the hypothesis that $\lambda_1=\cdots=\lambda_p$ against the alternative that at least two of the values λ_j are distinct. Derive the form of the likelihood ratio test statistic. Show that it may be approximated by

$$\frac{n}{\overline{X}} \sum_{j=1}^{p} (\overline{X}_j - \overline{X})^2$$

with

$$\overline{X}_j = \frac{1}{n} \sum_{i=1}^{n} X_{ij}, \quad \overline{X} = \frac{1}{np} \sum_{i=1}^{n} \sum_{j=1}^{p} X_{ij}.$$

Explain how you would test the hypothesis for large n.

Solution   See Example 4.13.   □

Problem 5.36   (MT-IB 1992-306D long) Let X_1, X_2, $\ldots$, X_n be an independent sample from a normal distribution with a known mean μ and an unknown variance σ^2 taking one of two values σ_1^2 and σ_2^2. Explain carefully how to construct a most powerful test of size α of the hypothesis $\sigma = \sigma_1$ against the alternative $\sigma = \sigma_2$. Does there exist an MP test of size α with power strictly less than α? Justify your answer.

Solution   By the NP Lemma, as both hypotheses are simple, the MP test of size $\leq \alpha$ is the likelihood ratio test

$$\text{reject } H_0: \ \sigma = \sigma_1 \text{ in favour of } H_1: \ \sigma = \sigma_2,$$

when

$$\frac{(2\pi\sigma_2^2)^{-n/2} \exp\left[-\sum_i (x_i - \mu)^2/(2\sigma_2^2)\right]}{(2\pi\sigma_1^2)^{-n/2} \exp\left[-\sum_i (x_i - \mu)^2/(2\sigma_1^2)\right]} > k,$$

where $k > 0$ is adjusted so that the TIEP equals α. Such test always exists because the normal PDF $f(x)$ is monotone in $|x|$ and continuous.
    By re-writing the LR as

$$\left(\frac{\sigma_1}{\sigma_2}\right)^n \exp\left[\frac{1}{2}\left(\frac{1}{\sigma_1^2} - \frac{1}{\sigma_2^2}\right)\sum(x_i - \mu)^2\right],$$

we see that if $\sigma_2 > \sigma_1$, we reject H_0 in the critical region

$$\mathcal{C}^+ = \left\{\sum_i (x_i - \mu)^2 \geq k^+\right\}$$

and if $\sigma_1 > \sigma_2$, in the critical region

$$\mathcal{C}^- = \left\{\sum_i (x_i - \mu)^2 \leq k^-\right\}.$$

Furthermore,

$$\frac{1}{\sigma_1^2} \sum_i (X_i - \mu)^2 \sim \chi_n^2 \text{ under } H_0$$

and

$$\frac{1}{\sigma_2^2} \sum_i (X_i - \mu)^2 \sim \chi_n^2 \text{ under } H_1.$$

The χ_n^2 PDF is

$$f_{\chi^2}(x) = \frac{1}{\Gamma(n/2)} \frac{1}{2^{n/2}} x^{n/2-1} e^{-x/2} I(x > 0).$$

Then if $\sigma_2 > \sigma_1$, we choose k^+ so that the size

$$\mathbb{P}(\mathcal{C}^+|H_0) = \int_{k^+/\sigma_1^2}^{\infty} \frac{1}{\Gamma(n/2)} \frac{1}{2^{n/2}} x^{n/2-1} e^{-x/2} dx = \alpha,$$

and if $\sigma_1 > \sigma_2$, k^- so that

$$\mathbb{P}(\mathcal{C}^-|H_0) = \int_0^{k^-/\sigma_1^2} \frac{1}{\Gamma(n/2)} \frac{1}{2^{n/2}} x^{n/2-1} e^{-x/2} dx = \alpha.$$

The power for $\sigma_2 > \sigma_1$ equals

$$\beta = \mathbb{P}(\mathcal{C}^+|H_1) = \int_{k^+/\sigma_2^2}^{\infty} \frac{1}{\Gamma(n/2)} \frac{1}{2^{n/2}} x^{n/2-1} e^{-x/2} dx$$

and for $\sigma_1 > \sigma_2$,

$$\beta = \mathbb{P}(\mathcal{C}^-|H_1) = \int_0^{k^-/\sigma_2^2} \frac{1}{\Gamma(n/2)} \frac{1}{2^{n/2}} x^{n/2-1} e^{-x/2} dx.$$

We see that if $\sigma_2 > \sigma_1$, then $k^+/\sigma_2^2 < k^+/\sigma_1^2$, and $\beta > \alpha$. Similarly, if $\sigma_1 > \sigma_2$, then $k^-/\sigma_2^2 > k^-/\sigma_1^2$, and again $\beta > \alpha$. Thus, $\beta < \alpha$ is impossible. $\square$

Gamma and Her Sisters

(From the series *'Movies that never made it to the Big Screen'*.)

Problem 5.37 (MT-IB 1992-406D long) Let ϵ_1, ϵ_2, ..., ϵ_n be independent random variables each with the $N(0, 1)$ distribution, and x_1, x_2, ..., x_n be fixed real numbers. Let the random variables Y_1, Y_2, ..., Y_n be given by

$$Y_i = \alpha + \beta x_i + \sigma \epsilon_i, \quad 1 \le i \le n,$$

where $\alpha, \beta \in \mathbb{R}$ and $\sigma^2 \in (0, \infty)$ are unknown parameters. Derive the form of the least squares estimator (LSE) for the pair (α, β) and establish the form of the distribution. Explain how to test the hypothesis $\beta = 0$ against $\beta \neq 0$ and how to construct a 95% CI for β.

(General results used should be stated carefully, but need not be proved.)

Solution   Define

$$\bar{x} = \frac{1}{n}\sum_{i=1}^{n} x_i, \ \ a = \alpha + \beta\bar{x} \text{ and } u_i = x_i - \bar{x},$$

with

$$\sum_i u_i = 0 \text{ and } Y_i = a + \beta u_i + \sigma\epsilon_i.$$

The LSE pair for (α, β) minimises the quadratic sum $R = \sum_{i=1}^{n}(Y_i - \mathbb{E}Y_i)^2 = \sum_{i=1}^{n}(Y_i - a - \beta u_i)^2$, i.e. solves

$$\frac{\partial}{\partial\alpha}R = \frac{\partial}{\partial\beta}R = 0.$$

This yields

$$\hat{a} = \bar{Y} \sim N\left(\alpha + \beta\bar{x}, \frac{\sigma^2}{n}\right), \ \ \hat{\alpha} \sim N\left(\alpha, \frac{\sigma^2}{n}\right),$$

and

$$\hat{\beta} = \frac{\mathbf{u}^{\mathsf{T}}\mathbf{Y}}{\mathbf{u}^{\mathsf{T}}\mathbf{u}} \sim N\left(\beta, \frac{\sigma^2}{\mathbf{u}^{\mathsf{T}}\mathbf{u}}\right),$$

independently. Also,

$$\hat{R} = \sum_i (Y_i - \hat{\alpha} - \hat{\beta}x_i)^2 = \sum_i Y_i^2 - n\bar{Y}^2 - \mathbf{u}^{\mathsf{T}}\mathbf{u}\hat{\beta}^2 \sim \sigma^2\chi_{n-2}^2,$$

since we have estimated two parameters. So, $R/(n-2)$ is an unbiased estimator for σ^2.
Now consider $H_0: \ \beta = \beta_0 = 0$, $H_1: \ \beta \neq \beta_0$. Then

$$\hat{\beta} \sim N\left(0, \frac{\sigma^2}{\mathbf{u}^{\mathsf{T}}\mathbf{u}}\right), \ \ \text{i.e.} \ \ \frac{\hat{\beta}\sqrt{\mathbf{u}^{\mathsf{T}}\mathbf{u}}}{\sigma} \sim N(0,1) \text{ iff } \beta = 0,$$

i.e. under H_0. Thus

$$T = \frac{\hat{\beta}\sqrt{\mathbf{u}^{\mathsf{T}}\mathbf{u}}}{\sqrt{\hat{R}/(n-2)}} \sim t_{n-2}.$$

So, given α, we reject H_0 when the value of $|T|$ exceeds $t_{n-2}(\alpha/2)$, the upper $\alpha/2$ point of t_{n-2}.
Finally, to construct an (equal-tailed) 95% CI for β, we take $t_{n-2}(0.025)$. Then from the inequalities

$$-t_{n-2}(0.025) < \frac{(\beta - \hat{\beta})\sqrt{\mathbf{u}^{\mathsf{T}}\mathbf{u}}}{\sqrt{\hat{R}/(n-2)}} < t_{n-2}(0.025)$$

we find that

$$\left(\widehat{\beta} - t_{n-2}(0.025)\sqrt{\frac{\widehat{R}}{(n-2)\mathbf{u}^{\mathsf{T}}\mathbf{u}}}, \ \widehat{\beta} + t_{n-2}(0.025)\sqrt{\frac{\widehat{R}}{(n-2)\mathbf{u}^{\mathsf{T}}\mathbf{u}}} \right)$$

is the required interval. $\square$

> Each of us has been doing statistics all his life,
> in the sense that each of us has been busily reaching
> conclusions based on empirical observations
> ever since birth.
>
> W. Kruskal (1919–), American statistician

We finish this volume with a story about F. Yates (1902–1994), a prominent UK statistician and a close associate of Fisher (quoted from [Wi], pp. 204–205). During his student years at St John's College, Cambridge, Yates had been keen on a form of sport which had a long local tradition. It consisted of climbing about the roofs and towers of the college buildings at night. (The satisfaction arose partly from the difficulty of the climbs and partly from the excitement of escaping the vigilance of the college porters.) In particular, the chapel of St John's College has a massive neo-Gothic tower adorned with statues of saints, and to Yates it appeared obvious that it would be more decorous if these saints were properly attired in surplices. One night he climbed up and did the job; next morning the result was generally much admired. But the College authorities were unappreciative and began to consider means of divesting the saints of their newly acquired garments. This was not easy, since they were well out of reach of any ordinary ladder. An attempt to lift the surplices off from above, using ropes with hooks attached, was unsuccessful, since Yates, anticipating such a move, had secured the surplices with pieces of wire looped around the saints' necks. No progress was being made and eventually Yates came forward and, without admitting that he had been responsible for putting the surplices up, volunteered to climb up in the daylight and bring them down. This he did to the admiration of the crowd that assembled.

The morale of this story is that maybe statisticians should pay more attention to self-generated problems. . . . (An observant passer-by may notice that presently two of the statues on the St John's chapel tower have staffs painted in a pale green colour, which obviously is not an originally intended decoration. Perhaps the next generation of followers of Fisher's school are practicing before shaping the development of twenty-first century statistics.)

Appendix 1. Tables of random variables and probability distributions

Table A1.1. *Some useful discrete distributions*

Family notation range	PMF $\mathbb{P}(X=r)$	Mean	Variance	PGF $\mathbb{E}s^X$
Poisson Po(λ) $0, 1, \ldots$	$\dfrac{\lambda^r e^{-\lambda}}{r!}$	λ	λ	$e^{\lambda(s-1)}$
Geometric Geom (p) $1, 2, \ldots$	$p(1-p)^r$	$\dfrac{1-p}{p}$	$\dfrac{1-p}{p^2}$	$\dfrac{p}{1-s(1-p)}$
Binomial Bin (n, p) $0, \ldots n$	$\dbinom{n}{r} p^r (1-p)^{n-r}$	np	$np(1-p)$	$[ps+(1-p)]^n$
Negative binomial NegBin (p, k) $0, 1, \ldots$	$\dbinom{r+k-1}{r} p^k (1-p)^r$	$\dfrac{k(1-p)}{p}$	$\dfrac{k(1-p)}{p^2}$	$\left[\dfrac{p}{1-s(1-p)}\right]^k$
Hypergeometric Hyp (N, D, n) $(n+D-N)_+,$ $\ldots, D \wedge n$	$\dfrac{\dbinom{D}{r}\dbinom{N-D}{n-r}}{\dbinom{N}{n}}$	$\dfrac{nD}{N}$	$\dfrac{nD(N-D)(N-n)}{N^2(N-1)}$	$_2F_1\begin{pmatrix} -D, & -n \\ -N; & 1-s \end{pmatrix}$
Uniform U[1, n] $1, \ldots, n$	$\dfrac{1}{n}$	$\dfrac{n+1}{2}$	$\dfrac{n^2-1}{12}$	$\dfrac{s(1-s^n)}{n(1-s)}$

Table A1.2. *Some useful continuous distributions 1*

Family, notation, range	PDF $f_X(x)$	Mean	Variance	MGF $\mathbb{E}e^{tX}$ CHF $\mathbb{E}e^{itX}$
Uniform $\mathrm{U}(a,b)$ (a,b)	$\dfrac{1}{b-a}$	$\dfrac{a+b}{2}$	$\dfrac{(b-a)^2}{12}$	$\dfrac{e^{bt}-e^{at}}{(b-a)t}$ $\dfrac{e^{ibt}-e^{iat}}{(b-a)t}$
Exponential $\mathrm{Exp}(\lambda)$ $\mathbb{R}_+$	$\lambda e^{-\lambda x}$	$\dfrac{1}{\lambda}$	$\dfrac{1}{\lambda^2}$	$\dfrac{\lambda}{\lambda-t}$ $\dfrac{\lambda}{\lambda-it}$
Gamma $\mathrm{Gam}(\alpha,\lambda)$ $\mathbb{R}_+$	$\dfrac{\lambda^\alpha x^{\alpha-1}e^{-\lambda x}}{\Gamma(\alpha)}$	$\dfrac{\alpha}{\lambda}$	$\dfrac{\alpha}{\lambda^2}$	$(1-t/\lambda)^{-\alpha}$ $(1-it/\lambda)^{-\alpha}$
Normal $\mathrm{N}(\mu,\sigma^2)$ $\mathbb{R}$	$\dfrac{\exp\left[-\dfrac{1}{2\sigma^2}(x-\mu)^2\right]}{\sqrt{2\pi\sigma^2}}$	μ	σ^2	$\exp\left(t\mu+\tfrac{1}{2}t^2\sigma^2\right)$ $\exp\left(it\mu-\tfrac{1}{2}t^2\sigma^2\right)$
Multivariate Normal $\mathrm{N}(\mu,\Sigma)$ $\mathbb{R}^n$	$\dfrac{\exp\left[-\tfrac{1}{2}(\mathbf{x}-\mu)^\top\Sigma^{-1}(\mathbf{x}-\mu)\right]}{\sqrt{(2\pi)^n\det\Sigma}}$	μ	Σ	$\exp\left(\mathbf{t}^\top\mu+\tfrac{1}{2}\mathbf{t}^\top\Sigma\mathbf{t}\right)$ $\exp\left(i\mathbf{t}^\top\mu-\tfrac{1}{2}\mathbf{t}^\top\Sigma\mathbf{t}\right)$
Cauchy $\mathrm{Ca}(\alpha,\tau)$ $\mathbb{R}$	$\dfrac{\tau}{\tau^2+(x-\alpha)^2}$	not defined	not defined	$\mathbb{E}(e^{tX})=\infty$ for $t\neq0$ $\mathbb{E}(e^{itX})=\exp(it\alpha-|t|\tau)$

Table A1.3. *Some useful continuous distributions 2*

Family, notation	Range	Distribution	Notes
Chi-square, χ_n^2	$\mathbb{R}_+$	$\chi_n^2 \sim \sum^n N(0,1)^2$	$\chi_n^2 \sim \text{Gam}\left(\frac{n}{2}, \frac{1}{2}\right)$
Student, t_n	$\mathbb{R}$	$t_n \sim \dfrac{N(0,1)}{\sqrt{\chi_n^2/n}}$	$t_1 \sim \text{Ca}(0,1)$, $\mathbb{E}X = 0$, $n > 1$ $$\text{Var } X = \frac{n}{n-2}, \ n > 2$$
Fisher, $F_{m,n}$	$\mathbb{R}_+$	$F_{m,n} \sim \dfrac{\chi_m^2/m}{\chi_n^2/n}$	$\mathbb{E}X = \dfrac{n}{n-2}, \ n > 2$; $$\text{Var } X = \frac{2n^2(m+n-2)}{m(n-2)^2(n-4)}, \ n > 4$$
Weibull, Weib (α)	$\mathbb{R}_+$	$f(x) = \alpha x^{\alpha-1} e^{-x^\alpha}$	$\text{Weib}(\alpha) \sim \text{Exp}(1)^{1/\alpha}$
Beta, Bet (r,s)	$[0,1]$	$f(x) = \dfrac{x^{r-1}(1-x)^{s-1}}{B(r,s)}$	$\text{Bet}(r,s) \sim \left[1 + \dfrac{\text{Gam}(s,\lambda)}{\text{Gam}(r,\lambda)}\right]^{-1}$; $$\mathbb{E}X = \frac{r}{r+s}$$
Logistic, Logist	$\mathbb{R}$	$f(x) = \dfrac{e^x}{(1+e^x)^2}$	$\text{Logist} \sim \log(\text{Par}(1) - 1)$; $\mathbb{E}(e^{tX}) = B(1+t, 1-t)$
Pareto, Par (α)	$[1, \infty]$	$f(x) = \alpha x^{-(\alpha+1)}$	$\text{Par}(\alpha) \sim \text{Bet}(\alpha, 1)^{-1}$; $\mathbb{E}X = \infty$ if $\alpha \le 1$ $\text{Var } X = \infty, \alpha \le 2$

Appendix 2. Index of Cambridge University Mathematical Tripos examination questions in IA Probability (1992–1999)

The references to Mathematical Tripos IA examination questions and related examples adopt the following pattern: 304H (short) and 310B stands for the short question 4H and the long question 10B from Paper 3 from the corresponding year. For example, Problem 1.14 is the short question 3H from Paper 3 of 1993.

IA sample questions 1992
1.8
1.16
1.40
1.72
1.73
2.29

IA specimen papers 1992
303B (short):   1.58
304B (short):   2.49
309B (long):   1.36
310B (long):   2.32
311B (long):   1.51
312B (long):   2.18

Mathematical Tripos IA 1992
303C (short):   1.24
304C (short):   2.4
309C (long):   2.50
310C (long):   1.70
311C (long):   1.17
312C (long):   1.80

Mathematical Tripos IA 1993
303H (short):   1.14
304H (short):   2.37
309H (long):   1.31

310D (long):   1.39
311D (long):   2.13
312D (long):   2.31

Mathematical Tripos IA 1994
303B (short):   1.29
304B (short):   2.17
309B (long):   1.53
310B (long):   1.61
311B (long):   1.15, 1.42
312B (long):   2.47

Mathematical Tripos IA 1995
303A (short):   1.34
304A (short):   1.59
309A (long):   1.64
310A (long):   1.27
311A (long):   2.22
312A (long):   1.77

Mathematical Tripos IA 1996
203A (short):   2.50
204A (short):   1.48
209C (long):   1.54
210C (long):   2.24
211B (long):   1.41
212B (long):   1.20

Mathematical Tripos IA 1997
203G (short):   1.9
204G (short):   1.50
209G (long):   1.47, 1.57
210G (long):   1.78
211G (long):   2.23 2.46
212G (long):   1.63

Mathematical Tripos IA 1998
203C (short):   1.21
204C (short):   1.56
209C (long):   1.62
210C (long):   1.46
211C (long):   2.45
212C (long):   2.16

Mathematical Tripos IA 1999
203C (short):   1.26
204C (short):   1.43
209C (long):   2.32
210C (long):   1.52
211C (long):   2.44
212C (long):   1.45, 2.33

Bibliography

[A] D. Applebaum. *Probability and Information: an Integrated Approach*. Cambridge: Cambridge University Press, 1996.

[ArBN] B.C. Arnold, N. Balakrishnan and H.N. Nagaraja. *A First Course in Order Statistics*. New York: Wiley, 1992.

[As] R.B. Ash. *Probability and Measure Theory*, with contributions from C. Doléans-Dade. 2nd ed. San Diego, CA: Harcourt/Academic; London: Academic, 2000.

[AūC] J. Auñón and V. Chandrasekar. *Introduction to Probability and Random Processes*. New York; London: McGraw-Hill, 1997.

[Az] A. Azzalini. *Statistical Inference: Based on the Likelihood*. London: Chapman and Hall, 1996.

[BE] L.J. Bain and M. Engelhardt. *Introduction to Probability and Mathematical Statistics*. 2nd ed. Boston, MA: PWS-KENT, 1992.

[BaN] N. Balakrishnan and V.B. Nevzorov. *A Primer on Statistical Distributions*. Hoboken, NJ/Chichester: Wiley, 2003.

[BarC] O.E. Barndorff-Nielsen and D.R. Cox. *Inference and Asymptotics*. London: Chapman and Hall, 1994.

[BartN] R. Bartoszynski and M. Niewiadomska-Bugaj. *Probability and Statistical Inference*. New York; Chichester: Wiley, 1996.

[Be] J.O. Berger. *Statistical Decision Theory and Bayesian Analysis*. 2nd ed. New York: Springer, 1985.

[BerL] D.A. Berry and B.W. Lindgren. *Statistics, Theory and Methods*. Pacific Grove, CA: Brooks/Cole, 1996.

[Bi] P. Billingsley. *Probability and Measure*. 3rd ed. New York; Chichester: Wiley, 1995.

[Bo] V.S. Borkar. *Probability Theory: an Advanced Course*. New York; London: Springer, 1995.

[Bor] A.A. Borovkov. *Mathematical Statistics*. Amsterdam: Gordon and Breach, 1998.

[BD] G.E.P. Box and N.R. Draper. *Evolutionary Operation: a Statistical Method for Process Improvement*. New York, Chichester : Wiley, 1998.

[BL] G.E.P. Box and A.Luceño. *Statistical Control by Monitoring and Feedback Adjustment*. New York; Chichester: Wiley, 1997.

[BTi] G.E.P. Box and G.C. Tiao. *Bayesian Inference in Statistical Analysis*. New York: Wiley, 1992.

[Box] *Box on Quality and Discovery: with Design, Control, and Robustness*. Ed.-in-chief G.C. Tiao; eds., S. Bisgaard *et al.*. New York; Chichester: Wiley, 2000.

[CK] M. Capinski and E. Kopp. *Measure, Integral and Probability*. London: Springer, 1999.

[CZ] M. Capinski and M. Zastawniak. *Probability through Problems*. New York: Springer, 2000.

[CaB] G. Casella and J.O. Berger. *Statistical Inference*. Pacific Grove, CA: Brooks/Cole, 1990.

[CaL] G. Casella and E.L. Lehmann. *Theory of Point Estimation*. New York; London: Springer, 1998.

[ChKB] C.A. Charalambides, M.V. Koutras and N. Balakrishnan. *Probability and Statistical Models with Applications*. Boca Raton, FL; London: Chapman and Hall/CRC, 2001.

[ChaHS] S. Chatterjee, M.S. Handcock and J.S. Simonoff. *A Casebook for a First Course in Statistics and Data Analysis*. New York; Chichester: Wiley, 1995.

[ChaY] L. Chaumont and M. Yor. *Exercises in Probability. A Guided Tour from Measure Theory to Random Processes, via Conditioning*. Cambridge: Cambridge University Press, 2003.

[ChoR] Y.S. Chow, H. Robbins and D. Siegmund. *The Theory of Optimal Stopping*. New York; Dover; London: Constable, 1991.

[ChoT] Y.S. Chow and H. Teicher. *Probability Theory: Independence, Interchangeability, Martingales*. 3rd ed. New York; London: Springer, 1997.

[Chu] K.L. Chung. *A Course in Probability Theory*. 3rd ed. San Diego, CA; London: Academic Press, 2001.

[ClC] G.M. Clarke and D. Cooke. *A Basic Course in Statistics*. 4th ed. London: Arnold, 1998.

[CoH1] D.R. Cox and D.V. Hinkley. *Theoretical Statistics*. London: Chapman and Hall, 1979.

[CoH2] D.R. Cox and D.V. Hinkley. *Problems and Solutions in Theoretical Statistics*. London: Chapman and Hall 1978.

[CouR] R. Courant and H. Robbins. *What Is Mathematics? An Elementary Approach to Ideas and Methods*. Oxford : Oxford University Press, 1996.

[CrC] J. Crawshaw and J. Chambers. *A Concise Course in Advanced Level Statistics: with Worked Examples*. 4th ed. Cheltenham: Nelson Thornes, 2001.

[DS] M.N. DeGroot and M.J. Schervish. *Probability and Statistics*. 3rd ed. Boston, MA; London: Addison-Wesley, 2002.

[De] J.L. Devore. *Probability and Statistics for Engineering and the Sciences*. 4th ed. Belmont, CA; London: Duxbury, 1995.

[Do] Y. Dodge Y. *et al.*, eds. *The Oxford Dictionary of Statistical Terms*. 6th ed. Oxford: Oxford University Press, 2003.

[DorSSY] A. Ya. Dorogovtsev, D.S. Silvestrov, A.V. Skorokhod, M.I. Yadrenko. *Probability Theory: Collection of Problems*. Providence, RI: American Mathematical Society, 1997.

[Du] R.M. Dudley. *Real Analysis and Probability*. Cambridge: Cambridge University Press, 2002.

[Dur1] R. Durrett. *The Essentials of Probability*. Belmont, CA: Duxbury Press, 1994.

[Dur2] R. Durrett. *Probability: Theory and Examples*. 2nd ed. Belmont, CA; London: Duxbury Press, 1996.

[E1] B. Efron. R.A. Fisher in the 21st century. Statistical Science, **13** (1998), 95–122.

[E2] B. Efron. Robbins, empirical Bayes and microanalysis. *Annals of Statistics*, **31** (2003), 366–378.

[ETh1] B. Efron and R. Thisted. Estimating the number of unseen species: How many words did Shakespeare know? *Biometrika*, **63** (1976), 435–447.

[ETh2] B. Efron and R. Thisted. Did Shakespeare write a newly-discovered poem? *Biometrika*, **74** (1987), 445–455.

[ETi] B. Efron and R. Tibshirani. *An Introduction to the Bootstrap*. New York; London: Chapman and Hall, 1993.

[EHP] M. Evans, N. Hastings and B. Peacock. *Statistical Distributions*. 3rd ed. New York; Chichester: Wiley, 2000.

[F] R.M. Feldman and C. Valdez-Flores. *Applied Probability and Stochastic Processes*. Boston, MA; London: PWS Publishing Company, 1996.

[Fe] W. Feller. *An Introduction to Probability Theory and Its Applications*. Vols 1 and 2. 2nd ed. New York: Wiley; London: Chapman and Hall, 1957–1971.

[FiB] J. Fisher Box. *R.A. Fisher. The Life of a Scientist*. New York; Chichester; Brisbane; Toronto: J. Wiley and Sons, 1978

[Fer] T.S. Ferguson. *Mathematical Statistics: a Decision Theoretic Approach*. New York; London: Academic Press, 1967.

[Fr] J.E. Freund. *Introduction to Probability*. New York: Dover, 1993.

[FreP] J.E. Freund and B.M. Perles. *Statistics: a First Course*. 7th ed. Upper Saddle River, NJ: Prentice Hall; London: Prentice-Hall International, 1999.

[FriG] B. Fristedt and L. Gray. *A Modern Approach to Probability Theory*. Boston, MA: Birkhäuser, 1997.

[G] J. Galambos. *Advanced Probability Theory*. 2nd ed., rev. and expanded. New York: M. Dekker, 1995.

[Gh] S. Ghahramani. *Fundamentals of Probability*. 2nd ed. International ed. Upper Saddle River, NJ: Pearson/Prentice-Hall, 2000.

[Gi] J.D. Gibbons. *Nonparametric Statistics: an Introduction*. Newbury Park, CA; London: Sage, 1993.

[Gn] B.V. Gnedenko. *Theory of Probability*. 6th ed. Amsterdam: Gordon and Breach, 1997.

[Go] H. Gordon. *Discrete Probability*. New York; London: Springer, 1997.

[Gr] A. Graham. *Statistics: an Introduction*. London: Hodder and Stoughton, 1995.

[GriS1] G. R. Grimmett, D. R. Stirzaker. *Probability and Random Processes: Problems and Solutions*. Oxford: Clarendon Press, 1992.

[GriS2] G.R. Grimmett and D. R. Stirzaker. *Probability and Random Processes*. 3rd ed. Oxford: Oxford University Press, 2001.

[GriS3] G. R. Grimmett, D. R. Stirzaker. *One Thousand Exercises in Probability*. Oxford: Oxford University Press, 2003.

[GrinS] C.M. Grinstead and J.L. Snell. *Introduction to Probability*. 2nd rev. ed. Providence, RI: American Mathematical Society, 1997.

[H] J. Haigh. *Probability Models*. London: Springer, 2002.

[Ha] A. Hald. *A History of Mathematical Statistics from 1750 to 1930*. New York; Chichester: Wiley, 1998.

[Has] K.J. Hastings. *Probability and Statistics*. Reading, MA; Harlow: Addison-Wesley, 1997.

[Haw] A.G. Hawkes. On the development of statistical ideas and their applications. Inaugural lecture at the University of Wales, Swansea, 1975.

[Hay] A.J. Hayter. *Probability and Statistics for Engineers and Scientists*. Boston, MA; London: PWS, 1996.

[He] L.L. Helms. *Introduction to Probability Theory: with Contemporary Applications*. New York: W.H. Freeman, 1997.

[Ho] J. Hoffmann-Jorgensen. *Probability with a View toward Statistics*. New York; London: Chapman and Hall, 1994.

[HogT] R.V. Hogg, E.A. Tanis. *Probability and Statistical Inference*. 6th ed. Upper Saddle River, NJ: Prentice Hall International, 2001.

[JP] J. Jacod, P. Protter. *Probability Essentials*. 2nd ed. Berlin, London: Springer, 2003.

[JaC] R. Jackson and J.T. Callender. *Exploring Probability and Statistics with Spreadsheets*. London: Prentice Hall, 1995.

[Je] H. Jeffreys. *Theory of Probability*. 3rd ed. Oxford: Oxford University Press, 1998.

[K] O. Kallenberg. *Foundations of Modern Probability*. 2nd ed. New York; London: Springer, 2002.

[Ki] S. Kim. *Statistics and Decisions: an Introduction to Foundations*. New York: Van Nostrand Reinhold; London: Chapman and Hall, 1992.

[Kinn] J.J. Kinney. *Probability: an Introduction with Statistical Applications*. New York; Chichester: J. Wiley and Sons, 1997.

[Kit] L.J. Kitchens. *Exploring Statistics: a Modern Introduction to Data Analysis and Inference*. 2nd ed. Pacific Grove, CA: Duxbury Press, 1998.

[KlR] D.A. Klain and G.-C. Rota. *Introduction to Geometric Probability*. Cambridge: Cambridge University Press, 1997.

[Ko] A.N. Kolmogorov. *Foundations of the Calculus of Probabilities*. New York : Chelsea Pub. Co., 1946.

[Kr] N. Krishnankutty. *Putting Chance to Work. A Life in Statistics*. A biography of C.R. Rao. State College, PA: Dialogue, 1996.

[LS] T.L. Lai and D. Siegmund. The contributions of Herbert Robbins to Mathematical Statistics. *Statistical Science*, **1** (1986), 276–284.

[La] J.W. Lamperti. *Probability: a Survey of the Mathematical Theory*. 2nd ed. New York; Chichester: Wiley, 1996.

[Lan] W.H. Lange. *Study Guide for Mason, Lind and Marchal's Statistics: an Introduction*. 5th ed. Pacific Grove, CA; London: Duxbury Press, 1998.

[LarM] R.J. Larsen and M.L. Marx. *An Introduction to Mathematical Statistics and its Applications*. 3rd ed. Upper Saddle River, NJ: Prentice Hall; London: Prentice Hall International, 2001.

[LawC] G. Lawler and L.N. Coyle. *Lectures on Contemporary Probability*. Providence, RI: American Mathematical Society, 1999.

[Le] P.M. Lee. *Bayesian Statistics: an Introduction*. 3rd ed. London: Arnold, 2004.

[Leh] E.L. Lehmann. *Testing Statistical Hypotheses*. 2nd ed. New York; London: Springer, 1997.

[LiS] D.V. Lindley and W.F. Scott. *New Cambridge Statistical Tables*. 2nd ed. Cambridge: Cambridge University Press, 1995.

[M] P. Malliavin. *Integration and Probability*. In cooperation with H. Airault, L. Kay and G. Letac. New York; London: Springer-Verlag, 1995.

[MaLM1] R.D. Mason, D.A. Lind and W.G. Marchal. *Statistics: an Introduction*. 5th ed. Pacific Grove, CA; London: Duxbury Press, 1998.

[MaLM2] R.D. Mason, D.A. Lind and W.G. Marchal. *Instructor's Manual for Statistics: an Introduction*. 5th ed. Pacific Grove, CA; London: Duxbury Press, 1998.

[McClS] J.T. McClave and T. Sincich. *A First Course in Statistics*. 7th ed. Upper Saddle River, NJ: Prentice Hall; London: Prentice-Hall International, 2000.

[McCo] J.H. McColl. *Probability*. London: Edward Arnold, 1995

[MeB] W. Mendenhall, R.J. Beaver and B.M. Beaver. *Introduction to Probability and Statistics*. 10th ed. Belmont, CA; London: Duxbury, 1999.

[MiM] I. Miller, M. Miller. *John Freund's Mathematical Statistics with Applications*. 7th ed. Upper Saddle River, NJ: Pearson Prentice-Hall Education, 2004.

[MT] F. Mosteller, J.W. Tukey. *Data Analysis and Regression: a Second Course in Statistics*. Reading, MA; London: Addison-Wesley, 1977.

[N] E. Nelson. *Radically Elementary Probability Theory*. Princeton, NJ: Princeton University Press, 1987.

[NiFY] Ye.P. Nikitina, V.D. Freidlina and A.V. Yarkho. *A Collection of Definitions of the Term 'Statistic(s)'* [in Russian]. Moscow: Moscow State University Publishing House, 1972.

[OF] A. O'Hagan and J. Forster. *Kendall's Advanced Theory of Statistics.* Vol. 2B. London: Hodder Arnold, 2004.

[Oc] M.K. Ochi. *Applied Probability and Stochastic Processes: in Engineering and Physical Sciences.* New York; Chichester: Wiley, 1990.

[OtL] R.L. Ott and M. Longnecker. *An Introduction to Statistical Methods and Data Analysis.* 5th ed. Australia; United Kingdom: Duxbury, 2001.

[PH] E. Pearson and H.O. Hartley (Eds). *Biometrika Tables for Statisticians.* Vol. 1.: Cambridge: Cambridge University Press, 1966.

[Pi] J. Pitman. *Probability.* New York: Springer-Verlag, 1993.

[Pol] D. Pollard. *A User's Guide to Measure Theoretic Probability.* Cambridge: Cambridge University Press, 2002.

[Por] S.C. Port. *Theoretical Probability for Applications.* New York; Chichester: Wiley, 1994.

[Ra] C.R. Rao. *Linear Statistical Inference and Its Applications.* 2nd ed. New York: Wiley, 2002.

[R] S. Rasmussen. *An Introduction to Statistics with data Analysis.* International student ed. Pacific Grove, CA: Brooks/Cole, 1992.

[Re] A.G. Rencher. *Linear Models in Statistics.* New York; Chichester: Wiley, 2000.

[Rei] C. Reid. *Neyman–from life.* New York, Hedelberg, Berlin: Springer-Verlag, 1982.

[Res] S.I. Resnick. *A Probability Path.* Boston, MA: Birkhäuser, 1999.

[Ri] J.A. Rice. *Mathematical Statistics and Data Analysis.* 2nd ed. Pacific Grove; CA; London: Duxbury Press, 1995.

[Ro] J. Rosenblat. *Basic Statistical Methods and Models for the Sciences.* Boca Raton, FL; London: Chapman and Hall/CRC, 2002.

[Ros1] S.M. Ross. *Solutions Manual for Introduction to Probability Models.* 4th ed. Boston, MA: Academic, 1989.

[Ros2] S.M. Ross. *Introductory Statistics.* New York; London: McGraw-Hill, 1996.

[Ros3] S.M. Ross. *Introduction to Probability Models.* 7th ed. San Diego, CA; London: Harcourt/Academic, 2000.

[Ros4] S.M. Ross. *Introduction to Probability and Statistics for Engineers and Scientists.* 2nd ed. San Diego, CA; London: Harcourt/Academic, 2000.

[Ros5] S. Ross. *A First Course in Probability.* 6th ed. Upper Saddle River, NJ; London: Prentice Hall, 2002.

[Ros6] S.M. Ross. *Probability Models for Computer Science.* San Diego, CA; London: Harcourt Academic Press, 2002.

[RotE] V.K. Rothagi and S.M. Ehsanes. *An Introduction to Probability and Statistics.* 2nd ed. New York, Chichester: Wiley, 2001.

[Rou] G.G. Roussas. *A Course in Mathematical Statistics.* 2nd ed. San Diego, CA; London: Academic, 1997.

[Roy] R.M. Royall. *Statistical Evidence: a Likelihood Paradigm.* London: Chapman and Hall, 1997.

[S] R.L. Scheaffer. *Introduction to Probability and Its Applications.* 2nd ed. Belmont, CA; London: Duxbury, 1995.

[Sc] R.B. Schinazi. *Probability with Statistical Applications.* Boston, MA: Birkhäuser, 2001.

[SeS] P.K. Sen and J.M. Singer. *Large Sample Methods in Statistics: an Introduction with Applications.* New York; London: Chapman and Hall, 1993.

[Sh] A.N. Shiryayev. *Probability.* 2nd ed. New York: Springer-Verlag, 1995.

[SiM] A.F. Siegel and C.J. Morgan. *Statistics and Data Analysis: an Introduction.* 2nd ed. New York; Chichester: Wiley, 1996.

[Sin] Y.G. Sinai. *Probability Theory: an Introductory Course*. Berlin; New York: Springer-Verlag, 1992.

[SpSS] M.R. Spiegel, J.J. Schiller and R.A. Srinivasan. *Schaum's Outline of Theory and Problems of Probability and Statistics*. 2nd ed. New York; London: McGraw-Hill, 2000.

[St1] D. Stirzaker. *Elementary Probability*. Cambridge: Cambridge University Press, 1994 (1996 [printing]).

[St2] D. Stirzaker. *Solutions Manual for Stirzaker's Elementary Probability*. Cambridge: Cambridge University Press, 1994.

[St3] D. Stirzaker. *Probability Vicit Expectation*. Chichester: Wiley, 1994.

[St4] D. Stirzaker. *Probability and Random Variables: a Beginner's Guide*. New York: Cambridge University Press, 1999.

[Sto] J.M. Stoyanov. *Counterexamples in Probability*. 2nd ed. Chichester: Wiley, 1997.

[Str] D.W. Strook. *Probability Theory: an Analytic View*. Cambridge: Cambridge University Press, 1993.

[T] K. Trivedi. *Probability and Statistics with Reliability, Queueing, and Computer Science Applications*. 2nd ed. New York; Chichester: Wiley, 2002.

[Tu] H.C. Tuckwell. *Elementary Applications of Probability Theory: with an Introduction to Stochastic Differential Equations*. 2nd ed. London: Chapman and Hall, 1995.

[Tuk] J.W. Tukey. *Exploratory Data Analysis*. Reading, MA: Addison-Wesley, 1977.

[WMM] R.E. Walpole, R.H. Myers and S.H. Myers. *Probability and Statistics for Engineers and Scientists*. 6th ed. Upper Saddle River, NJ; London: Prentice Hall International, 1998.

[We] R. Weber, IB Statistics, Lecture Notes, 2003. Available at www.statslab.cam.ac.uk/R.Weber.

[Wh] P. Whittle. *Probability via Expectation*. 4th ed. New York: Springer, 2000.

[Wi] M.V. Wilkes. *Memoirs of a Computer Pioneer*. Cambridge, MA: MIT Press, 1985.

[Wil] W. Willcox. Definitions of Statistics. *ISI Review*, **3**:4 (1935).

[Will] D. Williams. *Weighing the Odds: a Course in Probability and Statistics*. Cambridge: Cambridge University Press, 2001.

[WisH] G.L. Wise and E.B. Hall. *Counterexamples in Probability and Real Analysis*. New York; Oxford: Oxford University Press, 1993.

Index

asymptotic normality, 206

ballot problem, 31
branching process 96
    critical, 97
    subcritical, 97
    supercritical, 97

Cantor's staircase, 123
characteristic function (CHF) 62
    joint, 126
comparison of variances, 282
conjugate family (of PDFs/PMFs), 234
contingency tables, 271
correlation coefficient (of two RVs), 148
covariance (of two RVs), 39, 147
critical region (of a test) 242
    size of, 243

decision rule 236
    optimal Bayes, 236
distribution function, cumulative (CDF) 116
    joint, 126

error (in hypotheses testing) 242
    mean square (MSE), 218
    standard, 218
estimator
    best linear unbiased (BLUE), 291
    least squares (LSE), 290
    maximum likelihood (MLE), 213
    minimum MSE, 222
    optimal Bayes point, 235
    unbiased, 208
estimation
    interval, 229
    parametric, 206
    point, 229
exceedance, 166
expectation, or expected value (of an RV) 33
    conditional, 36
exponential families, 226
extinction probabilities, 98

factorisation criterion, 211
Fisher information, 222
formula
    Bayes, 8
    convolution, 40
    exclusion–inclusion, 27
    Stirling, 72
Fourier transform, 152
function
    Beta, 201
    concave, 76
    convex, 76
    Gamma, 114

Galton–Watson process, 104

hypothesis
    alternative, 242
    composite, 249
    conservative, 242
    null, 242
    one-side, 251
    simple, 242
hypothesis testing, 242

independence, 13
independent events, 14
independent identically distributed random variables
    (IID RVs), 37
independent observations, 165
independent random variables
    continuous, 133
    discrete, 37
inequality
    AM–GM, 91
    Cauchy–Schwarz (CS), 39
    Chebyshev, 75
    Chernoff, 76
    Cramér–Rao (CR), 222
    HM–GM, 91
    Hölder, 76
    Jensen, 76
    Markov, 75

interval
    confidence, 229
    prediction, 295

Jacobian, 122

Laplace transform, 61
Law of Large Numbers (LLN)
    weak, 78
    strong, 79
Lebesgue integration, 112
left-hand side (LHS), 27
Lemma
    Borel–Cantelli (BC), First, 132
    Borel–Cantelli (BC), Second, 132
    Neyman–Pearson (NP), 244
likelihood function, 245
likelihood ratio (LR) 245
    generalised (GLR), 262
    monotone (MLR), 249
linear regression
    simple, 249
log-likelihood function (LL), 213
loss
    posterior expected, 235
loss function 235
    quadratic, 235
    absolute error, 235

matrix
    inverse, 115
    invertible, 115
    orthogonal, 172
    positive-definite, 115
mean, mean value (of an RV)
    posterior, 235
    sample, 85
measure, 112
median 121
    posterior, 236
    sample, 136
memoryless property, 56, 132
method
    of moments, 220
    of maximum likelihood, 213
mode, 121
moment (of an RV), 61
moment generating function (MGF), 61

nuisance parameter, 263

outcome, outcomes, 3, 108

paradox
    Bertrand, 108
    Simpson, 275
percentile, 201

point (of a distribution)
    lower, 201
    upper, 201
posterior PDF/PMF (in Bayesian estimation), 180, 236
prior PDF/PMF (in Bayesian estimation), 179
probability
    conditional, 139
    posterior, 233
    prior, 233
probability density function (PDF) 112
    Beta, Cauchy, chi-square, exponential, Gamma, Gaussian or normal, Gaussian or normal bivariate, Gaussian or normal multivariate, jointly Gaussian or normal: *see* the list of continuous, probability distributions, 347, 348
    conditional, 132
    joint, 126
    support of, 123
    unimodal, 161
probability distribution, conditional, 7, 132
probability distribution, continuous
    Beta, 200
    Cauchy, 114
    chi-square, or χ^2, 197
    exponential, 112
    Fisher F-, 199
    Gamma, 114
    Gaussian, or normal, 112
    Gaussian or normal, bivariate, 127
    Gaussian or normal, multivariate, 115
    log-normal, 162
    Simpson, 115
    Student t, 198
    uniform, 112
probability distribution, discrete
    binomial, 54
    geometric, 56
    hypergeometric, 250
    multinomial, 271
    negative binomial, 57
    Poisson, 57
    uniform, 3
probability generating function (PGF), 58
probability mass function (PMF)
    binomial, geometric, multinomial, negative geometric, Poisson, uniform: *see* the list of discrete probability distributions, 346

quantile
    lower, 201
    upper, 201

random variable (RV), 33, 116
records, 45
reflection principle, 32
regression line, 290
relative entropy, 83
Riemann integration, 54
Riemann zeta-function, 54
right-hand side (RHS), 7
risk 236
    Bayes, 236

sample, 85
standard deviation 168
    sample, 218
standard error, 218
statistic (or sample statistic)
    Fisher F-, 282
    ordered, 155
    Pearson chi-square, or χ^2, 257
    Student t-, 253
    sufficient 209
        minimal, 212
statistical tables, 212
sum of squares
    between groups, 285
    residual (RSS), 293
    total, 285
    within group, 285

tail probabilities, 116
test
    analysis of variation (ANOVA), 284
    critical region of, 242

Fisher F-, 282
most powerful (MP), 243
Pearson chi-square, or χ^2-, 257
power of, 243
randomised, 245
significance level of, 243
size of, 243
Student t, 253
uniformly most powerful (UMP), 249
Theorem
    Bayes, 8
    Central limit (CLT)
        integral, 79
        local, 81
    De Moivre–Laplace (DMLT), 81
    Fisher, 216
    Rao–Blackwell (RB), 218
    Pearson, 257
    Wilks, 262
total set of outcomes, 6
type I
    error probability (TIEP), 242
type II
    error probability (TIIEP), 242

Unconscious Statistician, Law of the,
    35, 145
uncorrelated random variables, 39

variance 39
    sample, 209